Bioengineering and Biotechnology

Bioengineering and Biotechnology

Edited by **Joy Adam**

R CALLISTO REFERENCE

New York

Published by Callisto Reference,
106 Park Avenue, Suite 200,
New York, NY 10016, USA
www.callistoreference.com

Bioengineering and Biotechnology
Edited by Joy Adam

© 2016 Callisto Reference

International Standard Book Number: 978-1-63239-749-2 (Hardback)

Printed in the United States of America.

Contents

Preface

The purpose of the book is to provide a glimpse into the dynamics and to present opinions and studies of some of the scientists engaged in the development of new ideas in the field from very different standpoints. This book will prove useful to students and researchers owing to its high content quality.

Bioengineering is mainly concerned with developing cost effective diagnostic and curative technologies for problems related to human health. It combines the expertise of engineering with the rapidly advancing field of biology to provide innovative solutions. On the other hand biotechnology primarily focuses on making improvements in the living organisms by the means of genetic engineering, hybridization, tissue culture technologies, etc. They join hands to provide the best possible alterations in relation to living organisms. This book outlines the processes and applications of these disciplines in detail. A number of latest researches have been included to keep the readers up-to-date with the global concepts in this area of study. This book attempts to understand the multiple branches that fall under these disciplines and how such concepts have practical applications. It will serve as a complete source of knowledge for researchers and students alike.

At the end, I would like to appreciate all the efforts made by the authors in completing their chapters professionally. I express my deepest gratitude to all of them for contributing to this book by sharing their valuable works. A special thanks to my family and friends for their constant support in this journey.

Editor

Heterologous Soluble Expression of Recombinant OmpR of *Aeromonas hydrophila* and Its Immunogenic Potential

Sunita Kumari Yadav[1][#], Carmelita N. Marbaniang[2][#], Vibhuti Sharma[1], Aparna Dixit[1]*

[1]School of Biotechnology, Jawaharlal Nehru University, New Delhi, India
[2]RNA Biology Group, Institute for Molecular Infection Biology, University of Würzburg, Würzburg, Germany
[*]Email: adix2100@mail.jnu.ac.in, adixit7@yahoo.com, adixit7@gmail.com

Abstract

Aeromonas hydrophila, a gram negative bacterium is a major fish pathogen and causes major economic losses to aquaculture industry. Outer membrane proteins play a significant role in its survival during different environmental conditions and bacterial pathogenesis. The outer membrane protein R (OmpR) is a member of the two-component regulatory system of *Aeromonas hydrophila* which differentially regulates the expression of OmpF or OmpC depending on the osmolarity conditions. Role of OmpR has been demonstrated in its virulence in other infectious bacteria and it is found to be a potential drug target/vaccine candidate. However, the OmpR of *A. hydrophila* has not been characterized. In the present study, we report recombinant expression, purification of the OmpR of *A. hydrophila* strain Ah17 in salt inducible *E. coli* GJ1158 cells. Leaky expression of rOmpR was confirmed by Western blot analysis using anti-6 × His antibody. The histidine tagged recombinant OmpR (rOmpR) (~29 kDa) was purified using Ni-NTA affinity chromatography from the soluble fraction of induced *E. coli* cells. The rOmpR was found to be highly immunogenic with end point titres of greater than 1:80,000. The anti-rOmpR antisera were capable of agglutinating live *A. hydrophila* cells, thus showing vaccine potential of the rOmpR.

Keywords

OmpR, *Aeromonas*, Cross-Reactivity, Outer Membrane Protein, Expression, *E. coli* GJ1158

---

[*]Corresponding author.
[#]These authors made equal contributions.

1. Introduction

Aeromonads, a heterogenous group of gram-negative bacteria have been associated with various pathological syndromes in fish, causing a major economic loss to the aquaculture industry [1]-[4] of various species of Aeromonads. *Aeromonas hydrophila* is of significance as it is one of the major species that infects fresh water fish. The treatment with antibiotics results in antibiotic-resistant strains, and water treatment with chemicals raises a great concern for the residues in the environment and food fish. Therefore, vaccination against infectious agents continues to be one of the most effective methods to control *A. hydrophila* infection. Till date, no vaccine is available for the treatment of infection against these bacteria. Vaccination strategies in fish farms have mostly involved the use of formalin killed pathogen [5] or live attenuated vaccines. Immunization with polyvalent vaccines, heat-killed and live attenuated *Aeromonas hydrophila* has been reported [6]-[9]. Biofilm of *Aeromonas hydrophila* has been shown to offer protection against bacterial challenge [10]. However, studies on immunization with a specific antigen(s) have remained limited to laboratories only. The outer membrane/surface proteins have been considered to be ideal candidates for vaccine development. Recently, recombinant outer membrane proteins OmpTS, OmpA and OmpW of *A. hydrophila* have been investigated for their immunogenic potential [11]-[13].

Several outer membrane components have been reported to be associated with the virulence of *Aeromonas*. The pathogenicity of the bacteria depends on their ability to survive and proliferate in the stressful environmental conditions which may change markedly during invasion of the host cells. Upon infection, bacteria are faced with a change in environmental conditions in the host and the ability of bacterium to survive in the changing environment plays a key role in its pathogenicity. In response to these changing environmental conditions, all bacteria display a complex regulation of gene expression by utilizing the signal transduction system referred to as the two components regulatory system [14] [15]. The OmpF and OmpC are two major porins, constituting ~ 2% of the total cellular protein [16] in the outer membrane of bacteria whose expression is regulated in response to change in the osmolarity by two component regulatory system. The levels of these proteins are regulated by OmpR, a protein of two components regulatory systems [17].

The role of ompR and ompR-dependent genes in virulence has been well illustrated in *Salmonella typhimurium* [18] [19], *Shigella flexineri* [20] [21] *Yersinia enterolitica* [22] [23], and *Helicobacter pylori* [24]. It has been reported that the combined mutations in OmpF and OmpC result in avirulent *S. typhimurium*. Immunization with OmpR mutants of *S. typhimurium* fails to cause any mortality in BALB/c mice after oral challenge [19]. The OmpC protein has been indirectly reported to be involved in the adhesion and invasion of Crohn's disease-associated *Escherichia coli* strain LF82 [25]. Therefore, these proteins are important candidates for vaccine development. In addition, these proteins are present only in the bacterium and not in their hosts, and therefore can serve as a potential vaccine candidate or drug target for effective drug design. Since purification of the outer membrane protein to homogeneity from the bacteria is cumbersome due to significant similarity in their properties, recombinant route to produce the targeted protein is an attractive alternate. Present investigation reports production of soluble recombinant OmpR of *A. hydrophila* using heterologous *E. coli* expression system. The purified recombinant protein is then evaluated for its immunogenicity to assess its vaccine potential.

2. Materials and Methods

2.1. Materials

Expression vector pRSET-A was procured from Invitrogen USA. Reagents required for DNA modification (restriction enzymes and chemicals) were purchased from New England Biolabs, USA and Promega, USA. All other chemicals (analytical grade) used in the study were procured from Sigma-Aldrich Chemical Co., USA, unless otherwise stated. Oligonucleotides and primers used in the present study were synthesized by Microsynth, Switzerland.

2.2. Bacterial Strains

Aeromonas hydrophila (strain Ah17) was a kind gift from Dr. I. Karunasagar, College of Fisheries, Mangalore, India [26]. *Escherichia coli* DH5α and BL21 (λDE3) pLysS were procured from Gibco-BRL, USA and Novagen, USA respectively. *E. coli* GJ1158 cells were used for recombinant expression of the OmpR [27].

2.3. Cloning of the *ompR* Gene of *A. hydrophila* in Expression Vector

The *ompR* gene of Ah17 along with upstream and downstream region was cloned in PBCKS+ cloning vector [28]. The full length gene encoding mature OmpR of *A. hydrophila* Ah17 was PCR amplified using *pAhOmpR* (EMBL acc no. AM084417.1) (Locus CAJ29527.1) [28] as the template and gene specific primers (Forward: 5'-CCG**CTCGAG**ATGCGGATCCTGTTGGTGG-3'; Reverse: 5'-ACC**GGTACC**TCATGCCCGGCCGTCCCGG CGC-3') containing *Xho*I and *Kpn*I (in bold letters) in forward and reverse primers, respectively.

Polymerase chain reaction was performed at the following specified conditions: initial denaturation 94˚C for 5 min followed by 25 thermal cycles of denaturation at 94˚C for 45 s; annealing at 53˚C for 30 s, extension at 72˚C for 40 s with a final extension for 7 min at 72˚C. The *Xho*I- and *Kpn*I-digested amplified OmpR fragment was ligated to pRSET-A prokaryotic expression vector digested with the same enzymes. Putative recombinants were screened using restriction enzyme digestion analysis, and integrity of the OmpR gene was confirmed by automated DNA sequencing (DNA sequencing facility, University of Delhi South campus, New Delhi). The resulting recombinant was designated as *pRSETAh.ompR*.

2.4. Expression and Purification of Recombinant OmpR in *E. coli*

Initial attempts to express the recombinant OmpR from *pRSETAh.ompR* in standard *E. coli* BL21(DE3) cells did not result in any expression. Subsequently, *E. coli* GJ1158 cells were used for recombinant expression after transformation with the *pRSETAh.ompR*. *E. coli* GJ1158 cells harboring the *pRSETAh.ompR* were induced with the indicated concentrations of NaCl at 0.8 O.D$_{600nm}$ and expression of the recombinant protein was analyzed at 6 h post-induction by SDS-PAGE (12%) analysis. Optimization of inducer concentration was carried out by inducing the cells with different concentration of NaCl for 4 h. Time of induction was optimized by inducing the cells with optimum NaCl concentration (determined earlier) for different time intervals. Localization analysis of the expressed protein was performed in the induced cell lysates essentially as described earlier [26]. The recombinant OmpR protein harboring the 6 × -histidine tag at the N-terminus (rOmpR) was purified from the soluble fraction under native conditions using Ni^{2+}-NTA affinity chromatography. Briefly, LB (100 ml) containing 100 µg/ml ampicillin was inoculated with 1% of O/N grown culture of *E. coli* GJ1158 cells harboring *pRSE-TAh.ompR* and incubated at 37˚C with shaking (200 rpm) till A$_{600}$ reached 0.6 - 0.8. Induction of expression of the rOmpR was primarily done with 0.75 M NaCl for 6 h at 37˚C. The cells were harvested by centrifugation at 4000 rpm for 10 min at 4˚C. After washing the pellet with Washing Buffer (20 mM Tris-HCl, pH 8.5, 20% Sucrose), cells were lysed with 20 ml of lysis Buffer (20 mM Tris-HCl, pH 8.5, 1 mM EDTA, 1% Triton × 100, 200 µg/ml lysozyme) by stirring vigorously for 10 min at RT. All protein purification steps were performed at 4˚C. Soluble fraction obtained after lysis was collected by centrifugation at 12,000 rpm for 20 min. The rOmpR was purified from the soluble fraction using Ni^{2+} NTA affinity chromatography essentially as described [26] Briefly, the rOmpR present in the soluble fraction was incubated with Ni^{2+}-NTA slurry at 4˚C for 2 h in binding buffer (5 mM imidazole, 20 mM Tris- HCl pH 7.9, 0.5 M NaCl). Non-specific proteins were removed by washing the resin with 6 volumes of washing buffer (20 mM imidazole, 20 mM Tris- HCl pH 7.9, 0.5 M NaCl). The specifically bound rOmpR was eluted with elution buffer (200 mM imidazole, 20 mM Tris-HCl pH 7.9, 0.5 M NaCl). Different fractions were analyzed by SDS-PAGE. The fractions containing the purified rOmpR were pooled and subjected to dialysis in 20 mM Tris-HCl, pH 7.9 containing 10% glycerol. The purified rOmpR was stored at −20˚C in small aliquots until further use. Protein concentration was determined by the method of Lowry *et al.* [29] using bovine serum albumin as a standard.

2.5. Western Blot Analysis

The authenticity of the induced protein *i.e.* histidine tagged rOmpR in the induced cell lysates or the purified rOmpR was confirmed by Western blot analysis using anti-His or anti-rOmpR antibodies, respectively, essentially as described earlier [30]. Cell lysates or the rOmpR protein were resolved on 12% SDS-PAG and transferred onto nitrocellulose (NC) membrane. The blot was incubated with 2% BSA in 1 × PBS containing 0.05% Tween 20 (PBST) for 2 h at RT, followed by three subsequent washes with 1 × PBST for 10 min each at RT. The membrane was then incubated with the anti-6 × -His tag antibody (1:10,000) or anti-rOmpR antibody (1:5000) for 1 h, followed by incubation with the alkaline phosphatase conjugated secondary antibody (1:10,000) for 1 h at RT. The membrane was given 3 washes with 1 × PBST between each incubation. The immunoreactive bands were visualized by the Western blue stabilized substrate solution (Promega, USA).

2.6. Antibody Generation Against the rOmpR

Female BALB/c mice (4 - 6 weeks) were used for immunization to generate polyclonal antibodies against the recombinant protein. The use of animals for the study was approved by the Institutional Animal Ethics Committee and prescribed guidelines by the Institutional Animal Ethics Committee, JNU, New Delhi, India were followed while handling the animals.

Primary immunization was carried out using the rOmpR (20 μg/mouse) emulsified in complete Freund's adjuvant. Booster doses with the same amount of the rOmpR in incomplete Freund's adjuvant were given on day 14, and day 28 of primary immunization. Mice were bled on day 0 (pre-immune sera), and a week after each immunization *i.e.* on day 21 and day 35. Sera collected prior to immunization served as control pre-immune serum. End-point titers of the antisera were determined using ELISA. Ability of the polyclonal antibodies to specifically recognize the rOmpR was also confirmed by Western blotting.

2.7. Enzyme-Linked Immunosorbent Assay to Determine Antibody Titre

The rOmpR (500 ng/100μl/well) in 0.1 M carbonate-bicarbonate buffer, pH 9.8 was used for coating 96-well ELISA plate. After incubation overnight at 4°C, the plate was washed with 1 × PBST thrice. Blocking was done with 2% BSA made in 1 × PBST (100 μl each well) for 2 h at 37°C. Different dilutions of anti-sera (primary antibody) prepared in 1 × PBS were added (100 μl/well) and left for 2 h at 37°C followed by three washes with 1 × PBST (1 × PBS containing 0.05% Tween-20). Alkaline phosphatase-conjugated anti-mice antibody raised in goat (Santacruz, USA) was used at a dilution of 1:10,000 (in 1 × PBS) and incubated for 1 h at 37°C. The plate was again washed three times with 1 × PBST after every incubation. The substrate p-nitrophenylphosphate [100 μl of 1 mg/ml solution prepared in AP buffer (50 mM Na_2CO_3, 1 mM $MgCl_2$, pH 9.8)] was added to each well and left for 20 min. The absorbance was read at 450 nm in an ELISA reader (Tecan, USA).

2.8. Agglutination Assay

The assay was performed to assess the ability of anti-rOmpR antisera to agglutinate live *A. hydrophila* cells essentially as described by Yadav *et al.* [30]. Briefly, *A. hydrophila* (Ah17) cells were inoculated in LB broth from an overnight culture and grown for 5 - 6 h. For agglutination assay, 5×10^8 cfu from the log phase culture were taken and agglutination reaction set was made in 1 × PBS containing 1:250 dilution of the anti-rOmpR antisera. Equal numbers of *A. hydrophila* (Ah17) cells in 1 × PBS, with pre-immune sera, were included in the study as a control. The reaction mix was incubated for 1 h at 37°C followed by centrifugation at 5,000 rpm for 10 min. The pellet was resuspended in 1 × PBS. The resuspended cells were uniformly smeared on a clean glass slide and dried. The slide was heat fixed by passing through a flame transiently, and stained with methylene blue (Sigma-Aldrich Chemical Co, USA), followed by washing to remove the excess stain and visualised under microscope (Model Eclipse TE2000S, Nikon, USA).

3. Results and Discussion

A. hydrophila, a common fish pathogen, is responsible for causing major economic losses to the aquaculture industry. In addition to virulence factors expressed by the bacteria, its ability to adapt to different environments (outside and within the host) plays a critical role in its pathogenicity. This is conferred by the signal transduction system known as two components regulatory system. The OmpR is the key regulator of this system which controls the expression of OmpF and OmpC—the two major porins, differentially expressed under different osmolarity conditions. Role of OmpR and OmpR-dependent genes in bacterial virulence has been elucidated in a number of bacterial species and hence the proteins of this system become a potential drug target or vaccine candidates. We have earlier reported cloning and characterization of the OmpR gene of *A. hydrophila* [28] and *in silico* homology modelling of the OmpR of *A. hydrophila* for structure based drug design. However, for a protein to be evaluated as a vaccine candidate or drug target, large amounts of purified proteins are required. In the present study, we report soluble expression of recombinant OmpR of *A. hydrophila* and its purification, for evaluation of its immunogenic potential and ability to neutralize *A. hydrophila*.

3.1. Cloning of the *ompR* Gene *in E. coli* Expression Vector

Since the OmpR gene reported earlier contained 5'- and 3'-flanking sequences, including the signal sequence

[28], cloning of the gene region encoding the mature OmpR in expression vector was carried out. PCR amplification of the OmpR encoding the mature OmpR resulted in amplification of ~0.7 kb DNA fragment of the expected size. Successful cloning of the PCR amplified fragment after digestion with *Xho*I and *Kpn*I in pRSETA resulted in *in frame* cloning, confirmed by DNA sequencing. The *ompR gene* with a single open reading frame would encode an N-terminal 6 × -Histidine tagged recombinant OmpR of ~29.5 kDa, including the amino acid residues contributed by the vector.

3.2. Expression of the rOmpR in GJ1158 Cells

For evaluation as a drug target or a vaccine candidate, the purified protein in its native form is required in large amounts which would not be feasible to purify from their natural source. It is particularly more difficult with reference to membrane proteins due to their similar properties. We have reported large scale purification of OmpF of *A. hydrophila*, a member of the two component regulatory system [30]. However, the protein expressed as inclusion bodies which adds additional steps for purification and refolding to its native conformation. General strategies to express soluble proteins in standard *E. coli* expression hosts such as M15, BL21(DE3) etc, involve lowering the concentration of IPTG, induction temperature etc. This slows down the cellular metabolism and cell proliferation. As a result, the yield of soluble protein is often compromised. We have used an alternate *E. coli* expression host GJ1158 cells, in which the T7 RNA polymerase gene has been placed under the control of osmotically inducible *proU* promoter of *E. coli* [27]. These cells allow the use of standard expression plasmid in which the gene to be expressed is put under the control of T7 promoter. Successful expression of the rOmpR was obtained when the *E. coli* GJ1158 cells harboring the plasmid *pRSETAh.ompR* were induced with 0.3 M NaCl at 6 h post-induction. An intense band of approximately 29 kDa of the rOmpR was seen in the induced (**Figure 1(A)**, lane 2) as well as in uninduced cells (**Figure 1(A)**, lane 1). This could be due to leaky expression of expressed protein in GJ1158 cells as the expression of the recombinant protein is not under as tight a control as seen with inducible promoters such as *lac* or *tac*. Since the OmpR homologue is present in *E. coli*, the protein is native to the host cell and hence leaky expression of the rOmpR would not adversely affect the cellular metabolism of the host cell. Unlike other membrane protein, MerT, which could not be expressed in GJ1158 cells due to expression induced toxicity [31], OmpR of *A. hydrophila* could be successfully expressed in these *E. coli* host cells. Use of NaCl as an inducer has an advantage over IPTG with respect to its low cost and relatively no toxicity. Western blot analysis using anti-His antibodies confirmed the authenticity of the rOmpR as a clear band at the expected position could be observed in the induced cells (**Figure 1(B)**, lane 1). To obtain maximum expression of rOmpR, cultures induced with different concentrations of NaCl (0.1 M - 1 M; **Figure 1(C)**) were analysed for the rOmpR expression. As evident from the figure, maximum expression was obtained with 0.75 M NaCl (indicated on top of the panel). Time kinetics of expression revealed maximum expression of the rOmpR at 6 h post-induction (**Figure 1(D)**).

3.3. Localization and Purification of the rOmpR

Since the rOmpR is being expressed to ultimately evaluate its potential as a drug target or vaccine candidate, it is imperative to have the protein expressed as a soluble protein. Expression of soluble protein facilitates its structural characterization which is necessary for drug designing. Also, the soluble protein in its native form would retain conformational epitopes to bring about effective immune response. Analysis of the soluble and insoluble fractions of the induced cell lysates clearly showed that the rOmpR predominantly expressed in the soluble fraction (**Figure 1(E)**. lane 1), although little expression could also be seen in the insoluble fraction (**Figure 1(E)**, lane 2). Soluble expression of the rOmpR was achieved without addition of arginine, which is often used to direct the expression of aggregation-prone recombinant proteins in soluble fraction [32]. Our results are in agreement with Bhandari and Gowrishankar [27], who clearly demonstrated that induction with NaCl reduced sequestration of the overexpressed recombinant proteins within insoluble inclusion bodies. These authors attributed this to accumulation of glycine betaine, a component of yeast extract, upon osmotic shock with NaCl in GJ1158, which has earlier been reported to alleviate inclusion body formation of overexpressed dimethylallyl pyrophosphate: 5' AMP transferase of *agrobacterium* [33]. Soluble rOmpR was purified from the soluble fraction of induced cell lysates in a single step Ni^{2+}-NTA chromatography. The rOmpR eluted with 200 mM imidazole (**Figure 1(E)**, lane 3) was purified to near homogeneity (98%) based on the densitometric analysis. Approximately, 20 mg of purified soluble rOmpR could be obtained from 1 liter of culture at shake flask level. Soluble

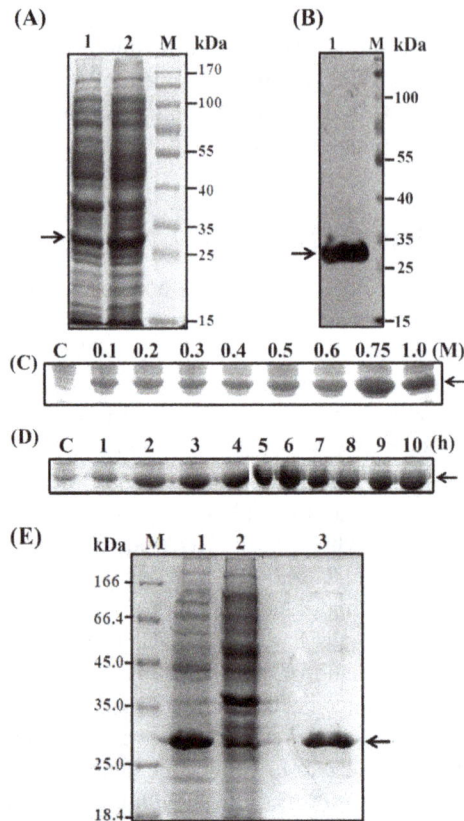

Figure 1. Recombinant expression of OmpR of *Aeromonas hydrophila* **(A)** *E. coli* GJ1158 cells harbouring *pRSETAh.ompR* were induced with 0.75 M NaCl. Total cell lysates of the uninduced cells and induced cells are shown in lanes 1 and 2, respectively. The arrow points to the rOmpR of ~29 kDa. **(B)** Western blot analysis of induced cell lysates (lane 1) using anti-His antibody **(C)** Optimization of inducer concentration for rOmpR expression. *E. coli* GJ1158 cells harboring *pRSE-TAh.ompR* were induced with different concentrations of NaCl (0.1 M to 1 M, indicated on top of the panel) for 6 h. Cell lysates (50 µg each) were analyzed on 12% SDS-PAGE. C refers to the uninduced cell lysates. The arrow points to the rOmpR. rOmpR expression could be seen at NaCl concentrations as low as 0.1 M. **(D)** Time kinetics of the rOmpR expression. Cell lysates of *E. coli* GJ1158 cells harboring *pRSETAh.ompR* induced with 0.75 M NaCl for different time periods (1 - 10 h, shown on top of the panel) were analyzed by SDS-PAGE (12%). Maximum expression of rOmpR is observed at 6 h. C refers to the uninduced cell lysates. The arrow points to the rOmpR **(E)** SDS-PAGE (12%) analysis of the soluble (lane 1) and pellet (lane 2) fractions of the induced cell lysates. Lane 3 shows the rOmpR purified from the soluble fraction of induced cell lysates using Ni-NTA chromatography. M denotes protein molecular weight markers in all the panels.

rOmpR thus produced can be used for its structural characterization for effective drug design. Comparative densitometric analysis showed that, ~1.5% - 1.8% fold purification has been achieved. This is due to the reason that majority (~80%) of the total protein in the soluble fraction of the induced culture appeared to be recombinant OmpR.

3.4. Specificity and Agglutination Ability of Anti-rOmpR Antisera

The outer membrane proteins have been identified as primary targets for vaccine development, as these are exposed on the cell surface and are primary contact molecules with the host cell that is involved in generation of immune response [34]. A number of outer membrane proteins have been shown to be highly immunogenic amongst different bacteria and immunization with these proteins has been reported to confer protection against the bacterial challenge [35]-[39]. Likewise, the rOmpR was found to be highly immunogenic and resulted in generation of efficient immune response (**Figure 2(A)**). Significant immune response was observed even after single booster. The anti-rOmpR antisera with the end point titers of ≥80,000 were obtained after first booster, which increased further after second booster. Immunoblot analysis of the induced cell lysates using anti-rOmpR antisera showed that the antisera was able to detect the expressed rOmpR as only a single band at the expected

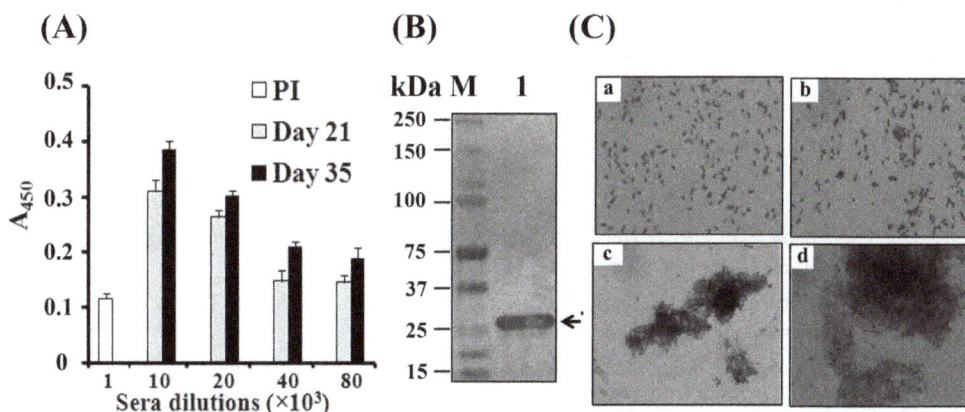

Figure 2. (A) End-point titer determination of the anti-rOmpR antisera. Antisera obtained from mice immunized with rOmpR at day 21 and day 35 was assessed for rOmpR-specific antibody titers by ELISA. Different dilutions of the anti-rOmpR antisera were added (in triplicate) to ELISA plates coated with purified rOmpR. Alkaline phosphatase-conjugated anti-mice antibody raised in goat were used as a detection tool. Aborbance at 450 nm measured 20 min after the substrate (p-nitrophenylphosphate) is plotted against the antisera dilution. Preimmune (PI) serum was included as control. (B) Immunoblot analysis for the Specificity of the anti-rOmpR sera: Induced cell lysate of *E. coli* GJ1158 cells harbouring *pRSE-TAh.ompR* were immunoblotted with anti-rOmpR antisera (1:5000) on nitrocellulose membrane. A dark immunoreactive band (pointed by arrow) visible only in the induced cell lysate (lane 1) confirms high specificity of the anti-rOmpR antisera. M indicates protein molecular weight (kDa) marker (C) Agglutination assay of anti-rOmpR antisera. Live *A. hydrophila* (strain Ah17), cells (5×10^8 CFU each) in 0.5 ml PBS were incubated with either pre-immune serum or anti-rOmpR antisera (1:250 dilution each). *A. hydrophila* cells pre-incubated with pre-immune sera shown in a, b whereas c, d show *A. hydrophila* incubated with anti-rOmpR antisera. Agglutination is evident only in *A. hydrophila* cells incubated with anti-rOmpR antisera. Images are taken at 40× magnification.

size in the induced cell lysate was observed (**Figure 2(B)**). Due to specific interaction of the antisera with the cell membrane, agglutination assays using serum have been routinely used for bacterial cell identification during an infection [40] [41]. Positive and specific agglutination indicates direct interaction of the antibodies present in the antisera with the bacterial cell. Therefore, neutralizing potential of the anti-rOmpR antisera was assessed by agglutination assay using live *A. hydrophila* cells. While incubation with pre-immune sera did not show any agglutination (**Figure 2(C)**, panels a and b), the live *A. hydrophila* incubated with anti-rOmpR antisera resulted in effective clumping of cells (**Figure 2(C)** panels c and d). Potential of anti-rOmpR antisera to agglutinate live *A. hydrophila* cells clearly indicates the neutralizing activity of the antisera and its vaccine potential. Earlier studies from laboratory have demonstrated agglutination ability of the antisera generated against another outer membrane protein of two component regulatory system namely rOmpF [30]. It is possible that immunization with these two proteins *i.e.* rOmpR and rOmpF would further augment immune response that would be able to curtail *Aeromonas* infection.

4. Conclusion

The present studies thus establish soluble expression of rOmpR using NaCl as an inducer which makes the production cost effective in comparison to use of IPTG as an inducer. We also demonstrated immunogenic and vaccine potential of rOmpR of *A. hydrophila* which can be evaluated further in animals susceptible to *A. hydrophila* infection followed by challenge studies.

Acknowledgements

Authors acknowledge the Department of Biotechnology, Ministry of Science and Technology, New Delhi for research grant to AD. SKY thanks the Council of Scientific and Industrial Research, New Delhi for providing research fellowship.

References

[1]　Abbott, S.L., Cheung, W.K.W. and Janda, J.M. (2003) The Genus *Aeromonas*: Biochemical Characteristics, Atypical Reactions, and Phenotypic Identification Schemes. *Journal of Clinical Microbiology*, **41**, 2348-2357.

http://dx.doi.org/10.1128/JCM.41.6.2348-2357.2003

[2]   Behera, B., Bhoriwal, S., Mathur, P., Sagar, S., Singhal, M. and Misra, M.C. (2011) Post-Traumatic Skin and Soft Tissue Infection Due to *Aeromonas hydrophila*. *Indian Journal of Critical Care Medicine*, **15**, 49-51. http://dx.doi.org/10.4103/0972-5229.78228

[3]   Cipriano, R.C. (2001) *Aeromonas hydrophila* and Motile Aeromonad Septicaemias of Fish. Revision of Fish Disease Leaflet 68, United States Department of Interior, Fish and Wildlife Service Division of Fishery Research, Washington DC, 1-25.

[4]   Janda, J.M. and Abbott, S.L. (2010) The Genus *Aeromonas*: Taxonomy, Pathogenicity, and Infection. *Clinical Microbiology Reviews*, **23**, 35-73. http://dx.doi.org/10.1128/CMR.00039-09

[5]   Michel, C. (1979) Furunculosis of Salmonids: Vaccination Attempts in Rainbow Trout (*Salmo gairdneri*) by Formalin Killed Germs. *Annales de Recherches Veterinaires*, **10**, 33-40.

[6]   Chandran, M.R., Aruna, B.V., Logambal, S.M. and Michael, R.D. (2002) Immunisation of Indian Major Carps against *Aeromonas hydrophila* by Intraperitoneal Injection. *Fish and Shellfish Immunology*, **13**, 1-9. http://dx.doi.org/10.1006/fsim.2001.0374

[7]   Karunasagar, I., Ali, A., Otta, S.K. and Karunasagar, I. (1997) Immunization with Bacterial Antigens: Infections with Motile *Aeromonads*. *Developments in Biological standardization*, **90**, 131-141.

[8]   Majumdar, T., Ghosh, D., Datta, S., Sahoo, C., Pal, J. and Mazumder, S. (2007) An Attenuated Plasmid-Cured Strain of *Aeromonas hydrophila* Elicits Protective Immunity in *Clarias batrachus*. *Fish and Shellfish Immunology*, **23**, 222-230. http://dx.doi.org/10.1016/j.fsi.2006.10.011

[9]   Moral, C.H., Castillo, E.F.D., Fieroo, P.L., Cortes, A.V., Castillo, J.A., Soriano, A.C., Salazar, M.S., Peralta, B.R. and Carrasco, G.N. (1998) Molecular Characterization of the *Aeromonas hydrophila* aroA Gene and Potential Use of an Auxotrophic aroA Mutant as a Live Attenuated Vaccine. *Infection and Immunity*, **66**, 1813-1821.

[10] Asha, A., Nayak, D.K., Shankar, K.M. and Mohan, C.V. (2004) Antigen Expression in Biofilm Cells of *Aeromonas hydrophila* Employed in Oral Vaccination of Fish. *Fish and Shellfish Immunology*, **16**, 429-436. http://dx.doi.org/10.1016/j.fsi.2003.08.001

[11] Khushiramani, R., Girisha, S.K., Karunasagar, I. and Karunasagar, I. (2007) Protective Efficacy of Recombinant OmpTs Protein of *Aeromonas hydrophila* in Indian Major Carp. *Vaccine*, **25**, 1157-1158. http://dx.doi.org/10.1016/j.vaccine.2006.10.032

[12] Maiti, B., Shetty, M., Shekar, M., Karunasagar, I. and Karunasagar, I. (2011) Recombinant Outer Membrane Protein A (OmpA) of *Edwardsiella tarda*, a Potential Vaccine Candidate for Fish, Common Carp. *Microbiological Research*, **167**, 1-7. http://dx.doi.org/10.1016/j.micres.2011.02.002

[13] Maiti, B., Shetty, M., Shekar, M., Karunasagar, I. and Karunasagar, I. (2012) Evaluation of Two Outer Membrane Proteins, Aha1 and OmpW of *Aeromonas hydrophila* as Vaccine Candidate for Common Carp. V*eterinary Immunology and Immunopathology*, **149**, 298-301. http://dx.doi.org/10.1016/j.vetimm.2012.07.013

[14] Mizuno, T. (1998) His-Asp Phosphotransfer Signal Transduction. *Journal of Biochemistry* (*Tokyo*), **123**, 555-563. http://dx.doi.org/10.1093/oxfordjournals.jbchem.a021972

[15] Robinson, V.L., Buckler, D.R. and Stock, A.M. (2000) A Tale of Two Components: A Novel Kinase and a Regulatory Switch. *Nature Structural and Molecular Biology*, **7**, 626-633. http://dx.doi.org/10.1038/77915

[16] Nikaido, H. Neidhardt, F.C., Curtiss III, R., Ingraham, J.L., Lin, E.C.C., Low Jr., K.B., Magasanik, B., Reznikoff, W.S., Riley, M., Schaechter, M. and Umbarger, H.E. (1996) Outer Membrane in *Escherichia coli* and *Salmonella*: Cellular and Molecular Biology. 2nd Edition, American Society for Microbiology Press, Washington DC, 29-47.

[17] Stock, J.B., Ninfa, A.J. and Stock, A.M. (1989) Protein Phosphorylation and Regulation of Adaptive Responses in Bacteria. *Microbiological Reviews*, **53**, 450-490.

[18] Chatfield, S.N., Dorman, C.J., Hayward, C. and Dougan, G. (1991) Role of OmpR-Dependent Genes in *Salmonella typhimurium* Virulence: Mutants Deficient in Both OmpC and OmpF Are Attenuated *in Vivo*. *Infection and Immunity*, **59**, 449-452.

[19] Dorman, C.J., Chatfield. S., Higgins, C.F., Hayward, C. and Dougan, G. (1989) Characterisation of Porin and OmpR Mutants of a Virulent Strain of *Salmonella typhimurium:* OmpR Mutants Are Attenuated *in Vivo*. *Infection and Immunity*, **57**, 2136-2140.

[20] Bernardini, M.L., Fontaine, A. and Sansonetti, P.J. (1990) The Two-Component Regulatory System OmpR-EnvZ Controls the Virulence of *Shigella flexneri*. *Journal of Bacteriology*, **172**, 6274-6281.

[21] Bernardini, M.L., Sanna, M.G., Fontaine, A. and Sansonetti, P.J. (1993) OmpC Is Involved in Invasion of Epithelial Cells by *Shigella flexneri*. *Infection and Immunity*, **61**, 3625-3635.

[22] Brzostek, K., Raczkowska, A. and Zasada, A. (2003) The Osmotic Regulator OmpR Is Involved in the Response of *Yersinia enterocolitica* O:9 to Environmental Stresses and Survival within Macrophages. *FEMS Microbiology Letters*,

228, 265-271. http://dx.doi.org/10.1016/S0378-1097(03)00779-1

[23] Raczkowska, A. and Brzostek, K. (2004) Identification of OmpR Protein and Its Role in the Invasion Properties of *Yersinia enterocolitica*. *Polish Journal of Microbiology*, **53**, 11-16.

[24] Bury-Mone, S., Thiberge, J.M., Contreras, M., Maitournam, A., Labigne, A. and De Reuse, H. (2004) Responsiveness to Acidity via Metal Ion Regulators Mediates Virulence in the Gastric Pathogen *Helicobacter pylori*. *Molecular Microbiology*, **53**, 623-638. http://dx.doi.org/10.1111/j.1365-2958.2004.04137.x

[25] Rolhion, N., Carvalho, F. and Darfeuille-Michaud, A. (2007) OmpC and the Sigma (E) Regulatory Pathway Are Involved in Adhesion and Invasion of the Crohn's Disease-Associated *Escherichia coli* Strain LF82. *Molecular Microbiology*, **63**, 1684-700. http://dx.doi.org/10.1111/j.1365-2958.2007.05638.x

[26] Agarwal, S., Gopal, K., Upadhyaya, T. and Dixit, A. (2007) Biochemical and Functional Characterization of UDP-Galactose 4-Epimerase from *Aeromonas hydrophila*. *Biochimica et Biophysica Acta*, **1774**, 828-837. http://dx.doi.org/10.1016/j.bbapap.2007.04.007

[27] Bhandari, P. and Gowrishankar, J. (1997) An *Escherichia coli* Host Strain Useful for Efficient Overproduction of Cloned Gene Products with NaCl as the Inducer. *Journal of Bacteriology*, **179**, 4403-4406.

[28] Chhabra, G., Upadhyaya, T. and Dixit, A. (2011) Molecular Cloning, Sequence Analysis and Structure Modeling of OmpR, the Response Regulator of *Aeromonas hydrophila*. *Molecular Biology Reports*, **39**, 41-50. http://dx.doi.org/10.1007/s11033-011-0708-3

[29] Lowry, O.H., Rosebrough, N.J., Farr, A.L. and Randall, R.J. (1951) Protein Measurement with the Folin Phenol Reagent. *Journal of Biological Chemistry*, **193**, 265-275.

[30] Yadav, S.K., Sahoo, P.K. and Dixit, A. (2014) Characterization of Immune Response Elicited by the Recombinant Outer Membrane Protein OmpF of *Aeromonas hydrophila*, a Potential Vaccine Candidate in Murine Model. *Molecular Biology Reports*, **41**, 1837-1848. http://dx.doi.org/10.1007/s11033-014-3033-9

[31] Senthil, K. and Gautam, P. (2010) Expression and Single-Step Purification of Mercury Transporter (merT) from *Cupriavidus metallidurans* in *E. coli*. *Biotechnology Letters*, **32**, 1663-1666. http://dx.doi.org/10.1007/s10529-010-0337-2

[32] Schaffner, J., Winter, J., Rudolph, R. and Schwarz, E. (2001) Co-Secretion of Chaperones and Low-Molecular-Size Medium Additives Increases the Yield of Recombinant Disulfide-Bridged Proteins. *Applied and Environmental Microbiology*, **76**, 3994-4000. http://dx.doi.org/10.1128/AEM.67.9.3994-4000.2001

[33] Blackwell, J.R. and Horgan, R. (1991) A Novel Strategy for Production of a Highly Expressed Recombinant Protein in an Active Form. *FEMS Microbiology Letters*, **295**, 10-12. http://dx.doi.org/10.1016/0014-5793(91)81372-F

[34] Osman, K.M. and Marouf, S.H. (2014) Comparative Dendrogram Analysis of OMPs of *Salmonella enteric* Serotype Enteritidis with *Typhimurium*, *Braendurup* and *Lomita* Isolated from Pigeons. *International Journal of Advanced Research*, **2**, 952-960.

[35] Khushiramani, R., Maiti, B., Shekar, M., Girisha, S.K., Akash, N., Deepanjali, A., Karunasagar, I. and Karunasagar, I (2012) Recombinant *Aeromonas hydrophila* Outer Membrane Protein 48 (Omp48) Induces a Protective Immune Response against *Aeromonas hydrophila* and *Edwardsiella tarda*. *Research in Microbiology*, **163**, 286-291. http://dx.doi.org/10.1016/j.resmic.2012.03.001

[36] Guan, R., Xiong, J., Huang, W. and Guo, S. (2011) Enhancement of Protective Immunity in European Eel (*Anguilla anguilla*) against *Aeromonas hydrophila* and *Aeromonas sobria* by a Recombinant *Aeromonas* Outer Membrane Protein. *Acta Biochimica et Biophysica Sinica (Shanghai)*, **43**, 79-88. http://dx.doi.org/10.1093/abbs/gmq115

[37] Pal, S., Theodor, I., Peterson, E.M. and de la Maza, L.M. (1997) Immunization with an Acellular Vaccine Consisting of the Outer Membrane Complex of *Chlamydia trachomatis* Induces Protection against a Genital Challenge. *Infection and Immunity*, **65**, 3361-3369

[38] Guo, S.L., Lu, P.P., Feng, J.J., Zhao, J.P., Lin, P. and Duan, L.H. (2015) A Novel Recombinant Bivalent Outer Membrane Protein of *Vibrio vulnificus* and *Aeromonas hydrophila* as a Vaccine Antigen of American Eel (*Anguilla rostrata*). *Fish and Shellfish Immunology*, **43**, 477-484. http://dx.doi.org/10.1016/j.fsi.2015.01.017

[39] Wright, J.C., Williams, J.N., Christodoulides, M. and Heckels, J.E. (2002) Immunization with the Recombinant PorB Outer Membrane Protein Induces a Bactericidal Immune Response against *Neisseria meningitides*. *Infection and Immunity*, **70**, 4028-4034. http://dx.doi.org/10.1128/IAI.70.8.4028-4034.2002

[40] Kronvall, G. (1973) Rapid Slide-Agglutination Method for Typing *Pneumococci* by Means of Specific Antibody Adsorbed to Protein A-Containing *staphylococci*. *Journal of Medical Microbiology*, **6**, 187-190. http://dx.doi.org/10.1099/00222615-6-2-187

[41] Svenungsson, B. and Linberg, A.A. (1978) Identification of *Salmonella* Bacteria by Co-Agglutination, Using Antibodies against Synthetic Disaccharide-Protein Antigens O2, O4 and O9, Adsorbed to Protein A-Containing *staphylococci*. *Acta Pathologica et Microbiologica Scandinavica Section B*, **86**, 283-290. http://dx.doi.org/10.1111/j.1699-0463.1978.tb00045.x

2

Prevalence and Intensity of Helminth Parasites of African Catfish *Clarias gariepinus* in Lake Manzala, Egypt

Rewaida Abdel-Gaber*, Manal El Garhy, Kareem Morsy

Zoology Department, Faculty of Science, Cairo University, Cairo, Egypt
Email: *rewaida@sci.cu.edu.eg

Abstract

The African catfish, *Clarias gariepinus*, is generally considered to be one of the most important tropical catfish species for aquaculture purposes. Parasitological investigation was performed in two hundred naturally collected fish samples during the period of February to December 2014. The prevalence of gastrointestinal helminth parasites infecting *C. gariepinus* was investigated. A total of 249 helminth parasites belonging to four genera were recovered from 130 (65%) examined fish samples. They were digenea *Orientocreadium batrachoides*, cestode *Polyonchobothrium clariae*, and nematode *Procamallanus laevionchus* and *Camallanus polypteri*. Majority of the recorded parasites were found in the intestine. Female fish samples had higher prevalence rate 72 (90%) than males 58 (48.33%), and there was no significant difference (P > 0.05) in infestation rate between the two sexes. The relationship of host size (weight/length) and parasite infection showed that there was no significant difference in the parasitic infection among three classes, although fish of larger sizes had more infections. In addition, this study determines the effect of fish age on the prevalence and intensity of gastrointestinal parasites.

Keywords

Lake Manzala, Fish, *Clarias gariepinus*, Helminth Parasites

1. Introduction

Fish is regarded as the cheapest source of protein among the urban and rural populace [1] [2]. The demand for fish as a source of protein increases as the human population grows [3]. In an attempt to increase fish supply as

---

*Corresponding author.

protein source, there has been tremendous increase in the development of fish farming [4]. *Clarias gariepinus*-Burchell 1822, the African catfish is generally considered to be one of the most important tropical catfish species for aquaculture in West Africa [5], with many names such as *C. mossambicus* Peters 1852 and *C. lazera*-Valenciennes 1840 being recognized as its junior synonyms [6]. This African catfish is widely distributed throughout Africa, inhabiting tropical swamps, lakes and rivers, some of which are subjected to seasonal drying [1]. However, farmers are constraint with massive fry and fingerling mortalities, especially in culture system due to the invasion of parasites [7] [8].

Akinsanya and Otubanjo [9] reported that fish from African freshwater were infected by a variety of adult helminth parasites ranging from monogenaen, digenean, cestodes, nematodes, acathocephalans and aspidogastrean. Paperna [10] reported different helminth parasite had varying degrees of been pathogenic. For example *Spirocamalllanus spirallis*, a common nematode parasite in the stomach of catfish which was reported to be non- pathogenic in spite of the form of attachment by their buccal capsule to the stomach mucosa of infected fish [11] [12], while species of Philometra and Acanthocephalans caused mild to severe pathology in fish. Parasites of fish could also constitute health hazards to humans when ingested with poorly cooked fish [13]. Therefore, this study reports the occurrence and prevalence for some helminth parasites inhabiting the gastrointestinal tract of *C. gariepinus* in Lake Manzala, Egypt.

2. Materials and Mthods

2.1. Study Area

Lake Manzala is the largest lake in the northern region of Egypt and the most productive for fisheries [14]. It extends between longitudes 31°45' - 32°22'E and latitudes 31°00' - 31°30'N. It extends to 64.5 km in the maximum length and 49 km in the maximum width and 239 km in total length of the shore line [15].

2.2. Parasitological Examination

A total of 200 freshwater *Clarias gariepinus* (Family Clariidae) were collected alive from Lake Manzala at Kafr El-Sheikh governorate by the aid of fishermen and transported a live to laboratory of Parasitological Research in large plastic bags partially filled with water and supplied with a good aeration according to Langdon and Jones [16]. The collection was made between February and December 2014.

The total length of each fish were measured in centimeters (cm) using measuring tape, while the weight of each fish was taken in grams (g) using a weighing balance. The sex of the fish was determined by examination of the papillae. Then, the collected fish samples were dissected and the mesenteric cavity examined for parasites. The gastrointestinal tract was then dissected from the rectum to the oesophagus and all parasitic helminth encountered were carefully detached from the stomach or intestinal mucosa. The internal organs of each fish were also examined for parasites or cysts. The helminth parasites from each fish were then fixed in 70% alcohol. The parasites were later stained and identified using identification keys of Yamaguti [17], Ukoli [18], Paperna [10]. Prevalence and mean intensity for selected parasites were determined according to Margolis *et al.* [19].

2.3. Statistical Analysis

The prevalence (%), and mean intensity were analyzed according to Bush *et al.* [20]. The relationships between factors such as host sex, weight, length, and parasitic infection were obtained from data using analysis of variance (ANOVA). All statistical analysis were done using SPSS version 15 for windows.

3. Ethical Considerations

Animal use followed a protocol approved and authorized by Institutional Animal Care and Use Committee (IACUC) in Faculty of science, Cairo University, Egypt.

4. Results

A total of two hundred fish were examined from Lake Manzala, 120 out of the 200 were males while the rest were female (**Table 1, Figure 1(a)**). The overall prevalence of helminth parasite in *C. gariepinus* was 65% (130/200). In addition, the prevalence of helminth infection in relation to host sex of *C. gariepinus*. Although, the prevalence in males (48.33%) was lower than in females (90%), and it was not statically significant ($\chi^2 =$ 0.85; P > 0.05) (**Table 2**). A total number of 249 parasitic helminths were recovered from 130 infected fish, while helminth intensity was higher in the intestine than in the stomach. **Table 2** and **Figure (1(b))** shown that helminth parasites belong to 4 genera which include nematodes, *Procamallanus laevionchus* Wedl 1862 (Camallanidae) and *Camallanus polypteri* Kabre and Petter 1997 (Camallanidae); one species of cestodes, *Polyonchobothrium clariae* Woodland 1925 (Ptychobothriidae); and one species of digenea, *Orientocreadium batrachoides* Tubangui 1931 (Orientocreadiidae). In addition, nematodes have the highest occurrence (33.33%), while, cestodes showed maximum prevalence (25.0%) as tapeworms dominated in the examined fish species.

The examined fish were categorized into three groups according to their length, which were I, II, III (I larger size up to 29 cm, II medium size from 19 to less than 29, and III smaller size less than 19 cm). **Table 3** and **Figure (1(c))** showed that the smallest fish are relatively less infected than the other length groups of the examined *C. gariepinus* and the percentage of infection increases with increasing fish lengths. The prevalence observed between sizes in relation to intestinal helminth was not significant ($\chi^2 = 5.14$; p > 0.05).

In addition, the weight of the normal and infected fish of *C. gariepinus* are grouped in three classes which were I, II, III (I larger weight up to 150 gm, II medium size from 132 gm to less than 150 gm, and III smaller size

Table 1. Overall prevalence of the intestinal helminth in *C. gariepinus*.

Fish sex	Different parameters			
	No. Examined fish	No. infected fish	Prevalence (%)	Percentage of infection
Male	120	58	29%	48.33%
Female	80	72	36%	90%
Total	200	130	65%	65%

Table 2. Prevalence, range and intensity for the recorded parasites of *Clarias gariepinus*.

Recorded parasites	Prevalence (%)		Range of parasites		No. of Parasites		Intensity	
	Stomach	Intestine	Stomach	Intestine	Stomach	Intestine	Stomach	Intestine
Orientocreadium batrachoides	--	8.33%	--	1 - 8	--	27	--	6.75 ± 0.2
Procamallanus laevionchus	--	23.33%	--	6 - 15	--	46	--	3.28 ± 0.1
Camallanus polypteri	10.0%	--	4-16	--	53	--	8.83±0.3	--
Polyonchobothrium clariae	--	25.0%	--	13 - 48	--	123	--	4.39 ± 0.1

Table 3. Relationship between body length and sex for *C. gariepinus* with the degree of parasitic infection.

Size categories of *C. gariepinus*	Body length (cm)		No. examined Fish		No. infected Fish		Prevalence of infection (%)	
	Non-Infected	Infected	Male	Female	Male	Female	Male	Female
Class I	29.0 - 45.4 (36.3 ± 0.2)	42.2 - 55.6 (50.6 ± 0.1)	35	55	23	40	65.71%	72.72%
Class II	19.0 - 24.9 (22.5 ± 0.1)	21.1 - 28.9 (25.1 ± 0.1)	30	40	18	26	60%	65%
Class III	11.9 - 16.9 (14.6 ± 0.1)	15.6 - 18.9 (17.1 ± 0.1)	16	24	10	13	62.5%	54.16%

Table 4. Relationship between body weight and sex for *C. gariepinus* with the degree of infection.

Weight categories of *C. gariepinus*	Body weight (gram)		No. examined Fish		No. infected Fish		Prevalence of infection (%)	
	Non-Infected	Infected	Male	Female	Male	Female	Male	Female
Class I	172.62 - 185.46 (80.32 ± 2.20)	150.32 - 160.10 (155.45 ± 1.30)	30	50	28	34	93.33%	68%
Class II	132.34 - 150.10 (145.54 ± 1.90)	128.34 - 142.10 (135.12 ± 1.90)	20	30	13	21	65%	70%
Class III	121.23 - 127.10 (125.92 ± 1.85)	115.79 - 122.29 (119.01 ± 1.55)	30	40	17	25	56.66%	62.5%

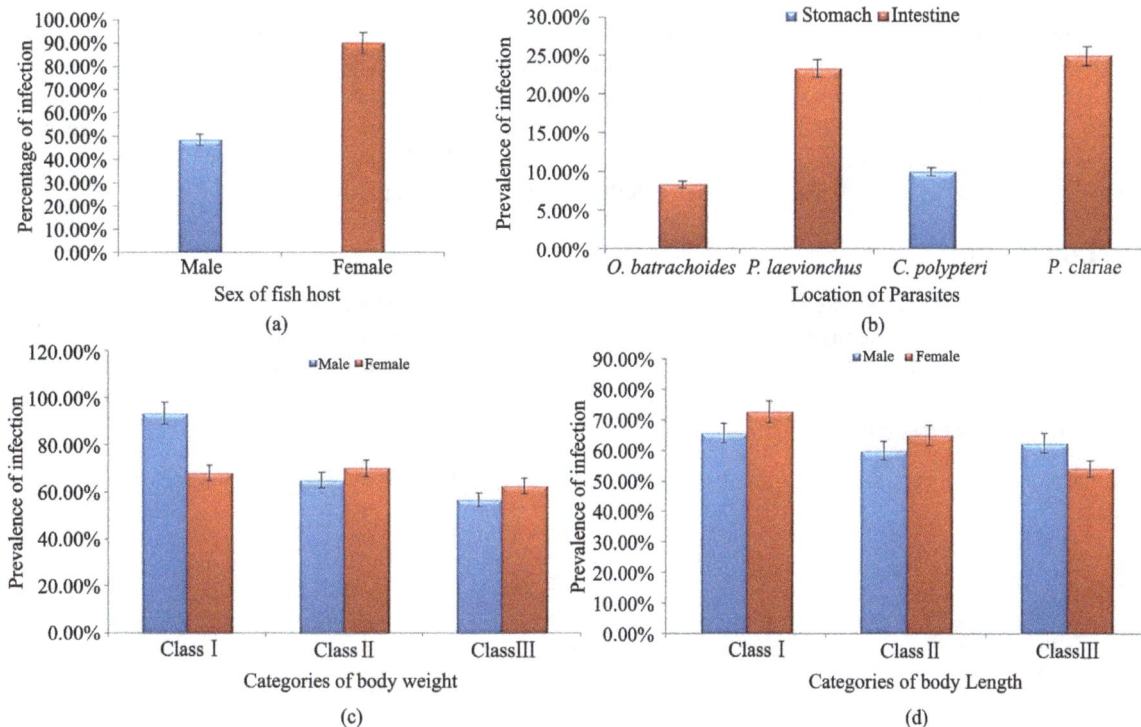

Figure 1. Different histograms for parasitic infection of *C. gariepinus* showing: (a) Percentage of infection in relation to sex fish host. (b) Distribution of parasitic infection on examined fish samples. ((c), (d)) Prevalence of infection in categorized three classes of the examined fish samples in relation to: c Body length. d Body weight.

from less than 132 gm). Parasitic helminthes were observed in all weight groups. Larger fish weights are heavily parasites than smaller ones (**Table 4**, **Figure 1(d)**).The prevalence observed between weight in relation to intestinal helminth was not significant ($\chi^2 = 8.91$; P > 0.05).

5. Discussion

African catfish is one of the most important fish species in Africa and the Middle East [21] [22]. However, parasitic infections are known to cause massive mortality in the fry and fingerling stages, especially in high-density aquaculture systems [23] [24]. In the present study, the high infection rate (65%) of *C. gariepinus* from Lake Manzala could therefore, be attributed to the contamination of the Lake by various pollutants and numerous tributaries coming from the River Nile. Khan and Thulin [25] reported that urban effluents promote aquatic pollution, thus making aquatic organism vulnerable to increased incidence of parasites. Although female fish were infected with more parasites species, infections were not significant in both males and females of *C. gariepinus*, these data are similar to the findings of Ayanda [26], who reported higher parasitic infestation in female *C. gariepinus* than the male as a result of their quest for survival, Omeji *et al.* [27], Emere and Egbe [28] who reported that due to physiological state of the female, most gravid females could have reduced resistance to infestation by parasites.

Of all helminth parasites recovered in the present study, they revealed that nematodes have the highest occurrence, while, cestodes showed maximum prevalence recovered from intestine, and these data are coincided by the results obtained by Paperna [10] and Eyo and Iyaji [29] who reported that the high infection of *C. gariepinus* by cestode parasites could be due to the ingestion of eggs, copepods and mollusks which serve as intermediate hosts of the larval stages of the cestodes.

Oniye *et al.* [30] reported that the increase in fish size is a reflection of increase in length and weight, which is hereby considered as a measure of age. In the present study, the high incidence of infestation obtained in bigger fish (>29 cm) is an indicator that size of the fish is important in determining the parasitic load compared to small fish, these data are similar to that obtained in previous reported by Mohammed [31] and Oniye and Aken'Ova

[32] who stated that the prevalence was found to be increased as the fish grow, and that could be attributed to the longer time of exposure to the environment by body size. In addition, the higher percentage of parasitic infection observed in the weight class (>150 g) indicated the increase in parasitism with increase in size. This could be due to the fact that bigger fish cover wider areas in search of food than the smaller ones and as a result, they take in more food than the smaller ones and this could expose them more to infestation by parasites. This agrees with the previous reports by Ayanda [26], Omeji *et al.* [27] [33] but disagrees with Tasawar *et al.* [34] who reported higher parasite load in smaller fish than the bigger counterparts.

6. Conclusion

It can be concluded that, there is a positive relationship between host age/size and increase in parasitism. In addition, the increased demand on fish as a source of protein should trigger further studies on fish species and their parasites to determine if there is a risk to humans by feeding more of a particular sex.

Acknowledgements

Authors extend their appreciation to Faculty of Science, Cairo University, Cairo, Egypt; which supported this work.

References

[1]   Olufemi, B.E., Akinlabi, D.A. and Agbede, S.A. (1991) Aerobic Bacterial Pathogens Isolated from the African Catfish *Clarias gariepinus*. *Tropical Veterinarian*, **9**, 177-180.

[2]   Khalil, L.F. and Polling, L. (1997) Check List of the Helminth Parasites of African Freshwater Fish. University of the North Republic of South Africa, 161 p.

[3]   Onyedineke, N.E., Obi, U., Ofoegbu, P.U. and Ukogo, I. (2009) Helminth Parasites of Some Freshwater Fish River Niger at Illushi, Edo State, Nigeria. *The Journal of American Science*, **6**, 16-21.

[4]   Usip, L.P.E., Udoidiong, O.M., Ekpo, I.E. and Ukut, I. (2014) Parasites of Cultured *Clarias gariepinus* (Burchell, 1822) from Three Fish Farms, Uyo Nigeria. *Global Advanced Research Journal of Food Science and Technology*, **3**, 84-89.

[5]   Skelton, P.H. and Teugels, G.G. (1992) Neotype Designation for the African Catfish *Clarias gariepinus* (Burchell 1822) (Pisces: Siluroidei: Clariidae). *Lchthyological Bulletin of the JLB Smith Institute of Ichthyology*, **56**, 1-8.

[6]   Skelton, P. (1993) A Complete Guide to the Freshwater Fishes of Southern Africa. Southern Book Publishers, Halfway House, 388 p.

[7]   Fagbenro, O.A.M., Adedire, C.O., Owoseeni, E.A. and Ayotunde, E.O. (1993) Studies on the Biology and Aquaculture Potential of Feral Catfish *Heterobranchus bidosalis* (Geoffry St. Hilarie 1809). *Tropical Zoology*, **16**, 67-79. http://dx.doi.org/10.1080/03946975.1993.10539209

[8]   Iyaji, F., Etim, L. and Eyo, J. (2009) Parasite Assemblages in Fish Host. *Bio-Research*, **7**, 561-570.

[9]   Akinsanya, B. and Otubanjo, O.A. (2006) Helminth Parasite of *Clarias gariepinus* (Clariidae) in Lekki Lagoon, Lagos. *Revista de Biología Tropical*, **54**, 93-99.

[10]   Paperna, I. (1996) Parasite, Infections and Disease of Fishes in Africa—An Update. *CIFA Technical Paper*, **31**, 1-220.

[11]   Paperna, I. (1964) The Metazoan Parasite Fauna of Israel Inland Water Fishes. *Bamidgeh* (*Bulletin Fish Culture Israel*), **16**, 3-66.

[12]   Khalil, L.F. (1969) Studies on the Helminth Parasites of Freshwater Fish of the Sudan. *Journal of Zoology*, **158**, 143-170. http://dx.doi.org/10.1111/j.1469-7998.1969.tb02132.x

[13]   Ibiwoye, T.I.I., Balogun, A.M., Ogunsisi, R.A. and Agbontale, J.J. (2004) Determination of the Infection Densities of Mudfish *Eustrongylides* in *Clarias gariepinus* and *Clarias anguillaris* from Bida Floodplain of Nigeria. *Journal of Applied Sciences and Environmental Management*, **8**, 39-44.

[14]   Bahnasawy, M., Khidr, A. and Dheina, N. (2011) Assessment of Heavy Metal Concentrations in Water, Plankton, and Fish of Lake Manzala, Egypt. *Turkish Journal of Zoology*, **35**, 271-280.

[15]   Donia, N. and Hussein, M. (2004) Eutrophication Assessment of Lake Manzala Using GIS Techniques. *Proceedings of the 8th International Water Technology Conference*, Alexandria, 2004, 393-308.

[16]   Langdon, J. and Jones, B. (2002) Design and Implementation of Health Testing Protocols for Fish with Special Reference to Sample Size, Test Sensitivity and Specify Predictive Value and Risk, Australian Standard Diagnostic Techniques for Fish Diseases.

[17] Yamaguti, S. (1959) Systema Helminth I. The Nematodes of Vertebrates Inter-Science Publishes Inc., New York.

[18] Ukoli, F.M.A. (1965) Occurrence, Morphology and Systematics of Caryophyllaeid cestode of the Genus Wenyonia Woodland 1923. From Fish in River Niger Nigeria. In the First Scientific Report of the Kainji Biological Research Team, 49-67.

[19] Margolis, L., Esch, G.W., Holmes, J.C., Kuris, A.M. and Scad, G.A. (1982) The Use of Ecological Terms in Parasitology (Report of an Ad-hoc committee of the American Society of Parasitology). *The Journal of Parasitology*, **68**, 131-133. http://dx.doi.org/10.2307/3281335

[20] Bush, A.O., Lafferty, K.D., Lotz, J.M. and Shostak,A.W. (1997) Parasitology Meets Ecology on Its Own Terms: Margolis *et al.* Revisited. *The Journal of Parasitology*, **83**, 575-583. http://dx.doi.org/10.2307/3284227

[21] Marcogliese, D.J. and Cone, D.K. (2001) Myxozoan Communities Parasitizing *Notropis hudsonius* (Cyprinidae) at Selected Localities on the St. Lawrence River, Quebec: Possible Effects of Urban Effluents. *The Journal of Parasitology*, **87**, 951-956. http://dx.doi.org/10.1645/0022-3395(2001)087[0951:MCPNHC]2.0.CO;2

[22] Hassan, A.A., Akinsanya, B. and Adegbaju, W.A. (2011) Impacts of Helminth Parasites on *Clarias gariepinus* and *Synodontis clarias* from Lekki Lagoon, Lagos, Nigeria. *Report and Opinion*, **2**, 42-48.

[23] Schludermann, C., Konecny, R., Laimgruber, S. and Auteurs, J.W. (2003) Fish Macroscopic Parasites as Indicator of Heavy Metal Pollution in River Sites in Austeria. *Parasitology*, **30**, 201-238.

[24] Williams, H.H. and Mackenzie, K. (2003) Marine Parasites as Pollution Indicators: An Update. *Parasitology*, **126**, S27-S41. http://dx.doi.org/10.1017/s0031182003003640

[25] Khan, R.A. and Thulin, J. (1991) Influence of Pollution on Parasites of Aquatic Animals. *Advances in Parasitology*, **30**, 201-238. http://dx.doi.org/10.1016/s0065-308x(08)60309-7

[26] Ayanda, O.I. (2009) Comparison of Parasitic Helminthes Infection between the Sexes of *Clarias gariepinus* from Asa Dam Ilorin, North-Central Nigeria. *Scientific Research and Essay*, **4**, 357-360.

[27] Omeji, S., Solomon, S.G. and Obande, R.A. (2010) A Comparative Study of the Common Protozoan Parasites of *Heterobranchus longifilis* from the Wild and Pond Environments in Benue State. *Pakistan Journal of Nutrition*, **9**, 865-872. http://dx.doi.org/10.3923/pjn.2010.865.872

[28] Emere, M.C. and Egbe, N.E.L. (2006) Protozoan Parasites of *Synodontis clarias* (A Freshwater Fish) in River Kaduna. *The Best Journal*, **3**, 58-64.

[29] Eyo, J.E. and Iyaji, F.O. (2014) Parasites of *Clarotes laticeps* (Ruppell 1832 Siluriformes, Bagridae) at Rivers Niger-Benus Confluence, Lokoja, Nigeria. *Journal of Fisheries and Aquatic Science*, **9**, 125-133. http://dx.doi.org/10.3923/jfas.2014.125.133

[30] Oniye, S.J., Adebote, D.A. and Ayanda, O.I. (2004) Helminth Parasite of *Clarias gariepinus* in Zaria, Nigeria. *African Journal of Aquatic Science*, **19**, 71-76. http://dx.doi.org/10.4314/jas.v19i2.20027

[31] Mohammed, A., Kalu, A.U., Yem, I.Y., Kolndacha, O., Nyaku, R.E. and Sanni, A.O. (2009) Bacteriological Quality of *Clarias gariepinus* (Burchell, 1822) in Lake Alau, Borno State, Nigeria. *The Best Journal*, **6**, 15-18.

[32] Oniye, S.J. and Aken'Ova, T.O. (1999) The Dynamics of Adult and Larval Stages of *Rhadinorhynchus* (horridus-Luhe,1912) in *Hyperopisusbebeoccidentalis* (hunther) from Zaria Dam. *Journal of Zoologic Society of Nigeria*, **1**, 7-8.

[33] Omeji, S., Solomon, S.G. and Uloko, C. (2013) Comparative Study on the Endo-Parasitic Infestation in *Clarias gariepinus* Collected from Earthen and Concrete Ponds in Makurdi, Benue State, Nigeria. *Journal of Agriculture and Veterinary Science*, **2**, 45-49. http://dx.doi.org/10.9790/2380-0214549

[34] Tasawar, Z., Umer, K. and Hayat, C.S. (2007) Observations on Lernaeid Parasites of Catlacatla from a Fish Hatchery in Muzaffargarh, Pakistan. *The Pakistan Veterinary Journal*, **27**, 17-19.

3

Laboratory Handling of *Didymosphenia geminata* (Lyngbye) Schmidt and the Effect of Control Efforts on Viability

Jorge Parodi[1]*, Pamela Olivares[1]*, Viviana Chavez[2], Matías Peredo-Parada[3,4]

[1]Laboratorio Fisiología de la Reproducción, Escuela de Medicina Veterinaria, Núcleo de Investigaciónen Producción Alimentaria, Facultad de Recursos Naturales, Universidad Católica de Temuco, Temuco, Chile
[2]Laboratorio de Investigación y Educación Tonalli Ltda, Temuco, Chile
[3]Departamento de Ingenieríaen Obras Civiles, Universidad de Santiago de Chile, Santiago, Chile
[4]Plataforma de Investigaciónen Ecohidrología y Ecohidráulica, EcoHyd Ltda, Santiago, Chile
Email: jparodi@uct.cl

Abstract

Didymosphenia geminata (Lyngbye) Schmidt is a type of diatom that exists in Chile as an introduced species, invading the country and its rivers. We collected samples of *D. geminata* from two sampling points in Chile, assessing their viability and response to control agents. Fresh *D. geminata* showed more than 90% of viable granular forms (containing granules in their cytoplasm); however, the dry form presents near 50% viability. By creating dry *D. geminata* through exposure to 38°C temperatures for 7 days, viability values of the granular form decreased to 20%. *D. geminata* kept at room temperature for more than 4 weeks reported values of granular forms at 50%, while samples that were refrigerated at 4°C maintained values of granular forms at 90% for 4 weeks. Previous studies suggest that high salt concentration affects the viability of *D. geminata*. When taking wet *D. geminata* samples and exposing them to a solution of 10% NaCl for 10 minutes, we observed no differences compared to the control samples, finding granular forms at 90%. When the *D. geminata* was exposed to a 5% soap solution, reductions of over 90% of the granular forms were observed. Our results suggest that the viability of *D. geminata* is associated with the granular content within their cytoplasm, and that it is possible to alter laboratory conditions for their study. These early studies are important in order to better manipulate the model in the laboratory, allowing us to obtain new evidence regarding the microalgae's biology through *in vitro* studies.

---

*Both authors collaborate in equal way.

Keywords

Plague, Viability, Laboratory, Didymo

1. Introduction

Known as "didymo", *Didymosphenia geminata* was reported by Rivera *et al.* (2013) [1] as present in the waters of the rivers of southern Chile. The microalgae are considered a pest in freshwater sources, and as a result control measures have been generated and implemented to prevent its expansion. From an environmental and social perspective, *D. geminata* may cause changes in the aquatic ecosystem [2] [3] and have negative impacts. Also, this invasive exotic alga generates economic problems for the tourism sector [4]. Its presence in aquatic ecosystems causes a loss of the ecological condition and a decrease in economic resources as a result of increased security measures being implemented and their effects on the landscape [4] [5]. *D. geminata* uses a great mass of stalks to adhere to the substrates of rivers, streams, and even lakes [6]. The species is an environmental problem that has led to the implementation of various strategies—without positive results—to control the microalgae in Chile and the rest of the world [7] [8]. In Chile, preliminary studies indicate that some rivers such as the Futaleufú, Biobío, and Puelo, are affected by *D. geminata*, and have implemented imported technology for control measures [4]. Studies indicate that *D. geminata* alters the microenvironment and reduces the fish population [9]. Furthermore, it reduces the aquatic macro-invertebrate communities and can block the filter systems used for the production of drinking water [10]-[12], and recently reports show that the *D. geminata* alters the normal function of salmon spermatozoa [13]. In Chile, there are no studies to confirm this, and the full effects that *D. geminata* has on the microflora and fish are unknown [1]. Laboratory research must be developed that replicates the conditions observed in the field, in order to understand the biology of the microalgae and test protocols for their control.

In Chile such studies have not yet been planned or carried out. Recently, a diagnosis of contaminated rivers was performed to determine factors that created conditions for *D. geminata* growth. Consequently, the handling of the microalgae in the laboratory is still incipient. We measured the mortality of *D. geminata* under different treatments commonly used for prevention or control, in order to propose a working model for *D. geminata* in the laboratory, which considered control measures that could be implemented in the field.

2. Methods

2.1. *D. geminata* Sample Collection

D. geminata was collected in the Futaleufú and Biobío rivers during the winter and spring of 2013. Samples were transported to the laboratory in plastic boxes, enclosed in darkness at 10°C. River water and colonized substrate was also collected for the microalgae.

2.2. Protocol for the Maintenance of *D. geminata* Samples

Samples collected were kept in aquariums for laboratory observation. The *D. geminata* contaminated rocks were dispersed in the aquariums, where 50% of the original river water and 50% distilled water (total volume of 14 liters) was added, making sure to leave a water column of 15 cm over them. The aquariums were maintained with insulated expanded polystyrene covers, reducing the temperature to an average of 12°C by using a cooling gel system. Water flow was kept constant using a 71,009 model Plaset-Italy 30 W power engine, and aeration. Macroscopic and microscopic changes in the aquarium with *D. geminata* were recorded daily as previous report [13].

2.3. *D. geminata* Mortality Studies

The mortality of *D. geminata* was observed by visual inspection with bright field microscopy, using an inverted Meijie (VT series, Techno Co. Ltd., Japan) microscope to observe the presence or absence of granules within the cytoplasm, denominating them as granular forms [14]. The number of *D. geminata* cells at 40× was counted, and the percentage of those that contained granules within the cytoplasm, indicating a viable form, was recorded. In order to improve the documentation with images, Nomarsky microscopy in a 40× objective (Olympus) was used for observation of intracellular structures, and was compared to the viability obtained using modified neutral red staining. To do this, samples were left 10 minutes in a neutral red solution 0.01% as a way to assess the viability of *D. geminata*. When a granular red coloration is observed, it indicates a viable form of *D. geminata* [15].

2.4. Treatments for Assessing *D. geminata* Mortality

Mortality was defined as: The percentage of *D. geminata* cells that were identified as unviable, as they did not present intracellular granules. Each sample, subjected to different treatments, and depending on the increase in non-granular forms, was assessed for its viability and compared to a control sample (untreated, fresh). Percent mortality was determined for each treatment. Fresh samples correspond to the samples collected from the aforementioned rivers, or the samples that were maintained in the laboratory for 2 months without being subjected to any treatment. *D. geminata* samples were subjected to 7 treatments in order to evaluate their efficiency on the mortality of the microalgae. These treatments included: dehydration or drying of *D. geminata* under ambient conditions; survival in seawater; treatment by acid digestion; and proposed biosafety treatments in New Zealand and Chile. Treatments with "didymo" are the following: 1) Dry; Dry samples from the sample of polluted rivers. Samples were subsequently hydrated as follows: 10 to 20 grams of algae in 50 ml of sterile distilled water. The samples were dried at room temperature and at 38°C. 2) Saline solution; NaCl concentrations between 1% and 25%. 3) Soap solutions; concentrations of 1% to 20%. 4) Sodium Hypochlorite; concentrations from 1% to 50%. 5) Natural seawater. 6) Acid; this standard treatment to prepare microalgae samples for electron microscopy involves exposing the samples to 2 or 3 ml of sulfuric acid for 35 minutes, then centrifuging them for 3 min at 4000 rpm. Afterwards, the supernatant is discarded and then 15 ml of distilled water is added. The sample is subsequently centrifuged for 3 min at 4000 rpm, repeating this wash 3 times before the samples are digested with acid. 7) Variation of temperature; microalgae samples were frozen at −20°C for 1 hr. and then thawed to observe the number of granular forms. Furthermore, they were exposed to high temperatures, using a solution of river water heated to 45°C, in which the microalgae were left for 30 minutes and then be observed under the microscope.

Finally, all treatments were subjected to visual inspection viability testing by light microscopy, Nomarsky, or neutral red staining, where the percentage of viable cells was obtained by observing the number of granular and non-granular forms in 10 different fields.

2.5. Statistical Analysis

The results are presented as the mean ± standard error of the mean (SEM). ANOVA analysis was performed, comparing all the observations. A post-test was applied, and the Bonferroni test was used for separation of means with $p < 0.05$. Levels of probability (p) less than 0.05 were considered statistically significant. All data was analyzed with the Prism 4.0 statistical program.

3. Results

3.1. Evaluation of Viability Using Intracellular Granules of *D. geminata*

We evaluated if the presence of granular forms in *D. geminata* was an indicator of viability. We compared the percentage of granular forms observed in a fresh sample using visual inspection. With a positive reaction for neutral red, we obtained similar rates for each procedure, 91% and 93%, respectively (**Figure 1(a)**). A similar situation occurs when the samples were dried at 38°C for 48 hrs. We obtained viability percentages equal to 76% by visual inspection, and 73% by the stain "neutral red". The results suggest that use of the number of granules in *D. geminata*is a good indicator of viability.

Mortality of *D. geminata*

We evaluated the effect of drying on the mortality of *D. geminata*. In **Figure 1(b)**, the variation of viability is shown in time, when the microalgae is dried at room temperature in the laboratory. The mortality samples dried at room temperature for 60 days showed a significant decline from day 15 (**Figure 1(b)**), reaching values of 41% ± 3%. In a constant temperature model in the laboratory we evaluated the mortality of *D. geminata*. In **Figure 1(c)**, the photomicrograph of the drying conditions is shown, at room temperature for 4 weeks, or 38°C for 7 days. In **Figure 1(d)**, the graph quantifying for viability under the conditions described

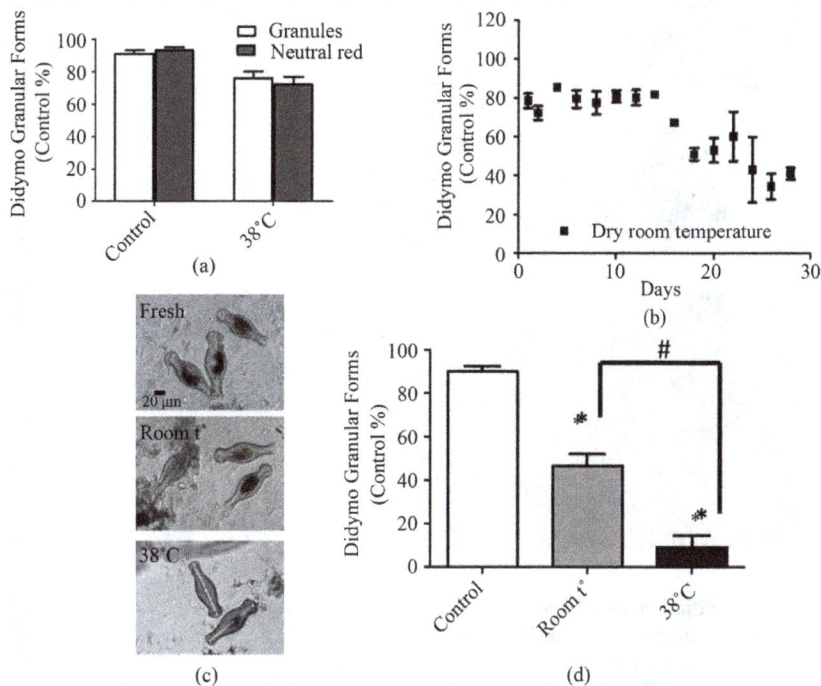

Figure 1. Maintained *Didymosphenia geminata*. (a) Comparison of neutral red staining and the observation of granular forms in fresh and 38˚C conditions; (b) Daily variation chart of dried *D. geminata* granular forms at room temperature; (c) Micrographs *of D. geminata* treated by drying. (d) Quantification of granular forms in drying conditions. The photomicrographs are representative of 5 independent observations. Each bars represents (mean ± SEM) the measurement of at least 5 independent experiments. The asterisk indicates $p < 0.05$ (ANOVA).

is shown, with an increase in mortality of *D. geminata* with values of 90% ± 5%, when the microalgae is dried at 38˚C for 7 days. The results of this treatment affirm the feasibility of quantifying death by counting granular forms, and the effect of physical treatment, such as drying, in *D. geminata* viability.

3.2. The Effect of Drying on *D. geminata* Mortality

Contaminated rivers present dried forms of the microalgae on their banks, which are usually described as dead microalgae. Hydration of these samples in the laboratory allowed us to observe that the number of granular forms is similar to the fresh samples handled in the laboratory, as demonstrated in the photomicrographs of **Figure 2(a)** and **Figure 2(b)**. This suggests that the dry material found on the banks of rivers is a potential pollutant. As a positive control for mortality, fresh samples were subjected to acid digestion, an effective method used to increase the mortality of *D. geminata* cells. **Figure 2(c)** shows the quantification of granular forms of samples that were acid-treated, fresh, or dried for 2 months and rehydrated. Acid treated samples showed 97% ± 1% mortality. In samples of fresh *D. geminata* granular forms presented at above 50%, and in the dry samples they presented at 58% ± 5%. These results suggest that drying of the material is not a highly effective control measure, even though it shows a significant change in the level of *D. geminata* mortality.

3.3. Effect of Laboratory Control Measures in *D. geminata* Mortality

The effects of salinity, surface tension, and change in the temperature have been used to decrease the viability of *D. geminata*. In our study, under the effect of 10% NaCl, we observed that at 30 minutes and 24 hours microalgae show a significant mortality, as shown in **Figure 3(a)**. **Figure 3(b)** shows the effect of increasing concentrations of NaCl, where solutions at 25% m/v reduce the viability of the samples to 34% ± 2.3%. **Figure 3(c)** shows the quantification of the effect of 10% NaCl solution when used for 30 minutes or 24 hours, demonstrating that only the 24-hour treatment is effective in increasing the mortality to 88% ± 2%, but it does not reach

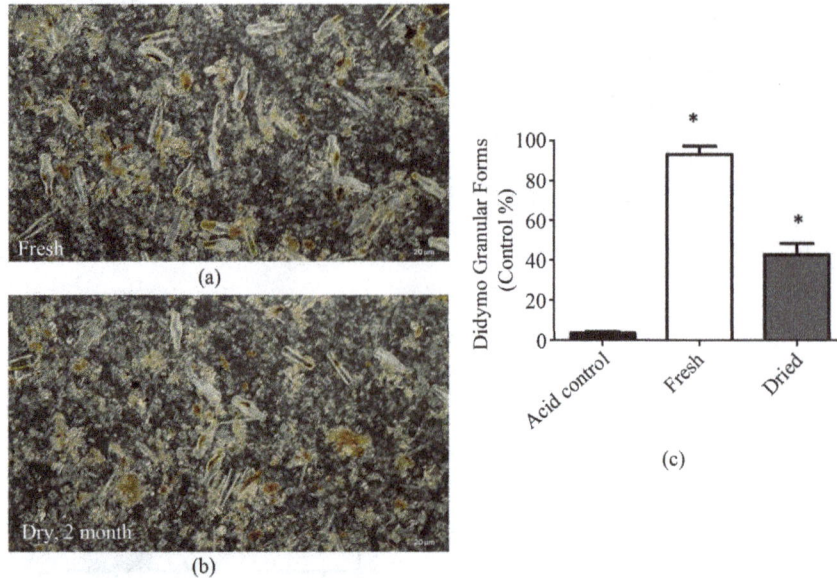

Figure 2. Hydration of *Didymosphenia geminata*. (a) Photomicrograph of fresh *D. geminata* maintained in an aquarium; (b) Photomicrograph of dry *D. geminata* hydrated in the laboratory; (c) Graph of the presence of granular forms under the conditions indicated above. Photomicrographs are representative of 5 independent experiments. Each bars represents (mean ± SEM) the measurement of at least 5 independent experiments. The asterisk indicates $p < 0.05$ (ANOVA).

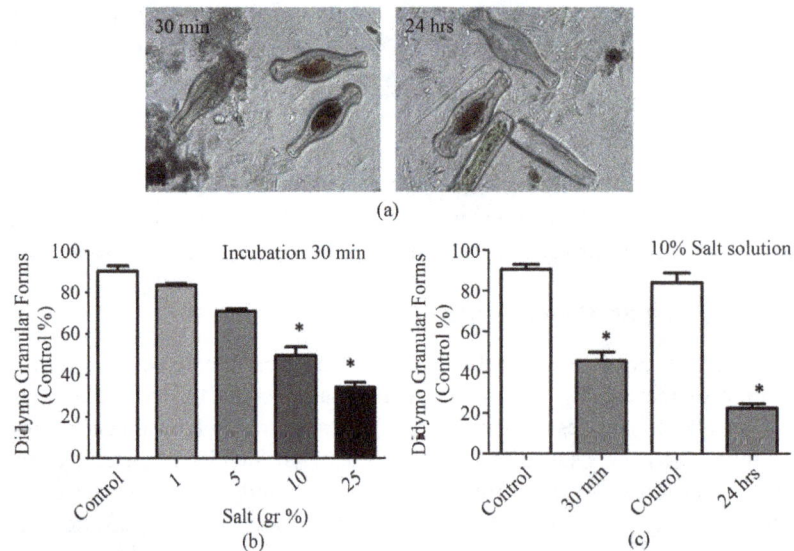

Figure 3. Effect of NaCl on *D. geminata* mortality. (a) Photomicrographs of *D. geminata* exposed 10% NaCl solution; (b) Curve of increasing NaCl concentrations; (c) Graph quantifying the effect of a 10% NaCl solution in time. Photomicrographs are representative of 5 independent experiments. Each bars represents (mean ± SEM) the measurement of at least 5 independent experiments. The asterisk indicates $p < 0.05$ (ANOVA).

100% mortality. These results suggest that the use of NaCl solutions is not effective in eliminating viable *D. geminata* cells in the laboratory. The use of other control measures described in research literature was also evaluated such as: Freezing, high temperatures, seawater, and soap solutions, to eliminate and/or control *D. geminata*. In **Figure 4(b)**, the quantification of mortality as a result of diverse biosecurity protocols used in different countries is shown. Treatments with salt or freezing are not efficient in eliminating the viability. However, the use of a 5% soap solution (soap, 5 min) was able to increase mortality of *D. geminata* cells up to a rate of 99% ±

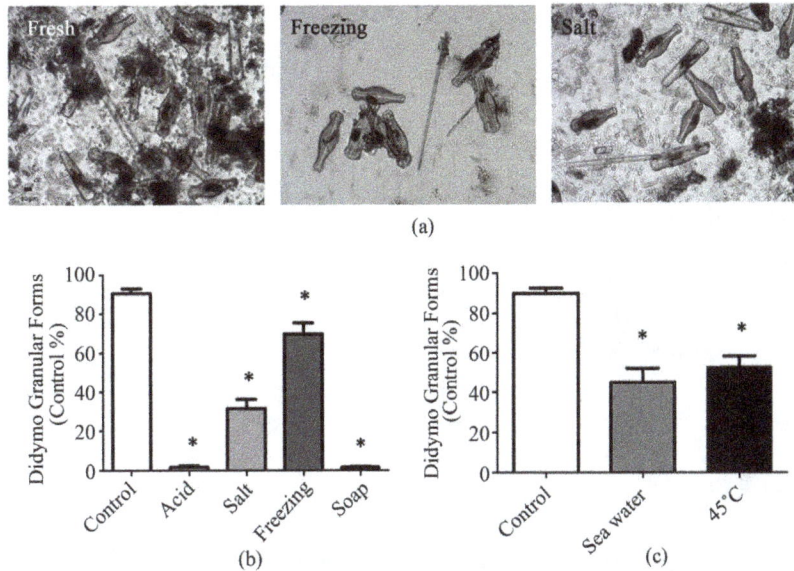

Figure 4. Other barrier methods. (a) Photomicrographs of fresh, frozen, and NaCl-treated *D. geminata*; (b) Graph of the diverse treatments that demonstrate *D. geminata* mortality; (c) Graph of unconventional treatments and their effect on *D. geminata* mortality. Photomicrographs are representative of 5 independent experiments. Each bars represents (mean ± SEM) the measurement of at least 5 independent experiments. The asterisk indicates $p < 0.05$ (ANOVA).

0.6%. **Figure 4(c)** shows the effect of seawater (for 30 min) and the subjection of algae to temperatures 45°C (for 30 min). The mortality of *D. geminata* cells reaches values close to 50% in both treatments. These results suggest that the use of soap solutions are effective in reducing the number of *D. geminata* granular forms, and are a good laboratory control measure.

3.4. Comparison of Sodium Hypochlorite and Soap Solutions in *D. geminata* Mortality

The mortality of *D. geminata* cells subjected to sodium hypochlorite is shown in **Figure 5(a)**. We observed a decrease in the number of granular forms when compared to the fresh sample that was subjected to soap solutions. **Figure 5(b)** shows the effect of increasing concentrations of chlorine and soap solutions, where the soap, at a lower concentration (10%), increases *D. geminata* cell mortality, reaching 88% ± 5%, compared to the same concentration of chlorine, which achieves 60% ± 3%. Chlorine achieved similar values at concentrations of 20% solution: 78% ± 5%. The findings of these experiments suggest that the use of sodium hypochlorite is effective in controlling the viability of *D. geminata*, but is less efficient than soap solutions.

4. Discussion

D. geminata has been studied for more than 20 years, and its problems are still present in several countries; especially in South America and New Zealand [16]. The results of this study suggest some control mechanisms for the handling of *D. geminata* in the laboratory, allowing us to establish the groundwork of protocols that will permit *in vitro* studies to be performed safely, therefore preventing the microalgae from spreading. Recently, a microculture model showed successful results; however, it has not been described as maintaining *D. geminata* or replicating the conditions of rivers, for studying their development [17]. Previous studies indicated that it is not possible to cultivate *D. geminata*, only to keep it viable for short periods of time, even when varying light or water quality conditions [6]. However, it is possible to preserve *D. geminata* when alive for diagnosis [18], and even move them in live or granular forms to collection sites before they are transported for analysis and study [19]. Yet, the description to keep them viable in an aquarium-type system is novel in this work, allowing laboratory studies to be done on the microalgae to understand their basic biology [13]. These observations indicate that it is possible to maintain *D. geminata* in the laboratory for a long time. Furthermore, and contradicting what

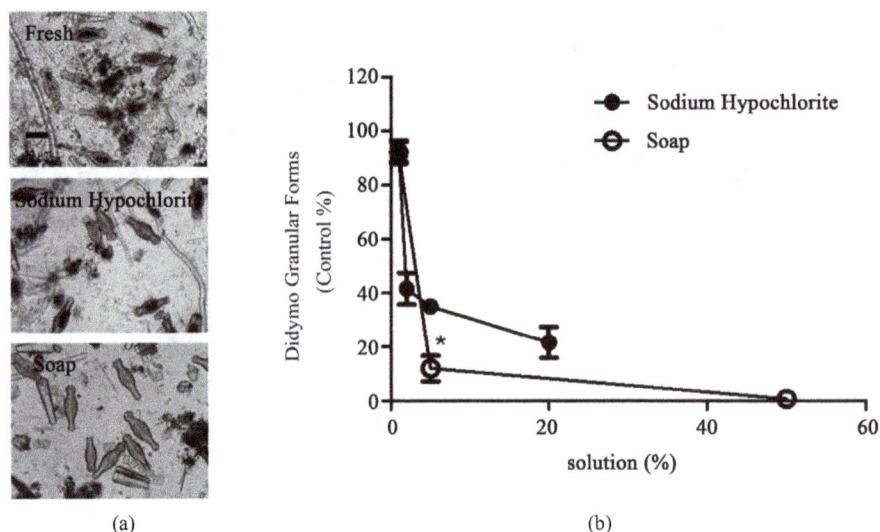

(a)   (b)

Figure 5. Comparative effects of detergent and sodium hypochlorite. (a) Photomicrographs of *D. geminata* under different conditions; (b) Graph of increasing concentrations for sodium hypochlorite and soap. Photomicrographs are representative of 5 independent experiments. Each point represents (mean ± SEM) the measurement of at least 5 independent experiments. The asterisk indicates $p < 0.05$ (ANOVA).

is shown in the literature, we have shown it is possible to study them *in vitro*. This study confirms that *D. geminata* can be maintained in the laboratory (**Figure 1(c)** and **Figure 2(a)**) using the protocols previously described in the methods section to keep them fresh, and achieving a viability of about 90% for several months. We used these methods for the fresh sample in our experiments (**Figure 3** and **Figure 4**). Thus, it seems possible to have a handling system for viable *D. geminata* in the laboratory using the protocol described above. This method can be improved to include river conditions, since the level of dissolved nutrients water has been described as favoring the growth of microalgae, but altering them would disturb the natural flora of rivers [20].

Many authors note that drying the samples of *D. geminata* would be a good way to control this pest [7], and it is widely used in various countries, including Chile. However, our findings suggest that while the method reduces the viability of the samples (**Figure 1(d)**), mortality only reaches about 50%. Mortality increases to 80% when the samples were subjected to constant temperatures of 38°C for 4 weeks (**Figure 1(d)**), an environment clearly not replicable in natural conditions, therefore requiring review of the procedure. In Chile, humid conditions would support the preservation of *D. geminata* when drying naturally; this idea is reinforced with our research, since only when using an oven-drying model (a closed system at a constant temperature) did we achieve a faster, and significant increase in sample mortality (**Figure 1(d)**). Taking this into consideration, dry *D. geminata* obtained from the rock of the rivers sampled was used, rehydrating them to assess their viability. We found that only 40% of the samples were viable, a value similar to that found in the samples maintained in aquariums (**Figure 2(b)**). The naturally dried samples retain a high viability; therefore they cannot be used as a single control method. Subjecting the "didymo" to a constant high temperature (38°C) show that it can be controlled, but this method requires other implementations and precautions. Other biosafety protocols are described as suitable for controlling *D. geminata* [20]-[22], such as the use of NaCl solutions [21]. The effect of increasing NaCl concentrations on *D. geminata* viability was evaluated (**Figure 3(b)**), showing that 25% solutions for 30 minutes achieved a *D. geminata* cell mortality of more than 80% (**Figure 3(b)**). A 10% NaCl solution was successful in reducing the viability by 50% after 30 min of incubation, and when used for longer the effect increased reaching values of 80% mortality (**Figure 3(c)**). It is suggested that the use of NaCl is a good way to control *D. geminata*; however, it leaves an important margin of viable forms, requiring an incubation time and NaCl concentration greater than what is recommended in research literature [21]. Our experiments show that using soap solution (5% soap for 5 min) reduces viability of *D. geminata* cells by more than 90% (**Figure 4(b)**), achieving values similar to those observed with acid digestion. This is an easy way to reduce *D. geminata* cell viability in a short period of time, and can be directly implemented to clean areas, equipment, and to ensure the low dispersion of samples being worked on in the laboratory. Additionally, we compared the effects of soap solution and sodium hypochlorite in controlling "didymo", finding that both treatments have significant effects on viability. However, the treatment with a 10% soap solution was more efficient (10% viability) than the 10% sodium hypochlorite solution (20% viability), and solutions with higher concentrations of soap were nearly 100% effective. Treatment with soap solution is a much more effective and inexpensive control for *D. geminata* (**Figure 5**).

5. Conclusion

Our findings show that it is possible to safely grow *D. geminata* under laboratory conditions, opening lines of research for better studying the biology and development of these microalgae. Our results indicate that the use of 5% soap solutions for 5 min is efficient in removing over 90% of the viable forms of *D. geminata*. Finally, dry samples of *D. geminata* are only efficient when they are induced at temperatures of at least 38°C for several weeks, suggesting that the drying implemented as a control mechanism up until now is not a good barrier to prevent the spread of "didymo" in rivers systems.

Acknowledgements

Funded by, the UCT Technical Assistance Agreement 278-2472 Didy 2013. Jorge Parodi has MECESUP UCT 0804 funding. We are indebted to Professor Ian Scott for his translation, revision and editing. Language editing services were provided by www.journalrevisions.com. We would like to thank Ms. Díaz, for the supply of fresh microalgae material.

References

[1]   Rivera, P., Basualto, S. and Cruces, F. (2013) On the Diatom *Didymosphenia geminata* (Lyngbye) M. Schmidt: Its Morphology and Distribution in Chile. *Gayana Botanica*, **70**, 154-158. http://dx.doi.org/10.4067/S0717-66432013000100015

[2]   Kelly, M.G. (2003) Short Term Dynamics of Diatoms in an Upland Stream and Implications for Monitoring Eutrophication. *Environmental Pollution*, **125**, 117-122. http://dx.doi.org/10.1016/S0269-7491(03)00075-7

[3]   Stoermer, E.F., Kreis Jr., R.G. and Andresen, N.A. (1999) Checklist of Diatoms from the Laurentian Great Lakes. II. *Journal of Great Lakes Research*, **25**, 515-566. http://dx.doi.org/10.1016/S0380-1330(99)70759-8

[4]   Reid, B.L., Hernandez, K.L., Frangopulos, M., Bauer, G., Lorca, M., Kilroy, C. and Spaulding, S. (2012) The Invasion of the Freshwater Diatom *Didymosphenia geminata* in Patagonia: Prospects, Strategies, and Implications for Biosecurity of Invasive Microorganisms in Continental Waters. *Conservation Letters*, **5**, 432-440. http://dx.doi.org/10.1111/j.1755-263X.2012.00264.x

[5]   Beville, S.T., Kerr, G.N. and Hughey, K.F.D. (2012) Valuing Impacts of the Invasive Alga *Didymosphenia geminata* on Recreational Angling. *Ecological Economics*, **82**, 1-10. http://dx.doi.org/10.1016/j.ecolecon.2012.08.004

[6]   Bothwell, M.L. and Kilroy, C. (2011) Phosphorus Limitation of the Freshwater Benthic Diatom *Didymosphenia geminata* Determined by the Frequency of Dividing Cells. *Freshwater Biology*, **56**, 565-578. http://dx.doi.org/10.1111/j.1365-2427.2010.02524.x

[7]   Clearwater, S.J., Jellyman, P.G., Biggs, B.J.F., Hickey, C.W., Blair, N. and Clayton, J.S. (2011) Pulse-Dose Application of Chelated Copper to a River for *Didymosphenia geminata* Control Effects on Macroinvertebrates and Fish. *Environmental Toxicology and Chemistry*, **30**, 181-195. http://dx.doi.org/10.1002/etc.369

[8]   Jellyman, P.G., Clearwater, S.J., Clayton, J.S., Kilroy, C., Blair, N., Hickey, C.W. and Biggs, B.J.F. (2011) Controlling the Invasive Diatom *Didymosphenia geminata*: An Ecotoxicity Assessment of Four Potential Biocides. *Archives of Environmental Contamination and Toxicology*, **61**, 115-127. http://dx.doi.org/10.1007/s00244-010-9589-z

[9]   Clearwater, S.J., Jellyman, P.G., Biggs, B.J., Hickey, C.W., Blair, N. and Clayton, J.S. (2010) Pulse-Dose Application of Chelated Copper to a River for *Didymosphenia geminata* Control: Effects on Macroinvertebrates and Fish. *Environmental Toxicology and Chemistry*, **30**, 181-195. http://dx.doi.org/10.1002/etc.369

[10] Kilroy, C., Larned, S.T. and Biggs, B.J.F. (2009) The Non-Indigenous Diatom *Didymosphenia geminata* Alters Benthic Communities in New Zealand Rivers. *Freshwater Biology*, **54**, 1990-2002. http://dx.doi.org/10.1111/j.1365-2427.2009.02247.x

[11] Bergey, E.A., Cooper, J.T. and Phillips, B.C. (2010) Substrate Characteristics Affect Colonization by the Bloom-Forming Diatom *Didymosphenia geminata*. *Aquatic Ecology*, **44**, 33-40. http://dx.doi.org/10.1007/s10452-009-9247-6

[12] Gillis, C.A. and Chalifour, M. (2010) Changes in the Macrobenthic Community Structure Following the Introduction

of the Invasive Algae *Didymosphenia geminata* in the Matapedia River (Québec, Canada). *Hydrobiologia*, **647**, 63-70.

[13] Olivares, P., Orellana, P., Guerra, G., Peredo-Parada, M., Chavez, V., Ramirez, A. and Parodi, J. (2015) Water Contaminated with *Didymosphenia geminata* Generates Changes in *Salmo salar* Spermatozoa Activation Times. *Aquatic Toxicology*, **163**, 102-108. http://dx.doi.org/10.1016/j.aquatox.2015.03.022

[14] Crippen, R.W. and Perrier, J.L. (1974) The Use of Neutral Red and Evans Blue for Live-Dead Determinations of Marine Plankton (with Comments on the Use of Rotenone for Inhibition of Grazing). *Stain Technology*, **49**, 97-104. http://dx.doi.org/10.3109/10520297409116949

[15] Jellyman, P.G., Clearwater, S.J., Clayton, J.S., Kilroy, C., Hickey, C.W., Blair, N. and Biggs, B.J.F. (2010) Rapid Screening of Multiple Compounds for Control of the Invasive Diatom *Didymosphenia geminata*. *Journal of Aquatic Plant Management*, **48**, 63-71.

[16] Bothwell, M.L., Taylor, B.W. and Kilroy, C. (2014) The Didymo Story: The Role of Low Dissolved Phosphorus in the Formation of *Didymosphenia geminata* Blooms. *Diatom Research*, **29**, 229-236. http://dx.doi.org/10.1080/0269249X.2014.889041

[17] Kuhajek, J.M., Lemoine, M., Kilroy, C., Cary, S.C., Gerbeaux, P. and Wood, S.A. (2014) Laboratory Study of the Survival and Attachment of *Didymosphenia geminata* (Bacillariophyceae) in Water Sourced from Rivers throughout New Zealand. *Phycologia*, **53**, 1-9. http://dx.doi.org/10.2216/13-145.1

[18] Agrawal, S.C. and Singh, V. (2002) Viability of Dried Filaments, Survivability and Reproduction under Water Stress, and Survivability Following Heat and UV Exposure in *Lyngbya martensiana*, *Oscillatoria agardhii*, *Nostoc calcicola*, *Hormidium fluitans*, *Spirogyra* sp. and *Vaucheria geminata*. *Folia Microbiologica*, **47**, 61-67. http://dx.doi.org/10.1007/BF02818567

[19] Aboal, M., Marco, S., Chaves, E., Mulero, I. and Garcia-Ayala, A. (2012) Ultrastructure and Function of Stalks of the Diatom *Didymosphenia geminata*. *Hydrobiologia*, **695**, 17-24. http://dx.doi.org/10.1007/s10750-012-1193-y

[20] Kilroy, C. and Bothwell, M. (2011) Environmental Control of Stalk Length in the Bloom-Forming, Freshwater Benthic Diatom *Didymosphenia geminata* (Bacillariophyceae). *Journal of Phycology*, **47**, 981-989. http://dx.doi.org/10.1111/j.1529-8817.2011.01029.x

[21] Root, S. and O'Reilly, C.M. (2012) Control: Increasing the Effectiveness of Decontamination Strategies and Reducing Spread. *Fisheries*, **37**, 440-448. http://dx.doi.org/10.1080/03632415.2012.722873

[22] Domozych, D.S., Toso, M. and Snyder, A. (2010) Biofilm Dynamics of the Nuisance Diatom, *Didymosphenia geminata* (Bacillariophyceae). *Nova Hedwigia*, **136**, 249-259.

4

Production of Endopolysaccharides from Malaysia's Local Mushrooms in Air-Lift Bioreactor

Shaiful Azuar Mohamad[1*], Mat Rasol Awang[1], Rusli Ibrahim[1], Choong Yew Keong[2], Mohd Yusof Hamzah[1], Rosnani Abdul Rashid[1], Sobri Hussein[1], Khairuddin Abdul Rahim[1], Fauzi Daud[3], Aidil Abdul Hamid[3], Wan Mohtar Wan Yusoff[3]

[1]Agrotechnology and Biosciences Division, Malaysian Nuclear Agency, Bangi, Malaysia
[2]Herbal Medicine Research Centre, Institute For Medical Research, Kuala Lumpur, Malaysia
[3]Faculty of Science and Technology, Universiti Kebangsaan Malaysia, Bangi, Malaysia
Email: *azuar@nuclearmalaysia.gov.my

Abstract

Four local mushroom species, viz. *Auricularis polytricha, Lentinus edodes, Agrocybe sp* and *Pleurotus flabellatus* were grown under submerged culture and screened for endopolysaccharides. The fermentation was done in 250 ml working volume Erlenmeyer flask and the fermentation curves for all species were established. *Pleurotus flabellatus* has the highest rate of biomass production at the rate of 0.180 g/L/day, at 10 days hence chosen for further investigation. Two additional media, viz. Mushroom Complete Media (MCM) and Yeast Malt (YM) were selected to be compared with potato extract(PE) media used initially. MCM media produced the highest biomass productivity at the rate of 0.311 g/L/day. *Pleurotus flabellatus* biomass was extracted using modified Mizuno method and the endopolysaccharide obtained was tested for β-glucan. The yield of β-glucan was 7.70 ± 1.11 g/100g. The polysaccharides were purified using column chromatography to yield four fractions. The fourth fraction F_4, gave the highest molecular weight at 3.058×10^6 Dalton (11.8%) and 1.282×10^4 Dalton (88.2%). The mushroom, *P. flabbelatus* was cultured using air-lift bioreactor, and the highest productivity was obtained at air-flowrate 2 L/min, yielding 2.25 g/L/day. The yield of biomass against substrate used (glucose consumption) $Y_{b/s}$ was 0.78 g/g.

Keywords

Submerged Culture Fermentation, Mushroom, β-Glucan, Column Chromatography, Molecular Weight

---

*Corresponding author.

1. Introduction

β-glucan obtained from mushrooms have been used as source of therapeutic agents functioning by modulating animal and human response and inhibiting certain tumor growth [1]-[3]. The mushroom derived polysaccharides can reduce the side effects significantly when take prior to and during radiotherapy/chemotherapy treatments [4]. Several polysaccharides including schizophyllan, lentinan, grifolan, krestin and polysaccharide-K (PSK) have been commercialized for clinical treatments of patients undergoing therapy [2]. Several reports about commercial products showed that Krestin which was derived from mycelium of *Trametes versicolor* had a molecular weight of 1.0×10^5 Dalton, Lentinan from fruit body of *Lentinus edodes* with 5.0×10^5 Dalton and Sonifilan from broth of *Schizopyllum commune* with 4.5×10^5 Dalton [3] [5]. This paper will focus on screening of β-glucan for local mushrooms grown under submerged culture fermentation.

The time taken to produce fruit bodies in solid state fermentation (SSF) often varies and especially for some medicinal mushrooms, the length tend to be longer. Submerged culture fermentation (SCF) has the advantage of producing higher quantity of mycelium, in a compact space, shorter incubation time and less contamination [6] [7]. The air-lift bioreactor will be used to compare biomass production in shake flasks.

β-glucan from local mushrooms will contribute to the development of the local industry if the productivity of SCF can be improved. This can be achieved by ensuring the productivity of the mycelium related to the endopolysaccharides production to produce at least 5% w/w of endopolysaccharides with different media.

2. Methodology

2.1. Biological Materials

The mushroom strains were collected by the Bioprocess Group, Agrotechnology and Biotechnology Division, Malaysian Nuclear Agency. The strains were maintained on potato-dextrose-agar (PDA) and subcultured every 3 months. Four local species of mushrooms tested were *Auricularia polytricha*, *Lentinus edodes*, *Agrocybe sp* and *Pleurotus flabellatus* due to its availability at the Nuclear Malaysia (NM) Mushroom Culture Collection and various reports showed the presence of β-glucan for all species.

2.2. Screening for High Biomass Species

The stock cultures of the species were transferred into the petri dish with PDA as the medium. Then 1 cm of the agar plate culture, was cut with a sterilized cutter and tranferred into a 500 ml Erlenmeyer flask containing 250 ml of media incubated using orbital shaker at 50 rpm, at room temperature of 25°C. The composition of media consists of potato extract (100 g/l) and glucose (30 g/l). The biomass was collected after 4, 6, 8, 10, 12, 14 and 16 days to obtain fermentation curves for all species.

2.3. Screening of Media

Based on literature, the media used were Mushroom Complete Media (MCM) which consists of 20 g/l glucose, 2 g/l meat peptone, 2 g/l yeast extract, 0.46 g/l KH_2PO_4, 1 g/l K_2HPO_4, and 0.5 g/l $MgSO_4 \cdot 7H_2O$. Yeast Malt (YM) consists of 10 g/l glucose, 3 g/l yeast extract, 3 g/l malt extract, and 5 g/l meat peptone. The fermentation curves of these media were compared to the initial media used.

2.4. Characterization

2.4.1. Hot Water Extraction to Produce Endopolysaccharides
The biomass (100 g) produced was extracted to obtain the endopolysaccharides using modified Mizuno method [8], involving hot water extraction for at least 2 h, filtration, concentration process and centrifugation. The supernatant was added to absolute ethanol (ratio 1:1) and kept overnight before lyophilization to get the polysaccharides.

2.4.2. Endopolysaccharide and β-Glucan Determination
The endopolysaccharide was tested using Mushroom and Yeast Beta Glucan Assay Procedure (Megazyme International Ireland Limited, 2008). The total beta glucan was obtained by hydrolysing the sample in concentrated HCl (37% v/v, ~10 M), followed by neutralization with KOH (2 M) and filtration with Whatman GF/A

glass fibre filter paper before enzymatic hydrolysis by exo-1,3 β glucanase and β-glucosidase. The α-glucan was obtained after the sample was hydrolysed with 2 M KOH followed by enzymatic hydrolysis using amyglucosidase and invertase, then filtration with Whatman No.1 filter paper. Both reactions above were reacted with Glucose Oxidase and Peroxidase (GOPOD) before measurement using UV Spectrophotometer at 510 nm.

2.4.3. Column Chromatography

The endopolysaccharides obtained from the extraction process were fractionated using Toyopearl DW-65F in column chromatography. Toyopearl DW-65F was diluted in phosphate buffer (0.05 M sodium dihidrogen phospate, 0.05 M of disodium hydrogen phosphate, and 0.1 M sodium chloride in 1 L of deionized water) and packed in a column. The fractions of endopolysaccharides sample obtained from the packed column were collected every 4 min and tested using phenol sulphuric acid test and its absorbance was measured at 490 nm.

2.4.4. Endopolysaccharides Molecular Weight Determination

Average weight of endopolysaccharides, M_w, was determined by GPC-MALLS (Gel permeation Chromatography-Multiangle Laser Light Scattering). The GPC system comprised an Agilent G1310A pump (Agilent Tecnologies, Santa Clara, USA), an Agilent G1329A auto-injector with an injection loop of 100 μL and a Wyatt 986 refractometer (Wyatt Technology, Santa Barbara, USA). The MALLS apparatus has a Wyatt Dawn-Heleos II laser photometer (Wyatt Technology, Santa Barbara, USA) equipped with a K5 flow cell and a He–Ne laser operating at k = 632.8 nm. An aqueous SEC column: Shodex OHpak SB-806 HQ (8.0 mm × 300 mm) (Showa Denko, Kawasaki, Japan) was used for the analysis.

The mobile phase consisted of a filtered (0.22 μm) phosphate buffer (0.05 M sodium dihidrogen phospate, 0.05 M of disodium hydrogen phosphate, and 0.1 M sodium chloride in 1 L deionized water) solution obtained using ultrapure water. The flow rate was 0.5 mL/min and analyses were performed at room temperature. The samples were dissolved in phosphate buffer solution and filtered (0.45 μm) to eliminate dust particles. The MALLS instrument was placed directly after the GPC columns and before the refractive index detector (DRI). Prior to measurements, a Dawn apparatus was calibrated using HPLC grade toluene and normalized using a 20 nm polystyrene latex standard (Thermo Scientific, Fremont, USA) in phosphate buffer solution. The performance of the HPSEC-MALLS system was checked with monodisperse pullulan of various molecular weights. A dn/dc value of 0.148 for β-glucan was used at wavelength 490 nm [9]. Data were collected from the DRI and MALLS and evaluated with the ASTRA software 5.3.4.14. Since β-glucans are polydisperse polysaccharides, average weights were compared. Results were estimated using second-order Zimm model.

2.5. Production in Air Lift Bioreactor (Submerged Culture Fermentation)

The 250 ml of mycelia biomass (500 ml shake flask) in MCM media was transferred to a 2.5 L working volume air lift bioreactor (5 L total volume) aseptically. The flow rates were varied from 0.5 L/min to 2.0 L/min. (vvm 0.2 to 0.8). The mycelia produced from the submerged culture fermentation were freeze-dried until constant weight. The mycelial biomass dry weights obtained were plotted against air flow rate inlet.

3. Results

3.1. Screening for High Biomass Species

The fermentation curves for all species were plotted and shown in **Figure 1**.

Figure 1 shows mycelia growth profiles in submerged culture fermentation for the four species selected using potato extract as the crude media. The biomass collected ranged from approximately 0.1 g to 0.7 g. As the fermentation duration increased, more media were consumed to produce more biomass. *P. flabellatus* produced the most consistent rate and highest biomass production whilst *L. edodes* species showed the lowest biomass production rate. The production rate of *Agrocybe sp* was slightly lower than *P. flabellatus* whilst the rate of *A. polytricha* was initially low but increased at the end of fermentation. The most consistent production was by *P. flabellatus*, with the highest rate of biomass production at 0.180 g/L/day, at the fermentation duration of 10 days.

3.2. Screening of Media

Figure 2 shows the mycelial growth profile for *P. flabellatus* using two additional media obtained from the lite-

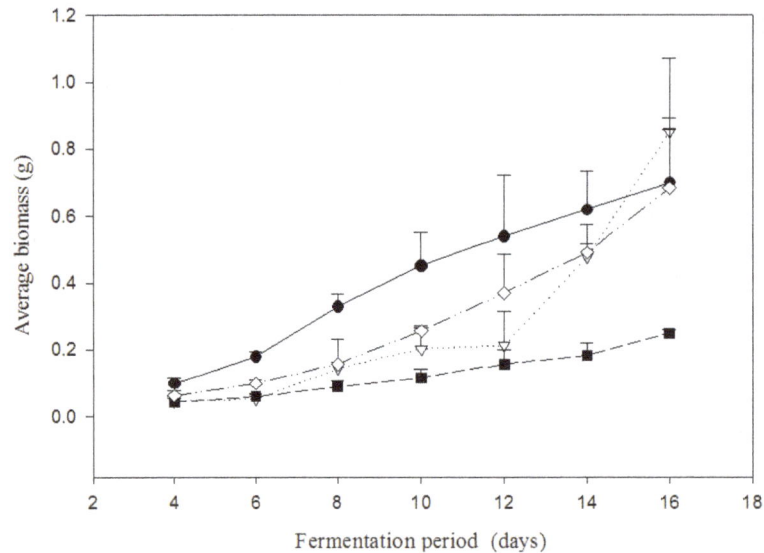

Figure 1. Fermentation curves for *Agrocybe sp*, *Auricularia polytricha*, *Lentinus edodes*, and *Pleurotus flabellatus*.

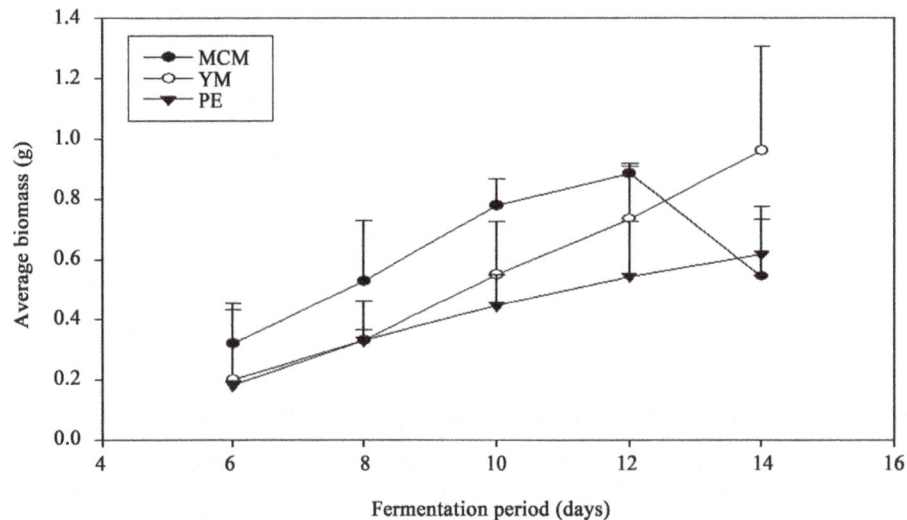

Figure 2. The fermentation curve of *P. flabellatus* using different media compared to the potato extract media.

rature compared to the initial potato extract media used. The two media produced more biomass compared to the crude media of potato extract. YM produced the most consistent rate but the production rate for MCM was higher than YM up to day 12. The calculation for each media and duration is shown in **Table 1**.

From the table, the highest production rate was 0.311 g/L/day using media MCM, and again the fermentation period of 10 days. Using this media the production rate increased by 72.7%. Hence, MCM was chosen for subsequent experiment.

3.3. Characterization of Polysaccharides

3.3.1. β-Glucan Determination

After the extraction process using Modified Mizuno method, 100 mg of polysaccharide from the mycelium of *P. flabellatus* sample was used to test the presence of beta glucan using the beta glucan assay kit (**Table 2**). The total glucan and α-glucan were obtained from the test done. The amount of β-glucan was obtained by subtraction of α-glucan from the total glucan. The total glucan in the biomass was 17.54 ± 2.91 g/100g whilst beta glucan yield was 7.70 ± 1.11 g/100g. No publication has reported this finding for the species studied.

3.3.2. Column Chromatography and Molecular Weight Determination

From the column chromatography, the value of absorbance from each fractions obtained from phenol sulphuric acid test were plotted against the number of bottles collected at 4 min interval from the column as shown in **Figure 3**. Samples from bottle 6 - 14, 15 - 21, 25 - 37 and 38 - 60 were combined to give fraction F_1, F_2, F_3 and F_4, respectively, to be analyzed further.

The four fractions were run in GPC-MALLS to determine the molecular weight. The F_4 has the highest molecular weight with two possible molecular weight 3.058×10^6 Dalton (11.8%) and 1.282×10^4 Dalton (88.2%). Other fractions indicated a lower molecular weight in the range of $\sim 10^3$ Dalton.

3.4. Production in Air-Lift Bioreactor

Table 3 showed the biomass and productivity in the air-lift bioreactor. The highest productivity of biomass in air-lift bioreactor with 2.5 L working volume of *Pleurotus flabellatus* is 2.25 g/L/day at volume per volume per min (vvm) 0.8. The air inlet flow rate did not seem to affect the productivity of the biomass very much. The yield of biomass against substrate used (glucose consumption) $Y_{b/s}$ was 0.78 g/g.

4. Discussion

The mushroom species *P. flabellatus* was chosen due to its consistency and highest production rate of mycelium at the rate of 0.180 g/L/day. For the media screening, MCM was chosen with highest productivity at the rate of

Table 1. The production rate calculation for different media.

Fermentation days	MCM (g/L/day)	YM (g/L/day)	PE (g/L/day)
6	0.213 ± 0.090	0.133 ± 0.156	0.122 ± 0.01
8	0.263 ± 0.100	0.165 ± 0.065	0.165 ± 0.018
10	0.311 ± 0.036	0.220 ± 0.070	0.179 ± 0.040
12	0.294 ± 0.008	0.244 ± 0.061	0.181 ± 0.061
14	0.155 ± 0.066	0.274 ± 0.099	0.176 ± 0.032

Table 2. The yield of β-glucan from crude extract of *P. flabellatus*.

Species	Total glucan (mg/100mg)	α-glucan (mg/100mg)	β-glucan (mg/100mg)
Mycelium of *P. flabellatus*	17.54 ± 2.91	9.84 ± 3.82	7.70 ± 1.11

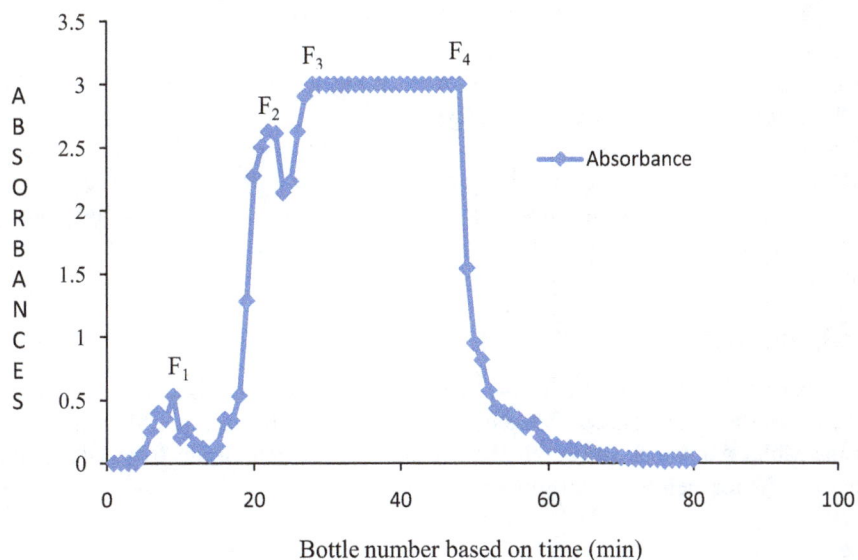

Figure 3. The reading of phenol sulphuric acid test from column chromatography of endopolysaccharides from *Pleurotus flabellatus*.

Table 3. The productivity of *Pleurotus flabellatus* biomass production in air-lift bioreactor.

Air inlet flowrate (L/min)	vvm	Mycelium biomass (g)	Reducing sugar (g/L)	Productivity (g/L/day)
0.5	0.2	20.40 ± 1.98	10.69 ± 1.71	2.04 ± 0.20
1.0	0.4	22.00 ± 2.12	8.30 ± 0.61	2.20 ± 0.21
1.5	0.6	20.85 ± 1.63	10.49 ± 2.36	2.09 ± 0.16
2.0	0.8	22.50 ± 4.10	8.49 ± 1.19	2.25 ± 0.41

0.311 g/L/day. In a similar study for *Ganoderma resinaceum* in 250 ml shake flask using MCM medium, the biomass production rate obtained was 0.333 g/L/day [10]. Another study indicated that the maximum biomass produced for *Pleurotus sajor caju* in shake flask was 6.5 g/l in 10 days (0.650 g/L/day) using deproteinized whey, diammonium phosphate and yeast extract as fermentation medium [11].

For the β-glucan content using assay kit by Megazyme, a paper reported that the β-glucan in endopolysaccharides of *Lentinus squarrosulus* was 11.36 ± 0.27 (%w/w) for the hot water extract in submerged culture fermentation [12]. Another paper reported the β-glucan content obtained from the fruit body of *G. applanatum*, *T. versicolor*, *L. edodes*, and *G. lucidum* to be 16.0, 33.4, 41.2 and 41.4 g/100g, respectively using dry weight of dialyzed crude extract [13]. This researcher used higher purity of crude extract using dialysis technique.

The high molecular weight in the order of 10^6 with 3.058×10^6 Dalton (11.8%) and 1.282×10^4 Dalton (88.2%) obtained from *P. flabellatus* indicated this compound has the potential to be explored for anti-tumor as reported by Akramiene and coworkers [14] regarding the application of high molecular weight β-glucan. Another report showed that the insoluble glucan obtained from yeast separated using size exclusion chromatography also has two peaks with molecular weight of 1×10^6 Da (1% of total mass) and 1.5×10^4 Da (99% of total mass) [15].

For the production of biomass using air-lift bioreactor, the value of 2.25 g/L/day reported in this experiment is in the same range as reported by Cho and coworkers [16]. The report showed that for *Tremella fuciformis* in 5 L airlift bioreactor, the maximum dry weight obtained was 10.30 g/l at days 5 (productivity 2.06 g/L/day). For the same species using stirred-tank bioreactor, the cell dry weight obtained was 8.83 g/L (productivity 1.77 g/L/day) [16].

In another study using stirred tank fermentor, the productivity of *Pleurotus sajor-caju* biomass was 0.648 g/L/day, with 3 L working volume, agitation speed of 150 rpm, and aeration rate of 2 vvm [17].

5. Conclusions

The species *P. flabellatus* has the highest biomass productivity (0.180 g/L/day) with potato extract as the crude media. Enhanced biomass productivity (0.311 g/L/day) was achieved with MCM. The yield of beta glucan from submerged culture fermentation of *P. flabellatus* was 7.70 ± 1.11 g/100g. The productivity of biomass in airlift bioreactor was 2.25 g/L/day and approximately 12.5 times higher compared to the initial value. The fourth fraction F_4 gave the highest molecular weight with 3.058×10^6 Dalton (11.8%) and 1.282×10^4 Dalton (88.2%).

The β-glucan (1.3:1.6) from *Pleurotus flabellatus* species has the potential to be produced in submerged culture fermentation at a higher productivity. The quantity and quality of the β-glucan can be purified further and tested for its effectiveness towards anti-tumor application. The high molecular weight produced from this research can be analyzed for a single compound and determine its exact molecular structure. The air-lift bioreactor can be custom-made and produced at a cheaper price locally.

Acknowledgements

I would like to thank MOSTI for the grant 02-03-01-SF0157 supporting this research and the staff from Nuclear Malaysia, Mr. Hassan Hamdani Hassan Mutaat, Mr. Mohd Meswan Maskom, Ms Nurul Shahnadz Amir Hamzah, Ms Liyana Mohd Ali Napia, Industrial Biotechnology Research Group from UKM and Herbal Medicine Research Centre, IMR for their technical support.

References

[1]   Wasser, S.P. (2002) Medicinal Mushrooms as a Source of Antitumor and Immunomodulating Polysaccharides. *Applied Microbiology and Biotechnology*, **60**, 258-274. http://dx.doi.org/10.1007/s00253-002-1076-7

[2]   Zhang, M., Cui, S.W., Cheung, P.C.K. and Wang, Q. (2007) Antitumor Polysaccharides from Mushrooms: A Review on Their Isolation Process, Structural Characteristics and Antitumor Activity. *Trends in Food Science & Technology*, **18**, 4-19. http://dx.doi.org/10.1016/j.tifs.2006.07.013

[3]   Ooi, V.E.C. and Liu, F. (2000) Immunomodulation and Anti-Cancer Activity of Polysaccharide-Protein Complexes. *Current Medicinal Chemistry*, **7**, 715-729. http://dx.doi.org/10.2174/0929867003374705

[4]   Smith, J.E., Rowan, N.J. and Sullivan, R. (2002) Medicinal Mushrooms: Their Therapeutic Properties and Current Medical Usage with Special Emphasis on Cancer Treatments. University of Strathclyde & Cancer Research, UK.

[5]   Mizuno, T. (1999) The Extraction and Development of Antitumor-Active Polysaccharides from Medicinal Mushrooms in Japan. *International Journal of Medicinal Mushroom*, **1**, 9-29. http://dx.doi.org/10.1615/IntJMedMushrooms.v1.i1.20

[6]   Bae, J.T., Sinha, J., Park, J.P., Song, C.H. and Yun, J.W. (2000) Optimization of Submerged Culture Conditions for Exobiopolymers Production by *Paecilomyces japonica*. *Journal of Microbiology and Biotechnology*, **10**, 482-487.

[7]   Choi, D.B., Lee, J.H., Kim, Y.S., Na, M.S., Choi, O.Y., Lee, H.D., Lee, M.K. and Cha, W.S. (2011) A Study of Mycelial Growth and Exopolysaccharides Production from a Submerged Culture of *Mycoleptodonoides aitchisonii* in Air-Lift Bioreactor. *Korean Journal of Chemical Engineering*, **28**, 1427-1432. http://dx.doi.org/10.1007/s11814-011-0109-2

[8]   Mizuno, T., Ando, M., Sugie, R., Ito, H., Shimura, K., Sumiya, T. and Matsuura, A. (1992) Antitumor Activity of Some Polysaccharides Isolated from an Edible Mushroom Ningyotake, the Fruiting Bodies and the Cultured Mycelium of *Polyporous confluens*. *Bioscience Biotechnology, Biochemistry*, **56**, 34-41. http://dx.doi.org/10.1271/bbb.56.34

[9]   Young, S.H. and Castranova, V. (2005) Toxicology of 1-3-Beta Glucans: Glucan as a Marker for Fungal Exposure. CRC Press, Boca Raton. http://dx.doi.org/10.1201/9780203020814

[10]   Kim, H.M., Paik, S.Y., Ra, K.S., Koo, K.B., Yun, J.W. and Choi, J.W. (2006) Enhanced Production of Exopolysaccharides by Fed-Batch Culture of *Ganoderma resinaceum*. *The Journal of Microbiology*, **44**, 233-242.

[11]   Mukhopadhyay, R., Chatterjee, S., Chatterjee, B.P. and Guha, A.K. (2005) Enhancement of Biomass Production of Edible Mushroom *Pleurotus sajor-caju* Grown in Whey by Plant Growth Hormones. *Process Biochemistry*, **40**, 1241-1244. http://dx.doi.org/10.1016/j.procbio.2004.05.006

[12]   Ahmad, R., Muniandy, S., Abdullah Shukri, N.I., Alias, S.M.U., Abdul Hamid, A., Wan Yusoff, W.M., Senafi, S. and Daud, F. (2014) Antioxidant Properties and Glucan Compositions in Various Crude Extract from *Lentinus squarrosulus* Mycelia Culture. *Advances in Bioscience and Biotechnology*, **5**, 805-814. http://dx.doi.org/10.4236/abb.2014.510094

[13]   Kozarski, M., Klaus, A., Niksic, M., Vrvic, M.M., Todorovic, N., Jakovljevic, D. and Van Griensven, L.J.L.D. (2012) Antioxidative Activities and Chemical Characterization of Polysaccharide Extracts from the Widely Used Mushrooms *Ganoderma applanatum, Ganoderma lucidum, Lentinus edodes* and *Trametes versicolor*. *Journal of Food Composition and Analysis*, **26**, 144-153. http://dx.doi.org/10.1016/j.jfca.2012.02.004

[14]   Akramiene, D., Kondrotas, A., Didziapetriene, J. and Kevelaitis, E. (2007) Effects of β-Glucans on the Immune System. *Medicina* (*Kaunas*), **43**, 597-606.

[15]   Tzianabos, A.O. (2000) Polysaccharide Immunomodulators as Therapeutic Agents: Structural Aspects and Biologic Function. *Clinical Microbiology Reviews*, **13**, 523-533. http://dx.doi.org/10.1128/CMR.13.4.523-533.2000

[16]   Cho, E.J., Oh, J.Y., Chang, H.Y. and Yun, J.W. (2006) Production of Exopolysaccharides by Submerged Mycelia Culture of a Mushroom *Tremella fuciformis*. *Journal of Biotechnology*, **127**, 129-140. http://dx.doi.org/10.1016/j.jbiotec.2006.06.013

[17]   Kim, S.W., Hwang, H.J., Park, J.P., Cho, Y.J., Song, C.H. and Yun, J.W. (2002) Mycelial Growth and Exo-Biopolymer Production by Submerged Culture of Various Edible Mushrooms under Different Media. *Letters in Applied Microbiology*, **34**, 56-61. http://dx.doi.org/10.1046/j.1472-765x.2002.01041.x

5

Effects of Media Composition and Auxins on Adventitious Rooting of *Bienertia sinuspersici* Cuttings

Jennifer Anne Northmore, Marie Leung, Simon Dich Xung Chuong*

Department of Biology, University of Waterloo, Waterloo, Canada
Email: *schuong@uwaterloo.ca

Abstract

An efficient *in vitro* method for rapid vegetative propagation of *Bienertia sinuspersici*, one of four terrestrial species of family Chenopodiaceae capable of performing C_4 photosynthesis within a single cell, was developed. Cuttings of *B. sinuspersici* were used to examine the effects of Murashige and Skoog (MS) media strength and auxins on adventitious root formation. Half-strength MS medium was determined to be ideal for adventitious root formation in Bienertia cuttings. Although cuttings cultured in medium containing 5.0 mg/L α-naphthalene acetic acid (NAA) promoted the highest number of adventitious roots, cuttings cultured in medium supplemented with 1.0 mg/L indole-3-butyric acid (IBA) produced the longest adventitious roots and had the highest survival rate upon transplanting to soil. Histological analysis revealed variations in the root anatomy generated by the various auxins which may affect adventitious root formation and subsequent establishment of cuttings in soil. Overall, the established procedure provides a simple and cost-effective means for the rapid propagation of the single-cell C_4 species *B. sinuspersici*.

Keywords

Bienertia sinuspersici, Single-Cell C_4 Photosynthesis, MS Media, Auxins, Adventitious Roots

1. Introduction

Bienertia sinuspersici, a halophytic shrub-like member of the Chenopodiaceae family indigenous to Southeast Asia, was discovered to possess a novel form of single-cell C_4 photosynthesis [1]. In this species, key photosynthetic enzymes and organelles such as chloroplasts and mitochondria are partitioned into two intracellular cy-

---

*Corresponding author.

toplasmic compartments within individual chlorenchyma cells achieving the equivalent functions of the mesophyll and bundle sheath cells in the Kranz-type C_4 plants [1]-[3]. The unique single-cell C_4 photosynthetic mechanism of *B. sinuspersici* has altered our understanding of how plants have evolved innovative solutions to improve photosynthetic efficiency. This discovery has also provided plant researchers with possibility of engineering C_3 crops to utilize the more efficient C_4 pathway to improve yields under extreme environmental conditions.

Conventional propagation of *B. sinuspersici* from seed-derived plants is insufficient in generating adequate plant material for basic cellular and molecular research purposes mainly due to its slow growth process. In addition, the limited seed stock and low seed production and viability further discourage its propagation via this method. Plants produced through vegetative asexual propagation are genetically identical to the parent and therefore contain the same desired phenotypes [4]. Moreover, there is an urgent need to maintain a healthy living stock so that sufficient material of this fascinating species can be rapidly generated for various research experiments. Therefore, an *in vitro* propagating method was established for *B. sinuspersici* using cuttings from seed-derived plants.

Vegetative asexual propagation provides a means for high-frequency replication of plant material through the use of cuttings. Vegetative asexual propagation relies on branches of an established mature plant as the starting material for creating a genetically identical daughter plant and that there is a sizeable increase in propagation efficiency as compared to growing plants from seeds. Adventitious root induction of cuttings has been established and employed as a routine method of propagating many plants for their agricultural, horticultural, industrial, or pharmaceutical properties [5] [6]. Vegetative asexual propagation is particularly advantageous for rare or attractive plants where a specific trait like flower size or colour is desired in daughter plants, which is of particular important value in the horticulture industry [4]. For example, vegetative cuttings are used to reproduce many grape vines to maintain a single cultivar with a particular appearance, and flavour within a vineyard [7]. Moreover, vegetative propagation is efficient because it reduces the amount of time required for seedling establishment and a large number of healthy living stock plants can be produced rapidly from a small amount of available original plant material. Rooted cuttings can also serve as potential rootstocks for grafting experiments for plant species that are recalcitrant to adventitious root induction process or for multiplication of transgenic shoots [8] [9]. In addition, findings on nutrient requirements and effects of auxins on rooting of *B. sinuspersici* cuttings may provide valuable insight into factors that influence the *in vitro* regeneration process.

In general, root induction can be initiated by adding an auxin to the growth medium [10]. The role of auxins in organ formation, especially root induction and adventitious root formation, has been invaluable in the field of plant tissue culture [4]. Rooting is primarily regulated by a high auxin to cytokinin ratio, therefore root formation can generally be induced by adding an auxin to the growth medium [10]. Common auxins used for root induction *in vitro* include 2,4 dichlorophenoxyacetic acid (2,4-D), indole-3-acetic acid (IAA), indole-3-butyric acid (IBA), and α-naphthalene acetic acid (NAA).

Except for two reports on the regeneration using indirect organogenesis as well as direct shoot micropropagation [11] [12], no additional *in vitro* studies on the rapid vegetative propagation of *B. sinuspersici* have been documented. The purpose of this study is to examine the effects of Murashige and Skoog (MS) basal medium strengths and various auxins on the induction of adventitious roots in *B. sinuspersici* cuttings so that an efficient and rapid method for vegetative propagation of massive plant material can be established.

2. Materials and Methods

2.1. Plant Materials

Cuttings (~5 - 8 cm) were obtained from six independent 6- to 8-month old *Bienertia sinuspersici* plants derived from seeds grown in a growth chamber (model GCW-15H; Environmental Growth Chambers, Ohio, USA) under 350 μmol/m^2/s^1 with a 14 h/10h light/dark photoperiod and a 25/18°C day/night temperature regime. Plants were fertilized with Miracle-Gro (24-8-16) and 150 mM NaCl once a week and watered when needed. Leaves were excised from the bottom 3 cm of the stem before cuttings were surface sterilized with 10% (v/v) bleach for 10 min and rinsed three times with sterile dH$_2$O for 5 min each.

2.2. Root Induction and Greenhouse Acclimation

Four sterilized cuttings were placed in each Magenta box containing various concentrations of MS medium (quarter-, half-, three quarter-, or full-strength). Full strength MS media consisted of 4.32 g/L MS basal medium with Gamborg's vitamins (Sigma-Aldrich, Ontario, Canada), 1.96 g/L MES, 30 g/L sucrose, and 7 g/L agar, adjusted to pH 5.8 and autoclaved at 121°C, 1.1 kg/cm^2 for 30 min. After autoclaving, the media was allowed to cool to approximately 60°C before hormones were added. All plant hormones were purchased from Sigma-Aldrich and prepared as 1 mg/ml working stock solutions in hydrochloric acid, filter sterilized through a 0.2 μm filter, and stored at −20°C.

Root induction of cuttings was also performed on half-strength MS medium supplemented with auxins (IAA, IBA, NAA, and 2,4-D) at 0.5, 1.0, 2.5, and 5.0 mg/L. Cuttings were maintained in a growth chamber (Percival Scientific, Iowa, USA) with a 14 h/10h light/dark photoperiod at 22°C under 25 μmol/m^2/s[1]. Rooted cuttings were carefully removed from the agar, gently washed in water, photographed, and the number and length of roots was recorded for each plant before transplanting to 10 cm pot containing Sunshine mix #4 potting soil. Transplanted cuttings were maintained in a humidifying dome and gradually acclimatized to greenhouse conditions. The survival rate of transplanted cuttings was determined 3 weeks after the plants were completely uncovered and transferred to greenhouse conditions.

2.3. Histological Analysis

Root samples were fixed in 2% (v/v) paraformaldehyde and 2% (v/v) glutaraldehyde (Electron Microscopy Sciences [EMS], Pennsylvania, USA) in 50 mM PIPES buffer, pH 7.2 overnight at 4°C, dehydrated with a graded ethanol series and gradually infiltrated with increasing concentrations of London Resin White (LR white; EMS, Pennsylvania, USA) acrylic resin. Sections (1 μm) were prepared on a Reichert Ultracut E ultramicrotome (Reichert-Jung, Heidelberg, Germany), dried onto glass slides, and stained with 0.1% (w/v) Toluidine blue (TBO) (Sigma-Aldrich, Ontario, Canada). Images were captured using a cooled CCD camera (Retiga 1350 Exi Fast, Qimaging, British Columbia, Canada) and OpenLab (OpenLab, Ontario, Canada) imaging software. Image processing was performed using Adobe Photoshop CS (Adobe, California, USA).

2.4. Statistical Analysis

Each treatment consisted of twenty cuttings and the experiment was repeated at least three times. To assess the treatment differences, the results were analyzed by analysis of variance (ANOVA) and the variation among means were analyzed by Duncan's multiple range test at $P < 0.05$.

3. Results

3.1. Effect of MS Concentration on Adventitious Rooting of Cuttings

After 3 weeks of culture, adventitious roots were observed on the cuttings in all treatments (**Table 1**). There was no significant difference in the number of roots formed among the different MS-containing media, however the presence of half- or three quarter-strength MS basal salts in the medium improved root length significantly. The highest number of roots were observed in the treatment containing quarter-strength MS, while three-quarter-strength MS produced the longest roots. Although media containing half-strength MS produced a compromise between root length and number, this treatment had the highest survival rate after transplantation. Therefore, half-strength MS was used for subsequent experiments to examine the effect of auxins on adventitious

Table 1. Effect of MS concentration on adventitious root induction from *B. sinuspersici* cuttings after 3 weeks of culture.

MS strength	Number of roots/cutting	Root length (cm)	Survival rate (%)
0	5.9 ± 2.5^a	3.2 ± 1.9^b	41.4 ± 12.0^c
1/4	10.0 ± 2.2^a	3.6 ± 2.4^b	100^a
1/2	9.4 ± 3.0^a	6.3 ± 3.8^{ab}	100^a
3/4	6.0 ± 3.1^a	9.8 ± 2.8^a	71.4 ± 9.0^b
1	8.6 ± 2.4^a	4.9 ± 2.2^{ab}	71.4 ± 11.0^b

Results represent means ± standard error mean (SEM) of three replicated experiments. For each column, numbers followed by the same letter are not significantly different using Duncan's Multiple Range Test, P < 0.05.

roots formation. Rooted cuttings were removed from media, transplanted to soil, and acclimatized to greenhouse conditions (**Figure 1**). Overall, the survival rates of rooted cuttings from MS treatments were significantly higher than that of the control (0 MS) with the quarter- and half-strength MS treatments providing the best rates (**Table 1**).

3.2. Effects of Auxin on Adventitious Rooting of Cuttings

After three weeks in culture, significant differences were found between many of the treatments cultured on half-strength MS media supplemented with various auxins (2,4-D, IAA, IBA, and NAA) (**Table 2**). **Table 2** shows that the addition of 1 mg/L IBA was most successful at inducing adventitious roots from cuttings. The addition of IBA to the growth medium produced several long roots with many lateral roots and root hairs (**Figure 2(c)**). These IBA-rooted cuttings had the highest survival rate after transplanting to soil. In general, cuttings

Table 2. Effect of different auxins on adventitious root induction from *B. sinuspersici* cuttings after 3 weeks of culture.

Hormone	Concentration (mg/L)	Number of roots/cutting	Root length (cm)	Survival rate (%)
2,4-D	0	11.8 ± 2.2^d	16.4 ± 7.4^{ab}	77.8 ± 10.1^b
	0.5	10.4 ± 3.2^d	1.8 ± 0.5^c	0.0^e
	1.0	13.4 ± 3.7^c	2.0 ± 0.7^c	50.0 ± 8.0^c
	2.5	5.6 ± 2.7^{de}	1.4 ± 0.5^c	25.0 ± 7.6^d
	5.0	1.7 ± 0.7^e	0.5 ± 0.4^e	17.0 ± 4.0^d
IAA	0.5	21.0 ± 5.0^b	2.7 ± 0.5^c	50.0 ± 10.4^c
	1.0	23.1 ± 4.1^b	3.4 ± 0.6^c	49.5 ± 9.4^c
	2.5	28.3 ± 8.2^b	2.9 ± 0.6^c	83 ± 8.7^b
	5.0	21.8 ± 7.6^{bc}	3.4 ± 0.6^c	17.0 ± 5.4^d
IBA	0.5	14.3 ± 1.9^c	27.3 ± 6.1^a	83.0 ± 5.4^b
	1.0	9.1 ± 1.6^d	29.4 ± 7.0^a	100^a
	2.5	10.7 ± 0.3^d	25.3 ± 6.7^a	100^a
	5.0	14.8 ± 3.7^{bc}	15.0 ± 5.6^{ab}	100^a
NAA	0.5	15.5 ± 4.4^{bc}	6.7 ± 2.6^b	100^a
	1.0	19.4 ± 6.0^{bc}	6.1 ± 3.0^b	50.0 ± 7.3^c
	2.5	16.5 ± 8.4^{bc}	6.4 ± 4.2^{bc}	66.0 ± 5.8^{bc}
	5.0	44.2 ± 13.6^a	1.3 ± 0.4^d	77.3 ± 9.3^b

Results represent means ± standard error mean (SEM) of three replicated experiments. For each column, numbers followed by the same letter are not significantly different using Duncan's Multiple Range Test, $P < 0.05$.

Figure 1. Induction of adventitious roots and plant regeneration from cuttings of *Bienertia sinuspersici*. (a) Sterilized cuttings of *B. sinuspersici* culture in rooting medium. (b) Cuttings with adventitious roots after 3 weeks of culture. (c) Cuttings with many long healthy roots from 1 mg/L IBA treatment. (d) Transplanted cutting fully acclimated to greenhouse conditions after 3 weeks. (e) *B. sinuspersici* plant derived from cuttings after 3 months in the greenhouse. Scale bars = 2 cm.

Figure 2. Effect of auxins on adventitious root formation and root anatomy in cuttings of *Bienertia sinuspersici*. Histological analysis of root morphology and anatomy. Cuttings were rooted on (a) and (f) hormone-free medium and medium supplemented with (b) and (g) IAA, (c) and (h) IBA, (d) and (i) NAA, or (e) and (j) 2,4-D. Scale bars = 2 cm.

with longer healthy roots were more predisposed to survival after transplantation compared to those with many short roots. For example, cuttings that were cultured on 5 mg/L NAA produced the highest number of roots, but the plants did not survive in soil, whereas cuttings treated with IBA produced fewer, longer roots and had higher survival rates after transplantation (**Table 2**). 2,4-D treatment produced a few short, feather-like aerial roots on cuttings just above the agar surface (**Figure 2(e)**). Similarly, cuttings cultured on media supplemented with IAA formed clusters of many short and thick roots at the base (**Figure 2(b)**). IAA-rooted cuttings demonstrated moderate success upon transplanting, due to the quantity of roots and the stability provided by the compact root structure. However, the cluster of roots made it difficult to remove the agar without causing physical damage. Cuttings cultured on media containing NAA developed many short and stubby root protrusions that were accompanied by a few feathery roots above the surface of the media similar to those observed in the 2,4-D treatment (**Figure 2(d)**). These cuttings with poorly developed roots were unable to establish in soil. It was generally observed that cuttings with many short roots had lower survival rates after transplantation, compared to those with fewer but longer roots. Even though cuttings cultured on 1 mg/L IBA produced fewer roots than some of the other treatments, the roots were the longest and had the highest rate of survival after transplanting. Roots induced by IBA appeared much healthier and more viable than those produced in the hormone-free media. In addition, high survival rates after transplantation to soil further suggest IBA as the favorable candidate for *in vitro* root induction.

The root anatomy was analyzed to examine the effect of auxins on the growth and development of adventitious roots in *B. sinuspersici*. Longitudinal sections of root tips on hormone-free media showed a thin root cap surrounding the root apical meristem that overlaps with the zone of elongation (**Figure 2(f)**). Despite this, these roots appeared thin and slightly brown in colour, raising the question of whether supplementing the media with auxins would produce healthier roots. Histological examination of root tips showed that IAA-induced roots consist of many smaller cells that also organized into distinct regions (**Figure 2(g)**). However, cells from the IAA-induced roots appeared to have a larger zone of cell division and were not elongating to produce long roots. Roots that were induced by IBA showed the best root anatomy as indicated by the presence of various distinctive regions (**Figure 2(h)**). Sections of root tips developed from the NAA treatment showed that each root tip consisted of many small cells organized into a condensed miniature root. The epidermal, cortical and root cap cells near the tip region appeared to be loosely packed and sloughed off prematurely (**Figure 2(i)**). Longitudinal sections of 2,4-D induced root tips showed thin roots with irregularly shaped elongated parenchyma cells, a poorly developed root cap, and sloughing of epidermal cells (**Figure 2(j)**).

4. Discussion

Murashige and Skoog basal medium is comprised of many different macro- and micronutrients initially optimized to the *in vitro* regeneration of tobacco. It is useful for the growth of most plant species, but its formulation is prepared in very high concentrations, particularly nitrogen, and may be desired in as much as a 10-fold dilution by a particular species and explant type [13]. Thus, lowering the strength of MS medium is often used *in vitro* culture for rooting of cuttings, as reported in "Delicious" apple, *Malus domestica* [14] and *Ophiorrhiza prostrata*, an anti-cancer drug producing plant [15]. This study further confirms that lowering the strength of MS medium was essential for inducing adventitious root formation in *B. sinuspersici* cuttings and that half-strength MS medium was determined to be the optimal concentration for this process.

Auxins are phytohormones critical to root growth and development. Plants use auxin gradients to promote cell division, determine cell fate, and establish environmental controls that regulate overall root development through multiple biosynthetic pathways [16]. Without the presence of exogenous auxins, *in vitro* root formation relies on endogenous auxin synthesized in the shoot apex and transported downwards to create an auxin gradient required for root induction [17]. This established gradient of auxin allows cells to maintain information about their growth and development past the initial signals that caused cell differentiation, while changes in the auxin gradient allow the plant to control its development [17]. These developmental changes can be induced artificially to promote root cell differentiation by adding exogenous auxins. Although root induction of cuttings was effective using hormone-free media (**Figure 2(a)**), auxins such as IBA induce healthier, longer, and more numerous adventitious roots. Moreover, examining the effect of different auxins on the induction and development of adventitious roots in *B. sinuspersici* cuttings would be resourceful for adventitious root induction experiments of indirect organogenesis-derived shoots.

Each of the auxins examined (2,4-D, IAA, IBA, and NAA) affected the *in vitro* induced roots differently. Medium supplemented with 1 mg/L IBA produced the longest roots and facilitated a high survival rate among plants after transplantation to soil. 2,4-D, IAA and NAA were determined to be poor rooting hormones for *B. sinuspersici* due to the formation of short, unhealthy roots and low survival rates when transplanted to soil. Studies comparing the effects all four examined auxins favoured IBA as a root induction agent in both *Plumbago zeylanica* [18] and *Eucalyptus* spp. [19]. Furthermore, root induction using IBA were also observed in African Blackwood, *Dalbergia melanoxylon* [20], Indian horsechestnut, *Aesculusindica* [21], and ponytail palm, *Nolinarecurvata* [22]. In general, high concentrations of auxins are required for root initiation, but may also inhibit its development [23]. It has been suggested that IBA is the most efficient auxin at inducing root growth because its longer side chain length makes it more difficult for plants to oxidize [24]. The ability of IBA to be metabolized to IAA may create a slow release to lower the concentration and allow root development to progress gradually, making it more suited to root development rather than root initiation compared to other auxins [19]. However, NAA is even more stable, and thus may be inhibitory to development due to its accumulated concentration. These results suggest that IBA has an impact on root growth by promoting and accelerating cell differentiation to form roots and increasing of rate of root growth, while 2,4-D, IAA, and NAA inhibit root development and elongation. These observations are consistent with our results showing that *B. sinuspersici* cuttings cultured on medium supplemented with IBA developed long roots and had a high survival rate upon transplanting to soil compared to those cultured on medium containing 2,4-D, IAA, or NAA.

5. Conclusion

In conclusion, we have developed an efficient and rapid method of propagating *B. sinuspersici* through cuttings. Although rooting of cuttings may be induced on hormone-free medium, half-strength MS medium containing

IBA appeared to be the most effective auxins at inducing adventitious roots in *B. sinuspersici* cuttings and may be used to facilitate root development *in vitro*-derived shoots. The single-cell C_4 species, *Bienertia sinuspersici*, can be successfully and efficiently propagated by this method.

Acknowledgements

This research was supported by the Natural Sciences and Engineering Research Council (NSERC) of Canada Research Grants and the University of Waterloo Start-Up Fund to SDXC.

References

[1]   Akhani, H., Barroca, J., Koteyeva, N., Voznesenskaya, E.V., Franceschi, V.R., Edwards, G.E., Ghaffari, S.M., Stichler, W. and Ziegler, H. (2005) *Bienertia sinuspersici* (Chenopodiaceae): A New Species from SW Asia and Discovery of a Third Terrestrial C_4 Plant without Kranz Anatomy. *Systematic Botany*, **30**, 290-301. http://dx.doi.org/10.1600/0363644054223684

[2]   Voznesenskaya, E.V., Franceschi, V.R., Kiirats, O., Artyusheva, E.G., Freitag, H. and Edwards, G.E. (2001) Proof of C_4 Photosynthesis without Kranz Anatomy in *Bienertia cycloptera* (Chenopodiaceae). *Plant Journal*, **31**, 649-662. http://dx.doi.org/10.1046/j.1365-313X.2002.01385.x

[3]   Chuong, S.D.X., Franceschi, V.R. and Edwards, G.E. (2006) The Cytoskeleton Maintains Organelle Partitioning Required for Single-Cell C_4 Photosynthesis in Chenopodiaceae Species. *Plant Cell*, **18**, 2207-2223. http://dx.doi.org/10.1105/tpc.105.036186

[4]   Kesari, V. and Krishnamacheri, A. (2007) Effect of Auxins on Adventitious Rooting from Stem Cuttings of Candidate plus Tree *Pongamia pinnata* (L.), a Potential Biodiesel Plant. *Trees*, **23**, 597-604. http://dx.doi.org/10.1007/s00468-008-0304-x

[5]   Thorpe, T.A. (2007) History of Plant Tissue Culture. *Molecular Biotechnology*, **37**, 169-180. http://dx.doi.org/10.1007/s12033-007-0031-3

[6]   Evan, B.R., Bali, G., Foston, M., Ragauskas, A.J., O'Neil, H.M., Shah, R., McGaughey, J., Reeves, D., Rempe, C.S. and Davison, B.H. (2015) Production of Deuterated Switchgrass by Hydroponic Cultivation. *Planta*, **242**, 215-222. http://dx.doi.org/10.1007/s00425-015-2298-0

[7]   This, P., Lacombe, T. and Thomas, M.R. (2006) Historical Origins and Genetic Diversity of Wine Grapes. *Trends in Genetics*, **22**, 511-519. http://dx.doi.org/10.1016/j.tig.2006.07.008

[8]   Belide, S., Hac, L., Singh, S.P., Green, A.G. and Wood, C.C. (2011) *Agrobacterium*-Mediated Transformation of Safflower and the Efficient Recovery of Transgenic Plants via Grafting. *Plant Methods*, **7**, 12. http://dx.doi.org/10.1186/1746-4811-7-12

[9]   Jaganath, B., Subramanyam, K., Subramanian, M., Karthik, S., Elayaraja, D., Udayakumar, R., Manickavasagam, M. and Ganapathi, A. (2014) An Efficient in Planta Transformation of *Jatrophacurcas* (L.) and Multiplication of Transformed Plants through *in Vivo* Grafting. *Protoplasma*, **251**, 591-601. http://dx.doi.org/10.1007/s00709-013-0558-z

[10] Skoog, F. and Miller, C.O. (1957) Chemical Regulation of Growth and Organ Formation in Plant Tissue Cultured *in Vitro*. *Symposia of the Society for Experimental Biology*, **11**, 118-131.

[11] Rosnow, J., Offermann, S., Park, J., Okita, T.W., Tarlyn, N., Dhingra, A. and Edwards, G.E. (2011) *In Vitro* Culture and Regeneration of *Bienertia sinuspersici* (Chenopodiaceae) under Increasing Concentrations of Sodium Chloride and Carbon Dioxide. *Plant Cell Reports*, **30**, 1541-1553. http://dx.doi.org/10.1007/s00299-011-1067-1

[12] Northmore, J.A., Zhou, V. and Chuong, S.D.X. (2012) Multiple Shoot Induction and Plant Regeneration of the Single-Cell C_4 Species *Bienertia sinuspersici*. *Plant Cell, Tissue and Organ Culture*, **108**, 101-109. http://dx.doi.org/10.1007/s11240-011-0018-4

[13] Huang, L. and Murashige, T. (1977) Plant Tissue Culture Media: Major Constitutents, Their Preparation and Some Applications. *Methods in Cell Science*, **3**, 539-548. http://dx.doi.org/10.1007/bf00918758

[14] Zimmerman, R.H. (1984) Rooting Apple Cultivars *in Vitro*: Interactions among Light, Temperature, Phloroglucinol and Auxin. *Plant Cell, Tissue and Organ Culture*, **3**, 301-311. http://dx.doi.org/10.1007/BF00043081

[15] Martin, K.P., Zhang, C.L., Hembrom, M.E., Slater, A. and Madassery, J. (2008) Adventitious Root Induction in *Ophiorrhiza prostrata*: A Tool for the Production of Camptothecin (an Anticancer Drug) and Rapid Propagation. *Plant Biotechnology Reports*, **2**, 163-169. http://dx.doi.org/10.1007/s11816-008-0057-4

[16] Overvoorde, P., Fukaki, H. and Beeckman, T. (2010) Auxin Control of Root Development. *Cold Spring Harbor Perspectives in Biology*, **2**, a001537. http://dx.doi.org/10.1101/cshperspect.a001537

[17] Grieneisen, V.A., Xu, J., Maréem, A.F.M., Hogeweg, P. and Scheres, B. (2007) Auxin Transport Is Sufficient to Generate a Maximum and Gradient Guiding Root Growth. *Nature*, **449**, 1008-1013. http://dx.doi.org/10.1038/nature06215

[18] Saxena, C., Samantaray, S., Rout, G.R. and Das, P. (2000) Effect of Auxins on *in Vitro* Rooting of *Plumbago zeylanica*: Peroxidase Activity as a Marker for Root Induction. *Biologia Plantarum*, **43**, 121-124. http://dx.doi.org/10.1023/A:1026519417080

[19] Fogaça, C.M. and Fett-Neto, A.G. (2004) Role of Auxin and Its Modulators in the Adventitious Rooting of *Eucalyptus* Species Differing in Recalcitrance. *Plant Growth Regulation*, **45**, 1-10. http://dx.doi.org/10.1007/s10725-004-6547-7

[20] Amri, E., Lyaruu, H.M.V., Nyomora, A.S. and Kanyeka, Z.L. (2009) Vegetative Propagation of African Blackwood (*Dalbergia melanoxylon* Guill. & Perr.): Effects of Age of Donor Plant, IBA Treatment and Cutting Position on Rooting Ability of Stem Cuttings. *New Forests*, **39**, 183-194.

[21] Majeed, M., Khan, M.A. and Mughal, A.H. (2009) Vegetative Propagation of *Aesculus indica* through Stem Cuttings Treated with Plant Growth Regulators. *Journal of Forest Research*, **20**, 171-173. http://dx.doi.org/10.1007/s11676-009-0031-1

[22] Bettaieb, T., Mhamdi, M. and Hajlaoui, I. (2008) Micropropagation of *Nolina recurvata* Hemsl.: β-Cyclodextrin Effects on Rooting. *Science Horticulturae*, **117**, 366-368. http://dx.doi.org/10.1016/j.scienta.2008.05.023

[23] De Klerk, G.J., Krieken, W.V.D. and Jong, J. (1999) The Formation of Adventitious Roots: New Concepts, New Possibilities. *In Vitro Cellular & Developmental Biology-Plant*, **35**, 189-199. http://dx.doi.org/10.1007/s11627-999-0076-z

[24] Fawcett, C.H., Wain, R.L. and Wightman, F. (1960) The Metabolism of 3-Indolylalkanecarboxylic Acids, and Their Amides, Nitriles and Methyl Esters in Plant Tissues. *Proceedings of the Royal Society London B*, **152**, 231-254. http://dx.doi.org/10.1098/rspb.1960.0035

6

Diversity and Origin of Indigenous Village Chickens (*Gallus gallus*) from Chad, Central Africa

Khadidja Hassaballah[1*], Vounparet Zeuh[2], Raman A. Lawal[3], Olivier Hanotte[3], Mbacké Sembene[4]

[1]Department of Biology, Faculty of Exact and Applied Sciences, University of N'Djamena, N'Djamena, Chad
[2]Livestock Polytechnic Institute of Moussoro, Moussoro, Chad
[3]School of Life Sciences, The University of Nottingham, University Park, Nottingham, UK
[4]Department of Animal Biology, Faculty of Sciences and Techniques, Cheikh Anta Diop University of Dakar, Dakar, Senegal
Email: [*]hassaballaro@yahoo.fr

Abstract

In this study we assess the maternal genetic diversity and origin of indigenous village chickens from Chad complementing previous phenotypic and biometric measurements studies. We analysed a 387 bp fragment of the mitochondrial DNA (mtDNA) D-loop region of 181 village chickens from three populations of western Chad (Lake Chad/Hadjer Lamis), central Chad (Guera) and south-west Chad (Pala) and at different poultry markets in N'Djamena. Twenty-five polymorphic sites and 20 haplotypes are identified. Phylogenetic and network analyses group all chicken into a single mtDNA haplogroup D. Comparison with reference sequences shows that this haplogroup is the commonest one observed in chicken and it supports the Indian subcontinent as the maternal center of origin for the village chicken in Chad. Little genetic variation was found within and between populations which is in agreement with a recent and a maternal founding effect for the chicken in the country.

Keywords

mtDNA, D-Loop, Genetic Variation, Origin, Village Chickens, Chad

---

[*]Corresponding author.

1. Introduction

The domestic chicken is one of the most common and widespread domestic animals species with an estimated population in 2010 of more than 1.6 billion in Africa [1]. It is also the most abundant species of domestic bird in the world [2]. Human keeps chickens primarily as a source of food, consuming both their meat and their eggs. Many authors think that the main maternal ancestor of village chicken is the red jungle fowl *Gallus gallus* ssp. which was domesticated in South and South-East Asia at least since 5400 BC [3].

The introduction of village chicken in Africa remains relatively little documented [4]. The earliest evidence of chicken on the African continent is from Egypt and it dates from around 2000 BC. Chickens were most likely imported at the time as a curiosity and an addition to exotic menageries. Chicken did not become a regular feature of the Egyptian farmyard before the Ptolemaic period of 304 - 330 BC [5]. The oldest recognizable pictorial evidence of birds was reported by Haller in 1954 [6] and was assumed to have dated from the second half of the fourteenth century BC. In West Africa the earliest evidences are from Mali *circa* in AD 450 - 850, while archeological evidences show the presence of chicken in Chad in post-1700 AD [7].

Typically, today most African households will keep 5 to 20 indigenous birds essentially for eggs and meat production [8] [9]. However, the initial trigger for the adoption of domestic chickens by African communities, as it has been assumed for the initial domestication of the chicken in Asia, could have been for socio-culture and/ or recreation (e.g. cockfighting) rather than as a new source of food. Today indigenous chickens represent an important valuable animal genetic resource and the conservation, sustainable exploitation and improvement of local breeds are therefore important issues. Many of the populations developed over hundreds of years were selected for morphological and appearance characteristics as much as for production purposes. This is illustrated by the very large numbers of chicken breeds and ecotypes found across the world.

Recent genetic studies have pointed to multiple maternal origins of domestic chicken in Asia [10]. Also, the African continent likely witnessed several major introductions of chickens from different Asian centers of origin including South Asia and islands South East Asia. In East Africa two distinct major mtDNA haplogroups, A and D *sensu* [11], are present. It has been proposed that African haplogroup A originates from Southeast and/or East Asia [11]-[14]; and using phylogeographic information on the modern geographic distribution across Europe and Asia [10], the Indian subcontinent has been proposed as the initial center of origin for haplogroup D [11] [13] [14]. Other halogroups occur on the continent but at a very low frequency. Haplogroups B, C may have reached Africa following recent introductions of improved commercial chickens [11]. This suggestion is supported by the presence of identical or closely related haplotypes belonging to these haplogroups in European and commercial birds [13] [15] [16]. However, a more ancient and direct introduction of one or both of these haplogroups from their centers of origins in Asia also remains possible. The center of origin of haplogroup E, also observed at very low frequency on the Africa continent, remains speculative so far; it has only been observed in the north of the Equator in Sudan and Ethiopia [11].

From the five on the African continent haplogroups reported by Mwacharo *et al.* [4], two haplogroups, A and D, dominate the continent. Haplogroup D is found in all the African countries studied and is the most common in all countries with the exception of Madagascar, Zimbabwe and eastern Kenya. It is the only haplogroup so far identified in West Africa, represented by Nigeria [17]. The next commonest haplogroup is A, which is absent from Uganda, Sudan, Nigeria and Ethiopia but is the commonest in Madagascar, Malawi, Zimbabwe and eastern Kenya. The other haplogroups are observed at very low frequencies in all studies, being present in only one or two birds.

This study completes information from phenotypes and biometric measurements of the domestic chickens from Chad [18] [19]. These previous studies examined the phenotypic diversity and took biometric measurements of three different populations (Hadjer-Lamis/Lac Chad, Guera and West Mayo-Kebbi). The findings of these studies suggest that these populations living in different geographic areas may represent different ecotypes with different phenotypic characteristics. We now address the issue of the maternal genetic origins of the chicken from Chad by examining sequence information from the mitochondrial DNA D-loop and by comparing our findings with sequences of reference from Asia. Our results are adding new information on the geographic distribution of chicken mitochondrial DNA haplogroups in Africa and more particularly in Central Africa.

2. Materials and Methods

2.1. Sample Collection and DNA Extraction

Genomic DNA was extracted from air dried blood preserved on FTA classic cards (Whatman Biosciences), using the recommended manufacturer protocol, from 186 unrelated local village chickens from three geographical regions in Chad (Central Africa). The samples include 50 birds from the western region (Hadjer-Lamis/Lake Chad) representing population I (Chad I), 50 birds from the central region (Guera) representing population II (Chad II), 50 birds from Southern region (West Mayo-Kebbi) representing population III (Chad III) and 36 birds samples from various poultry markets of N'Djamena representing population IV (Chad IV). The DNA concentration and purity, A260/A280 ratio between 1.8 and 2.0, were assessed using a NanoDrop® 1000 Spectrophotometer. Potential DNA degradation was visualized on 1% agarose gel. Five samples (2 from Chad III and 3 from Chad IV) were removed from the analysis following poor DNA extraction yield and/or bad quality sequence information. To address the possible Asian origin of Chad village chicken, 9 Asian reference haplotypes were included in the analyses including the chicken mtDNA reference downloaded from the National Centre for Biotechnology Information (NCBI) (GenBank accession number AB098668).

2.2. PCR Amplification and Sequencing

Five hundred and forty nine base pairs of the mtDNA D-loop region were amplified using AV1F2 (5'-AGGACTACGGCTTGAAAAGC-3' [11] as the forward primer and H547 (5'-ATGTGCCTGACCGAGGAACCAG-3', accession number AB098668, Komiyama et al., 2003) as the reverse primer. PCR amplifications were carried out in a 20 µl reaction volumes containing 40 ng genomic DNA, 5× Phire reaction buffer (containing 1.5 mM MgCl$_2$ at final reaction concentration), 200 µM of each dNTP, 0.5 µM of each primer and 0.4 µl of Phire Hot Start II DNA Polymerase (Thermo Scientific Ltd). The thermo-cycling conditions were Lid (110°C), hot start 98°C (2 min), denaturation 98°C (5 sec), annealing 63°C (10 sec), elongation 72°C (15 sec), 35 cycles and final extension step at 72°C for 1 min [20]-[22]. PCR products were purified using the NucleoSpin® Gel and PCR Clean-UP kit [23]. Purified products were sequenced using the AV1F2 (5'-AGGACTACGGCTTGAAAAGC-3') as the forward primer and H547 (5'-ATGTGCCTGACCGAGGAACCAG-3') as the reverse sequence primers.

2.3. Sequence and Phylogenetic Analysis

For each sample, two sequences were generated. The forward and reverse primers were trimmed and the two sequences were compared for consistency generating a 549 bp consensus sequence using CodonCode Aligner version 5.1.3 (www.codoncode.com/). The consensus sequence was aligned against the reference (GenBank accession number AB098668) [23] using Clustal × version 2.1 [25]. Subsequent analyses were restricted to the first 397 bp of the sequence which includes the hypervariable region (HV1) of the D-loop [17]. A Neighbour-Joining (NJ) tree was constructed for all samples (sequences from Chad and reference sequences) with a 1000 boostrap replicates using *MEGA* version 6.0 [26]. The Median Joining (MJ) network was constructed using NETWORK 4.6.1.2 [27]. Sequence variation (nucleotide diversity, haplotype diversity and average number of nucleotide at each population) were calculated using DnaSP v5 [28].

3. Results

The analysis involved the first 397 bp of mtDNA D-loop sequences including the hypervariable region (HV1). A total of 20 haplotypes (3 from the population Chad I, 9 from Chad II, 10 from Chad III and 8 from Chad IV) defined by 25 polymorphic sites from 181 sequences was found (**Figure 1**). The individual haplotypes and haplogroups used in this study are defined by CD (Chad), H1 to H20 (Haplotype 1 to 20) while HapA, HapA01i03, HapA02, HapB, HapC, HapD, HapE, and HapF all represents different haplogroups from the Asian continent [11]. Haplotype and nucleotide diversities range from (0.500 ± 0.074 - 0.740 ± 0.058) and (0.0048 ± 0.0013 - 0.0074 ± 0.0017) across populations (**Table 1**), with Chad I and Chad II showing the lowest values for both. In particular despite the analysis of 50 birds only three haplotypes were observed in populations Chad while at the other extreme we do observe eight different haplotypes out of 33 birds from population Chad IV.

```
                    1111222222222222233333333 3
                    43791133445669990114456 99
                     379272336617678605244716 N

Reference ATATATCCTTTCATACTCTAATTCT 1
CDH1      ....GC..CCCT.C...TC...... 127
CDH2      ....GC.TCCCT.C...TC...... 10
CDH3      ....GC..CCCT.C...TC...C.. 2
CDH4      ..T.GC..CCCT.C...TC...... 1
CDH5      ....GC..CCCTGC...TC...... 6
CDH6      ....GC..CCCT.C...TC.G.... 3
CDH7      ....GC..C.CT.C...TC...... 2
CDH8      ....GC..CCC..C...TC..C... 1
CDH9      ....GC..CCCT.C.T.TC...... 1
CDH10     .C..GCTTCCCT.C...TC...... 1
CDH11     ...CGC..CCCT.C...TC...... 1
CDH12     ....GC.TCCCT.C...TCG..... 2
CDH13     ....GC..CCCT.C...TC.....C 6
CDH14     ....GC..CCCT.C..CTC...... 1
CDH15     ....GC..CCCT.C...TC....TC 7
CDH16     T...GC..CCCT.C...TC...... 1
CDH17     ....GC..C.CT.C...TC..C... 1
CDH18     ....GC..CCCT.CG..TC....T. 1
CDH19     .C..GC.TCCCT.C...TC...... 1
CDH20     ....GC..CCCT.C...TC....T. 6
```

Figure 1. Polymorphic sites of the 20 haplotypes observed in the mtDNA D-loop region from 181 village chicken sequences from Chad. The number of individuals within each haplotype is indicated by "N". The dots (.) indicate identity with reference sequence (GenBank accession number AB098668) [24].

Table 1. Sampling population, sample size, population genetic diversity measures, standard deviation (SD) for each population and Tajima's D (P value).

Population	Sample size	Haplotype CDH1 (number of individuals observed)	Number of haplotypes	Haplotype diversity (SD)	Nucleotide diversity per site (SD)	Average number of nucleotide differences (k)	Tajima's D (P value)
*Chad I	50	40	3	0.500 (0.074)	0.00484 (0.00129)	1.92344	−2.10352 ($P < 0.05$)
*Chad II	50	37	9	0.583 (0.076)	0.00542 (0.00129)	2.15020	−2.32076 ($P < 0.01$)
*Chad III	48	27	10	0.740 (0.058)	0.00656 (0.00130)	2.59962	−2.00892 ($P < 0.05$)
*Chad IV	33	21	8	0.714 (0.073)	0.00744 (0.00166)	2.90128	−2.09414 ($P < 0.05$)
All populations	181	127	20	0.541 (0.044)	0.00312 (0.00049)	1.21270	−2.37429 ($P < 0.01$)

*Chad I = Western region (Hadjer-Lamis/Lake Chad); Chad II = Central region (Guera); Chad III = Southern region (West Mayo-Kebbi); Chad IV = Various poultry markets of N'Djamena.

The Neighbour-joining tree and it bootstraps values support close relationships between all haplotypes with reference haplotype D belonging to the same group (**Figure 2**). In particular, all haplotypes identified in this study belong to haplogroup D (refer as haplogroup E in [10]), with one major CDH1 haplotype presents in 127 observations out of 181 (**Figure 1**). This haplotype (CDH1) is identical to our haplotype of reference HapD.

Haplotype network analysis (**Figure 3** and **Figure 4**) further confirms the close relationship between all haplotypes. **Figure 3** clearly illustrates that all these haplotypes belongs to a single expansion event centred on haplotype HapD (CDH1). This is also the case when the populations are analysed separately (**Figure 4**).

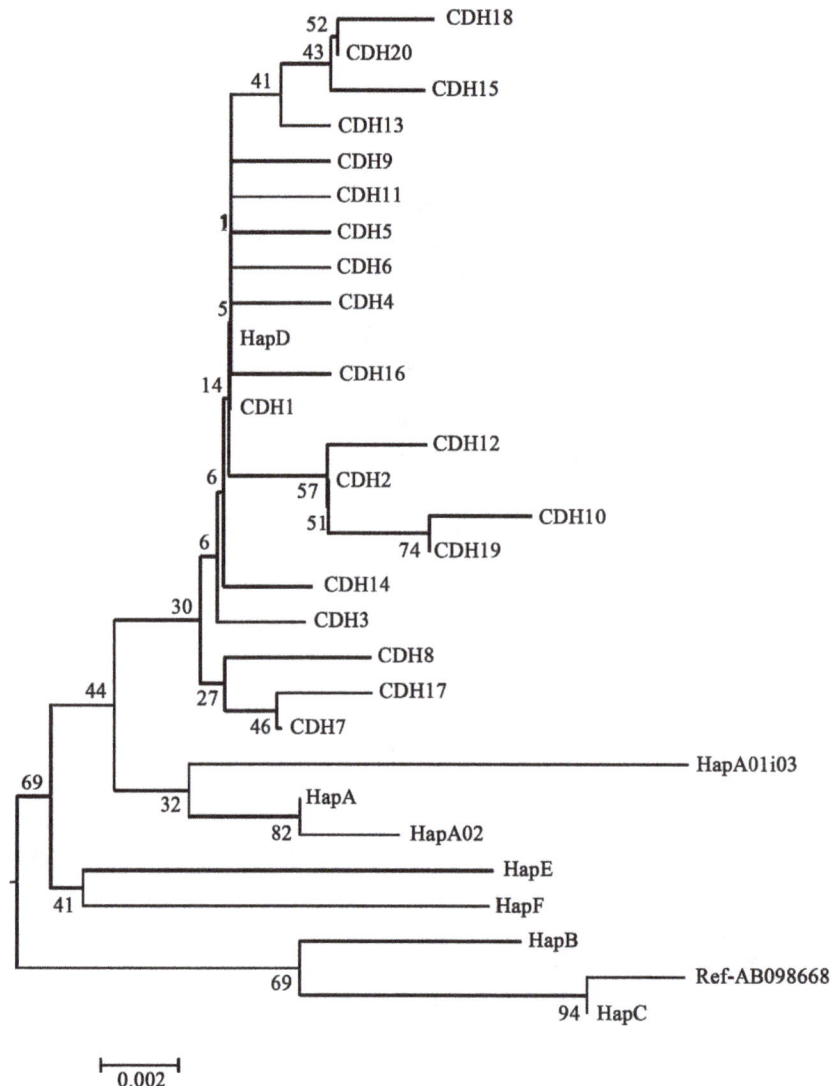

Figure 2. Neighbour-joining tree reconstructed from the 20 haplotypes identified in the 181 Chad village chicken sequences and the nine haplotypes of references using MEGA 6.0. The percent bootstrap value is represented by the numbers at the node after 1000 replication.

4. Discussion

The 20 haplotypes detected from a total of 25 polymorphic sites belong to a single haplogroup D (**Figure 2**). On the African continent two major haplogroups, A and D, have been reported previously [11]-[14]. Haplogroup D was found in all African countries studied and it is the commonest in all countries with the exception of Madagascar, Zimbabwe as well as the Eastern part of Kenya. Like in Chad, haplogroup D is the only haplogroup found so far in Nigeria [17], it also by far the commonest in Sudan and South Sudan (being refer as Clade IV in [29] in agreement with a possible common origin for the chicken from West and Central Africa. Haplogroup A was not observed in Chad. This haplogroup is absent from Uganda, Sudan, Nigeria, Sudan and South Sudan but is the commonest in Madagascar, Malawi, Zimbabwe and East Kenya. The absence of haplogroup A in Chad indicates that the likely introduction of chicken along the coast of East Africa did not contribute on the maternal side to the today genetic pool of the modern indigenous chicken from Chad. None of the rare haplogroups previously reported on the African continent (Mwacharo *et al.*, 2013 [4]) were found in Chad.

The Indian subcontinent has been proposed as the initial center of origin for the haplogroup D found in Africa [11] [13] [14]. A possible Indian subcontinent origin of some African chickens is further supported by the

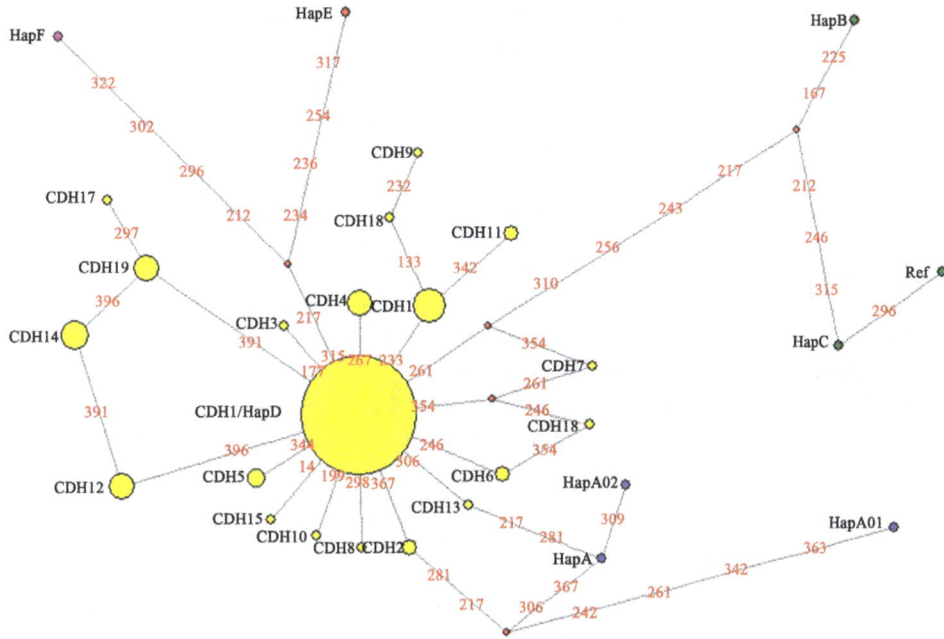

Figure 3. Median-joining network based on the mtDNA D-loop HV1 region for the 20 haplotypes of Chad indigenous chicken derived from 181 sequences and the nine reference haplotypes. Inferred ancestral haplotypes not sampled here are represented in red. The size of the circles is proportional to the frequency of the frequency of each haplotype. The positions of nucleotide mutations, compared to the reference sequence (GenBank accession number AB098668), correspond to the numbers between haplotype nodes.

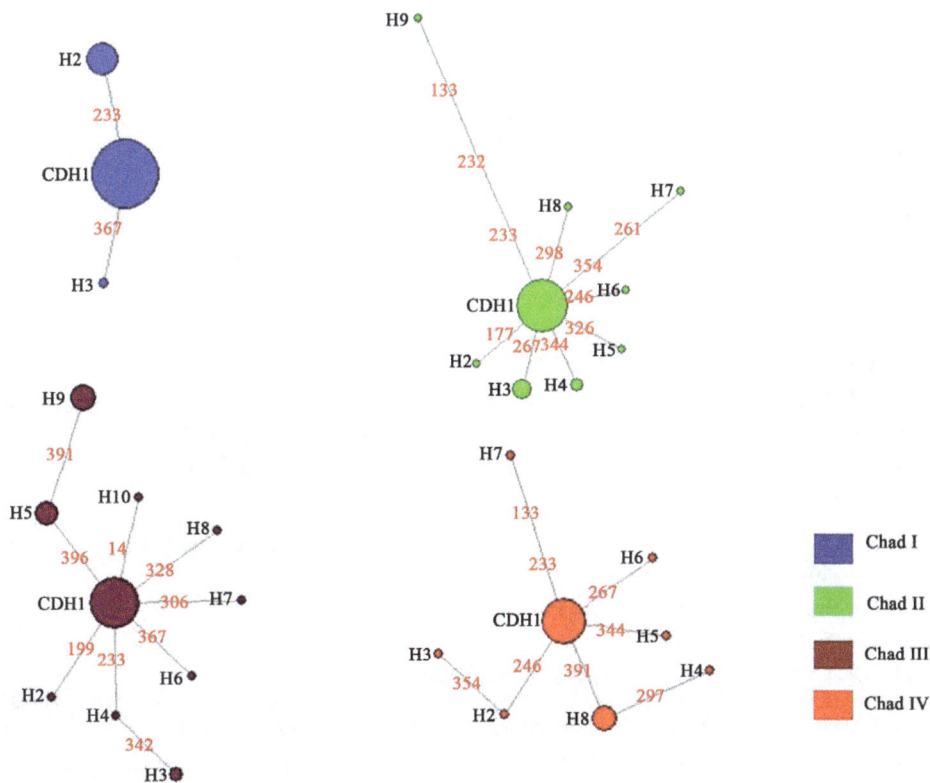

Figure 4. Median-joining network based on the mtDNA D-loop HV1 region for individual population. The number of individual in haplotype CDHI is referring in **Table 1**.

commonly observed yellow skin phenotype across African village chickens [15] [18] [30]-[33]. This phenotype has recently been shown at the molecular level to be a legacy of successful introgression of the grey jungle fowl *Gallus sonneratii*, a yet to be domesticated *Gallus* species with restricted geographic range on the Indian subcontinent, into domestic chicken [34]. The yellow skin phenotype is also observed in Libya [35], and Egypt where haplogroup D has been shown to be present in indigenous chicken [36]. Together all these findings support that haplogroup D observed in local chicken from Chad originated from a single source, which is likely the Indian subcontinent. More specifically, we propose that the ancient maternal ancestor of indigenous chicken from Chad would have entered the country through trans-Saharan trading from Egypt or Sudan rather than through migration along the Sahelian belt from East Africa.

Within Chad, we do observe difference between populations (**Table 1**). The smallest number of haplotype was observed in Chad I while at the other end we do observe the largest number of haplotype in population Chad IV. In this later population, the birds were sampled at N'Djamena markets and it may be expected to find in the city birds that may have been imported from different parts of the countries or even neighbouring countries. This is further support by the large average nucleotide divergence between haplotypes observed in this region (**Table 1**). At the opposite we have found a very small number haplotypes in population CHD1 (regions of Hadjer-Lamis and Lake Chad) and CHD2 (region of Guera) two populations from the Sahel zone.

5. Conclusion

The aim of this study was to assess the possible origin of local chickens' populations in Chad and their genetic variation. A total of 20 haplotypes defined by 25 polymorphic sites from 181 sequences were identified. Only haplogroup D was observed. Haplogroup D is also the only haplogroup found in Sudan, Nigeria and now with the results of our study in Chad suggesting a common route and entry point of chicken to Central and West Africa involving trans-Saharan trading from Egypt or Sudan. Combined with phenotypic information, these results provide important baselines information to guide poultry improvement programs aiming to conserve and utilise indigenous animal genetic resources from Chad.

Acknowledgements

This work was financially supported by a National Fund (CONFOFOR) for Scientific Research and Training of Trainers (Chad). This article was produced as part of ongoing effort aiming to better characterize the local chickens populations of Chad for conservation, improvement, sustainable utilization purposes.

References

[1]   FAOSTAT (2012) FAO Statistical Yearbook 2012. FAO, Rome. http://faostat.fao.org/

[2]   Garrigus, W.P. (2007) Poultry Farming. Encyclopaedia Britannica. http://www.britannica.com/eb/article-9111040

[3]   Crawford, R.D. (1990) Origin and History of Poultry Species. In: Crawford. R.D., Ed., *Poultry Breeding and Genetics*. Elsevier, Amsterdam

[4]   Mwacharo, J.M., Bjørnstad, G., Han, J.L. and Hanotte, O. (2013) The History of African Village Chickens: An Archaeological and Molecular Perspective. *African Archaeological Review*, Published Online.

[5]   MacDonald, K.C. and Edwards, D.N. (1993) Chicken in Africa: The Importance of Qasr Ibrim. *Antiquity*, **67**, 584-590.

[6]   Haller, A. (1954) Die Gräber und Grüfte von Assur. Berlin, Germany.

[7]   Rivallain, J. and Van Neer, W. (1983) Inventory of the Archaeological and Faunal Material at Koyom, South Chad. *Anthropologie*, Masson editor Nr. 88, 441-448.

[8]   Guèye, E.F. (1998) Village Egg and Fowl Meat Production in Africa. *World's Poultry Science Journal*, **54**, 73-86. http://dx.doi.org/10.1079/WPS19980007

[9]   Guèye, E.F. and Bessei, W. (1995) La poule locale sénégalaise dans le contexte villageois et les possibilités d'amélioration de ses performances. In: *Sustainable Rural Poultry Production. Proceedings of an International Workshop*, 13-16 June 1995, the International Livestock Research Institute, Addis Ababa, Ethiopia, 112-123.

[10] Liu, Y.P., Wu, G.S., Yao, Y.G., Miao, Y.W., Luikart, G., Baig, M., Beja-Pereira, A., Ding, Z.L., Alanichamy, M.G. and Zhang, Y.P. (2006) Multiple Maternal Origins of Chickens: Out of the Asian Jungles. *Molecular Phylogenetics and Evolution*, **38**, 12-19. http://dx.doi.org/10.1016/j.ympev.2005.09.014

[11] Mwacharo, J.M., Bjørnstad, G., Mobegi, V., Nomura, K., Hanada, H., Amano, T., Jianlin, H. and Hanotte, O. (2011)

Mitochondrial DNA Reveals Multiple Introductions of Domestic Chicken in East Africa. *Molecular Phylogenetics and Evolution*, **58**, 374-382. http://dx.doi.org/10.1016/j.ympev.2010.11.027

[12] Razafindraibe, H., Mobegi, V.A., Ommeh, S.C., Rakotondravao, G., Bjørnstad, G., Hanotte, O. and Jianlin, H. (2008) Mitochondrial DNA Origin of Indigenous Malagasy Chicken. *Annals of the New York Academy of Sciences*, **1149**, 77-79. http://dx.doi.org/10.1196/annals.1428.047

[13] Muchadeyi, F.C., Eding, H., Simianer, H., Wollny, C.B.A., Groeneveld, E. and Weigend, S. (2008) Mitochondrial DNA D-Loop Sequences Suggest a Southeast Asian and Indian Origin of Zimbabwean Village Chicken. *Animal Genetics*, **39**, 615-622. http://dx.doi.org/10.1111/j.1365-2052.2008.01785.x

[14] Mtileni, B.J., Muchadeyi, F.C., Maiwashe, A., Chimonyo, M., Groeneveld, E., Weigend, S. and Dzama, K. (2011) Diversity and Origin of South African Chicken. *Poultry Science*, **90**, 2189-2194. http://dx.doi.org/10.3382/ps.2011-01505

[15] Dana, N., Megens, H.J., Crooijmans, R.P.M.A., Hanotte, O., Mwacharo, J.M., Groenen, M.A. and Van Arendonk, J.A. (2010) East Asian Contributions to Dutch Traditional and Western Commercial Chickens Inferred from mtDNA Analysis. *Animal Genetics*, **42**, 125-133. http://dx.doi.org/10.1111/j.1365-2052.2010.02134.x

[16] Ceccobelli, S., Di Lorenzo, P., Hovirag, L., Monteagudo, L., Tejedor, M.T. and Castellini, C. (2015) Genetic Diversity and Phylogeographic Structure of Sixteen Mediterranean Chicken Breeds Assessed with Microsatellites and Mitochondrial DNA. *Livestock Science*, **175**, 27-36. http://dx.doi.org/10.1016/j.livsci.2015.03.003

[17] Adebambo, A.O., Mobegi, V., Mwacharo, A., Oladejo, J.M., Adewale, B.M., Ilori, R.A., Makanjuola, L.O., Afolayan, B.O., Bjørnstad, O., Jianlin, H. and Hanotte, O. (2010) Lack of Phylogeographic Structure in Nigerian Village Chickens Revealed by Mitochondrial DNA D-Loop Sequence Analysis. *International Journal of Poultry Science*, **9**, 503-507. http://www.pjbs.org/ijps/ab1670.htm http://dx.doi.org/10.3923/ijps.2010.503.507

[18] Hassaballah, K., Zeuh, V. and Sembene, M. (2014) Phenotypic Diversity of Local Chickens (*Gallus domesticus*) in Three Ecological Zones of Chad. *International Journal of Current Research in Biosciences and Plant Biology*, **1**, 1-8.

[19] Hassaballah, K., Zeuh, V., Mopate, L.Y. and Sembene, M. (2015) Caractérisation morpho-biométrique des poules (*Gallus gallus*) locales dans trois zones agro-écologiques du Tchad. *Livestock Research for Rural Development*, **27**, Article No.: 53. http://www.lrrd.org/lrrd27/3/hass27053.html

[20] Chester, N. and Marshak, D.R. (1993) Dimethyl Sulfoxide-Mediated Primer T_m Reduction: A Method for Analyzing the Role of Renaturation Temperature in the Polymerase Chain Reaction. *Analytical Biochemistry*, **209**, 284-290. http://dx.doi.org/10.1006/abio.1993.1121

[21] Nord, K., Gunneriusson, E., Ringdahl, J., Ståhl, S., Uhlén, M. and Nygren, P.A. (1997) Binding Proteins Selected from Combinatorial Libraries of an α-Helical Bacterial Receptor Domain. *Nature Biotechnology*, **15**, 772-777. http://dx.doi.org/10.1038/nbt0897-772

[22] Wikman, M., Steffen, A.C., Gunneriusson, E., Tolmachev, V., Adams, G.P., Carlsson, J. and Ståhl, S. (2004) Selection and Characterization of HER2/Neu-Binding Affibody Ligands. *Protein Engineering*, **17**, 455-462. http://dx.doi.org/10.1093/protein/gzh053

[23] Vogelstein, B. and Gillespie, D. (1979) Preparative and Analytical Purification of DNA from Agarose. *Proceedings of the National Academy of Sciences of the United States of America*, **76**, 615-619. http://dx.doi.org/10.1073/pnas.76.2.615

[24] Komiyama, T., Ikeo, K. and Gojobori, T. (2003) Where Is the Origin of the Japanese Gamecocks? *Gene*, **317**, 195-202. http://dx.doi.org/10.1016/S0378-1119(03)00703-0

[25] Thompson, J.D., Gibson, T.J., Plewniak, F., Jeanmougin, F. and Higgins, D.G. (1997) The ClustalX Windows Interface: Flexible Strategies for Multiple Sequence Alignment Aided by Quality Analysis Tools. *Nucleic Acids Research*, **25**, 4876-4882. http://dx.doi.org/10.1093/nar/25.24.4876

[26] Tamura, K., Stecher, G., Peterson, D., Filipski, A. and Kumar, S. (2013) MEGA6: Molecular Evolutionary Genetics Analysis Version 6.0. *Molecular Phylogenetics and Evolution*, **30**, 2725-2729. http://dx.doi.org/10.1093/molbev/mst197

[27] Bandelt, H.J., Forster, P. and Röhl, A. (1999) Median-Joining Networks for Inferring Intraspecific Phylogenies. *Molecular Phylogenetics and Evolution*, **16**, 37-48. http://dx.doi.org/10.1093/oxfordjournals.molbev.a026036

[28] Librado, P. and Rozas, J. (2009) DnaSP v5: A Software for Comprehensive Analysis of DNA Polymorphic Data. *Bioinformatics*, **25**, 1451-1452. http://dx.doi.org/10.1093/bioinformatics/btp187

[29] Wani, C.E., Yousif, I.A., Ibrahim, M.E. and Musa, H.H. (2014) Molecular Characterization of Sudanese and Southern Sudanese Chicken Breeds Using mtDNA D-Loop. *Genetics Research*, **2014**, Article ID: 928420. http://dx.doi.org/10.1155/2014/928420

[30] FAO (2009) Characterization of Domestic Chicken and Duck Production Systems in Egypt. Prepared by Haitham H., Mohamed Kosba and Olaf Thieme. *AHL-Promoting Strategies for Prevention and Control of HPAI*, Rome.

[31] Youssao, I.A.K., Tobada, P.C., Koutinhouin, B.G., Dahouda, M., Idrissou, N.D., Bonou, G.A., Tougan, U.P., Ahounou, S., Yapi-Gnaoré, V., Kayang, B., Rognon, X. and Tixier-Boichard, M. (2010) Phenotypic Characterisation and Molecular Polymorphism of Indigenous Poultry Populations of the Species *Gallus gallus* of Savannah and Forest Ecotypes of Benin. *African Journal of Biotechnology*, **9**, 369-381. http://www.academicjournals.org/AJB

[32] Daikwo, I.S., Okpe, A.A. and Ocheja, J.O. (2011) Phenotypic Characterization of Local Chickens in Dekina. *International Journal of Poultry Science*, **10**, 444-447. http://www.pjbs.org/ijps/ab1905.htm http://dx.doi.org/10.3923/ijps.2011.444.447

[33] Melesse, A. and Negesse, T. (2011) Phenotypic and Morphological Characterization of Indigenous Chicken Populations in Southern Region of Ethiopia. *Animal Genetic Resources Information*, **49**, 19-31. http://dx.doi.org/10.1017/S2078633611000099

[34] Eriksson, J., Larson, G., Gunnarsson, U., Bed'hom, B., Tixier-Boichard, M., Strömstedt, L., Wright, D., Jungerius, A., Vereijken, A., Randi, E., Jensen, P. and Andersson, L. (2008) Identification of the *Yellow Skin* Gene Reveals a Hybrid Origin of the Domestic Chicken. *PLoS Genetics*, 4, e1000010. http://dx.doi.org/10.1371/journal.pgen.1000010

[35] El-Safty, S.A. (2012) Determination of Some Quantitative and Qualitative Traits in Libyan Native Fowls. *Egypt. Poultry Science*, **32**, 247-258.

[36] Elkhaiat, I., Kawabe, K., Saleh, K., Younis, H., Nofal, R., Masuda, S. and Shimogori, T. (2014) Genetic Diversity of Egyptian Native Chickens Using mtDNA D-Loop Region. *Journal of Poultry Science*, **51**, 359-363. http://dx.doi.org/10.2141/jpsa.0130232

7

Nutrient Requirements and Fermentation Conditions for Mycelia and Crude Exo-Polysaccharides Production by *Lentinus squarrosulus*

Felicia N. Anike[1], Omoanghe S. Isikhuemhen[1]*, Dietrich Blum[1], Hitoshi Neda[2]

[1]Mushroom Biology and Fungal Biotechnology Laboratory, North Carolina A&T State University, Greensboro, USA
[2]Forestry and Forest Products Research Institute (FFPRI), Tsukuba, Japan
Email: *omon@ncat.edu

Abstract

Lentinus squarrosulus Mont. is an emerging tropical white rot basidiomycete, with nutritional and medicinal benefits. Low levels of commercial cultivation of the mushrooms limit their availability for use as food and medicine. Mycelia from submerged fermentation are a suitable alternative to the mushroom from *L. squarrosulus*. Three strains, 340, 339 and 218, were studied to determine optimum growth conditions for mycelia mass and crude exo-polysaccharides (CEPS) production. The experiments were conducted in a completely randomized design (CRD) with a factorial structure. Nutrients involving 8 carbon and 8 nitrogen sources were screened, and concentrations of the best sources were optimized. Optimized nutrients, interaction between strains and other parameters such as agitation and medium volume were investigated to obtain optimum fermentation conditions for biomass and CEPS production. Biomass yield varied among strains depending on carbon or nitrogen nutrient sources. Starch and yeast extract at 30 and 25 g/L were identified as the most important nutrients in mycelia and CEPS production. Nutrient optimization resulted in a 3-fold increase in mycelia mass: 12.8, 10.0 and 15.3 g/L in strains 340, 339 and 218 respectively. There was a significant interaction between strain, agitation, and volume ($p < 0.001$). Mycelia mass increased with volume under shake conditions, while polysaccharides decreased. There was a weak and negative correlation between mycelia mass and polysaccharides ($p = 0.02$). Static conditions favored more polysaccharide production. Optimized fermentation conditions resulted in very high increase in biomass: 238.1, 266.9 and 185.0 g/L in strains 340, 339 and 218 respectively. Results obtained could be useful in modeling fermentation systems for large-scale production of mycelia mass, CEPS and other bio-products from *L. squarrosulus*.

---

*Corresponding author.

Keywords

Mycelia Mass, Carbon and Nitrogen Requirements, Exopolysaccharides, *Lentinus squarrosulus*, Submerged Fermentation

1. Introduction

Lentinus squarrosulus Mont. is a white-rot basidiomycete found in many countries in African and Asia. Its fruit bodies are consumed as food in Sub-Saharan Africa and Southeast Asia [1]. The fruit body, if harvested within 3 days of fruiting, is used as a meat substitute. It was not until recently that research started to emerge on its application in food [2], medicine [3] [4] and bioremediation [5]. Proximate analysis shows that both the cap and stipe are rich in proteins, sugars, fiber, lipids, amino acids, vitamins B, C, and D, and minerals [6]. The water soluble glucans from *L. squarrosulus* have immune-enhancing properties [7], while water soluble extract from mycelia eliminated ulcer in rats within 72 hours [6]. Other investigators reported its potential as a biocontrol agent against *Rigidoporus lignosus*, a fungal pathogen of rubber, and an antimicrobial against *Bacillus subtilis*, *Mucor ramannianus* and yeast [4].

Submerged fermentation has been used widely in the production of mycelia and bioactive compounds in other basidiomycetes [8]-[10]. The process offers several advantages including: high productivity, compact and controlled environment for quality and consistency of products, and shortened production time [8]. Therefore it is reasonable to explore some factors that could enhance mycelia yield and CEPS secretion in *L. squarrosulus*. Nutritional and physiological factors are critical for bioprocess optimization during fermentation [11]. However, there is scant information on how these factors affect mycelia yield and crude exopolysaccharide (CEPS) secretions in *L. squarrosulus*. Carbon and nitrogen as nutrients are identified to be essential for cell survival, and can greatly influence cell proliferation, metabolite biosynthesis and secretion in submerged fermentation [12]. Variability in utilization of different carbon sources by basidiomycetes within the same species and across genera is widely reported [13]. Therefore it was necessary to include various sources of these nutrients. Eight different commonly used and affordable carbon and nitrogen sources were selected and tested. Since basidiomycetes require oxygen for aerobic respiration, agitation was necessary for homogenous distribution of available oxygen [12] [14].

To date there are only two reports on the effect of nutrient factors on vegetative mycelia growth in *L. squarrosulus* [15] [16]. The authors studied the effect of medium components on only one strain, but did not consider agitation and scalability. The present study investigates the influence of different sources and concentrations of carbon and nitrogen, as well as agitation, on mycelia growth and CEPS secretion in three strains of *L. squarrosulus*. The study also presents data on optimized medium for the growth of *L. squarrosulus* under static and shake fermentation conditions. Furthermore, the optimized conditions were studied for applicability in scale up fermentation.

2. Methods

2.1. Microorganisms and Inocula

Three strains of *L. squarrosulus*, 340, 339 and 218, used in the study were from the culture collection of the Mushroom Biology and Fungal Biotechnology Laboratory at North Carolina A&T State University in Greensboro. Strain 218 originated from Ghana, while strains 339 and 340 are from Okinawa Japan. The strains were maintained on PDA slants at 4°C until used. The cultures were activated by subculturing unto PDA media and incubated at 30°C for 3 days. Inocula for all experiments were prepared by blending a plate culture (60 × 15 mm) of each strain with 200 ml of sterile distilled, de-ionized water, using a Warring blender. All experimental media were seeded with 2% v/v inoculum.

Experiments were conducted using a basal medium composed of the following compounds, per liter: KH_2PO_4, 1 g; NaH_2PO_4, 0.4 g; $MgSO_4 \cdot 7H_2O$, 0.5 g; $CuSO \cdot 7H_2O$, 0.5 g; $CaCl_2 \cdot H_2O$, 74 mg; $ZnSO_4 \cdot 7H_2O$, 6 mg; $FeSO_4 \cdot 7H_2O$, 5 mg; $MnSO_4 \cdot 4H_2O$, 3.79 mg; $CoCl_2 \cdot 6H_2O$, 1 mg; Thiamine HCl, 0.1 mg; Pyridoxine HCl, 0.1 mg; Nicotinic acid, 0.1 mg. Carbon and nitrogen sources, and their concentrations, varied according to the parameter tested. The media were sterilized at 121°C for 15 min except for the fructose-containing media, which were autoclaved at 110°C.

2.2. Nutrient Requirements for Mycelia Production

2.2.1. Carbon Sources
To determine the best carbon source for optimum yield of mycelia, a 3 × 8 factorial experiment was set up in 50 mL basal medium (in 250 ml Erlenmeyer flasks) containing 10 g/L of one of the following carbon sources: dextrose, fructose, mannose, mannitol, sorbitol, sucrose, starch or xylose. The control medium had no sugar. Each treatment flask was separately inoculated with either strain 340, 339 or 218 and each treatment was replicated five times. Inoculated media were incubated at 30°C for 14 days. Mycelia biomass were harvested by filtration through Whatman No. 1 filter paper, washed three times and dried to a constant weight at 65°C.

2.2.2. Nitrogen Sources
To determine the best nitrogen source for mycelia yield, a 3 × 8 factorial experiment (strain and nitrogen source) was set up. Eight grams per liter of organic nitrogen sources—peptone, yeast extract, corn steep liquor and urea, and nitrogen contents corresponding to 0.96 g/L from inorganic sources—potassium sulfate, ammonium nitrate and ammonium sulfate—were separately used in the growth medium. Starch (10 g/L) identified from previous experiments was used as the carbon source. Sterilization, inoculation, incubation and mycelia yield determination were done as described in Section 2.2.1.

2.2.3. Concentrations of Selected Carbon and Nitrogen Sources
To determine the most appropriate concentration of selected carbon (starch) and nitrogen (yeast extract) sources to support high mycelia yield, two separate 3 × 9 factorial experiments of strain and starch or yeast extract was conducted. The strains were cultivated at different concentrations ranging from 0 - 30 g/L of starch or yeast extract. Mycelia was harvested and determined as described in Section 2.2.1.

2.3. Scale up Fermentation under Shake and Static Conditions for Mycelia and Crude Exopolysaccharide Production

Series of 3 × 5 × 2 experiments were conducted to determine the effect of strain, volume and agitation on mycelia and exopolyssacharide yield. The strains were incubated for six days in an optimized medium composed of starch at 30 g/L, and yeast extract at 25 g/L in 5 different scale-up medium volumes of 50, 100, 250, 500, and 1000 mL under static and shake ~150 rpm (Barnstead/Lab-Line Max Q4000) conditions. Mycelia was harvested by filtration with Whatman no. 1 filter paper and washed three times. Crude exopolysaccharide (CEPS) was precipitated as previously reported [17]. The culture filtrates were stirred with 4 volumes of absolute ethanol, mixed vigorously and stored at 4°C overnight. Precipitated CEPS were pelleted at 10,000 g (Eppendorf Centrifuge 5430R) at 4°C for 10 min. Harvested mycelia and CEPS were dried at 65°C to constant weight. Mycelia mass and CEPS were expressed as g dry weight/L of culture liquid.

2.4. Experimental Design and Statistical Analysis

The experimental design is a completely randomized design (CRD) with 2 or 3 factorial structure depending on the experiment. All experiments were carried out in 5 replicates, and the results are expressed as mean values. Differences among means were compared using Duncan's multiple range test. Statistical Analysis Software (SAS) version 9.3 was used for analysis of variance (ANOVA). Mean comparisons, and regression and correlation analysis were carried out where applicable. Under scale-up fermentation experiments, analysis of variance was performed to determine whether there is main effect or interaction between any two or all of the variables in the 3 factors tested. Regression and correlation analysis was used to estimate the variability in mycelia/polysaccharide yield due to culture volume, while correlation measured the strength of the relationship between mycelia yield and polysaccharide secretion among strains.

3. Results and Discussion

3.1. Utilization of Different Carbon Sources

The effect of eight different carbon sources on mycelia growth in strains 340, 339 and 218 of *L. squarrosulus* was studied. Analysis of variance (ANOVA) shows that there is a highly significant interaction between strain and carbon source ($p < 0.0001$). Mean comparison using Duncan's multiple range test shows that mycelia yield was highest when starch and mannose were used as carbon sources regardless of strain (**Table 1**). In these two carbon sources, strains 340 and 218 had similar mean mycelia yield of 4.31 g/L that was significantly different from strain 339 (3.29 g/L). Beyond these two sugars, strains vary significantly in their growth response to other carbon sources studied. Modest mycelia yield was achieved with dextrose only in strain 218. The sugar alcohols (mannitol and sorbitol) produced higher mycelia mass in strain 218 than strains 339 and 340. The least mycelia yield was produced with fructose, xylose and sorbitol in strains 218, 339 and 340 respectively. The result from the study is consistent with other work that reported starch and mannose as being widely utilized by many basidiomycetes [10] [18]. Starch utilization is possible because some basidiomycetes synthesize effective amylolytic enzymes for hydrolysis of starch. Literature shows that mannose and dextrose are good substrates for cellular respiration [19]. Therefore, it is not surprising that the strains studied produced high biomass with mannose. Similar response is seen in other basidiomycetes including *Cordycceps militaris* [18], *Pleurotus tuber-regium* [20], and *Grifola frondosa* [21]. Intermediate utilization of mannitol by strain 218 and other basidiomycetes is associated with substrate oxidation or dehydrogenase enzyme activity [22]. In contrast, Gbolagade *et al.* [15] reported that other sugars such as fructose and maltose stimulated the most growth in *L. subnudus* (syn. *Lentinus squarrosulus*). The highest mean mycelia mass reported in their work was remarkably lower, 0.19 g/L, compared to 4.42 g/L observed in the present study. Since mean mycelia yield did not differ between starch and mannose within the strains studied, starch was used in further experiments, since it is more readily available and cheap compared to mannose.

3.2. Utilization of Different Nitrogen Sources

Analysis of variance shows that there is interaction between nitrogen source and strain ($p < 0.001$), resulting in differences in mycelia yield among strains (**Table 2**). Organic nitrogen sources were generally preferred over inorganic sources except urea. Strain 218 did not grow on urea supplemented medium. Among organic sources, yeast extract was clearly superior to peptone and corn steep liquor ($p < 0.0001$). With yeast extract there was no significant difference between strains 340 and 218, which had higher mean mycelia yield of 6.17 g/L compared to 5.61 g/L in strain 339 ($p < 0.0001$). Mycelia yield was significantly reduced with inorganic nitrogen sources. KNO_2 and the control treatments inhibited growth of all strains. The preference of basidiomycetes for organic nitrogen sources has been reported by others in experiments with *Lentinus subnudus* [15], *Agaricus cinnamomea* [23], and *Hericium erinaceus* [24]. Most basidiomycetes prefer complex organic nitrogen sources in submerged fermentations, probably because certain essential amino acid(s) are not readily synthesized from inorganic sources during fermentation [25]. The result is supported by Jennison *et al.* [26], who reported that white and brown rot fungi failed to grow when potassium nitrate, potassium nitrite, and ammonium chloride were used as nitrogen sources. Growth was significantly higher with ammonium nitrate than ammonium sulfate ($p < 0.0001$).

Table 1. Mycelia yield (g/L) of three strains of *L. squarrosulus* grown in 8 different carbohydrate sources.

Carbon sources	Strain 218	Strain 339	Strain 340
	Mycelia dry weight g/L*		
Starch	4.30[a]	3.30[bcd]	4.39[a]
Mannose	4.42[a]	3.27[cd]	4.30[a]
Dextrose	4.30[a]	2.48[fgh]	3.54[bc]
Mannitol	2.82[def]	1.47[kl]	1.86[ijk]
Sorbitol	2.65[efg]	1.35[klm]	1.73[ijkl]
Xylose	2.00[hij]	1.07[m]	3.36[bcd]
Fructose	1.604[jklm]	1.82[ijk]	3.06[cde]
Sucrose	2.00[hij]	2.24[ghi]	3.85[ab]
Control	1.24[lm]	1.07[lm]	1.24[lm]

*Means with the same letters within columns and rows are not significantly different ($p < 0.0001$).

Table 2. Mycelia yield (g/L) of 3 strains of *L. squarrosulus* grown in 8 different nitrogen sources.

Nitrogen sources	Strain 218	Strain 339	Strain 340
	Mycelia dry weight g/L*		
Organic			
Yeast Extract	6.12[a]	5.61[b]	6.22[a]
Peptone	5.24[b]	4.47[c]	5.24[b]
Corn Steep Liquor	2.85[de]	3.02[d]	2.59[e]
Urea	0.63[i]	0.95[hi]	1.06[gh]
Inorganic			
$(NH_4)_2SO_4$	2.17[f]	0.57[i]	2.07[f]
NH_4NO_3	1.35[g]	0.91[j]	1.38[g]
KNO_3	0.057[j]	0.07[hi]	0.06[j]
KNO_2	0.00[j]	0.00[j]	0.00[j]
Control	0.00[j]	0.00[j]	0.00[j]

*Means with the same letters within columns and rows are not significantly different ($p < 0.0001$).

The apparent differences in utilization of various ammonium compounds could arise in part from differences in hydrogen ion concentrations produced in aqueous solutions by the compounds. Ammonium sulfate supplies twice the acidity as ammonium nitrate in solution. Mean mycelia yields from yeast extract was significant among strains, and therefore it was used in further studies.

3.3. Effect of Different Concentrations of Selected Carbon and Nitrogen Sources on Mycelia Yield

In previous experiments, starch and yeast extract were selected as the best carbon and nitrogen sources that support cell growth and high biomass yield in the strains studied. It was important to optimize the concentration of these nutrients in order to maximize mycelia yield. ANOVA shows that strain and concentration interact with both nutrients at a highly significant level ($p < 0.0001$). Regression analysis shows that there is a non-linear relationship between starch or yeast extract concentration and mycelia yield; the equation of the model is displayed in **Figure 1** and **Figure 2** respectively. The strains had variable yield in mycelia ($p < 0.05$) depending on concentration of starch or nitrogen within the range tested. Starch accounted for more variability in mycelia yield than yeast extract in strains 218 (96% versus 87%) and 339 (95% versus 79%) while they seem to be of equal strength in strain 340 (84% versus 85%). Mean mycelia yield was highest at elevated starch concentrations of between 20 - 30 g/L. Strains 218 and 340 produced the most mycelia at 30 g/L, although there was no difference in mycelia yield between 25 and 30 g/L in strain 339 ($p = 0.02$). Based on statistical data, 30 g/L of starch was selected and used in further experiments.

Strain response to yeast extract is different from starch. Strain 218 and 340 produced the most mycelia with no difference in mean yield between 20, 25 and 30 g/L yeast extract. Therefore 25 g/L was selected as the median concentration for further research. Biomass was significantly lower ($p = 0.05$) in strain 339 compared to others, however its response to yeast extract concentration was similar to other strains.

3.4. Scale up Fermentation for Mycelia Yield and Crude Exopolysaccharide (CEPS) Production in Optimized Medium under Static and Shaken Conditions

The result (**Table 3**) from Analysis of Variance (ANOVA) shows that there is highly significant interaction ($p < 0.0001$) in all possible combinations of the three factors tested (strain versus agitation, strain versus volume, volume versus agitation, strain versus volume versus agitation). Since the factors considered particularly influence fermentation, data interpretation follows a 3-way ANOVA. Regardless of strain type and culture volume, shake

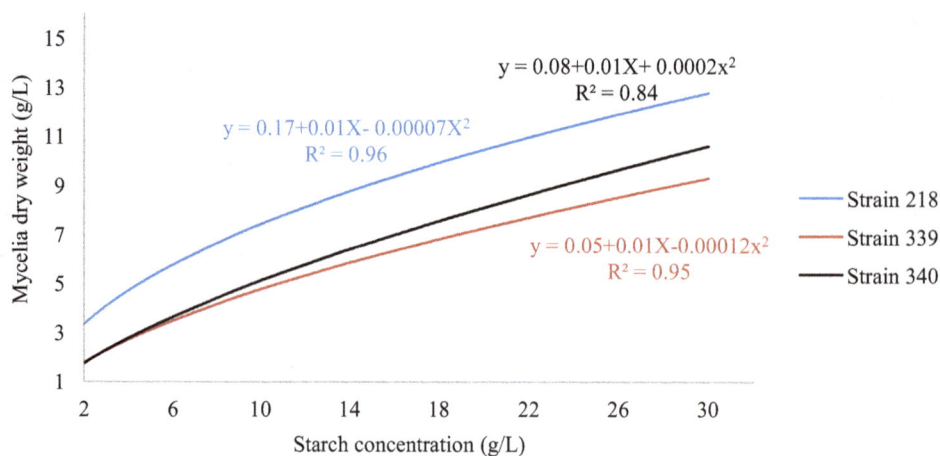

$$y = 0.08+0.01X+ 0.0002x^2$$
$$R^2 = 0.84$$

$$y = 0.17+0.01X- 0.00007X^2$$
$$R^2 = 0.96$$

$$y = 0.05+0.01X-0.00012x^2$$
$$R^2 = 0.95$$

Figure 1. Regression of mycelia yield in 3 strains of *L. squarrosulus* cultured in different concentrations of starch as sole carbon source.

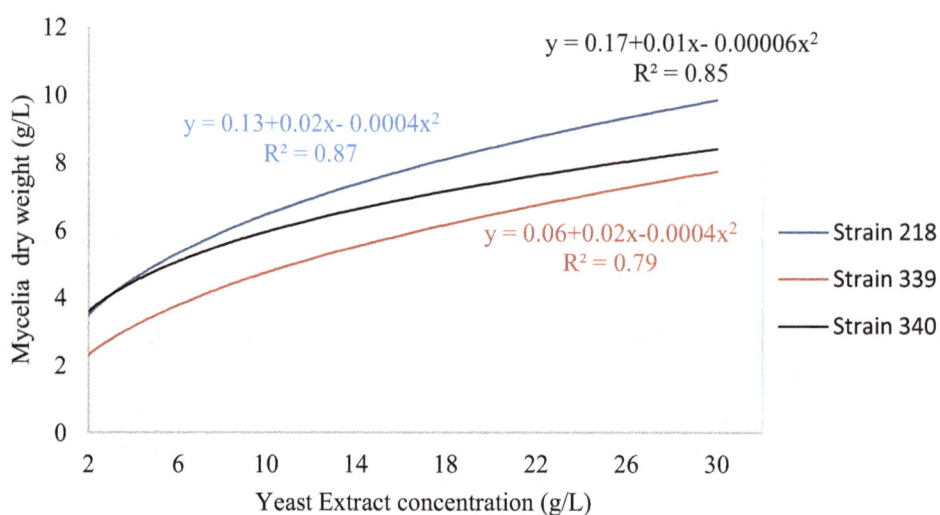

$$y = 0.17+0.01x- 0.00006x^2$$
$$R^2 = 0.85$$

$$y = 0.13+0.02x- 0.0004x^2$$
$$R^2 = 0.87$$

$$y = 0.06+0.02x-0.0004x^2$$
$$R^2 = 0.79$$

Figure 2. Regression of Mycelia yield in 3 strains of *L. squarrosulus* cultured in different concentrations of yeast extract as solenitrogen source.

Table 3. Analysis of variance of mycelia yield in 3 strains of *L. squarrosulus* showing interaction between 3 factors (strain, volume and agitation).

Sources of variation	df	Sum of squares	Mean square	F value	Pr > F
Treatment	29	682,495.72	23534.33	555.60	<0.0001
Strain	2	5099.89	2549.94	60.20	<0.0001
Volume	4	215,214.55	53,803.63	1270.20	<0.0001
Strain × volume	8	8747.07	1093.38	25.81	<0.0001
Agitation	1	214,880.50	5072.90	142.32	<0.0001
Strain × agitation	2	3134.04	1567.02	36.99	<0.0001
Volume × agitation	4	228,147.83	57,036.96	1346.53	<0.0001
Strain × volume × agitation	8	7271.82	908.97	21.46	<0.0001
Error	120	5083.02	42.36		
Total	149	687,578.75			

fermentation resulted in a highly significant increase in mycelia mass ($p < 0.0001$) than static fermentation (**Table 4**). The difference in mycelia yield under the two conditions ranged from 19.45 - 255 g/L, depending on volume of culture medium. Under agitation, mycelia mass increased with volume, whereas the reverse was observed with static fermentation, and mycelia mass decreased with volume. This pattern is consistent in the three strains tested. In similar observations, *Cordyceps jianxiensis* had a reduced growth when the culture volume increased from 50 to 300 mL under static fermentation condition [27]. In all strains, the order of increase in mycelia yield was 50 mL < 100 mL < 250 mL < 500 mL <1000 mL. The three strains produced the most mycelia mass in the highest volume tested (1000 mL). At this volume, mycelia mass produced by strain 339 (266 g/L) is higher and differs significantly from strain 340 (239.8 g/L) and strain 218 ((185 g/L) (**Table 4**). These values are higher than 25.8 g/L reported by Ahmad *et al.* [16]. Regression analysis of mycelia yield as a function of volume (**Figure 2**), was significant with an R^2 of 0.29. Since 29% of variability in mycelia mass is explained by volume, it is possible that other parameters such as nutrient and agitation may have played major roles in influencing mycelia yield. There is a highly significant interaction ($p < 0.0001$) between strain, volume and agitation in crude exopolysaccharide (CEPS) secretion. The influence of agitation on CEPS production in submerged fermentation by other mushroom have been reported [28] [29]). High mycelia biomass did not necessarily lead to high CEPS secretion in strains studied (**Figure 3** and **Figure 4**). In fact, CEPS secretion decreased with increase in mycelia yield. This is consistent with the work of Nour El-Dein *et al.* [30] on *Pleurotus pulmonarius*, Lin and Sung [31] on *Antrodia cinnamomea*, Isikhuemhen *et al.* [17] on *L. squarrosulus* under Solid State Fermentation (SSF) and Diamantopoulou *et al.* [32] [33].

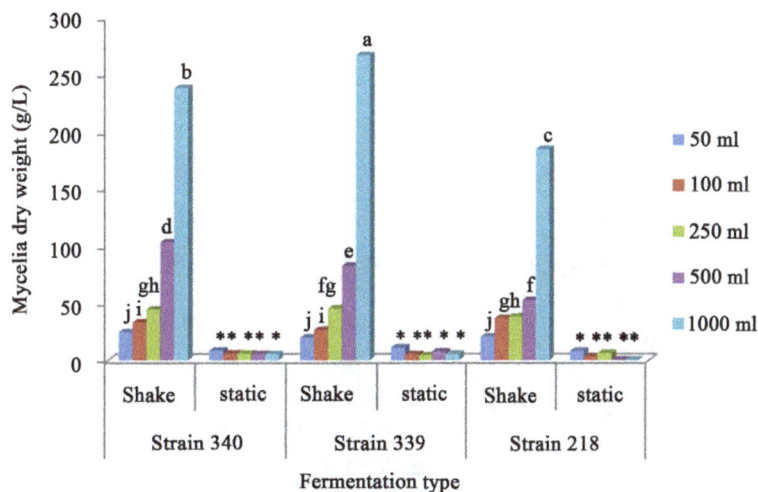

Figure 3. Mycelia dry weight (g/L) of 3 strains of *L. squarrosulus* grown in 5 different volumes of optimized medium (50 - 1000 mL) with or without agitation. Bars with same letter or symbol (*) indicate that the means are not significantly different ($p < 0.0001$).

Table 4. Mycelia dry weight resulting from submerged fermentation in 3 strains of *L. squarrosulus* using optimized medium, at different volumes with or without agitation.

Volume (mL)	Shake			Static		
	Strain 218	Strain 339	Strain 340	Strain 218	Strain 339	Strain 340
	Mycelia dry weight g/L*					
50	20.60[K]	19.97[K]	24.66[K]	8.18[lm]	11.02[l]	8.50[lm]
100	36.50[HI]	26.90[JK]	33.4I[J]	3.21[lm]	5.33[lm]	5.78[lm]
250	38.06[ghi]	45.42[fg]	44.76[fgh]	6.30[lm]	4.14[lm]	5.78[lm]
500	52.63[f]	82.71[e]	103.32[d]	0.60[m]	7.59[lm]	5.47[lm]
1000	185.00[c]	266.08[a]	238.80[b]	0.53[m]	5.60[lm]	5.34[lm]

*Means with same letters are not significantly different ($p < 0.0001$).

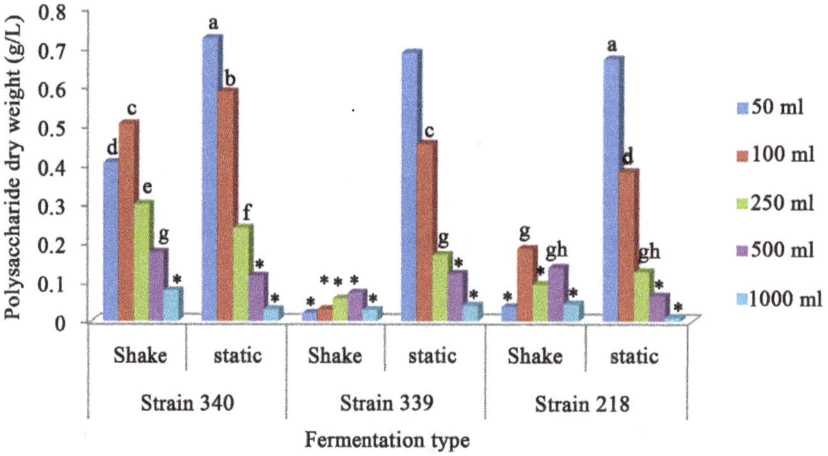

Figure 4. Crude exopolysaccharide in 3 strains of *L. squarrosulus* grown in 5 different volumes of optimized medium (50 - 1000 mL) with or without agitation. Bars with same letter or symbol (*) indicate that the means are not significantly different (*p* < 0.0001).

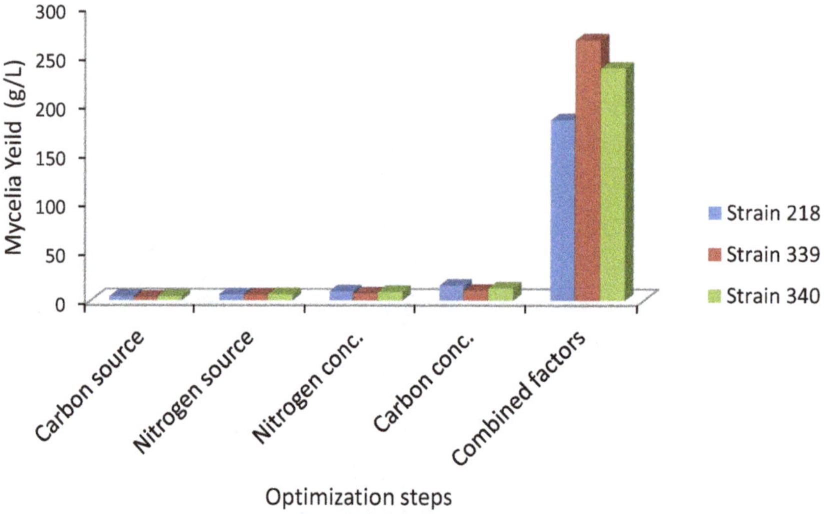

Figure 5. Increase in biomass dry weight at each stage of fermentation optimization process in 3 strains of *L. squarrosulus*. C = carbon; N = nitrogen; conc. = concentration; comb. = combined; scale up was done in 1L medium volume.

Since mycelia mass increases with medium volume as discussed above, it is logical that CEPS will decrease as medium volume increases. These results do not agree with Xiao *et al.* [27] who recorded higher CEPS in higher medium than low medium volumes. Higher CEPS was also reported by Ahmad *et al.* [16]. The highest CEPS of 0.72 g/L was observed in 50 mL medium capacity in strain 340 while the lowest (0.01 g/L) was recorded in 1000 mL culture volume in strain 218. Regression analysis shows that 25% of variability in CEPS is due to volume. Since biomass and polysaccharide production are inverse to each other, it was necessary to determine if there is a relationship between them. A regression and correlation analysis shows a weak negative relationship (−0.25) between biomass yield and polysaccharide secretion (*p* = 0.002). This result is in line with earlier report of, Diamantopoulou *et al.* [34], who performed statistical analysis regarding correlation between mycelia mass and EPS for five mushroom strains in static and agitated cultures, and indicated a significant negative relationship between mycelia production with EPS synthesis. Isikhuemhen *et al.* [17] recorded high yield of CEPS (up to 5.13 mg/mL) in the same mushroom (*L. squarrosulus*), albeit in Solid State Fermentation on cornstalk substrate. However in this study, CEPS was generally low in all strains, and maxed out at 0.72 mg/mL in strain 340, 0.69 mg/mL in strain 339 and 0.3 mg/mL in strain 218. In future studies, we will consider other factors that could increase CEPS yield, such as pH and aeration. Each step of the fermentation optimization process was an improvement on the previous step and mycelia mass increased accordingly (**Figure 5**). The highest mycelia mass was achieved when individual optimized factors were combined together in a single experiment. This is the first report on mycelia and CEPS production by *L. squarrosulus* in submerged fermentation under static and shake conditions.

4. Conclusion

The overall goal of the experiment was to optimize fermentation conditions for mycelia mass production and crude exopolysaccharide secretion in *L. squarrosulus*. There are several factors that can affect fermentation, and the factors considered in the present study do not cover all, but the results suggest that they are particularly influential. We conclude that all the factors studied: strain, nutrient, volume, and agitation affect product yield. Moreover, there is interaction between combinations of these factors. The practical result of the study yields information for the fermentation and biotechnology industry, which use mycelia, polysaccharides, and/or their products for food and health promoting benefits, and for the research community interested in bioactive compounds for further research. Optimally, biomass and exopolysaccharides should be produced in culture medium containing 30 g/L of starch as the carbon source, and 25 g/L of yeast extract as the nitrogen source. During fermentation, if mycelia are desired, a higher volume of 1000 mL is recommended, with agitation at 150 rpm. However if exopolysaccharide is the product of interest, lower volumes of 50 - 100 mL should be used with no agitation. Using such medium and physiological conditions, mycelia yield and/or polysaccharide secretion can be scaled up or down based on desired product.

Acknowledgements

The research in this paper was funded through the National Institute for Food and Agriculture of the United States Department of Agriculture, Project No. NC.X-999-5-09-120-1, in the Agricultural Research Program, North Carolina Agricultural and Technical State University.

References

[1] Isikhuenhen, O.S., Adenipekun, C.O. and Ohimain, E. (2010) Preliminary Studies on Mating and Improved Strain Selection in the Tropical Mushroom *Lentinus squarrosulus* Mont. *International Journal of Medicinal Mushrooms*, **12**, 177-183. http://dx.doi.org/10.1615/IntJMedMushr.v12.i2.80

[2] Okhuoya, J.A., Akpaja, E.O. and Abot, O. (2005) Cultivation of *Lentinus squarrosulus* (Mont) Singer on Sawdust of Selected Tropical Tree Species. *International Journal of Medicinal Mushrooms*, **7**, 213-218. http://dx.doi.org/10.1615/IntJMedMushr.v7.i3.790

[3] Isaka, M., Sappan, M., Rachtawee, P. and Boonpratuang, T. (2011) A Tetrahydrobenzofuran Derivative from the Fermentation Broth of *Lentinus squarrosulu* BCC 22366. *Phytochemistry Letters*, **2**, 106-108. http://dx.doi.org/10.1016/j.phytol.2010.12.002

[4] Sudirman, L.I., Lefebvre, G., Kiffer, E. and Botton, B. (1994) Purification of Antibiotics Produced by *Lentinus squarrosulus* and Preliminary Characterization of a Compound Active against *Rigidoporus lignosus*. *Current Microbiology*, **29**, 1-6. http://dx.doi.org/10.1007/BF01570183

[5] Adenikpekun, C.O. and Isikhuemhen, O.S. (2008) Bioremediation of Engine Oil Polluted Soil by the Tropical White Rot Fungus *Lentinus squarrosulus* Mont. (Singer). *Pakistan Journal of Biological Sciences*, **11**, 1634-1637. http://dx.doi.org/10.3923/pjbs.2008.1634.1637

[6] Omar, N.A.M., Abdullah, N., Kuppusamy, Abdulla, M.A. and Sabaratnam, V. (2011) Nutritional Composition, Antioxidant Activities, and Antiulcer Potential of *Lentinus squarrosulus* (Mont.) Mycelial Extract. *Evidence-Based Complementary and Alternative Medicine*, **2011**, Article ID: 539356.

[7] Bhunia, S.A., Dey, B., Maity, K.K., Patra, S., Maiti, S., Maiti, T., Sikdar, S.R. and Islam, S.S. (2011) Isolation and Characterization of an Immunoenhancing Extracts of an Edible Mushroom, *Lentinus squarrosulus* (Mont.) Singer. *Carbohydrate Research*, **346**, 2039-2044. http://dx.doi.org/10.1016/j.carres.2011.05.029

[8] Tang, Y.Z., Zhu, L.W., Li, H.M. and Li, D.S. (2007) Submerged Culture of Mushrooms in Bioreactors—Challenges, Current-State-of-the-Art, and Future. *Food Technology*, **45**, 221-229.

[9] Castro, J.I. and Taylor, E.R. (2002) Use of Different Nitrogen Sources by the Ectomycorrhizal Mushroom *Cantharellus cibarius*. *Mycorrhiza*, **12**, 131-137. http://dx.doi.org/10.1007/s00572-002-0160-2

[10] Chen, W., Zhao, Z. and Li, Y. (2011) Simultaneous Increase of Mycelia Biomass and Intracellular Polysaccharide from *Formes formentarius* and Its Biological Function of Gastric Cancer Intervention. *Carbohydrate Polymers*, **65**, 369-375. http://dx.doi.org/10.1016/j.carbpol.2011.02.035

[11] Zhong, J.J. and Tang, Y.J. (2004) Submerged Cultivation of Medicinal Mushrooms for Production of Valuable Bioactive Metabolites. *Advances in Biochemical Engineering/Biotechnology*, **87**, 25-59. http://dx.doi.org/10.1007/b94367

[12] Wasser, S.P., Elisashvili, V.I. and Tan, K.K. (2003) Effects of Carbon and Nitrogen Sources in the Medium on *Tremella mesenterica* Retz: Fr. (Heterobasidiomycetes) Growth and Polysaccharide Production. *International Journal of Medicinal Mushrooms*, **5**, 49-56. http://dx.doi.org/10.1615/InterJMedicMush.v5.i1.70

[13] Manjunathan, J. and Kaviyarasan, V. (2010) Studies on the Growth Requirements of *Lentinus tuber-regium* (Fr.), An Edible Mushroom. *Middle-East Journal of Scientific Research*, **5**, 81-85.

[14] Sinha, J., Bae, J.T., Park, J.P., Kim, K.H., Song, C.H. and Yun, J.W. (2001) Changes in Morphology of *Paecilomyces japonica* and Their Effect on Broth Rheology during Production of Exo-Biopolymers. *Applied Microbiology and Biotechnology*, **56**, 88-92. http://dx.doi.org/10.1007/s002530100606

[15] Gbolagade, J.S., Fasidi, I.O., Ajayi, E.J. and Sobowale, A.A. (2006) Effect of Physico-Chemical Factors and Semi-Synthetic Media on Vegetative Growth of *Lentinus subnudus* (Berk.), an Edible Mushroom from Nigeria. *Food Chemistry*, **99**, 742-747. http://dx.doi.org/10.1016/j.foodchem.2005.08.052

[16] Ahmad, R., Al-Shorgani, N.K.N., Hamid, A.A., Yusoff, W.M.W. and Daud, F. (2013) Optimization of Medium Components Using Response Surface Methodology (RSM) for Mycelium Biomass and Exopolysaccharide Production by *Lentinus squarrosulus*. *Advances in Bioscience and Biotechnology*, **4**, 1079-1085. http://dx.doi.org/10.4236/abb.2013.412144

[17] Isikhuenhen, O.S., Mikiashvili, N.A., Adenikpekun, C.O., Ohimain, E. and Shahbazi, G. (2012) The Tropical White Rot Fungus, *Lentinus squarrosulus* Mont.: Lignocellulolytic Enzymes Activities and Sugar Release from Cornstalks under Solid State Fermentation. *World Journal of Microbiology and Biotechnology*, **28**, 1961-1966. http://dx.doi.org/10.1007/s11274-011-0998-6

[18] Kwon, J.S., Lee, J.S., Shin, W.C., Lee, K.E. and Hong, E.K. (2009) Optimization of Culture Conditions and Medium Compositions for the Production of Mycelial Biomass and Exo-Polysaccharides with *Cordyceps militaris* in Liquid Culture. *Biotechnology and Bioprocess Engineering*, **14**, 756-762. http://dx.doi.org/10.1007/s12257-009-0024-0

[19] Griffin, D.H. (1994) Fungal Physiology. 2nd Edition, John Wiley and Sons, New York.

[20] Wu, J.Z., Cheung, P.C.K., Wong, K.H. and Huang, N.L. (2003) Studies on the Submerged Fermentation of *Pleurotus tuber-regium* (Fr.) Singer: 1. Physical and Chemical Factors Affecting the Rate of Mycelial Growth and Bioconversion Efficiency. *Food Chemistry*, **81**, 389-393. http://dx.doi.org/10.1016/S0308-8146(02)00457-0

[21] Shih, I., Chou, B., Chen, C., Wu, J. and Hsieh, C. (2008) Study of Mycelia Growth and Bioactive Polysaccharide Production in Batch and Feed-Batch Culture of *Grifola frondosa*. *Bioresource Technology*, **99**, 785-793. http://dx.doi.org/10.1016/j.biortech.2007.01.030

[22] Solomon, P.S., Water, O.D.C. and Oliver, R.P. (2007) Decoding the Mannitol Enigma in Filamentous Fungi. *Trends in Microbiology*, **15**, 257-262. http://dx.doi.org/10.1016/j.tim.2007.04.002

[23] Shih, I., Pan, K. and Hsieh, C. (2006) Influence of Nutritional Components and Oxygen Supply on Mycelia Growth and Bioactive Metabolites Production in Culture of *Anthrodia cinnamomea*. *Process Biochemistry*, **41**, 1129-1135. http://dx.doi.org/10.1016/j.procbio.2005.12.005

[24] Huang, D., Cui, F., Li, Y., Zhang, Z., Zhao, J., Han, X., Xiao, X., Qian, J., Wu, Q. and Guan, G. (2007) Nutritional Requirement for the Mycelia Biomass and Exopolymer Production by *Hericum erinaceus* CZ-2. *Food Technology and Biotechnology*, **45**, 389-395.

[25] Jung, I.C., Kim, S.H., Kwon, Y.I., Kim, S.Y., Lee, J.S., Park, S., Park, K.S. and Lee, J.S. (1997) Cultural Condition for the Mycelial Growth of *Phelinus igniarius* on Chemically Defined Medium and Grains. *The Korean Journal of Mycology*, **25**, 133-142.

[26] Jennison, M.W., Newcomb, M.D. and Henderson, R. (1955) Physiology of the Wood-Rotting Basidiomycetes. Growth and Nutrition in Submerged Culture in Synthetic Media. *Mycologia*, **47**, 275-304. http://dx.doi.org/10.2307/3755451

[27] Xiao, J., Chen, D., Wan, W., Hu, X., Qi, Y. and Liang, Z. (2006) Enhanced Simultaneous Production of Mycelia and Intracellular Polysaccharide in Submerged Cultivation of *Cordyceps jiangxiensis* Using Desirability Functions. *Process Biochemistry*, **41**, 1887-1893. http://dx.doi.org/10.1016/j.procbio.2006.03.031

[28] Park, J.P., Kim, Y.M., Kim, S.W., Hwang, H.J., Cho, Y.J., Lee, Y.S., Song, C.H. and Yun, J.W. (2002) Effect of Agitation Intensity on the Exo-Biopolymer Production and Mycelia Morphology in *Cordyceps militaris*. *Enzyme and Microbial Technology*, **34**, 433-438.

[29] Babitskaya, V.G., Shcherba, V.V., Puchkova, T.A. and Smirnov, D.A. (2005) Polysaccharides of *Ganoderma lucidum*: Factors Affecting Their Production. *Applied Biochemistry and Microbiology*, **41**, 169-173.

http://dx.doi.org/10.1007/s10438-005-0029-1

[30] El-Dein, M.M.N., El-Fallal, A.A., Toson, E.S.A. and Hereher, F.E. (2004) Exopolysaccharides Production by *Pleurotus pulmonarius*: Factors Affecting Formation and Their Structures. *Pakistan Journal of Biological Sciences*, **7**, 1078-1084. http://dx.doi.org/10.3923/pjbs.2004.1078.1084

[31] Lin, E. and Sung, S. (2006) Cultivation Conditions Influence Exopolysaccharide Production by the Edible Basidiomycete *Antrodia cinnamomea* in Submerged Culture. *International Journal of Food Microbiology*, **108**, 182-187. http://dx.doi.org/10.1016/j.ijfoodmicro.2005.11.010

[32] Diamantopoulou, P., Papanikolaou, S., Katsarou, E., Komaitis, M., Aggelis, G. and Philippoussis, A. (2012) Mushroom Polysaccharides and Lipids Synthesized in Liquid Agitated and Static Cultures. Part I: Screening Various Mushroom Species. *Applied Biochemistry and Biotechnology*, **167**, 536-551. http://dx.doi.org/10.1007/s12010-012-9713-9

[33] Diamantopoulou, P., Papanikolaou, S., Katsarou, E., Komaitis, M., Aggelis, G. and Philippoussis, A. (2012) Mushroom Polysaccharides and Lipids Synthesized in Liquid Agitated and Static Cultures. Part II: Study of *Volvariella volvacea*. *Applied Biochemistry and Biotechnology*, **167**, 1890-1906. http://dx.doi.org/10.1007/s12010-012-9714-8

[34] Diamantopoulou, P., Papanikolaou, S., Komaitis, M., Aggelis, G. and Philippoussis, A. (2014) Patterns of Major Metabolites Biosynthesis by Different Mushroom Fungi Grown on Glucose-Based Submerged Cultures. *Bioprocess and Biosystems Engineering*, **37**, 1385-1400. http://dx.doi.org/10.1007/s00449-013-1112-2

8

Characterization and Antioxidant Properties of OJP2, a Polysaccharide Isolated from *Ophiopogon japonicus*

Sairong Fan[1*], Junjun Wang[1,2*#], Yingge Mao[1,2], Yuan Ji[1,2], Liqin Jin[1,2†], Xiaoming Chen[1,2†], Jianxin Lu[1,2]

[1]Key Laboratory of Laboratory Medicine, Ministry of Education of China, School of Laboratory Medicine & Life Science, Wenzhou Medical University, Wenzhou, China
[2]Institute of Glycobiological Engineering, School of Laboratory Medicine & Life Science, Wenzhou Medical University, Wenzhou, China
Email: [†]liqinjin@126.com, [†]xmchen01@163.com

Abstract

A water-soluble polysaccharide (OJP2) obtained from the roots of *Ophiopogon japonicas*, was precipitated with 95% ethanol and purified by DEAE-52 cellulose anion-exchange and Sephadex G-100 gel filtration chromatography. The characteristics of OJP2 were determined by chemical analysis, high performance gel permeation chromatography (HPGPC), and gas chromatography-mass spectrometry (GC-MS). The results showed that the average molecular weight (Mw) of OJP2 was 35.2 kDa, and five kinds of monosaccharides including rhamnose, arabinose, xylose, glucose and galactose in a molar ratio of 0.5:5:4:1:10. Furthermore, the antioxidant activity of OJP2 was evaluated in H_2O_2-treated HaCaT cells and glucose-treated LO2 cells. The results show that OJP2 can increase the activity of SOD and NO production, and decrease the level of MDA in these two kinds of injury cells. OJP2 should be explored as a novel and potential natural antioxidant agent for use in functional foods or medicine.

Keywords

Polysaccharide, *Ophiopogon japonicas*, Characterization, Antioxidant

---

[*]These authors contributed equally to this work.
[#]Present address: The People's Hospital of Pingyang.
[†]Corresponding authors.

1. Introduction

Free radicals, the highly reactive molecules, play an important positive physiological role and, at the same time, they may exert toxic effects [1] [2]. There is increasing evidence that free radicals are able to damage cell membranes and numerous biological substances [3] [4], resulting in various diseases including cardiovascular diseases, cancer, aging, Parkinson's and Alzheimer's disease, atherosclerosis, and impairment of immune function [5]-[8]. Antioxidants can scavenge free radicals and help to reduce oxidative damage. Polysaccharides extracted from natural sources have been found to have variety of biological activities, such as antioxidant, immunobiological and antitumor activity, which have attracted lots of attention in the biochemical and medical area [9]-[13].

Ophiopogon japonicus (Thunb.) Ker-Gawl is a well known traditional Chinese medicine used to treat cardiovascular and chronic inflammatory diseases for thousands of years, widely distributed in south-east Asia, and has been confirmed in various experiments as having anti-inflammatory, anti-arrhythmia, and microcirculation improvement etc. [14] [15]. Chemical studies have shown that this plant includes saponins, polysaccharide and homoisoflavonoidal compounds [16]. In recent years, the polysaccharides isolated from the roots of *O. japonicus* have drawn the attention of researchers and consumers due to their nutritional and health protective value in hypoglycemic, anti-ischaemia, immunostimulation, inhibiting platelets aggregation, etc. [17]-[19]. However, the structure and function of these polysaccharides have not been well characterized.

In the present study, we isolated and purified a polysaccharide (designated OJP2 below) from *O. japonicas* using DEAE-cellulose anion-exchange and a Sephadex G-150 column chromatography. In addition, the characteristics and antioxidant activity of the polysaccharides are also investigated. It will be helpful to better find its functional properties for the wide application in food and pharmaceutical industries.

2. Material and Methods

2.1. Materials and Chemicals

The roots of *O. japonicus* were collected in Dongtou, Zhejiang province (China). Sephadex G-150, DEAE-cellulose, fucose, arabinose, rhamnose, xylose, glucose, galactose, mannose, trifluoroacetic acid (TFA) was purchased from Sigma. RPMI-1640 and DMEM medium was purchased from Gibco. The assay kits for superoxide dismutase (SOD), malondialdehyde (MDA) and nitric oxide (NO) were purchased from Jiancheng Biologic Project Company, Nanjing, Jiangsu Province. Cell line HaCaT and LO2 was purchased from Shanghai Institute of Cell Biology, Chinese Academy of Sciences. All the other chemicals used were of analytical grade.

2.2. General Methods

Gas chromatography (GC) and Gas chromatography-mass spectrometry (GC-MS) were described previously [20]. The products were dried by lyophilization. Gas chromatography (GC) (Shimadzu GC-2010) equipped with RTX-50 column (30.0 m × 0.25 mm × 0.25 μm) and flame-ionization detector (FID). Evaporation was performed at around 45°C under reduced pressure. The operation was performed using the following conditions: column temperature was programmed from 140°C (maintained for 2 min) to 170°C at a rate of 6°C/min, and increased to 173°C at a rate of 0.2°C/min, then increased to 233°C at a rate of 6°C/min, held for 40 min at 233°C; the rate of N_2 carrier gas was 1.0 ml/min; injection temperature was 250°C; detector temperature was 300°C. Gas chromatography-mass spectrometry (GC-MS) was run on the instrument Shimadzu GCMS-QP2010 (Shimadzu, Japan) and equipped with RTX-50 column (30 m × 0.25 mm × 0.25 μm), and at temperatures programmed from 140°C (maintained for 2 min) to 250°C (kept for 20 min) at a rate of 3°C /min. Nitrogen was the carrier gas.

2.3. Cell Culture and MTT Assay

HaCaT cell were maintained in DMEM medium supplemented with heat-inactivated fetal bovine serum (10%) at 37°C in humidified air containing 5% CO_2. The cells were plated into triplicate wells (100 μL/well) in 96-well ((5 × 10^5 cells/well) flat bottom tissue culture plates and co-cultured with samples at indicated concentrations and 250 μM H_2O_2. Serum free DMEM was used as control. Following 24 h incubation at 37°C and 5% CO_2, the supernatants were collected for detection of SOD activity, NO and MDA levels using commercial assay kits.

LO2 Cell were maintained in DMEM medium supplemented with heat-inactivated fetal bovine serum (10%)

at 37°C in humidified air containing 5% CO_2. The cells were plated into triplicate wells (200 μL/well) in 96-well flat bottom tissue culture plates and co-cultured with samples at indicated concentrations and 30 mmol/L glucose. Serum free DMEM was used as control. Following 24 h incubation at 37°C and 5% CO_2, the cell was measured by MTT assay. The absorbance at 570 nm test wavelength was translated into inhibition ratio for comparison.

$$\text{Inhibition ratio} = \left(A_0 - A_s \right) / A_0 \times 100\%$$

where A_0 and A_s are absorbance of blank and sample respectively.

2.4. Isolation and Purification of the Polysaccharide

The polysaccharide was prepared as described previously [20]. The roots of *O. japonicus* were soaked with 95% ethanol to remove the pigments and small lipophilic molecules. The residue was then extracted with 10 vol. of distilled water at 90°C for 3 h thrice. All water-extracts were combined, filtrated, concentrated, and precipitated with 95% EtOH (1:4, v/v) at 4°C for overnight. The precipitate was collected by centrifugation and deproteinated by Sevag method [21]. Finally the supernatant was lyophilized to give crude polysaccharides.

The crude polysaccharides were dissolved in distilled water and filtered through a membrane (0.45 μm). Then the solution was applied to a DEAE-52 cellulose column. Fractions were eluted with increased concentration of NaCl (0.01 - 1.0 M). The polysaccharide fractions were collected, concentrate, dialysed and finally lyophilized. Then, the sample was further purified by Sephadex G-150, and the second fraction was collected and lyophilized to give a polysaccharide named OJP2.

The polysaccharide was monitored by the phenol-sulfuric acid method [22].

2.5. Homogeneity and Molecular Weight

The homogeneity and molecular weight of OJP2 was evaluated and determined by high performance gel permeation chromatography (HPGPC) as described previously [20]. The sample solution was applied to Waters High Performance Liquid Chromatography (HPLC) equipped with a TSK-GEL G5000 SWXL column (7.8 × 300 mm), eluted with 0.1 mol/L Na_2SO_4 solution at a flow rate of 0.4 ml/min and detected by a Waters 2414 Refractive Index Detector. The columns were calibrated with Dextran T-series standard of known molecular weight (200,000, 70,000, 40,000, 10,000, 5000 Da). The molecular weight of OJP2 was estimated by reference to the calibration curve made above.

2.6. Analysis of Monosaccharide Composition

The monosaccharide of OJP2 was analyzed by GC as described previously [20]. OJP2 was hydrolyzed with 2 M TFA (2 ml) at 120°C for 2 h. After removing TFA with methanol, the hydrolyzed product was reduced with $NaBH_4$ (50 mg), followed by neutralization with dilute acetic acid and evaporated at 45°C. The reduced products (alditols) were added with 1 ml pyridine and 1 ml acetic anhydride in a boiling water bath for 1 h. The acetylated products were analyzed by GC.

2.7. Methylation Analysis

OJP2 (20 mg) was methylated three times according to the method of Needs and Selvendran [23]. The methylated products were extracted by chloroform and examined by IR spectroscopy. The absence of the absorption peak corresponding to hydroxyl indicated the complete methylation. The product was hydrolyzed using 2 M TFA, followed by reduction using $NaBH_4$ and finally acetylated with acetic anhydride. The partially methylated alditol acetates were analyzed by GC-MS.

2.8. Biochemical Assays

Lipid peroxidation was determined by quantifying MDA concentrations, which was spectrophotometrically measured by the absorbance of a red-colored product with thiobarbituric acid [24]. SOD and NO were measured with commercial kits, and the test method was done according to the reagent protocol prepared by the manufacturing firm.

2.9. Statistical Analysis

Results were presented as mean ± standard deviation (S.D.). Data were analyzed by one-way ANOVA using Student's *t-test*. *P*-values less than 0.05 were considered significant.

3. Results

3.1. Isolation and Characterization of OJP2

The polysaccharides extracted from the root of *O. japonicus* were purified through with DEAE-cellulose column and Sephadex G-100 column and named OJP2. OJP2 has no absorption at 280 and 260 nm in the UV spectrum, indicating the absence of protein and nucleic acid.

The average molecular weight of OJP2 was determined as 88.1 kDa by HPGPC. Results from phenol-sulfuric acid assay showed that OJP2 contained 95.3% carbohydrate.

OJP2 was hydrolyzed by TFA into individual monosaccharides that were further reduced and acetylated for GC analysis. The results showed that OJP2 is composed of Rha, Ara, Xyl, Glc, Gal with a relative molar ratio of 0.5:5:4:1:10 (**Figure 1**).

The IR spectrum of OJP2 was shown in **Figure 2**. The IR spectrum revealed a typical major broad stretching peak around 3400 - 3500 cm^{-1} for the hydroxyl group, and the small band at around 2950.21 cm^{-1} was attributed to the C-H stretching and bending vibrations. The relatively strong absorption peak at around 1614.55 cm^{-1} reflects the absorption of the C=O group that is part of glycosides [20] [25]. The absorptions at 1020.91 and 1099.76 cm^{-1} indicated a pyranose form of sugars [26]. The region between 950 and 1200 cm^{-1} is dominated by ring vibrations overlapped with stretching vibrations of (C-OH) side groups and the (C-O-C) glycosidic band vibration [20] [27].

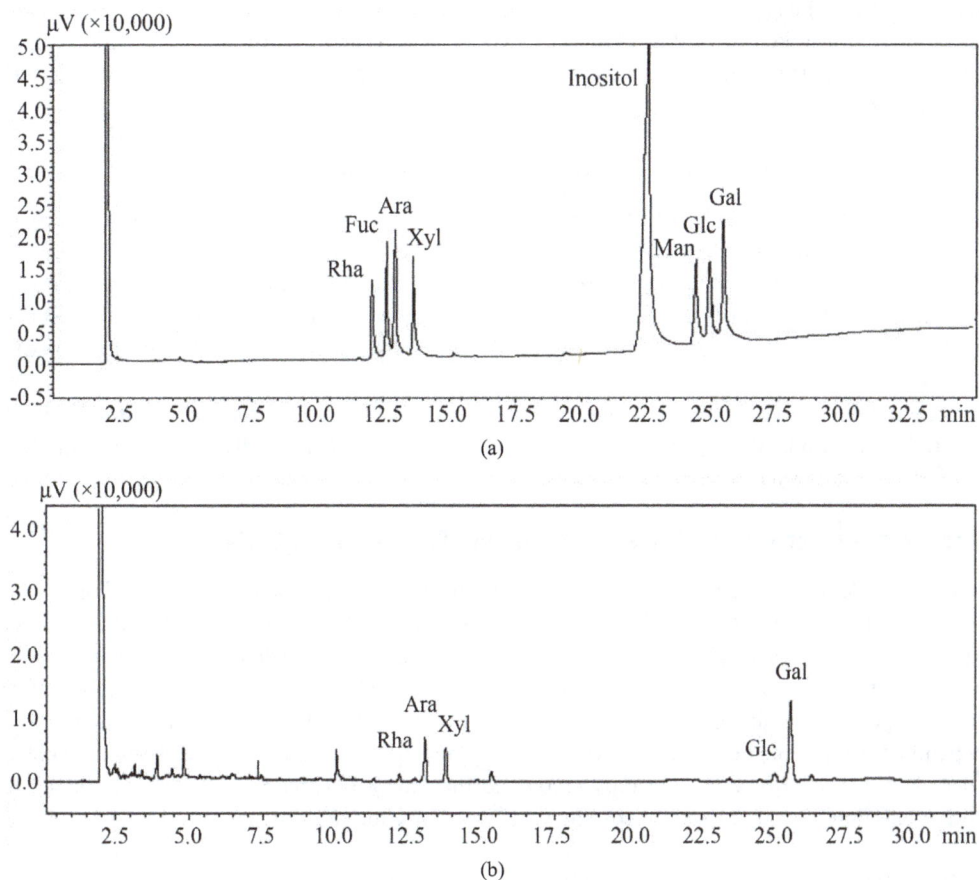

Figure 1. GC profile of OJP2. (a) Standard monosaccharides; (b) Monosaccharide composition of OJP2; Rha = Rhamnose, Fuc = Fucose, Ara = Arabinose, Xyl = Xylose, Man = Mannose, Glc = Glucose, Gal = Galactose.

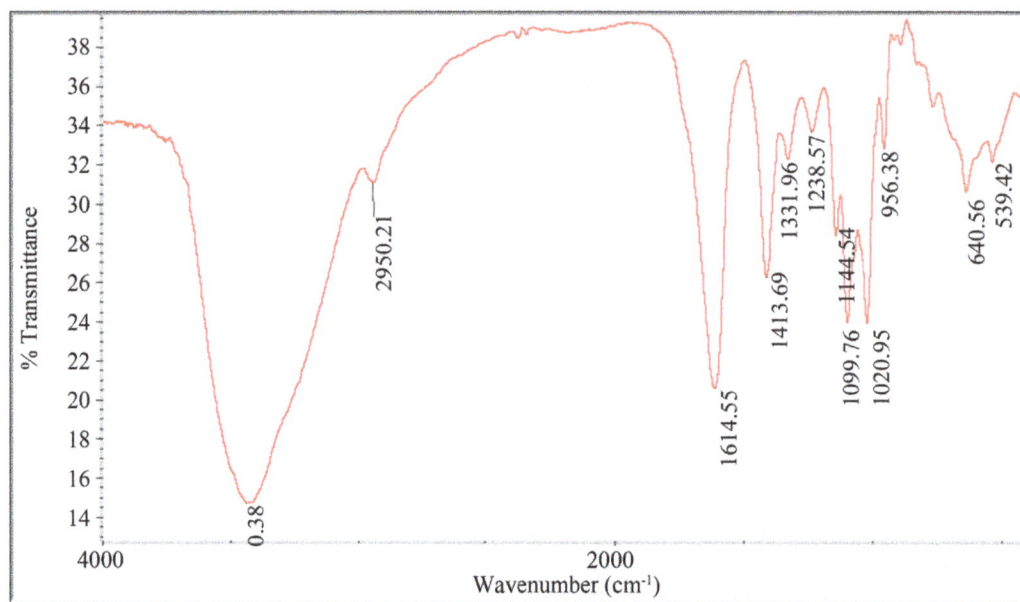

Figure 2. IR spectrum of the polysaccharide of OJP2 isolated from the roots of *O. japonicas*.

The fully methylated OJP2 was hydrolyzed with acid, converted into alditol acetates and analyzed by GC-MS. As shown in **Table 1**, the ratios of methylated fragments were calculated based on the areas of the methylated products and corrected using the effective-carbon response method [28]. The GC-MS results (**Table 1**) indicated that 2,5-Me-Xyl (1,3-linked Xyl), 2,3,4-Me-Gal (1,6-linked Gal) and part of Ara were major components of the backbone structure, part of Ara and Gal were distributed in branches, and residues of branches terminated with Rha, Glc and Gal.

3.2. Antioxidant Activities of OJP2 in H_2O_2-Treated HaCaT Cells

As shown in **Table 2**, the activity of SOD was significantly ($P < 0.01$) decreased in H_2O_2-treated group compared to the normal control group. OJP2-treated (250, 500, 1000 µg/mL) significantly ($P < 0.01$) increased the activities of SOD in H_2O_2-treated HaCaT cells with a dose-dependent manner. The treatment of HaCaT cells with H_2O_2 afforded the decrease in the NO production. However, incubation of the H_2O_2-treated cells with OJP2 polysaccharide markedly reduced in a dose-dependent manner the decrements in NO production of H_2O_2-treated cells ($P < 0.01$).

The levels of MDA in H_2O_2-treated groups was significantly ($P < 0.01$) increased (**Table 2**). Incubation of the H_2O_2-treated cells with OJP2 polysaccharide significantly decreased ($P < 0.01$ or $P < 0.05$) the levels of MDA, although it did not act as a dose-dependent manner.

3.3. Antioxidant Activities of OJP2 in Glucose-Treated LO2 Cells

As shown in **Table 3**, the treatment of LO2 cells with 30 mmol/L glucose caused inhibition of the cells growth and the inhibition ratio up to 52.8%. However, incubation of the glucose-treated cells with OJP2 polysaccharide markedly decreased in a dose-dependent manner the inhibition ratio of glucose-treated cells.

The treatment of LO2 cells with 30 mmol/L glucose afforded the decrease in the activity of SOD and NO production. However, incubation of the glucose-treated cells with OJP2 polysaccharide markedly reduced in a dose-dependent manner the decrements in SOD activity and NO production of glucose-treated cells ($P < 0.01$ or $P < 0.05$). The treatment of OJP2 not only recovered the glucose-decreased SOD activity and NO production, but also increased SOD activity and NO production, especially for NO production.

4. Discussion

Oxidation phenomena have been implicated in many illnesses, such as diabetes mellitus, arteriosclerosis,

Table 1. The results of methylation analysis of OJP2.

No.	Methylated sugar[a]	Molar ratio	Linkages types
1	2,3,4-Me-Rha	1	Rha-(1→
2	2,3-Me-Ara	3	→5)-Ara-(1→
3	2,3,4,6-Me-Glc	2	Glc-(1→
4	2,3,4,6-Me-Gal	3	Gal-(1→
5	3,5-Me-Ara	2	→2)-Ara-(1→
6	2,5-Me-Ara	3	→3)-Ara-(1→
7	2,5-Me-xyl	7	→3)-Xyl-(1→
8	2,3,4-Me-Gal	15	→6)-Gal-(1→
9	2-Me-Ara	1	→3,5)-Ara-(1→
10	2,4-Me-Gal	3	→3,6)-Gal-(1→

[a]2,3,4,6-Me-Glc = 1,5-di-O-acetyl-2,3,4,6-tetra-O-methyl-glucose, etc.

Table 2. Effects of OJP2 on SOD activity, NO and MDA levels of H_2O_2-treated HaCaT cells[a].

	SOD (U/mL)	NO (μmol/L)	MDA (nmol/L)
Normal control	69.79 ± 2.68	35.24 ± 1.21	0.47 ± 0.04
H_2O_2-treated control	33.15 ± 3.25[b]	27.23 ± 1.03[b]	0.69 ± 0.03[b]
OJP2-250	46.73 ± 1.89[d]	32.01 ± 0.85[d]	0.58 ± 0.04[c]
OJP2-500	56.12 ± 3.17[d]	35.09 ± 0.89[d]	0.51 ± 0.03[d]
OJP2-1000	58.35 ± 3.08[d]	36.05 ± 1.09[d]	0.52 ± 0.04[d]

[a]Normal control: without H_2O_2 injury; H_2O_2-treated control: incubated with 250 μM H_2O_2; OJP2-250: incubated with 250 μg/mL OJP2 and 250 μM H_2O_2; OJP2-500: incubated with 500 μg/mL OJP2 and 250 μM H_2O_2; OJP2-1000: incubated with 1000 μg/mL OJP2 and 250 μM H_2O_2; OJP2 was added to the culture 24 h prior to H_2O_2 addition; Three independent experiments were carried out in triplicates. Data represent mean ± S.D. [b]$P < 0.01$ compared with the normal control. [c]$P < 0.05$ compared with the H_2O_2-treated control. [d]$P < 0.01$ compared with the H_2O_2-treated control.

Table 3. Effects of OJP2 on NO production and the activity of SOD in glucose-treated LO2 cells.

.	BC[a]	GC[a]	OJP2 (1.5 mg/mL)[a]	OJP2 (1 mg/mL)[a]	OJP2 (0.5 mg/mL)[a]	OJP2 (0.25 mg/mL)[a]
Inhibition (%)		52.8 ± 2.5	39.8 ± 3.7[b]	40.2 ± 2.6[c]	42.9 ± 3.2[c]	45.6 ± 2.5[c]
SOD (IU)	18.56 ± 1.23	15.21 ± 1.21[d]	20.25 ± 1.05[c]	19.18 ± 0.97[c]	19.06 ± 0.83[c]	17.52 ± 1.07[b]
NO (μmol/L)	5.89 ± 0.14	4.24 ± 0.08[d]	11.23 ± 0.43[c]	10.87 ± 0.39[c]	8.53 ± 0.56[c]	7.81 ± 0.32[c]

[a]BC: Blank control (without glucose treated); GC: Glucose (30 mmol/L) treated control; the OJP2 groups: LO2 cell co-cultured with samples at indicated concentrations and 30 mmol/L glucose; [b]$P < 0.05$, [c]$P < 0.01$ (compared with the Glucose treated); [d]$P < 0.01$ (compared with the Blank control).

nephritis, Alzheimer's disease and cancer [29] [30]. SOD, one of the major antioxidant enzymes, decompose superoxide peroxide, against reactive oxygen species generated *in vivo* during oxidative stress, and being involved in the cellular defense mechanisms [31]. Our study shows that OJP2 treatment markedly restores the activity of SOD in H_2O_2-treated cells, and also increases the activity of SOD in glucose-treated LO2 cells. These finding suggest that OJP2 could considerably improve cellular antioxidative defense against oxidative stress.

MDA, generated under high levels of un-scavenged free radicals, is regarded as an index of cellular damage and cytotoxicity [32]. The study showed that the levels of MDA in H_2O_2-treated HaCaT cells or glucose-treated LO2 cells were increased. On the contrary, OJP2 treatment could decrease the MDA content elevation in heart tissues of diabetic rats. It was quite possible that the free radicals were effectively neutralized or scavenged, resulting in antioxidant effect of OJP2.

Nitric oxide (NO), a biologically active unstable radical, is generated from the metabolism of L-arginine by the enzyme nitric oxide synthase (NOS), and may quench the superoxide anion [33]. NO is also known to regulate immune responses and plays an important role in the protection against the onset and progression of cardiovascular disease, and decreased NO bioavailability has been proposed as one of the determinants of vascular damage [34] [35]. Our results show that OJP2 treatment increased the NO production both in H_2O_2-treated HaCaT cells and glucose-treated LO2 cells.

5. Conclusions

In conclusion, according to our experiment, the polysaccharide fraction (OJP2), with the MW 88.1 KDa, isolated from the root of *O. japonicus*. We demonstrated that OJP2 was a heteropolysaccharide consisting of Rha, Ara, Xyl, Glc, and Gal with a relative molar ratio of 0.5:5:4:1:10; 2,5-Me-Xyl (1,3-linked Xyl), 2,3,4-Me-Gal (1,6-linked Gal) or part of Ara were major components of the backbone structure; part of Ara and Gal were distributed in branches; and residues of branches terminated with Rha, Glc and Gal.

Biologically this polysaccharide exhibits significant antioxidant activity *in vitro*, and protective effects on H_2O_2-induced injury in HaCaT cells or glucose-induced injury LO2 cells. It might be a great potential source for the development of antioxidant agent for use on functional foods or medicine.

Acknowledgements

This work was supported by the Natural Science Foundation of Zhejiang Province of China (No. LY12B02009), and Key Science and Technology Innovation Team of Zhejiang Province (2010R50048).

References

[1]   Kozarski, M., Klaus, A., Niksic, M., Jakovljevic, D., Helsper, J.P.F.G. and Van Griensven, L.J.L.D. (2011) Antioxidative and Immunomodulating Activities of Polysaccharide Extracts of the Medicinal Mushrooms *Agaricus bisporus*, *Agaricus brasiliensis*, *Ganoderma lucidum* and *Phellinus linteus*. *Food Chemistry*, **129**, 1667-1675. http://dx.doi.org/10.1016/j.foodchem.2011.06.029

[2]   Wu, X.J. and Hansen, C. (2008) Antioxidant Capacity, Phenol Content, and Polysaccharide Content of *Lentinus edodes* Grown in Whey Permeate-Based Submerged Culture. *Journal of Food Science*, **73**, 1-8. http://dx.doi.org/10.1111/j.1750-3841.2007.00595.x

[3]   Tsai, M., Song, T., Shih, P. and Yen, G. (2007) Antioxidant Properties of Water-Soluble Polysaccharides from *Antrodia cinnamomea* in Submerged Culture. *Food Chemistry*, **104**, 1115-1122. http://dx.doi.org/10.1016/j.foodchem.2007.01.018

[4]   Thetsrimuang, C., Khammuang, S., Chiablaem, K., Srisomsap, C. and Sarnthima, A. (2011) Antioxidant Properties and Cytotoxicity of Crude Polysaccharides from *Lentinus polychrous* Lév. *Food Chemistry*, **128**, 634-639. http://dx.doi.org/10.1016/j.foodchem.2011.03.077

[5]   Wu, X.J. and Hansen, C. (2008) Antioxidant Capacity, Phenol Content, and Polysaccharide Content of *Lentinus edodes* Grown in Whey Permeate-Based Submerged Culture. *Journal of Food Science*, **73**, 1-8. http://dx.doi.org/10.1111/j.1750-3841.2007.00595.x

[6]   Xu, W., Zhang, F., Luo, Y., Ma, L., Kou, X. and Huang, K. (2009) Antioxidant Activity of a Water-Soluble Polysaccharide Purified from *Pteridium aquilinum*. *Carbohydrate Research*, **344**, 217-222. http://dx.doi.org/10.1016/j.carres.2008.10.021

[7]   Shin, K.H., Lim, S.S., Lee, S.H., Lee, Y.S. and Cho, S.Y. (2001) Antioxidant and Immunostimulating Activities of the Fruiting Bodies of *Paecilomyces japonica*, a New Type of Cordyceps sp. *Annals of the New York Academy of Sciences*, **928**, 261-273. http://dx.doi.org/10.1111/j.1749-6632.2001.tb05655.x

[8]   Finkel, T. and Holbrook, N.J. (2000) Oxidants, Oxidative Stress and the Biology of Aging. *Nature*, **408**, 239-247. http://dx.doi.org/10.1038/35041687

[9]   Xue, S.X., Chen, X.M., Lu, J.X. and Jin, L.Q. (2009) Protective Effect of Sulfated Achyranthes Bidentata Polysaccharides on Streptozotocin-Induced Oxidative Stress in Rats. *Carbohydrate Polymers*, **75**, 415-419. http://dx.doi.org/10.1016/j.carbpol.2008.08.003

[10]   Chen, R.Z., Liu, Z.Q., Zhao, J.M., Chen, R.P., Meng, F.L., Zhang, M. and Ge, W.C. (2011) Antioxidant and Immunobiological Activity of Water-Soluble Polysaccharide Fractions Purified from *Acanthopanax senticosu*. *Food Chemistry*, **127**, 434-440. http://dx.doi.org/10.1016/j.foodchem.2010.12.143

[11]   Chen, X.M., Nie, W.J., Fan, S.R., Zhang, J.F., Wang, Y.X., Lu, J.X. and Jin, L.Q. (2012) A Polysaccharide from *Sar-*

gassum fusiforme Protects against Immunosuppression in Cyclophosphamide-Treated Mice. *Carbohydrate Polymers*, **90**, 1114-1119. http://dx.doi.org/10.1016/j.carbpol.2012.06.052

[12] Chen, R.Z., Meng, F.L., Liu, Z.Q., Chen, R.P. and Zhang, M. (2010) Antitumor Activities of Different Fractions of Polysaccharide Purified from *Ornithogalum caudatum* Ait. *Carbohydrate Polymers*, **80**, 845-851. http://dx.doi.org/10.1016/j.carbpol.2009.12.042

[13] Chen, X.M., Nie, W.J., Yu, G.Q., Li, Y.L., Hu, Y.S., Lu, J.X. and Jin, L.Q. (2012) Antitumor and Immunomodulatory Activity of Polysaccharides from *Sargassum fusiforme*. *Food and Chemical Toxicology*, **50**, 696-700. http://dx.doi.org/10.1016/j.fct.2011.11.015

[14] Zhou, Y.H., Xu, D.S., Feng, Y., Fang, J.N., Xia, H.L. and Liu, J. (2003) Effects on Nutrition Blood Flow of Cardiac Muscle in Mice by Different Extracts in Radix Ophiopogonis. *China Journal of Experimental Traditional Medical Formulae*, **9**, 22-23.

[15] Huang, H.C. and Ni, Z. (2003) Effect of Ophiopogonis on the Auricular Microcirculation in Mice. *Shanghai Laboratory Animal Science*, **23**, 57-58.

[16] Kou, J., Sun, Y., Lin, Y., Cheng, Z., Zheng, W., Yu, B. and Xu, Q. (2005) Anti-Inflammatory Activities of Aqueous Extract from Radix *Ophiopogon japonicus* and Its Two Constituents. *Biological & Pharmaceutical Bulletin*, **28**, 1234-1238. http://dx.doi.org/10.1248/bpb.28.1234

[17] Zheng, Q., Feng, Y., Xu, D.S., Lin, X. and Chen, Y.Z. (2009) Influence of Sulfation on Anti-Myocardial Ischemic Activity of *Ophiopogon japonicus* Polysaccharide. *Journal of Asian Natural Products Research*, **11**, 306-321. http://dx.doi.org/10.1080/10286020902727363

[18] Chen, X.M., Tang, J., Xie, W.Y., Wang, J.J., Jin, J., Ren, J., Jin, L.Q. and Lu, J.X. (2013) Protective Effect of the Polysaccharide from *Ophiopogon japonicus* on Streptozotocin-Induced Diabetic Rats. *Carbohydrate Polymers*, **94**, 378-385. http://dx.doi.org/10.1016/j.carbpol.2013.01.037

[19] Fan, J. and Zhang, X. (2006) Research Progress in Pharmacology of *Ophiopogon japonicus* Polysaccharides. *Chinese Archives of Traditional Chinese Medicine*, **24**, 626-627.

[20] Chen, X.M., Jin, J., Tang, J., Wang, Z.F., Wang, J.J., Jin, L.Q. and Lu, J.X. (2011) Extraction, Purification, Characterization and Hypoglycemic Activity of a Polysaccharide Isolated from the Root of *Ophiopogon japonicus*. *Carbohydrate Polymers*, **83**, 749-754. http://dx.doi.org/10.1016/j.carbpol.2010.08.050

[21] Staub, A.M. (1965) Removal of Proteins from Polysaccharides. *Methods in Carbohydrate Chemistry*, **5**, 5-7.

[22] Dubois, M., Gilles, K.A., Hamilton, J.K., Rebers, P.A. and Smith, F. (1956) Colorimetric Method for Determination of Sugars and Related Substance. *Analytical Chemistry*, **28**, 350-356. http://dx.doi.org/10.1021/ac60111a017

[23] Needs, P.W. and Selvendran, R.R. (1993) Avoiding Oxidative Degradation during Sodium Hydroxide/Methyl Iodide-Mediated Carbohydrate Methylation in Dimethyl Sulfoxide. *Carbohydrate Research*, **245**, 1-10. http://dx.doi.org/10.1016/0008-6215(93)80055-J

[24] Yagi, K. (1994) Lipid Peroxides and Related Radicals in Clinical Medicine. In: Armstrong, D., Ed., *Free Radicals in Diagnostic Medicine*, Plenum Press, New York, 1-14. http://dx.doi.org/10.1007/978-1-4615-1833-4_1

[25] Zhu, M.Y., Mo, J.G., He, C.S., Xie, H.P., Ma, N. and Wang, C.J. (2010) Extraction, Characterization of Polysaccharides from *Lycium barbarum* and Its Effect on Bone Gene Expression in Rats. *Carbohydrate Polymers*, **80**, 672-676. http://dx.doi.org/10.1016/j.carbpol.2009.11.038

[26] Zhao, G.H., Kan, J.Q., Li, Z.X. and Chen, Z.D. (2005) Structural Features and Immunological Activity of a Polysaccharide from *Dioscorea opposita* Thunb Roots. *Carbohydrate Polymers*, **61**, 125-131. http://dx.doi.org/10.1016/j.carbpol.2005.04.020

[27] Yang, L. and Zhang, L.M. (2009) Chemical Structural and Chain Conformational Characterization of Some Bioactive Polysaccharides Isolated from Natural Sources. *Carbohydrate Polymers*, **76**, 349-361. http://dx.doi.org/10.1016/j.carbpol.2008.12.015

[28] Sweet, D.P., Shapiro, R.H. and Albersheim, P. (1975) Quantitative Analysis by Various g.l.c. Response-Factor Theories for Partially Methylated and Partially Ethylated Alditol Acetates. *Carbohydrate Research*, **40**, 217-225. http://dx.doi.org/10.1016/S0008-6215(00)82604-X

[29] Misthos, P., Katsaragakis, S., Milingos, N., Kakaris, S., Sepsas, E., Athanassiadi, K., Theodorou, D. and Skottis, I. (2005) Postresectional Pulmonary Oxidative Stress in Lung Cancer Patients. The Role of One-Lung Ventilation. *European Journal of Cardiothoracic Surgery*, **27**, 379-383. http://dx.doi.org/10.1016/j.ejcts.2004.12.023

[30] Dadé, M.M., Schinella, G.R., Fioravanti, D.E. and Tournier, H.A. (2011) Antioxidant and Cytotoxic Properties of an Aqueous Extract from the Argentinean Plant *Hedeoma multiflorum*. *Pharmaceutical Biology*, **49**, 633-639. http://dx.doi.org/10.3109/13880209.2010.526949

[31] Yao, D.C., Shi, W.B., Gou, Y.L., Zhou, X.R., Tak, Y.A. and Zhou, Y.K. (2005) Fatty Acid-Mediated Intracellular Iron

Translocation: A Synergistic Mechanism of Oxidative Injury. *Free Radical Biology and Medicine*, **39**, 1385-1398. http://dx.doi.org/10.1016/j.freeradbiomed.2005.07.015

[32] Karthikeyan, K., Bai, B.R. and Devaraj, S.N. (2007) Cardioprotective Effect of Grape Seed Proanthocyanidins on Isoproterenol-Induced Myocardial Injury in Rats. *International Journal of Cardiology*, **115**, 326-333. http://dx.doi.org/10.1016/j.ijcard.2006.03.016

[33] Wattanapitayakul, S.K., Weinstein, D.M., Holycross, B.J. and Bauer, J.A. (2000) Endothelial Dysfunction and Peroxynitrite Formation Are Early Events in Angiotensin-Induced Cardiovascular Disorders. *The FASEB Journal*, **14**, 271-278.

[34] Hamed, S., Brenner, B., Aharon, A., Daoud, D. and Roguin, A. (2009) Nitric Oxide and Superoxide Dismutase Modulate Endothelial Progenitor Cell Function in Type 2 Diabetes Mellitus. *Cardiovascular Diabetology*, **8**, 56. http://dx.doi.org/10.1186/1475-2840-8-56

[35] Mendez, I.I., Chung, Y.H., Jun, H.S. and Yoon, J.W. (2004) Immunoregulatory Role of Nitric Oxide in Kilham Rat Virus-Induced Autoimmune Diabetes in DR-BB Rats. *The Journal of Immunology*, **173**, 1327-1335. http://dx.doi.org/10.4049/jimmunol.173.2.1327

9

Agrobacterium-Mediated Transformation of Mexican Lime (*Citrus aurantifolia* Swingle) Using Optimized Systems for Epicotyls and Cotyledons

Maria Luiza P. de Oliveira[1]*, Gloria Moore[2], James G. Thomson[3], Ed Stover[1]

[1]USDA-ARS Subtropical Insects and Horticulture Research Unit, Fort Pierce, USA
[2]Horticultural Science Department, Institute of Food and Agricultural Science, University of Florida, Gainesville, USA
[3]USDA-ARS Crop Improvement and Utilization, Albany, USA
Email: *Maria.Oliveira@ars.usda.gov

Abstract

Transgenic Mexican lime (*Citrus aurantifolia* Swingle) was produced through two explant sources, each using systems previously optimized for each source. One used epicotyls segments, which was the predominant explant for transgenic *Citrus* production following co-cultivation with *Agrobacterium*, and has a well-established protocol. The other procedure used embryo cotyledons from mature seeds, which was developed in our lab as an alternative for stable *Citrus* transformation. Cotyledon transformation and regeneration protocols were optimized by comparing variables in culture medium composition on shoot regeneration and four parameters in transient transformation. The optimized protocols were compared, and frequency of regeneration, frequency of transgenic plant-recovery and stable transformation efficiency indicated the superiority of the cotyledon protocol for *Agrobacterium*-mediated genetic transformation in Mexican lime. The tissue choice resulted in marked improvement in shoot regeneration (14.1% of explants producing shoots in epicotyls; 55.8% in cotyledons), stable transformation frequency (11.4% of epicotyls explants; 40.2% in cotyledons), and frequency of transgenic plant-recovery (37.9% in epicotyl explants; 92.6% in cotyledons). Thus, easy availability of explants using embryo cotyledons from mature seeds, technical simplicity, shortening of transformation time-course, and higher transformation and regeneration frequencies makes this new system an attractive alternative over the previously published *Citrus* transformation protocols. In the course of this project, we generated Mexican lime with a Recombinase Mediated Exchange Cassette landing pad, which was designed for stacking transgenes.

---

*Corresponding author.

Keywords

Agrobacterium tumefaciens, Embryo Cotyledons, Epicotyls Segments, Recombinase Mediated Exchange Cassette, Tissue Culture

1. Introduction

The genetic transformation of *Citrus* has been widely studied as a tool to generate transgenic plants with enhanced tolerance of biotic [1]-[4] and/or abiotic stresses [5]. These protocols have relied on somatic embryogenesis from nucellar calli [6] or from protoplast-derived cultures [7], or more commonly, shoot organogenesis from epicotyl or internodal stem segments [8]-[11]. However, most of these *A. tumefaciens*-based systems have shown low transformation efficiencies, and/or transgenic plant-recovery.

Physiological state, ontogeny of explants and the explant-source position on mother plants greatly affects *in vitro* development. Different explant sources have different growth potential due to differences in age, and endogenous metabolic status [12] [13]. Likewise, Barcelo-Munoz *et al.* 1999 [14] reported that the choice of an appropriate explant is critical for success in morphogenesis and the ability of various types to respond for regeneration on their inheritance capabilities. Some efforts have been undertaken to explore other explants to develop an improved transformation system, with high capacity for regeneration and a shorter transformation time. Gene transfer into readily available tissue explants and regeneration without substantial *in vitro* culture are desirable characteristics to consider in evaluating alternative techniques for genetic transformation [15]. The resources required to maintain a continuous supply of explants at the correct developmental stage for transformation to be substantial.

Previous reports using epicotyls segments from Mexican lime have shown low affinity for *Agrobacterium tumefaciens* infection which results in low transformation efficiency [16] [17]. Therefore we decided to explore the use of cotyledons as an explant source. In the present study, we examined the timing and frequency of transformation between the conventional protocol, using epicotyls as explants, and an alternative optimized protocol for *Agrobacterium*-mediated transformation in Mexican lime. The alternative protocol using cotyledons from mature seeds provided an improvement in regeneration capacity, longer shoots and robust rooting, with a very simple transformation procedure, reducing the timing for transgenic Mexican lime plant-recovery. Application of this procedure may also accelerate the efficient recovery of other *Citrus* recalcitrant genotypes from *in vitro* transformation.

2. Materials and Methods

2.1. Plant Material and Explants Preparations

Seeds of Mexican lime were extracted from mature fruits and stored in a refrigerator at 4°C until used. They were surface sterilized under aseptic conditions for 1 min in 70% (v/v) ethanol, and further immersed in a solution containing 2.5% (v/v) sodium hypochlorite (Clorox, USA) and 0.02% (v/v) tween 20, then rinsed three times with sterile distilled water. For seedling epicotyl explant transformation, sterilized seeds were placed onto *Citrus* seed germination medium (MS medium solidified with 2.4 g·L^{-1} gelrite) in tall Magenta vessels, and incubated at 27°C for 30 days in the darkness, followed by 7 - 10 days on a 16/8h light/dark cycle. The light green seedling epicotyls were aseptically cut for co-culture with *Agrobacterium*. For cotyledon isolation, the sterilized seeds were dried overnight, and the two seed coats were peeled off aseptically, the embryonic axes were aseptically removed, and their cotyledons were used directly as explants.

2.2. Adventitious Shoot Induction Medium for Cotyledon Explants

For regeneration experiments from cotyledons, explants were placed on MS salts with 30 g·L^{-1} sucrose, and supplemented with different combinations of 6-benzylaminopurine (BAP), α-naphthalene acetic acid (NAA), Indole acetic acid (IAA) and 6-furfuryl-aminopurine (Kin), using a factorial design. The medium was , brought to pH 5.7 ± 0.1 prior to the addition of agar, solidified with 6 g·L^{-1} agar (Sigma), and autoclaved at 121°C for 25 min. The cotyledons were placed exterior surface (previously in contact with seed coat) down in contact with the medium. For each treatment at least 50 explants (5 plates) were cultured. Cultures were initially incubated in darkness at $27°C \pm 1°C$ for 3 weeks and then transferred to a 16/8-h (light/dark) photoperiod for 3 weeks with 36 $\mu mol \cdot m^{-2} \cdot s^{-1}$ radiation provided by fluorescent tubes. All treatments were repeated twice.

2.3. Sensitivity Test of "Mexican Lime" Cotyledons to Kanamycin

Prior to transformation, an effective concentration of kanamycin was determined for selection of transformants by culturing non-transgenic cotyledon explants on regeneration medium (MS medium supplemented with 2 $mg \cdot L^{-1}$ BAP, 1 $mg \cdot L^{-1}$ Kin, 1 $mg \cdot L^{-1}$ NAA, 3% sucrose and 0.6% agar) containing different concentrations of kanamycin (0, 15, 20, 25, 30, 40, 50, 60 $mg \cdot L^{-1}$), and incubated in darkness for 3 weeks followed by 16 h photoperiod with fluorescent light (36 $\mu mol \cdot m^{-2} \cdot s^{-1}$ light radiation) at $27°C \pm 1°C$ for 3 weeks. Each treatment had 5 petri dishes with 5 explants. Frequency of explants forming shoots and the number of shoots per explants were recorded. All experiments were repeated twice.

2.4. Optimized Conditions Evaluated for Transient *Agrobacterium* Transformation of Cotyledon Explants

The transformation parameters were optimized for cotyledon explants using the bacterial strain EHA 105 harboring the pCTAGDV-KCN3 binary vector [18]. Parameters were tested, one at a time, in a sequential order. The optimized conditions determined in earlier experiments were used in subsequent experiments. The following four parameters (and treatments) were tested in the order stated: 1) Agro infiltration method (use of vacuum chamber during inoculation and/or dipping); 2) density of bacterial culture; 3) duration of co-cultivation period; and 4) concentration of acetosyringone transformation enhancer.

For testing the first parameter, the cotyledon explants were placed in 50 ml Falcon tubes with 15 ml of *Agrobacterium* suspension ($OD_{600} = 0.5$) and were subject to different vacuum durations (5, 10, 15, 20 and 25 min). After vacuum inoculation the explants were maintained in *Agrobacterium* solution for an additional 10 min. These treatments were compared with only immersion of the explants into *Agrobacterium* inoculum ("dipping"). To compare the effect of *Agrobacterium* concentration (0.25, 0.50, 0.75 and 1.0 OD_{600}), explants were placed in 50 ml Falcon tubes with 10 ml *Agrobacterium* suspension and kept under vacuum for 15 minutes, followed by an additional 10 min in *Agrobacterium* inoculum. The effects of co-culture period and the presence of acetosyringone (AS) during co-cultivation period were evaluated, and the cotyledons inoculated with *Agrobacterium* were placed on co-cultivation medium for 0, 1, 2, 3 or 4 days, and the best co-culture period was selected to analyze the effect of transformation enhancer acetosyringone during the co-culture (concentrations at 0, 50, 100 or 150 μM). All parameters were evaluated using transient DsRed expression as described below.

2.5. DsRed Expression Assay

The expression of DsRed was observed in cotyledon explants under a fluorescent microscope (Olympus BX3) at 10x magnification and the appropriate filter for detection of the red fluorescence of the DsRed protein, which has an excitation maximum at 545 nm and emission maximum at 600 nm. Comparisons of the transient DsRed expression levels were made by counting DsRed expressing explants 10 days after co-cultivation. After *Agrobacterium* inoculation, explants were blotted on filter paper, transferred to co-culture medium for 3 days, which was then transferred to the selection medium for an additional 10 days. Analysis for each parameter used twenty five cotyledon explants and was repeated thrice.

2.6. Stable Transformation and Selection of Transgenic Plants

Agrobacterium tumefaciens strain EHA 105/pCTAGDV-KCN3, described in De Oliveira *et al.* 2015 [18], was cultured overnight in an orbital shaker at 28°C in 50 ml of YEP medium containing 100 $mg \cdot L^{-1}$ kanamycin. Bacterial cells were harvested by centrifugation at 3500 rpm for 5 min, resuspended to the designated OD_{600} in liquid MS basal medium containing 100 μM acetosyringone for explant inoculation. The *A. tumefaciens* suspension described above was used for epicotyl and cotyledon transformation.

Figure 1 outlines the procedure for epicotyl and cotyledon explant transformation in Mexican lime. An epicotyl transformation protocol was previously described by De Oliveira *et al.*, 2015 [18]. The optimized protocol was established for *Agrobacterium*-mediated transformation of *Citrus* using cotyledons as described below. The

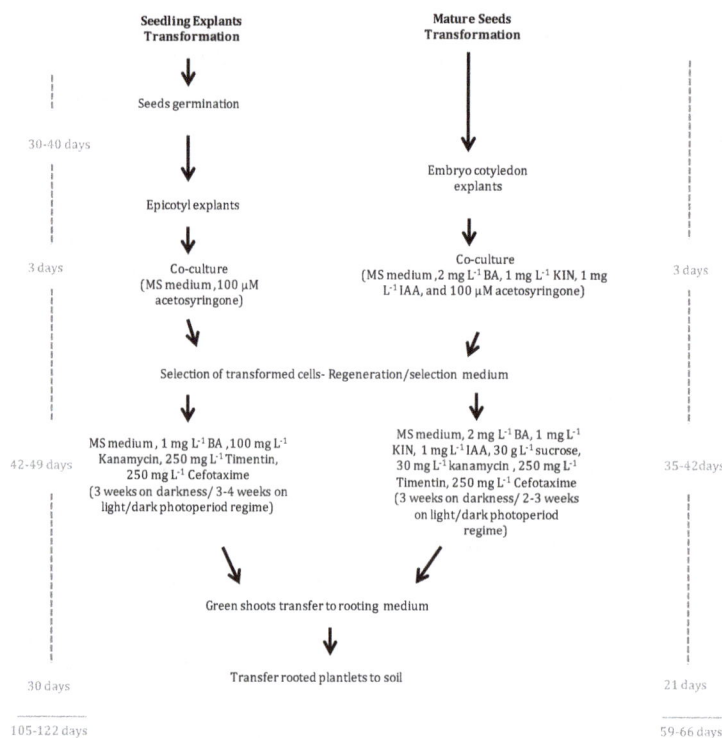

Figure 1. Outline of *A. tumefaciens*-mediated transformation protocols for epicotyls and cotyledon explants for Mexican lime.

cotyledons were isolated, placed into 50 ml Falcon tubes containing 10 ml *Agrobacterium* suspension (OD_{600} = 0.50) and exposed to vacuum infiltration for 15 min (vacuum pump at 0.25 MPa = 75 in. of Hg). Following incubation, explants were blotted dry on sterile filter paper and placed exterior side down on semi-solid co-cultivation medium (MS basal medium containing 2 mg·L^{-1} BAP, 1 mg·L^{-1} Kin and 1 mg·L^{-1} IAA, 100 µM acetosyringone and 30 g·L^{-1} sucrose). Co-cultivation was conducted in the dark at 27°C ± 1°C, for 3 days. The explants were then transferred to selection medium (MS medium containing 2 mg·L^{-1} BAP, 1 mg·L^{-1} Kin, 1 mg·L^{-1} IAA, 30 mg·L^{-1} sucrose, 250 mg·L^{-1} timentin, 250 mg·L^{-1} cefotaxime, and 30 mg·L^{-1} kanamycin) and maintained in darkness for 3 weeks, followed by incubation under a 16/8h light/dark photoperiod regime at 27°C ± 1°C, for an additional 3 weeks.

2.7. Molecular Analysis

Genomic DNA was isolated from leaves of transgenic and non-transgenic plants as previously described by Štorchová *et al.* (2000) [19]. For each sample of DNA, 100 ng of genomic DNA in 25 µL volume was used per PCR reaction. CodAORF70F60 and CodAORF1137R60 primers 5' CAAGACCCTTCCTCTATATAAG-3' and 5'-CGAGTTCATAGAGATAACCTTC-3', were used to detect the negative selectable marker sequence. The expected fragment size of the amplified DNA segment is 1067 bp. As a control, DNA from non-transgenic cv. Carrizo was included in each PCR screening. PCR positive transgenic plants were subsequently confirmed by Southern Blot. Ten micrograms of genomic DNA of non-transgenic (NT) and transgenic plants were digested with EcoRI for 6 h at 37°C and separated by electrophoresis on a 0.8% (w/v) agarose gel. The DNA was then transferred to a Hybond-N membrane (Amersham) and hybridized with the ^{32}P-labelled *cod*A sequence (Produced by Promega TaqTM polymerase using primers CodAORF70F60 5' CAAGACCCTTCCTCTATAT AAG-3' and CodAORF1137R60 5'-CGAGTTCATAGAGATAACCTTC-3') as described by De Oliveira *et al.* (2015) [18].

2.8. Statistical Analysis

All data were analyzed by analysis of variance (ANOVA) and Tukey's multiple comparison tests at 5% probability for all the variables studied.

3. Results and Discussion

3.1. Comparison of PGR Composition on Adventitious Shoot Regeneration

Most of the published protocols for *Agrobacterium*-mediated transformation in *Citrus* have used epicotyls as target cells for incorporation of the T-DNA [20]. Here, the cotyledons were extracted from mature seeds and conditions for regeneration and *Agrobacterium* transformation were optimized.

In preliminary studies, the cotyledons from seeds of Mexican lime were cultured on different media formulations containing different concentrations of BAP, Kin, NAA, and IAA (**Figure 2**). The results indicated that the frequency of explants with shoots and subsequent number of the shoots per explants were influenced by the type of plant growth regulator as well as concentration and combination of cytokinin or auxin. In the present study, the two cytokinins, BAP and Kin exerted a synergistic effect on the proportion of cotyledons producing shoots and the number of shoots per explants. Both parameters were higher for the cytokinin combination than for either cytokinin alone. The number of cotyledon segments forming shoots and the number of shoots per cotyledon increased with increasing cytokinin concentration and decreased with the addition of auxin as either NAA or IAA to the optimal high cytokinin medium. Addition of NAA and IAA did not cause significant differences on cotyledon morphogenesis in some treatments but in media with a high cytokinin levels the presence of auxin did enhance root formation (data not showed). Therefore, greatest regeneration capacity was achieved with 2 mg·L^{-1} BAP, 1 mg·L^{-1} Kin and 1 mg·L^{-1} IAA resulting in the highest regeneration efficiency (52%) and number of regenerated shoots per explant (1.89) although several other growth regulator combinations were not different at the $p < 0.05$ level. Inclusion of auxin enhanced root formation compared with the same treatment with similar levels of cytokinin (data not shown). Therefore this plant growth regulator regime was used in the selection media throughout genetic transformation experiments.

It has been shown that BAP is a requisite for optimal shoot regeneration in epicotyls, cotyledons, shoot tips, nodal and internodal stem segments of *Citrus*, while the effect of auxins seems to be marginal [21]-[25]. Our data demonstrate that in addition to BAP, inclusion of either Kin or IAA enhances shoot organogenesis from cotyledon explants.

3.2. Kanamycin Sensitivity

After the establishment of the optimal regeneration medium, we evaluated the response of cotyledons of Mexican lime cultured for 45 days in the presence of kanamycin at various concentrations (0, 15, 20, 30, 40, 50 and 60 mg·L^{-1}) to determine a suitable level for selection of transgenic plant cells within the non-transgenic explant tissue. **Figure 3** indicated that Mexican lime cotyledons were extremely sensitive to kanamycin. Kanamycin at the relatively low concentration of 30 mg·L^{-1} was detrimental to non-transformed organogenic potential without causing necrosis from antibiotic stress, such as visible browning of the cotyledon and reduced survival rate. In concentration of 40 mg·L^{-1} kan or more, all explants turned brown and stopped growing. These results suggested that the concentration of 30 mg·L^{-1} kanamycin would be effective for selection of transformants throughout transformation experiments.

3.3. Optimization of Transformation Conditions for T-DNA Delivery into Cotyledons of Mexican Lime

To determine optimal conditions for genetic transformation using cotyledons as explants, four studies were performed to establish optimal conditions for T-DNA delivery into tissue by *Agrobacterium*-mediated transformation (variables were vacuum infiltration and/or dipping, *Agrobacterium* concentration, co-cultivation period, and concentration of acetosyringone in co-cultivation medium). *DsRed* gene expression was used to monitor early transformation events.

The first experiment compared the vacuum infiltration and/or standard dipping (**Table 1**). DsRed expression increased with infiltration time up to 10 minutes, but longer infiltration hindered bacterial elimination and increased *Agrobacterium* overgrowth, resulting in reduced explants survival. Therefore, 10 minutes of vacuum infiltration was used in subsequent experiments. It has been reported that vacuum infiltration during *Agrobacterium* incubation period enhanced transformation efficiency in several species [9] [26]-[29] by improving pene-

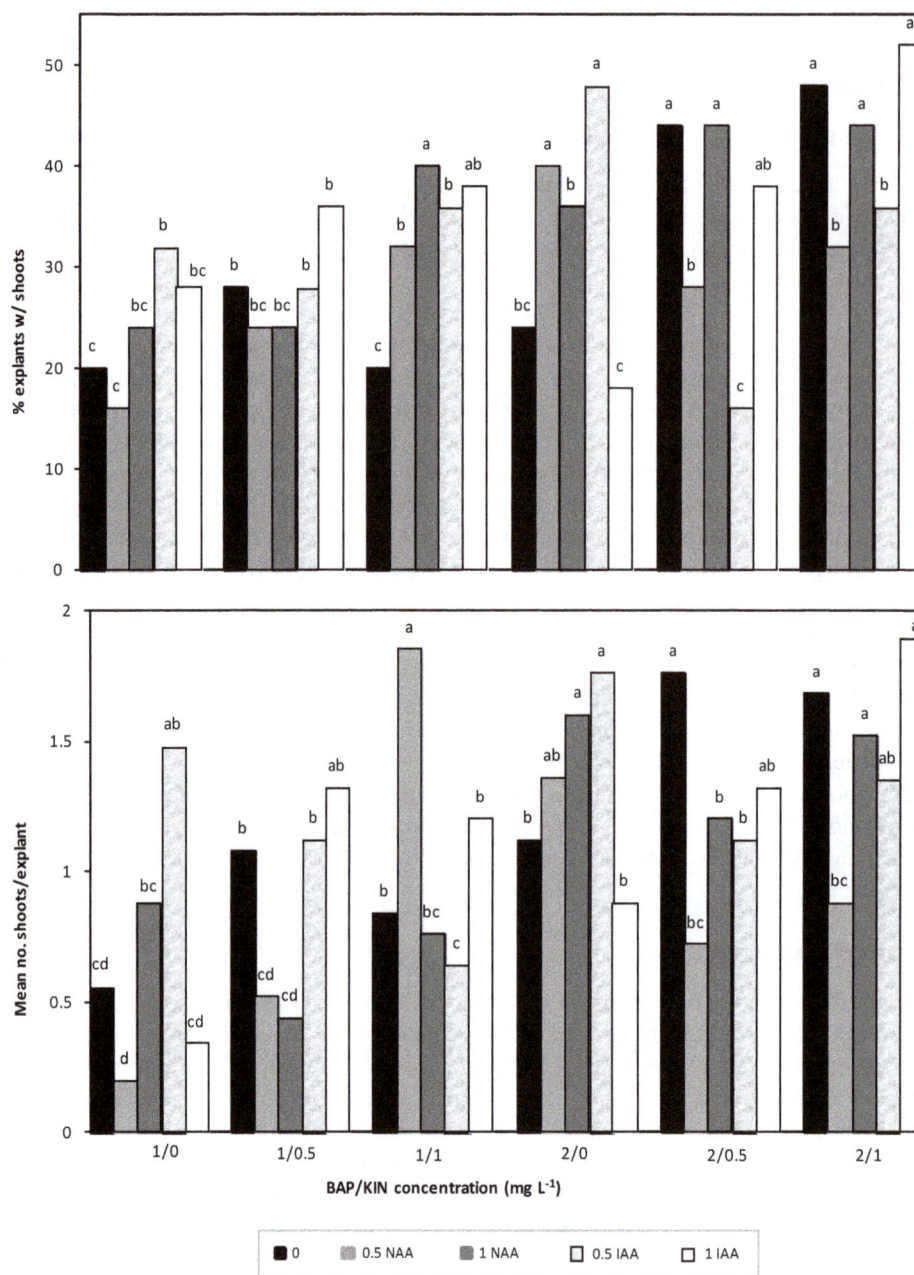

Figure 2. Organogenic response of cotyledon explants from Mexican lime to different concentrations and combinations of 6-benzyladenine (BAP), 6-furfurylaminopurine (Kin), α-naphthaleneacetic acid (NAA) and indole-3-acetic acid (IAA) in the shoot induction medium. Data are from two independent experiments. Means followed by the same letter do not differ significantly by Tukey HSD test (p ≤ 0.05).

tration of *Agrobacterium* into the explants.

Transient expression was enhanced at bacterial density higher than OD_{600} of 0.25 (**Table 1**) but $OD_{600} > 1.0$ increased *Agrobacterium* overgrowth. Consequently, OD between 0.50 and 0.75 were chosen for subsequent transformation experiments.

Increase in co-culture period resulted in an increased rate of DsRed expression. Co-culture for 3 or 4 days resulted in the highest proportion of explants expressing DsRed (**Table 1**), in spite of 4 days resulted in *Agrobacterium* overgrowth on the explant surface and a subsequent decrease in the explants survival. Therefore, we used a 3 day co-cultivation for subsequent experiments.

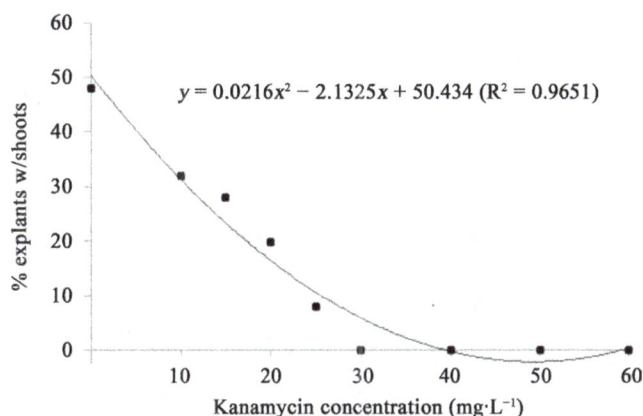

Figure 3. Shoot regeneration from cotyledon explants of Mexican lime as affected by different concentrations of kanamycin after 45 days in culture.

Table 1. DsRed transient expression assay measuring the effect of four parameters (and treatments) in *Agrobacterium*-mediated transient transformation of Mexican lime using embryo cotyledons: 1) Agro infiltration methods (use of vacuum chamber during *Agrobacterium* inoculation and/or dipping); 2) density of bacterial culture (0.25, 0.50, 0.75 and 1.0 OD_{600}); 3) duration of co-cultivation period (0, 1, 2, 3 and 4 days); and 4) concentration of transformation enhancer (acetosyringone 0, 50, 100 and 150 µM). Twenty five explants were analyzed by fluorescence stereoscopy (Olympus BX3) and the red explants were counted in each experiment. Values are the mean of three replicates ± SE. Data are from two independent experiments. Means followed by the same letter within a column do not differ significantly by Tukey HSD test ($p \leq 0.05$).

Parameters		Explants tested	Reps	*DsRed* expressing explants (t = 10 dac)
Agro infiltration methods	Dipping 10 min	25	3	5 ± 3^c
	Vacuum 5 min	25	3	8 ± 1^{bc}
	Vacuum 10 min	25	3	17 ± 2^a
	Vacuum 15 min	25	3	11 ± 4^{ab}
Density of bacterial culture (OD_{600})	0.25	25	3	8 ± 3^b
	0.50	25	3	16 ± 2^a
	0.75	25	3	15 ± 1^a
	1.0	25	3	17 ± 2^a
Duration of co-cultivation period (days)	1	25	3	5 ± 2^c
	2	25	3	11 ± 1^b
	3	25	3	18 ± 2^a
	4	25	3	17 ± 3^a
Transformation enhancer/acetosyringone (µM)	0	25	3	12 ± 2^b
	50	25	3	14 ± 2^b
	100	25	3	22 ± 1^a
	150	25	3	21 ± 3^a

Inclusion of acetosyringone at 100 - 150 µM in the co-culture medium resulted in the greatest proportion of explants displaying transient DsRed expression. Acetosyringone at 150 µM did not increase expression compared to 100 µM, therefore acetosyringone was used at 100 µM in all subsequent experiments. The use of acetosyringone (AS) during co-culture increases *Agrobacterium*-mediated transformation frequencies has previously been seen in several *Citrus* genotypes, such as, sweet orange [30] [31], "Carrizo" citrange [32], and Rio Red grapefruit [33]. AS is a phenolic compound naturally produced during wounding of plant cells that induces the transcription of the virulence genes of *Agrobacterium* [34].

From these experiments we concluded that *Agrobacterium*-mediated transformation using cotyledon of *Citrus* was near optimal with the following protocol: cotyledons were vacuum infiltrated in inoculation suspension at OD_{600} of 0.5 - 0.75 for 10 min followed by "resting" in the inoculum suspension for additional 10 min, blotting on filter paper, and co-culture in the dark for 3 days with supplementation of 100 μM of acetosyringone. Therefore, these optimized conditions of T-DNA delivery were evaluated on stable transformation efficiency. **Figure 4** shows transient and stable DsRed transformation in Mexican lime, confirming the transgene integration using the optimized cotyledon transformation protocol.

3.4. Molecular Analysis of Transgenic Plants

PCR verified the stable transformation of Mexican lime (**Table 2**). Twenty-nine PCR positive shoots were recovered following transformation from 254 inoculated epicotyls (11.4% of transformation efficiency), compared to 41 PCR positive shoots from 102 inoculated cotyledons (40.2 % of transformation efficiency). Only 37.9% of PCR positive plantlets from epicotyls were rooted and successfully acclimated under greenhouse conditions, one third those obtained using the cotyledon transformation protocol where 92.6% of transgenic shoots were successfully established. This may largely be due to the shorter length of shoots obtained from epicotyl explants (**Figure 5(A)**, **Figure 5(B)**), compared to longer shoots obtained from cotyledons (**Figure 5(B)**, **Figure 5(C)**). The root systems produced from cotyledon-derived transgenic shoots (**Figure 5(D)**) were more vigorous compared to those derived from epicotyls (**Figure 5(E)**).

The results from Southern blot analysis done at 120 days after shoot establishment in the greenhouse, shows strong transgene amplification and hybridization in cotyledon-derived transgenic plants (**Figure 5(F)**), consistent with stable transformation.

It is important to note that in this study, putative transgenic shoots were directly induced from cotyledons without the intervening step of seed germination which does not require weeks of preparation and staging of materials necessary when using epicotyls explants. The transgenic shoots resulting from cotyledon transformation were longer and had more vigorous root systems than those from epicotyl transformation, further reducing the time from initiating the transformation protocol to successfully establishing rooted plants *ex vitro*. The shoots produced per explants, transformation frequency, and transgenic plant establishment per explant were significantly improved using cotyledon transformation compared with epicotyl transformation. The epicotyl ex-

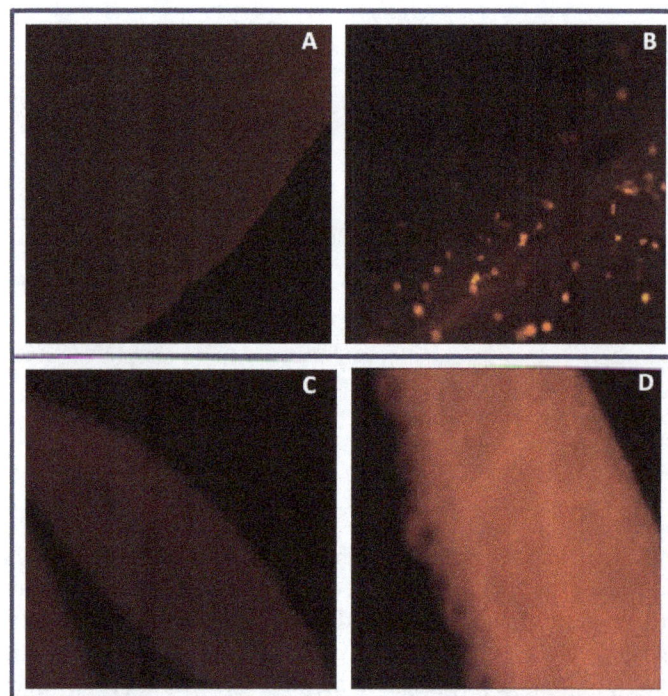

Figure 4. Transient and stable expression of DsRed fluorescent marker gene in Mexican lime in the embryo cotyledon and shoots under Olympus BX3 fluorescent microscope at 10× magnification using DsRed-specific filters. ((A)-(B)) *In situ* detection of DsRed expression in Mexican lime cotyledon co-cultivated with pCTAGDV-KCN3 for 10 days. (A) Non-transgenic cotyledon; (B) Cotyledon transgenic with optimized protocol. ((B)-(C)) Regenerating plant leaf with red fluorescence. (C) Non-transgenic plant; (D) transgenic plant line #9.

Table 2. Regeneration frequency, frequency of transgenic plant-recovery, and stable transformation efficiency of Mexican lime using epicotyls and embryo cotyledons explants.

Source of Explants	No. of explants inoculated	No. of regenerated shoots (Frequency of shoot regeneration %)[z]	No. of PCR+ shoots	No. of rooted PCR+ plantlets (Frequency of transgenic plant-recovery %)[y]	Transformation efficiency (%)[x]
Epicotyl	254	36 (14.1)	29	11 (37.9)	11.4
Cotyledon	102	57 (55.8)	41	38 (92.6)	40.2

[z]Frequency of shoot regeneration was calculated as number of regenerated shoots divided by the number of explants inoculated;
[y]Frequency of transgenic plant-recovery was calculated as number of rooted PCR+ plantlets transfer to the soil by the number of PCR+ shoots;
[x]Transformation efficiency was calculated as the number of PCR+ shoots divided by the number of explants inoculated in *Agrobacterium* co-cultivation media.

Figure 5. *In vitro* morphogenesis and *Agrobacterium tumefaciens*-mediated transformation of Mexican lime from epicotyls and cotyledon explants. (A). Adventitious bud formation on epicotyl explants after 6 weeks of culture on regeneration/selection MS medium containing 1 mg·L^{-1} BAP, 250 mg·L^{-1} timentin, 250 mg·L^{-1} cefotaxime, and 100 mg·L^{-1} kanamycin; (B) Comparison side by side of adventitious bud on cotyledon and epicotyl explants after 6 weeks of culture on regeneration/selection medium; (C). Adventitious bud formation on cut edge of a cotyledon explant after 6 weeks of culture on regeneration/selection MS medium containing 2 mg·L^{-1} BAP, 1 mg·L^{-1} Kin, 1 mg·L^{-1} IAA, 250 mg·L^{-1} timentin, 250 mg·L^{-1} cefotaxime, and 30 mg·L^{-1} kanamycin; (D). Rooted shoots from cotyledon after 21 days in MS medium containing NAA at 0.5 mg·L^{-1}; (E) Rooted shoots from epicotyl after 30 days in MS medium containing NAA at 0.5 mg·L^{-1}; (F) Schematic representation of the *Agrobacterium* binary vector T-DNA carrying the *codA* gene, and Southern Blot analysis of non-transgenic and transgenic Mexican lime *codA* lines derived from cotyledon transformation experiments. Total plant genomic DNA was digested with *Eco*RI and hybridized with a ^{32}P-labeled *codA* probe (gray bar). Samples are listed as above. NT, negative control (non-transgenic *Citrus* plant) that has been regenerated from cotyledon segments; Lanes #1 - 11 are *codA* transgenic lines.

plants transformation and establishment efficiencies generated in this study were similar to those previously reported for Mexican lime using a slightly different protocol [16] [17]. It must be noted that only two cotyledon explants are generated from each seed while as many as ten epicotyls explants are produced from each seed. However, production of epicotyl explants results in the loss of some seedlings to contamination and frequent discarding of some seedlings when initial staging does not correctly anticipate need for explants.

4. Conclusion

In conclusion, it has been demonstrated that cotyledons from Mexican lime are easily transgenic using *A. tume-*

faciens. The use of cotyledon explants may be widely applicable for genetic transformation of *Citrus.* Transforming cotyledons directly rather than seedlings or juvenile plants is technically simpler, less time-consuming, and at least in Mexican lime produces transgenic plants at a greater efficiency than the standard *Citrus* transformation protocols. In addition, this shoot regeneration system based on cotyledons may facilitate the use of particle-bombardment technology for transformation of *Citrus* species that are difficult to transform via *Agrobacterium.* During the course of this project, we utilized the pCTAGDV-KCN3 vector for Mexican lime transformation [18]. This vector was designed to provide a genomic landing pad or founder line for Recombinase Mediated Exchange Cassette (RMCE) genome targeting. The system is designed specifically for stacking multiple transgenes in a site-specific and sequential manner and also removes unneeded selectable markers when completed [35]. Analysis of the RMCE targeted integration protocol is currently being investigated.

Acknowledgements

This research was financially supported by the California *Citrus* Research Board under project # 5200-140A. This research was also supported by USDA Agricultural Research Service CRIS projects 5325-21000-018, 5325-21000-020, 6618-21000-014-00 and by the Biotechnology Risk Assessment Program competitive grant number 2010-33522-21773 from the USDA—National Institute of Food and Agriculture. Mention of trade names or commercial products is solely for the purpose of providing specific information and does not imply recommendation or endorsement by the US Department of Agriculture. USDA is an equal opportunity provider and employer.

References

[1]   Cardoso S.C., Barbosa-Mendes, J.M., Boscariol-Camargo, R.L., Christiano, R.S.C., Filho, A.B., Vieira, M.L.C., Mendes, B.M.J. and Mourão Filho, F.A.A. (2010) Transgenic Sweet Orange (*Citrus sinensis* L. Osbeck) Expressing the Attacin A Gene for Resistance to *Xanthomonas citri* subsp. citri. *Plant Molecular Biology Reports*, **28**, 185-192. http://dx.doi.org/10.1007/s11105-009-0141-0

[2]   Orbovic, V., Soria, P., Moore, G.A. and Grosser, J.W. (2011) The Use of *Citrus* Tristeza Virus (CTV) Containing a Green Fluorescent Protein Gene as a Tool to Evaluate Resistance/Tolerance of Transgenic *Citrus* Plants. *Crop Protection*, **30**, 572-576. http://dx.doi.org/10.1016/j.cropro.2011.01.001

[3]   He, Y., Chen, S., Peng, A., Zou, X., Xu, L., Lei, T., Liu, X. and Yao, L. (2011) Production and Evaluation of Transgenic Sweet Orange (*Citrus sinensis* Osbeck) Containing Bivalent Antibacterial Peptide Genes (Shiva A and Cecropin B) via a Novel *Agrobacterium*-Mediated Transformation of Mature Axillary Buds. *Scientia Horticulturae*, **18**, 99-107. http://dx.doi.org/10.1016/j.scienta.2011.01.002

[4]   Ali, S., Mannan A., Oirdi M.E., Waheed A. and Mirza B. (2012) *Agrobacterium*-Mediated Transformation of Rough Lemon (*Citrus jambhiri* Lush) with Yeast HAL2 Gene. *BMC Research Notes*, **5**, 285. http://dx.doi.org/10.1186/1756-0500-5-285

[5]   Bunnag S. and Tangpong, D. (2012) Genetic Transformation of *Citrus sinensis* L. with an Antisense ACC Oxidase Gene. *American Journal of Plant Sciences*, **3**, 1336-1340. http://dx.doi.org/10.4236/ajps.2012.39161

[6]   Dutt, M. and Grosser, J.W. (2010) An Embryogenic Suspension Cell Culture System for *Agrobacterium*-Mediated Transformation of *Citrus. Plant Cell Reports*, **29**, 1251-1260. http://dx.doi.org/10.1007/s00299-010-0910-0

[7]   Guo, W., Duan, Y., Olivares-Fuster, O., Wu, Z., Arias, C.R., Burns, J.K. and Grosser, J.W. (2005) Protoplast Transformation and Regeneration of Transgenic Valencia Sweet Orange Plants Containing a Juice Quality-Related Pectin Methylesterase Gene. *Plant Cell Reports*, **24**, 482-486. http://dx.doi.org/10.1007/s00299-005-0952-x

[8]   Costa, M.G.C., Otoni, W.C. and Moore, G.A. (2002) An Evaluation of Factors Affecting the Efficiency of *Agrobacterium*-Mediated Transformation of *Citrus paradisi* (Macf.) and the Production of Transgenic Plants Containing Carotenoid Biosynthetic Genes. *Plant Cell Reports*, **21**, 365-373. http://dx.doi.org/10.1007/s00299-002-0533-1

[9]   De Oliveira, M.L.P., Febres, V.J., Costa, M.G.C., Moore, G.A. and Otoni, W.C. (2009) High-Efficiency *Agrobacterium*-Mediated Transformation of *Citrus* via Sonication and Vacuum Infiltration. *Plant Cell Reports*, **28**, 387-395. http://dx.doi.org/10.1007/s00299-008-0646-2

[10] Favero, P., Mourão Filho, F.A.A., Stipp, L.C.L. and Mendes, B.M.J. (2012) Genetic Transformation of Three Sweet Orange Cultivars from Explants of Adult Plants. *Acta Physiologiae Plantarum*, **34**, 471-477. http://dx.doi.org/10.1007/s11738-011-0843-4

[11] Marutani-Hert, M., Bowman, K.D., McCollum, G.T., Mirkov, E., Evens, T.J. and Niedz, R.P. (2012) A Dark Incubation Period Is Important for *Agrobacterium*-Mediated Transformation of Mature Internode Explants of Sweet Orange,

Grapefruit, Citron, and a Citrange Rootstock. *PLoS ONE*, **7**, e47426. http://dx.doi.org/10.1371/journal.pone.0047426

[12] Chern, A., Hosskawa, Z., Cherubini, C. and Cline, M. (1993) Effects of Node Position on Lateral Bud out Growth in the Decapitation Shoot of *Ipomoea nil*. *Ohio Journal of Sciences*, **93**, 11-13.

[13] Litz, R.E., Raharjo, S., Efendi, D., Pliego-Alfaro, F. and Barcelo-Munoz, A. (2005) *Persea americana* Avocado. In: Litz, R., Ed., *Biotechnology of Fruit and Nut Crops*, Cromwell Press, Trowbridge, 331-335. http://dx.doi.org/10.1079/9780851996622.0326

[14] Barcelo-Munoz, A., Encina, C.L., Simon-Perez, E. and Pliego-Alfaro, F. (1999) Micropropagation of Adult Avocado. *Plant Cell Tissue and Organ Culture*, **58**, 11-17. http://dx.doi.org/10.1023/A:1006305716426

[15] Birch, R.G. (1997) Plant Transformation: Problems and Strategies for Practical Applications. *Annual Reviews of Plant Physiology and Plant Molecular Biology*, **48**, 297-326. http://dx.doi.org/10.1146/annurev.arplant.48.1.297

[16] Dutt, M. and Grosser, J.W. (2009) Evaluation of Parameters Affecting *Agrobacterium*-Mediated Transformation of Citrus. *Plant Cell Tissue and Organ Culture*, **98**, 331-340. http://dx.doi.org/10.1007/s11240-009-9567-1

[17] Dutt, M., Vasconcellos, M. and Grosser, J.W. (2011) Effects of Antioxidants on *Agrobacterium*-Mediated Transformation and Accelerated Production of Transgenic Plants of Mexican Lime (*Citrus aurantifolia* Swingle). *Plant Cell Reports*, **107**, 79-89. http://dx.doi.org/10.1007/s11240-011-9959-x

[18] De Oliveira, M.L.P., Stover, E. and Thomson, J.G. (2015) The codA Gene as a Negative Selection Marker in Citrus. *SpringerPlus*, **4**, 264. http://dx.doi.org/10.1186/s40064-015-1047-y

[19] Štorchová, H., Hrdlièková, R., Chrtek Jr., J., Tetera, M., Fritze, D. and Fehrer, J. (2000) An Improved Method of DNA Isolation from Plants Collected in the Field and Conserved in Saturated NaCl/CTAB Solution. *Taxon*, **49**, 79-84. http://dx.doi.org/10.2307/1223934

[20] Donmez, D., Simsek, O., Izgu, T., Aka, K.Y. and Yesim, Y.M. (2013) Genetic Transformation in Citrus. *The Scientific World Journal*, **2013**, Article ID: 491207. http://dx.doi.org/10.1155/2013/491207

[21] Bordón, Y., Guardiola, J.L. and García-Luis, A. (2000) Genotype Affects the Morphogenic Response *in Vitro* of Epicotyl Segments of *Citrus* Rootstocks. *Annals of Botany*, **86**, 159-166. http://dx.doi.org/10.1006/anbo.2000.1177

[22] Paudyal, K.P. and Haq, N. (2000) *In Vitro* Propagation of Pummel (*Citrus grandis* L. Osbeck). *In Vitro Cell and Developmental Biology*, **36**, 511-516. http://dx.doi.org/10.1007/s11627-000-0091-6

[23] Moreira-Dias, J.M., Molina, R.V., Bordon, Y., Guardiola, J.L. and Garcia-Luis, A. (2000) Direct and Indirect Shoot Organogenic Pathways in Epicotyl Cuttings of *Troyer citrange* Differ in Hormone Requirements and in Their Response to Light. *Annals of Botany*, **85**, 103-110. http://dx.doi.org/10.1006/anbo.2000.1001

[24] Cervera, M., Navarro, A., Navarro, L. and Peña, L. (2008) Production of Transgenic Adult Plants from Clementine Mandarin by Enhancing Cell Competence for Transformation and Regeneration. *Tree Physiology*, **28**, 55-66. http://dx.doi.org/10.1093/treephys/28.1.55

[25] Rodríguez, A., Cervera, M., Peris, J.E. and Peña, L. (2008) The Same Treatment for Transgenic Shoot Regeneration Elicits Opposite Effect in Mature Explants from Two Closely Related Sweet Orange (*Citrus sinensis* (L.) Osb. Genotypes. *Plant Cell Tissue and Organ Culture*, **93**, 97-106. http://dx.doi.org/10.1007/s11240-008-9347-3

[26] Amoah, B.K., Wu, H., Sparks, C. and Jones, H.D. (2001) Factors Influencing *Agrobacterium*-Mediated Transient Expression of *uidA* in Wheat Inflorescence Tissue. *Journal of Experimental Botany*, **52**, 1135-1142. http://dx.doi.org/10.1093/jexbot/52.358.1135

[27] Charity, J.A., Holland, L., Donaldson, S.S., Grace, L. and Walter, C. (2002) *Agrobacterium*-Mediated Transformation of *Pinus radiata* Organogenic Tissue Using Vacuum-Infiltration. *Plant Cell Tissue Organ Culture*, **70**, 51-60. http://dx.doi.org/10.1023/A:1016009309176

[28] Acereto-Escoffié, P.O.M., Chi-Manzanero, B.H., Echeverría-Echeverría, S., Grijalva, R., Kay, A.J., González-Estrada, T., Castaño, E. and Rodrígues-Zapata, L.C. (2005) *Agrobacterium*-Mediated Transformation of *Musa acuminata* cv. "Grand Nain" Scalps by Vacuum Infiltration. *Scientia Horticulturae*, **105**, 359-371. http://dx.doi.org/10.1016/j.scienta.2005.01.028

[29] Canche-Moo, R.L.R., Ku-Gonzalez, A., Burgeff, C., Loyola-Vargas, V.M., Rodríguez-Zapata, L.C. and Castaño, E. (2006) Genetic Transformation of *Coffea canephora* by Vacuum Infiltration. *Plant Cell Tissue and Organ Culture*, **84**, 373-377. http://dx.doi.org/10.1007/s11240-005-9036-4

[30] Bond, J.E. and Roose, M.L. (1998) *Agrobacterium*-Mediated Transformation of the Commercially Important Citrus Cultivar Washington Navel Orange. *Plant Cell Reports*, **18**, 229-234. http://dx.doi.org/10.1007/s002990050562

[31] Mendes, B.M.J., Boscariol, R.L., Mourão Filho, F.A.A. and De Almeida, W.A.B. (2002) *Agrobacterium*-Mediated Genetic Transformation of "Hamlin" Sweet Orange. *Pesquisa Agropecuaria Brasileira*, **37**, 955-961. http://dx.doi.org/10.1590/S0100-204X2002000700009

[32] Cervera, M., Pina, J.A., Juárez, J., Navarro, L. and Peña, L. (1998) *Agrobacterium*-Mediated Transformation of Ci-

trange: Factors Affecting Transformation and Regeneration. *Plant Cell Reports*, **18**, 271-278. http://dx.doi.org/10.1007/s002990050570

[33] Yang, Z.N., Ingelbrecht, I.L., Louzada, E., Skaria, M. and Mirkov, T.E. (2000) *Agrobacterium*-Mediated Transformation of the Commercially Important Grapefruit Cultivar Rio Red (*Citrus paradisi* Macf.). *Plant Cell Reports*, **19**, 1203-1211. http://dx.doi.org/10.1007/s002990000257

[34] De la Riva, G.A., González-Cabrera, J., Vázquez-Padrón, R. and Ayra-Pardo, C. (1998) *Agrobacterium tumefaciens*: A Natural Tool for Plant Transformation. *Electronic Journal of Biotechnology*, **1**, 24-25.

[35] Wang, Y., Yau, Y.-Y., Perkins-Balding, D. and Thomson, J. (2011) Recombinase Technology: Applications and Possibilities. *Plant Cell Reports*, **30**, 267-285. http://dx.doi.org/10.1007/s00299-010-0938-1

10

Nanomedicine: Tiny Particles and Machines, from Diagnosis to Treatment of Cardiovascular Disease, Provides Huge Achievements

Md. Ismail[1], Md. Faruk Hossain[2], Md. Fazlul Karim[3], Hossain Uddin Shekhar[1]*

[1]Department of Biochemistry and Molecular Biology, University of Dhaka, Dhaka, Bangladesh
[2]Department of Biological Sciences, St John's University, New York, USA
[3]Department of Biological Sciences, Eastern Illinois University, Charleston, USA
Email: *hossainushekhar@gmail.com

Abstract

Cardiovascular disease is one of many reverberating ailments that affect and kill hundreds of thousands of people around the world. To date treatments that offer improvement in the health condition of diseased people include the most promising nanomedicine although it is in its infancy, yet attaining attention from researchers of top notch day by day. In this current review importance is given on the application of nanomedicine in the diagnosis as well as treatment of cardiovascular disease.

Keywords

Nanomedicine, Cardiovascular Disease, Diagnosis, Treatment

1. Introduction

Technological application of nanometer sized molecule in medicine with the purpose of fighting and curing ailments is by default the definition of nanomedicine [1]. Currently, nanomedicine is a rapidly-flourishing as well as hectic sector of research activity that attains the focus of researchers of top notch reputation around the globe. In order to get rid of many present challenges in the treatment of cardiovascular, cancer as well as many other diseases, nanomedicines provide excellent solutions by virtue of its unique characteristics [2]-[8]. Cardiovascu-

---

*Corresponding author.

lar disease (CVD) is the name of a forerunner ailment that inflicts and kills millions of people around the globe and is expected to be continued as one of the top most indorsers to healthcare expenditure. The stronghold of CVD is the developed countries, but, it is also disseminating with a quick pace, among developing countries of the world. Information from World Health Organization (WHO) validates this claim, which states that approximately 17 million people's lives have been snatched away by CVD each year throughout the world [9]. Mostly attributed to the significant developments in surgical interventions, diagnostics and consciousness as well as concomitant lifestyle amendments, cardiovascular-related morbidity and mortality, in a time frame of 30 years in the later parts of the twentieth century, have resulted in a more than twofold reduction [10] [11]. The unconscionable number of cardiovascular diseases-related morbidity and mortality is thought to be the reason, behind the pressing requirement of more efficacious schemes to ameliorate the patient's condition. The last most prominent breakthrough in technology to affect CVD took place over a decade ago when Palmaz & Schatz introduced coronary stent which got its approval from the FDA in 1994. Since then, the new smash hit curatives (statins, beta blockers, and etc.) and the subtleties of surgical processes, have become the hallmark of reliance. But now the emerging and ever evolved nanomedicine is expected to confront and deal efficiently the present challenges in cardiovascular disease as well as to bring about a breakthrough in the identification and treatment of cardiovascular disease effectively (**Figure 1**). In this review, our discussion is related to the recent developments in the arena of nanobiotechnology for the diagnosis and treatment of cardiovascular disease, in light of nanoparticles, *ex vivo* biomarkers, *in vivo* sensors and programmable bio-nanochip (P-BNC) system for the diagnosis purpose as well as in light of theranostic and therapeutic nanoparticles, innovative liposomal platforms and tissue regenerating devices for the treatment purpose of cardiovascular disease.

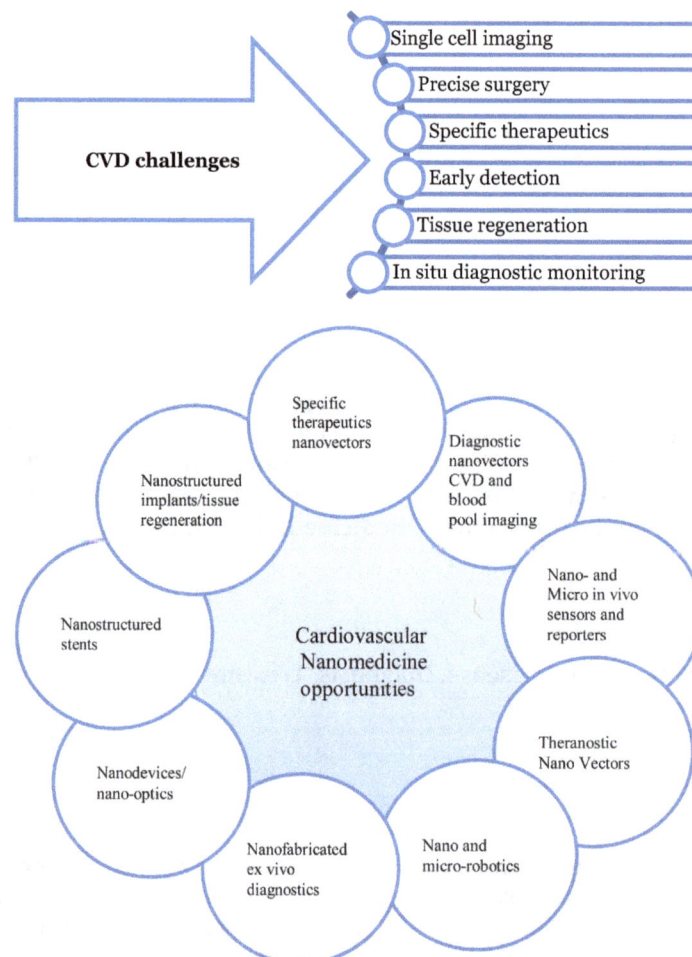

Figure 1. Outline of challenges in diagnosis and treatment of CVD and scopes to intervene with nanomedicine.

2. Nanomedicine in the Diagnosis of Cardiovascular Disease

2.1. Advanced and Sophisticated Diagnosis of CVD with Nanoparticles

A bulk number of drug delivery systems based on particles of nanometer size have recently been evolved, with variegated features and multiple functionalities [12]-[15] showing variegations in 1) sizes 2) shapes and 3) surface functionalization. Identification and characterization of initial disease stages before the appearance of gross disease manifestations, are the abilities of multifaceted nanoscale contrast agents. Nanoparticles that have the capacity to generate contrast, can be useful in cardiovascular imaging. Paramagnetic, fluorescent and other particles are contrast generating nanoparticles those can be used in the detection as well as characterization of initial ailment stages before the disease become more conspicuous and fatal. Internal structures images are now obtainable in which process there is a prerequisite of certain radiofrequency waves as well as magnetic fields in magnetic resonance imaging (MRI) mediated cardiovascular imaging. The sole purpose of using contrast agents is to magnify the subtle changes in the energy level of tissues in MRI. Gadolinium-diethylenetriaminopentaacetic acid is an example of a paramagnetic contrast agent which provides a bright contrast in MR images [16] [17]. A recent example of T1 enhancing contrast agent is manganese nanoparticles [18]. An eminent example of nanoparticles that can give off light is quantum dots. Quantum dots are special types of nanoparticle having fluorescence within itself. There is a positive correlation between an increase in particle size and an increase in emission wavelength [19]. For imaging purpose, micro particle-based contrast molecules are in use and example of this includes porous silicon particles which have encapsulated iron oxide nanoparticle. Improvement in contrast, has come from this iron oxide nanoparticle [20]. Macrophages engulf these multistage particles through phagocytosis, hence providing the chance of imaging of inflamed portions where macrophages aggregate [21]. An example of this macrophage aggregation is atherosclerotic plaque. Multimodal imaging is not an illusion, rather an assertion in the age of nanoscience, where nanoparticles have more than one contrast agents [22]-[24]. An illustrious example of these types of molecule is 18F-CLIO (18F-cross-linked iron oxide). Markers of angiogenesis, macrophages [25], collagen III [26], as well as fibrin [27] all together are important in atherosclerotic plaque image targeting. Plaque rupture provides many signals and one of the earliest signals is fibrin deposition. Through ultrasound [28] and magnetic resonance imaging [29] arterial thrombi can be imaged by targeting fibrin and other tissue factor. Specific interaction between nanoparticles conjugated to ligands and $\alpha v\beta 3$-integrin can be exploited in angiogenesis targeting [30]. In MRI and computed tomography (CT) various contrast agents, for instance iodine can be carried within nanoliposomes and the benefit is that the contrast agents have a significant reduction in body clearance, ameliorating potentialities of blood as well as cardiac imaging in study models [31] [32].

2.2. CVD Diagnosis with Programmable Bio-Nanochips (P-BNCs)

Assessment of CVD in quick time and in a reliable way is the prerequisite of point-of-care treatment, where the clinicians can have significant relief in discharging personnel with benign etiology while giving proper treatment to CVD inflicted personnel. Hence the scheme termed programmable bio-nanochip (P-BNC), a novel medical device with the potentiality of providing high functioning as well as trimmed cost, comes under light. The capacity to immediately ensure sensitive, authentic simultaneous assessment of multiple prime biomarkers of cardiology at the point-of-care pledges to change the scenario of clinical nosology. To attain this target, researchers have worked to ameliorate the present state of point-of-care IVD through the improvement, substantiation, as well as the effectuation of P-BNCs [33]-[40]. Programmable bio-nanochips gets its name from its ability to work when reprogrammed while maintaining standard platform in assessing the bio-markers affiliated with specific ailments. High performance at reduced cost is possible if the similarity between microelectronics industry and the ability to mass-produce the sensor elements, can be maintained, where the "chip" term emphasizes. It contains incorporated nano-nets and quantum dots for the purpose of efficacious and instant biomarker capture and augmented signal development respectively. P-BNCs provide output within minutes while methods like ELISA (enzyme-linked immunoassay) provide output within hours. P-BNCs have detection capacity in more magnitude lower than the traditional. Researchers have found that saliva has got many biomarkers of acute myocardial infarction diagnosis [41] and best 10 biomarkers with the most significant information are tumor necrosis factor-alpha (TN F-α), C-reactive protein (CRP), RANTES, soluble intercellular adhesion molecule-1 (sICAM-1), myeloperoxidase (MPO), myoglobin (MYO), matrix metalloproteinase-9 (MMP-9), interleukin-1 beta (IL-1β), adiponectin, and soluble CD40 Ligand (sCD40L) [42] [43].

2.3. *Ex Vivo* Biomarkers for CVD Diagnosis

Individual biomarkers of CVD that have drawn the attention of researchers include the levels of, fibrinogen [44], D-dimer [45], B-type natriuretic peptide [46], C-reactive protein [47], as well as homocysteine [48] for the identification of high-risk population. To specifically select low molecular weight proteins, nanoporous materials can be useful because they can monitor as well as find new circulating biomarkers in body fluids [49] [50]. pH variations measurement, detection of small quantity of molecules, whether chemical or biological, all are possible with the advent of nanowires [51]. Myoglobin, CK-MB (creatine kinase MB isoenzyme), and cardiac troponins have got FDA approval, so automatically they are in the focus of commercial cardiac biomarker point-of-care devices [52] [53]. Troponin antibodies and nickel nanohairs have been combined with modified viral nanoparticles to identify troponin in serum where the identification limit of troponin is six to seven orders of magnitude lower than traditional immunological assays [54].

2.4. *In Vivo* Sensors in CVD Diagnosis

In situ quick identification of ions such as, H^+, Na^+, K^+ and Ca^{2+} have been possible through the development of nanosensors. H^+ and K^+ ion activity can serve as an important marker in case of acute myocardial infarction onset [55]. Analysis at in vivo is possible because nanosensors have been implanted in epicardial and the arterial region [56]. Field effect transistors (FET) technology has already been exploited for the development of a silicon needle with multi-nanosensor for the identification of myocardial infarction [57]. Real-time detection of Ca^{2+} ions is also possible, through functionalized nanowires [51]. Shin KH, *et al.* [58] described the *in-vitro* development of bio-MEMS pressure sensors which can be used for the purpose of *in-situ* assessment of blood pressure. Kim J-H, *et al.* [59] described the development of near-infrared fluorescence sensors for NO with single-walled carbon nanotube technology.

3. Nanomedicine in the Treatment of Cardiovascular Disease

3.1. Intervention in Cardiovascular Disease with Liposome

Delivery of therapeutics to targeted tissues is possible through the design and construction of nanoscale particles while reduced toxicity as well as higher efficiency is obtainable [60] [61]. Liposomes have huge skillfulness with regard to physicochemical characteristics, permitting acclimatization of this tiny particle to dovetail to the exact application in biology [62]. Liposomes circulatory half life has been extended through the emergence of polyethylene glycol (PEG) which assist in the avoidal of phagocytic cells of the body. Coupling of liposomes with peptides or proteins for increased targeting to specific tissues is also possible [63] [64]. Although initiatives have been carried out to exploit the benefits of liposomal approach but currently, there are no approved liposomal formulations for the treatment of CVD in human. In an attempt to treat chronic myocardial ischemia, during angioplasty and stenting, Hedman *et al.* [65] delivered vascular endothelial growth factor encoding plasmid through liposomes, with the aim to prevent in-stent restenosis and postangioplasty but the treatment doomed to change the incidence of restenosis, while it was demonstrated that gene transfer using liposomes was a viable and well tolerated approach. In a different clinical trial paclitaxel nanoparticles fixed to albumin was used to prevent in-stent restenosis. Reports claimed that without major complication the innovative approaches were well tolerated at 10 or 30 mg/m^2 [66] [67]. Zhang *et al.* [68] described acquirement of the surface modified liposome with peptide which has an arginine-rich sequence (CRPPR) to coronary endothelial in myocardial infarction and ischemia models. In comparison to nontargeted liposomes, the CRPPR-conjugated nanostructure attains a 47-fold increase in accruement in the injured tissue vasculature. It is worth mentioning that healthy tissue vasculature has accruement to a lesser extent. Modified liposomes can be used to minimize accidental or unwanted damage, to tissues that are healthy and adjacent, after myocardial infarction by reducing inflammatory responses from macrophages. Macrophages exist at the nearby area of infarction and thus can serve as the basis for cell-based targets for therapy. Modified liposomes have been developed by Harel-Adar *et al.* [69] that have surface phosphatidylserine (PS). Apoptotic cells surface phosphatidylserine act as a trigger to initiate the inflammatory circuit of macrophages. Phosphatidylserine containing surface attenuated liposomes upon engulfment by macrophages brings in the release of more anti-inflammatory cytokines, upregulation of CD206 as well as the accompaniment of TNF-alpha and CD86 downregulation. Fabricated fluorescent PEGylated liposome has the ability to distribute therapeutics to the infarcted heart [70]. Liposomes loaded with oligodextran surfactants as well as RGD can serve the purpose of targeted delivery and reduced RES uptake, respectively [71]. The surfactants of oligosaccharide class imitate cell glycocalyx of restricted opsonization, resulting in reduced RES uptake [72]. Research activity of Lestini *et al.* validated the surreptitious nature through which liposomes can reduce opsonization and RES clearance. Activated platelets have P-selectin on their surface which is the target of glycoliposomes with negative charge [73]. This delivery system exploited the imitation properties of activated leukocytes of containing P-selectin glycoprotein ligand 1 in order to ease the specific attachment between P-selec- tin receptor on activated platelets and liposomes of the target. TMR-484 is a special liposomal formulation of prednisolone, developed by Joner *et al.* [74]. The injury site high with chondroitin sulfate proteoglycans, which can bind with great affinity with prednisolone. After 24 hours of administration, reports claimed, the lesion had liposome concentration that had increased by 100-fold in comparison to nonstented arteries, when tested in a laboratory animal model of atheroma. Solubilization of aggregated cholesterol in atherosclerosis is one of the approaches to the treatment of atherosclerosis. Liposomes loaded with phosphatidylcholine (PC) were developed by Cho *et al.* [75] with the purpose of lesion enrichment with high-density lipoprotein (HDL). The process of atherosclerosis includes the accumulation and subsequent oxidization of LDL in the artery wall and finally their picked up by foam cells. HDL can limit the process of inflammation that leads to atherosclerosis. Liposomal phosphatidylcholine could be a good therapy choice for the regression of atherosclerotic plaque because cholesterol-fed rabbits have undergone infusion of liposomal formulations and have resulted in a reduced cholesterol content as well as atherosclerotic plaque volume in the aortic walls. Activated macrophages have been shown to be attracted by liposomal components in atheroma components that are metabolically active. The liposome platform contained, nanogold, lipoprotein-associated phospholipase A2 and rhodamine, which has a surface potential of negative value. Walton *et al.* [76] demonstrated that in lesion of rabbits with Watanabe heritable hyperlipidemia have positive macrophage targeting, which has been assured through transmission electron microscopy, that have indicated the presence of liposome in high amount within the atheromas and the basis of identification was nanogold component.

3.2. Nanoparticles with Theranostic and Therapeutic Properties

Drugs that are cytotoxic in nature and can prevent smooth muscle cell growth are being in use to inhibit restenosis. Example of these drugs includes etoposides, paclitaxel, doxorubicin etc. Immunomodulators or inflammatory response inhibitory molecules like Cyclosporine A, steroids etc. and platelet derived growth factor receptor antagonists are also in use of preventing restenosis. Encapsulation of these therapeutic molecules in nanoparticles gives sanctuary from degradation of enzymes as well as permit sustained release profile [77] [78]. As an important advancement in drug delivery, there is a growing interest for theranostic agents [79]. Treatment results and therapeutic molecules, both of them can be precisely observed through the combination of therapeutic nanoparticle and diagnostic imaging modality. Theranostic nanoparticles imaging abilities can serve numerous purpose, including the verification of the delivery of specific molecules to its target, designing of dose patterns, as well as identification of personnel who either respond or not responding in a specifically designed therapy. For instance in hyperlipidemic animals a sustained antiangiogenesis therapy was carried out through the use of theranostic $\alpha v\beta 3$-integrin targeted paramagnetic nanoparticles. Data from MRI studies has proven a decline (between 50% and 75%) in neovascular signal for 21 days. Histological evaluation was acceptable. These outcomes have pointed towards the broad window of this strategy for efficacious therapy that is antiangiogenic in nature [80] [81].

3.3. Nanomedicine in Tissue Regeneration Devices

Nanotechnology finds itself in the field of cardiovascular device development and research and it has already devoted itself to the improvement of stent technology. Narrowing of the blood vessel is termed as stenosis, which hamper normal blood flow. The mainstream challenges in using stents in an initiative of revascularization of the narrowed arteries are in-stent restenosis that results from intimal hyperplasia [82] as well as activated platelets mediated thrombosis in the later stage [83]. Establishment of the stent as the drug delivery platform is the following degree of device sophistication. Johnson & Johnson, Guidant, Boston Scientific as well as Medtronic are the industry giants that started the production of drug eluting stents (DES) which are illustrious for releasing drugs, for instance sirolimus as well as paclitaxel for the exploitation of their anti-proliferative benefits. Drug eluting stents have manifested lower occurrence of restenosis after six months of procedure in comparison to their usual counterparts (metal), but recent long term sophisticated researches have raised concerns about whether DES can deliver long term benefits or not [84]-[87]. Incorporation of stent within the vessel wall is possible just because of the fact that the eluted drugs are anti-proliferative in character. Cell cycle inhibition results from eluted drugs that suppress the proliferation. As a result the normal vessel cannot remodel themselves. Incomplete neonatal coverage causes exposure of stent structure that leads to the formation of a thrombus, which is very fatal and complicated even results in increased death that can be attributed to the thrombosis of later stage. This whole scenario usually occurs if there is a premature surcease of dual antiplatelet therapy [85] [86]. Currently, researchers are trying to deliver tacrolimus [88] by exploiting the ability of aluminum oxide stent surfaces. Paclitaxel elution by using matrixes of carbon-carbon nanomaterials has also been attempted [89]. Nanoporous TiO_2 films have been tested for various drug delivery [90]. Inflicted vessels problem can be solved by increasing the interaction between endothelial cells and stent surfaces, while stents provide the platform of revascularization process. This enhancement in interaction exploits surface nanotexturing. Vascular tissue has specific structures and imitation of this structure is possible, for improving the adhesion of cells, through nanoscale topography on hydroxyapatite substrates [91] and nickel titanium [83]. Afterwards there is an increased endothelialization of the stent and decreased thrombosis [92].

4. Conclusion

Transcending expansion and understanding of sectors like molecular biology, material science, genetics, cellular biology, bioengineering and proteomics construct nanobiotechnology which acts as a fixative to connect the dots, between interactions on the microscopic and molecular levels, to form an outstanding and inclusive platform from where it can become the major potential actors in the race towards the progress of CVD diagnosis and treatment. Scientists have made us optimistic about nanomedicine with their outstanding research outcomes. The day is not so far when, currently seeming fledgling, nanomedicine will become the hegemon in the diagnosis and treatment of cardiovascular disease.

Conflict of Interests

The authors declare that there is no conflict of interests regarding the publication of this paper.

References

[1]   Kim, B.Y.S., Rutka, J.T. and Chan, W.C.W. (2010) Nanomedicine. *New England Journal of Medicine*, **363**, 2434-2443. http://dx.doi.org/10.1056/NEJMra0912273

[2]   Bharali, D.J. and Mousa, S.A. (2010) Emerging Nanomedicines for Early Cancer Detection and Improved Treatment: Current Perspective and Future Promise. *Pharmacology & Therapeutics*, **128**, 324-335. http://dx.doi.org/10.1016/j.pharmthera.2010.07.007

[3]   Sajja, H.K., East, M.P., Mao, H., Wang, Y.A., Nie, S. and Yang, L. (2009) Development of Multifunctional Nanoparticles for Targeted Drug Delivery and Noninvasive Imaging of Therapeutic Effect. *Current Drug Discovery Technologies*, **6**, 43-51. http://dx.doi.org/10.2174/157016309787581066

[4]   Ledet, G. and Mandal, T.K. (2012) Nanomedicine: Emerging Therapeutics for the 21st Century. *U.S. Pharmacist*, **37**, 7-11.

[5]   Godin, B., Sakamoto, J.H., Serda, R.E., Grattoni, A., Bouamrani, A. and Ferrari, M. (2010) Emerging Applications of Nanomedicine for the Diagnosis and Treatment of Cardiovascular Diseases. *Trends in Pharmacological Sciences*, **31**, 199-205. http://dx.doi.org/10.1016/j.tips.2010.01.003

[6]   Chhatriwalla, A.K. and Bhatt, D.L. (2008) Should Dual Antiplatelet Therapy after Drug-Eluting Stents Be Continued for More Than 1 Year? *Circulation Cardiovascular Interventions*, **1**, 217-225. http://dx.doi.org/10.1161/CIRCINTERVENTIONS.108.811380

[7]   Galvin, P., Thompson, D., Ryan, K.B., McCarthy, A., Moore, A.C., Burke, C.S., *et al.* (2012) Nanoparticle-Based Drug Delivery: Case Studies for Cancer and Cardiovascular Applications. *Cellular and Molecular Life Sciences*, **69**, 389-404. http://dx.doi.org/10.1007/s00018-011-0856-6

[8]   Bhaskar, S., Tian, F., Stoeger, T., Kreyling, W., de la Fuente, J.M., Grazú, V., *et al.* (2010) Multifunctional Nanocarriers for Diagnostics, Drug Delivery and Targeted Treatment Across Blood-Brain Barrier: Perspectives on Tracking and Neuroimaging. *Particle and Fibre Toxicology*, **7**, 3. http://dx.doi.org/10.1186/1743-8977-7-3

[9]   World Health Organization (2011) Programmes and Projects: Global Atlas on Cardiovascular Disease Prevention and

Control. World Health Organization, Geneva. http://www.who.int/cardiovascular_diseases/en/

[10] Hoyert, D. and Xu, J. (2012) Deaths: Preliminary Data for 2011. National Vital Statistics Reports. National Center for Health Statistics, Hyattsville, **61**, 1-65.

[11] Jemal, A., Siegel, R., Ward, E., Hao, Y., Xu, J. and Thun, M.J. (2009) Cancer Statistics, 2009. *CA: A Cancer Journal for Clinicians*, **59**, 225-249. http://dx.doi.org/10.3322/caac.20006

[12] Riehemann, K., Schneider, S.W., Luger, T.A., Godin, B., Ferrari, M. and Fuchs, H. (2009) Nanomedicine—Challenge and Perspectives. *Angewandte Chemie International Edition*, **48**, 872-897. http://dx.doi.org/10.1002/anie.200802585

[13] Peer, D., Karp, J.M., Hong, S., Farokhzad, O.C., Margalit, R. and Langer, R. (2007) Nanocarriers as an Emerging Platform for Cancer Therapy. *Nature Nanotechnology*, **2**, 751-760. http://dx.doi.org/10.1038/nnano.2007.387

[14] Ferrari, M. (2008) Nanogeometry: Beyond Drug Delivery. *Nature Nanotechnology*, **3**, 131-132. http://dx.doi.org/10.1038/nnano.2008.46

[15] Ferrari, M. (2005) Cancer Nanotechnology: Opportunities and Challenges. *Nature Reviews Cancer*, **5**, 161-171. http://dx.doi.org/10.1038/nrc1566

[16] Smith, R.C. and McCarthy, S. (1992) Physics of Magnetic Resonance. *Journal of Reproductive Medicine*, **37**, 19-26.

[17] Sosnovik, D.E., Nahrendorf, M. and Weissleder, R. (2008) Magnetic Nanoparticles for MR Imaging: Agents, Techniques and Cardiovascular Applications. *Basic Research in Cardiology*, **103**, 122-130. http://dx.doi.org/10.1007/s00395-008-0710-7

[18] Pan, D., Senpan, A., Caruthers, S.D., Williams, T.A., Scott, M.J., Gaffney, P.J., *et al.* (2009) Sensitive and Efficient Detection of Thrombus with Fibrin-Specific Manganese Nanocolloids. *Chemical Communications*, 3234-3236. http://dx.doi.org/10.1039/b902875g

[19] Michalet, X., Pinaud, F.F., Bentolila, L.A., Tsay, J.M., Doose, S., Li, J.J., *et al.* (2005) Quantum Dots for Live Cells, *in Vivo* Imaging, and Diagnostics. *Science*, **307**, 538-544. http://dx.doi.org/10.1126/science.1104274

[20] Serda, R.E., Godin, B., Tasciotti, E., Liu, X. and Ferrari, M. (2009) Mitotic Trafficking of Silicon Microparticles. *Nanoscale*, **1**, 250-259. http://dx.doi.org/10.1039/b9nr00138g

[21] Kooi, M.E., Cappendijk, V.C., Cleutjens, K.B., Kessels, A.G., Kitslaar, P.J., Borgers, M., *et al.* (2003) Accumulation of Ultrasmall Superparamagnetic Particles of Iron Oxide in Human Atherosclerotic Plaques Can Be Detected by *in Vivo* Magnetic Resonance Imaging. *Circulation*, **107**, 2453-2458. http://dx.doi.org/10.1161/01.CIR.0000068315.98705.CC

[22] Devaraj, N.K., Keliher, E.J., Thurber, G.M., Nahrendorf, M. and Weissleder, R. (2009) 18F Labeled Nanoparticles for *in Vivo* PET-CT Imaging. *Bioconjugate Chemistry*, **20**, 397-401. http://dx.doi.org/10.1021/bc8004649

[23] Nahrendorf, M., Zhang, H., Hembrador, S., Panizzi, P., Sosnovik, D.E., Aikawa, E., *et al.* (2008) Nanoparticle PET-CT Imaging of Macrophages in Inflammatory Atherosclerosis. *Circulation*, **117**, 379-387. http://dx.doi.org/10.1161/CIRCULATIONAHA.107.741181

[24] Chen, W., Vucic, E., Leupold, E., Mulder, W.J., Cormode, D.P., Briley-Saebo, K.C., *et al.* (2008) Incorporation of an apoE-Derived Lipopeptide in High-Density Lipoprotein MRI Contrast Agents for Enhanced Imaging of Macrophages in Atherosclerosis. *Contrast Media & Molecular Imaging*, **3**, 233-242. http://dx.doi.org/10.1002/cmmi.257

[25] Amirbekian, V., Lipinski, M.J., Briley-Saebo, K.C., Amirbekian, S., Aguinaldo, J.G., Weinreb, D.B., *et al.* (2007) Detecting and Assessing Macrophages *in Vivo* to Evaluate Atherosclerosis Noninvasively Using Molecular MRI. *Proceedings of the National Academy of Sciences of the United States of America*, **104**, 961-966. http://dx.doi.org/10.1073/pnas.0606281104

[26] Cyrus, T., Abendschein, D.R., Caruthers, S.D., Harris, T.D., Glattauer, V., Werkmeister, J.A., *et al.* (2006) MR Three-Dimensional Molecular Imaging of Intramural Biomarkers with Targeted Nanoparticles. *Journal of Cardiovascular Magnetic Resonance*, **8**, 535-541. http://dx.doi.org/10.1080/10976640600580296

[27] Botnar, R.M., Buecker, A., Wiethoff, A.J., Parsons Jr., E.C., Katoh, M., Katsimaglis, G., *et al.* (2004) *In Vivo* Magnetic Resonance Imaging of Coronary Thrombosis Using a Fibrin-Binding Molecular Magnetic Resonance Contrast Agent. *Circulation*, **110**, 1463-1466. http://dx.doi.org/10.1161/01.CIR.0000134960.31304.87

[28] Lanza, G.M., Trousil, R.L., Wallace, K.D., Rose, J.H., Hall, C.S., Scott, M.J., *et al.* (1998) *In Vitro* Characterization of a Novel, Tissue-Targeted Ultrasonic Contrast System with Acoustic Microscopy. *The Journal of the Acoustical Society of America*, **104**, 3665-3672. http://dx.doi.org/10.1121/1.423948

[29] Morawski, A.M., Winter, P.M., Crowder, K.C., Caruthers, S.D., Fuhrhop, R.W., Scott, M.J., *et al.* (2004) Targeted Nanoparticles for Quantitative Imaging of Sparse Molecular Epitopes with MRI. *Magnetic Resonance in Medicine*, **51**, 480-486. http://dx.doi.org/10.1002/mrm.20010

[30] Winter, P.M., Morawski, A.M., Caruthers, S.D., Fuhrhop, R.W., Zhang, H., Williams, T.A., *et al.* (2003) Molecular Imaging of Angiogenesis in Early-Stage Atherosclerosis with Alpha(v) Beta3-Integrin-Targeted Nanoparticles. *Circu-*

lation, **108**, 2270-2274. http://dx.doi.org/10.1161/01.CIR.0000093185.16083.95

[31]   Kao, C.Y., Hoffman, E.A., Beck, K.C., Bellamkonda, R.V. and Annapragada, A.V. (2003) Long-Residence-Time Nano-Scale Liposomal Iohexol for X-Ray-Based Blood Pool Imaging. *Academic Radiology*, **10**, 475-483. http://dx.doi.org/10.1016/S1076-6332(03)80055-7

[32]   Mukundan Jr., S., Ghaghada, K.B., Badea, C.T., Kao, C.Y., Hedlund, L.W., Provenzale, J.M., *et al.* (2006) A Liposomal Nanoscale Contrast Agent for Preclinical CT in Mice. *American Journal of Roentgenology*, **186**, 300-307. http://dx.doi.org/10.2214/AJR.05.0523

[33]   Christodoulides, N., Dharshan, P., Wong, J., Floriano, P.F., Neikirk, D. and McDevitt, J.T. (2007) A Microchip-Based Assay for Interleukin-6. *Methods in Molecular Biology*, **385**, 131-144. http://dx.doi.org/10.1007/978-1-59745-426-1_10

[34]   Goodey, A., Lavigne, J.J., Savoy, S.M., Rodriquez, M.D., Curey, T., Tsao, A., *et al.* (2001) Development of Multianalyte Sensor Arrays Composed of Chemically Derivitized Polymeric Microspheres Localized in Micromachined Cavities. *Journal of the American Chemical Society*, **123**, 2559-2570. http://dx.doi.org/10.1021/ja0033411

[35]   Christodoulides, N., Tran, M., Floriano, P.N., Rodriquez, M., Goodey, A., Ali, M., *et al.* (2002) A Microchip-Based Multianalyte Assay System for the Assessment of Cardiac Risk. *Analytical Chemistry*, **74**, 3030-3036. http://dx.doi.org/10.1021/ac011150a

[36]   Ali, M.F., Kirby, R., Goodey, A.P., Rodriguez, M.D., Ellington, A.D., Neikirk, D.P., *et al.* (2003) DNA Hybridization and Discrimination of Singlenucleotide Mismatches Using Chip-Based Microbead Arrays. *Analytical Chemistry*, **75**, 4732-4739. http://dx.doi.org/10.1021/ac034106z

[37]   Jokerst, J.V., Raamanathan, A., Christodoulides, N., Floriano, P.N., Pollard, A.A., Simmons, G.W., *et al.* (2009) Nano-Bio-Chips for High Performance Multiplexed Protein Detection: Determinations of Cancer Biomarkers in Serum and Saliva Using Quantum Dot Bioconjugate Labels. *Biosensors and Bioelectronics*, **24**, 3622-3629. http://dx.doi.org/10.1016/j.bios.2009.05.026

[38]   Lavigne, J.J., Savoy, S., Clevenger, M.B., Ritchie, J.E., McDoniel, B., Yoo, S.J., *et al.* (1998) Solution-Based Analysis of Multiple Analytes by a Sensor Array: Toward the Development of an "Electronic Tongue". *Journal of the American Chemical Society*, **120**, 6429-6430. http://dx.doi.org/10.1021/ja9743405

[39]   Christodoulides, N., Floriano, P.N., Mohanty, S., Dharshan, P., Griffin, M., Lennart, A., *et al.* (2007) Lab-on-a-Chip Methods for Point of Care Measurements of Salivary Biomarkers of Periodontitis. *Annals of the New York Academy of Sciences*, **1098**, 411-428. http://dx.doi.org/10.1196/annals.1384.035

[40]   Christodoulides, N., Mohanty, S., Miller, C.S., Langub, M.C., Floriano, P.N., Dharshan, P., *et al.* (2005) Application of Microchip Assay System for the Measurement of C-Reactive Protein in Human Saliva. *Lab on a Chip*, **5**, 261-269. http://dx.doi.org/10.1039/b414194f

[41]   Christodoulides, N., Floriano, P.N., Sanchez, X., Li, L.Y., Hocquard, K., Patton, A., *et al.* (2012) Programmable Bio-Nanochip Technology for the Diagnosis of Cardiovascular Disease at the Point of Care. *Methodist Debakey Cardiovascular Journal*, **8**, 6-12. http://dx.doi.org/10.14797/mdcj-8-1-6

[42]   Floriano, P.N., Christodoulides, N., Miller, C.S., Ebersole, J.L., Spertus, J., Rose, B.G., *et al.* (2009) Use of Saliva-Based Nano-Biochip Tests for Acute Myocardial Infarction at the Point of Care: A Feasibility Study. *Clinical Chemistry*, **55**, 1530-1538. http://dx.doi.org/10.1373/clinchem.2008.117713

[43]   Vasan, R.S. (2006) Biomarkers of Cardiovascular Disease: Molecular Basis and Practical Considerations. *Circulation*, **113**, 2335-2362. http://dx.doi.org/10.1161/CIRCULATIONAHA.104.482570

[44]   Danesh, J., Lewington, S., Thompson, S.G., Lowe, G.D., Collins, R., Kostis, J.B., *et al.* (2005) Plasma Fibrinogen Level and the Risk of Major Cardiovascular Diseases and Nonvascular Mortality: An Individual Participant Meta-Analysis. *JAMA*, **294**, 1799-1809.

[45]   Cushman, M., Lemaitre, R.N., Kuller, L.H., Psaty, B.M., Macy, E.M., Sharrett, A.R., *et al.* (1999) Fibrinolytic Activation Markers Predict Myocardial Infarction in the Elderly. The Cardiovascular Health Study. *Arteriosclerosis, Thrombosis, and Vascular Biology*, **19**, 493-498. http://dx.doi.org/10.1161/01.ATV.19.3.493

[46]   Wang, T.J., Larson, M.G., Levy, D., Benjamin, E.J., Leip, E.P., Omland, T., *et al.* (2004) Plasma Natriuretic Peptide Levels and the Risk of Cardiovascular Events and Death. *New England Journal of Medicine*, **350**, 655-663. http://dx.doi.org/10.1056/NEJMoa031994

[47]   Danesh, J., Wheeler, J.G., Hirschfield, G.M., Eda, S., Eiriksdottir, G., Rumley, A., *et al.* (2004) C-Reactive Protein and Other Circulating Markers of Inflammation in the Prediction of Coronary Heart Disease. *New England Journal of Medicine*, **350**, 1387-1397. http://dx.doi.org/10.1056/NEJMoa032804

[48]   Mangoni, A.A. and Jackson, S.H. (2002) Homocysteine and Cardiovascular Disease: Current Evidence and Future Prospects. *The American Journal of Medicine*, **112**, 556-565. http://dx.doi.org/10.1016/S0002-9343(02)01021-5

[49]   Gaspari, M., Cheng, M., Terracciano, R., Liu, X., Nijdam, A.J., Vaccari, L., *et al.* (2006) Nanoporous Surfaces as

Harvesting Agents for Mass Spectrometric Analysis of Peptides in Human Plasma. *Journal of Proteome Research*, **5**, 1261-1266. http://dx.doi.org/10.1021/pr050417+

[50] Luchini, A., Geho, D.H., Bishop, B., Tran, D., Xia, C., Dufour, R.L., *et al.* (2008) Smart Hydrogel Particles: Biomarker Harvesting: One-Step Affinity Purification, Size Exclusion, and Protection against Degradation. *Nano Letters*, **8**, 350-361. http://dx.doi.org/10.1021/nl0721741

[51] Cui, Y., Wei, Q., Park, H. and Lieber, C.M. (2001) Nanowire Nanosensors for Highly Sensitive and Selective Detection of Biological and Chemical Species. *Science*, **293**, 1289-1292. http://dx.doi.org/10.1126/science.1062711

[52] Yang, Z. and Zhou, D.M. (2006) Cardiac Markers and Their Point-of-Care Testing for Diagnosis of Acute Myocardial Infarction. *Clinical Biochemistry*, **39**, 771-780. http://dx.doi.org/10.1016/j.clinbiochem.2006.05.011

[53] Brogan Jr., G.X. and Bock, J.L. (1998) Cardiac Marker Point-of-Care Testing in the Emergency Department and Cardiac Care Unit. *Clinical Chemistry*, **44**, 1865-1869.

[54] Park, J.S., Cho, M.K., Lee, E.J., Ahn, K.Y., Lee, K.E., Jung, J.H., *et al.* (2009) A Highly Sensitive and Selective Diagnostic Assay Based on Virus Nanoparticles. *Nature Nanotechnology*, **4**, 259-264. http://dx.doi.org/10.1038/nnano.2009.38

[55] Vogt, S., Troitzsch, D., Späth, S. and Moosdorf, R. (2004) Efficacy of Ion-Selective Probes in Early Epicardial *in Vivo* Detection of Myocardial Ischemia. *Physiological Measurement*, **25**, N21-N26. http://dx.doi.org/10.1088/0967-3334/25/6/N02

[56] Ji, T., Rai, P., Jung, S. and Varadan, V.K. (2008) *In Vitro* Evaluation of Flexible pH and Potassium Ion-Sensitive Organic Field Effect Transistor Sensors. *Applied Physics Letters*, **92**, Article ID: 233304. http://dx.doi.org/10.1063/1.2936296

[57] Barhoumi, H., Haddad, R., Maaref, A., Bausells, J., Bessueille, F., Léonard, D., *et al.* (2001) New Technology for Multi-Sensor Silicon Needles for Biomedical Applications. *Sensors and Actuators B: Chemical*, **78**, 279-284. http://dx.doi.org/10.1016/S0925-4005(01)00826-7

[58] Shin, K.H., Moon, C.R., Lee, T.H., Lim, C.H. and Kim, Y.J. (2005) Flexible Wireless Pressure Sensor Module. *Sensors and Actuators A*, **123-124**, 30-35. http://dx.doi.org/10.1016/j.sna.2005.01.008

[59] Kim, J.-H., Heller, D.A., Jin, H., Barone, P.W., Song, C., Zhang, J.Q., *et al.* (2009) The Rational Design of Nitric Oxide Selectivity in Single-Walled Carbon Nanotube Nearinfrared Fluorescence Sensors for Biological Detection. *Nature Chemistry*, **1**, 473-481. http://dx.doi.org/10.1038/nchem.332

[60] Lammers, T., Kiessling, F., Hennink, W.E. and Storm, G. (2010) Nanotheranostics and Image-Guided Drug Delivery: Current Concepts and Future Directions. *Molecular Pharmaceutics*, **7**, 1899-1912. http://dx.doi.org/10.1021/mp100228v

[61] Barenholz, Y. (2012) Doxil®—The First FDA-Approved Nano-Drug: Lessons Learned. *Journal of Controlled Release*, **160**, 117-134. http://dx.doi.org/10.1016/j.jconrel.2012.03.020

[62] Mufamadi, M.S., Pillay, V., Choonara, Y.E., Du Toit, L.C., Modi, G. and Naidoo, D. (2011) A Review on Composite Liposomal Technologies for Specialized Drug Delivery. *Journal of Drug Delivery*, **2011**, Article ID: 939851. http://dx.doi.org/10.1155/2011/939851

[63] Maurer, N., Fenske, D.B. and Cullis, P.R. (2001) Developments in Liposomal Drug Delivery Systems. *Expert Opinion on Biological Therapy*, **1**, 923-947. http://dx.doi.org/10.1517/14712598.1.6.923

[64] Immordino, M.L., Dosio, F. and Cattel, L. (2006) Stealth Liposomes: Review of the Basic Science, Rationale, and Clinical Applications, Existing and Potential. *International Journal of Nanomedicine*, **1**, 297-315.

[65] Hedman, M., Hartikainen, J. and Syvänne, M. (2003) Safety and Feasibility of Catheter-Based Local Intracoronary Vascular Endothelial Growth Factor Gene Transfer in the Prevention of Postangioplasty and In-Stent Restenosis and in the Treatment of Chronic Myocardial Ischemia: Phase II Results of the Kuopio Angiogenesis Trial (KAT). *Circulation*, **107**, 2677-2683. http://dx.doi.org/10.1161/01.CIR.0000070540.80780.92

[66] Margolis, J., McDonald, J., Heuser, R., Klinke, P., Waksman, R., Virmani, R., *et al.* (2007) Systemic Nanoparticle Paclitaxel (Nab-Paclitaxel) for In-Stent Restenosis I (SNAPIST-I): A First-in-Human Safety and Dose-Finding Study. *Clinical Cardiology*, **30**, 165-170. http://dx.doi.org/10.1002/clc.20066

[67] McDowell, G., Slevin, M. and Krupinski, J. (2011) Nanotechnology for the Treatment of Coronary in Stent Restenosis: A Clinical Perspective. *Vascular Cell*, **3**, 8. http://dx.doi.org/10.1186/2045-824X-3-8

[68] Zhang, H., Li, N. and Sirish, P. (2012) The Cargo of CRPPR-Conjugated Liposomes Crosses the Intact Murine Cardiac Endothelium. *Journal of Controlled Release*, **163**, 10-17. http://dx.doi.org/10.1016/j.jconrel.2012.06.038

[69] Harel-Adar, T., Ben Mordechai, T., Amsalem, Y., Feinberg, M.S., Leor, J. and Cohen, S. (2011) Modulation of Cardiac Macrophages by Phosphatidylserine-Presenting Liposomes Improves Infarct Repair. *Proceedings of the National Academy of Sciences of the United States of America*, **108**, 1827-1832. http://dx.doi.org/10.1073/pnas.1015623108

[70] Dvir, T., Bauer, M., Schroeder, A., Tsui, J.H., Anderson, D.G., Langer, R., *et al.* (2011) Nanoparticles Targeting the Infarcted Heart. *Nano Letters*, **11**, 4411-4444. http://dx.doi.org/10.1021/nl2025882

[71] Lestini, B.J., Sagnella, S.M., Xu, Z., Shive, M.S., Richter, N.J., Jayaseharan, J., *et al.* (2002) Surface Modification of Liposomes for Selective Cell Targeting in Cardiovascular Drug Delivery. *Journal of Controlled Release*, **78**, 235-247. http://dx.doi.org/10.1016/S0168-3659(01)00505-3

[72] Holland, N.B., Qiu, Y., Ruegsegger, M. and Marchant, R.E. (1998) Biomimetic Engineering of Non-Adhesive Glyco-calyx-Like Surfaces Using Oligosaccharide Surfactant Polymers. *Nature*, **392**, 799-801. http://dx.doi.org/10.1038/33894

[73] Zhu, J., Xue, J., Guo, Z., Zhang, L. and Marchant, R.E. (2007) Biomimetic Glycoliposomes as Nanocarriers for Targeting P-Selectin on Activated Platelets. *Bioconjugate Chemistry*, **18**, 1366-1369. http://dx.doi.org/10.1021/bc700212b

[74] Joner, M., Morimoto, K., Kasukawa, H., Steigarwald, K., Meri, S., Nakazawa, G., *et al.* (2008) Site-Specific Targeting of Nanoparticle Prednisolone Reduces In-Stent Restenosis in a Rabbit Model of Established Atheroma. *Arteriosclerosis, Thrombosis, and Vascular Biology*, **28**, 1960-1966. http://dx.doi.org/10.1161/ATVBAHA.108.170662

[75] Cho, B.H., Park, J.R., Nakamura, M.T., Odintsov, B.M., Wallig, M.A. and Chung, B.H. (2010) Synthetic Dimyris-toylphosphatidylcholine Liposomes Assimilating into High-Density Lipoprotein Promote Regression of Atherosclerot-ic Lesions in Cholesterol-Fed Rabbits. *Experimental Biology and Medicine*, **235**, 1194-1203. http://dx.doi.org/10.1258/ebm.2010.009320

[76] Walton, B.L., Leja, M., Vickers, K.C., Estevez-Fernandez, M., Sanguino, A., Wang, E., *et al.* (2010) Delivery of Ne-gatively Charged Liposomes into the Atheromas of Watanabe Heritable Hyperlipidemic Rabbits. *Vascular Medicine*, **15**, 307-313. http://dx.doi.org/10.1177/1358863X10374118

[77] Danenberg, H.D., Fishbein, I., Gao, J., Mönkkönen, J., Reich, R., Gati, I., *et al.* (2002) Macrophage Depletion by Clo-dronate-Containing Liposomes Reduces Neointimal Formation after Balloon Injury in Rats and Rabbits. *Circulation*, **106**, 599-605. http://dx.doi.org/10.1161/01.CIR.0000023532.98469.48

[78] Buxton, D.B. (2009) Nanomedicine for the Management of Lung and Blood Diseases. *Nanomedicine*, **4**, 331-339. http://dx.doi.org/10.2217/nnm.09.8

[79] Cyrus, T., Zhang, H., Allen, J.S., Williams, T.A., Hu, G., Caruthers, S.D., *et al.* (2008) Intramural Delivery of Rapa-mycin with Alphavbeta3-Targeted Paramagnetic Nanoparticles Inhibits Stenosis after Balloon Injury. *Arteriosclerosis, Thrombosis, and Vascular Biology*, **28**, 820-826. http://dx.doi.org/10.1161/ATVBAHA.107.156281

[80] Winter, P.M., Caruthers, S.D., Zhang, H., Williams, T.A., Wickline, S.A. and Lanza, G.M. (2008) Antiangiogenic Synergism of Integrin-Targeted Fumagillin Nanoparticles and Atorvastatin in Atherosclerosis. *JACC: Cardiovascular Imaging*, **1**, 624-634. http://dx.doi.org/10.1016/j.jcmg.2008.06.003

[81] Winter, P.M., Neubauer, A.M., Caruthers, S.D., Harris, T.D., Robertson, J.D., Williams, T.A., *et al.* (2006) Endothelial Alpha(v)Beta3 Integrin-Targeted Fumagillin Nanoparticles Inhibit Angiogenesis in Atherosclerosis. *Arteriosclerosis, Thrombosis, and Vascular Biology*, **26**, 2103-2109. http://dx.doi.org/10.1161/01.ATV.0000235724.11299.76

[82] Hoffmann, R., Mintz, G.S., Dussaillant, G.R., Popma, J.J., Pichard, A.D., Satler, L.F., *et al.* (1996) Patterns and Me-chanisms of In-Stent Restenosis. A Serial Intravascular Ultrasound Study. *Circulation*, **94**, 1247-1254. http://dx.doi.org/10.1161/01.CIR.94.6.1247

[83] Samaroo, H.D., Lu, J. and Webster, T.J. (2008) Enhanced Endothelial Cell Density on NiTi Surfaces with Sub-Micron to Nanometer Roughness. *International Journal of Nanomedicine*, **3**, 75-82.

[84] Kastrati, A., Mehilli, J., Pache, J., Kaiser, C., Valgimigli, M., Kelbaek, H., *et al.* (2007) Analysis of 14 Trials Compar-ing Sirolimus-Eluting Stents with Bare-Metal Stents. *New England Journal of Medicine*, **356**, 1030-1039. http://dx.doi.org/10.1056/NEJMoa067484

[85] Lagerqvist, B., James, S.K., Stenestrand, U., Lindbäck, J., Nilsson, T., Wallentin, L., *et al.* (2007) Long-Term Out-comes with Drug-Eluting Stents versus Bare-Metal Stents in Sweden. *New England Journal of Medicine*, **356**, 1009-1019. http://dx.doi.org/10.1056/NEJMoa067722

[86] Mauri, L., Hsieh, W.H., Massaro, J.M., Ho, K.K., D'Agostino, R. and Cutlip, D.E. (2007) Stent Thrombosis in Ran-domized Clinical Trials of Drug-Eluting Stents. *New England Journal of Medicine*, **356**, 1020-1029. http://dx.doi.org/10.1056/NEJMoa067731

[87] Stone, G.W., Moses, J.W., Ellis, S.G., Schofer, J., Dawkins, K.D., Morice, M.C., *et al.* (2007) Safety and Efficacy of Sirolimus- and Paclitaxel-Eluting Coronary Stents. *New England Journal of Medicine*, **356**, 998-1008. http://dx.doi.org/10.1056/NEJMoa067193

[88] Wieneke, H., Dirsch, O., Sawitowski, T., Gu, Y.L., Brauer, H., Dahmen, U., *et al.* (2003) Synergistic Effects of a Nov-el Nanoporous Stent Coating and Tacrolimus on Intima Proliferation in Rabbits. *Catheterization and Cardiovascular Interventions*, **60**, 399-407. http://dx.doi.org/10.1002/ccd.10664

[89] Bhargava, B., Reddy, N.K., Karthikeyan, G., Raju, R., Mishra, S., Singh, S., *et al.* (2006) A Novel Paclitaxel-Eluting

Porous Carbon-Carbon Nanoparticle Coated, Nonpolymeric Cobalt-Chromium Stent: Evaluation in a Porcine Model. *Catheterization and Cardiovascular Interventions*, **67**, 698-702. http://dx.doi.org/10.1002/ccd.20698

[90] Ayon, A.A., Cantu, M., Chava, K., Agrawal, C.M., Feldman, M.D., Johnson, D., *et al.* (2006) Drug Loading of Nanoporous TiO$_2$ Films. *Biomedical Materials*, **1**, L11-L15. http://dx.doi.org/10.1088/1748-6041/1/4/L01

[91] Liu, D.M., Yang, Q. and Troczynski, T. (2002) Sol-Gel Hydroxyapatite Coatings on Stainless Steel Substrates. *Biomaterials*, **23**, 691-698. http://dx.doi.org/10.1016/S0142-9612(01)00157-0

[92] Caves, J.M. and Chaikof, E.L. (2006) The Evolving Impact of Microfabrication and Nanotechnology on Stent Design. *Journal of Vascular Surgery*, **44**, 1363-1368. http://dx.doi.org/10.1016/j.jvs.2006.08.046

11

Simultaneous Use of Entomopathogenic Fungus *Beauveria bassiana* and Diatomaceous Earth against the Larvae of Indian Meal Moth, *Plodia interpunctella*

Mohsen Arooni-Hesari, Reza Talaei-Hassanloui*, Qodrat Sabahi

Department of Plant Protection, College of Agriculture and Natural Resources, University of Tehran, Karaj, Iran
Email: *rtalaei@ut.ac.ir

Abstract

The suppressive ability of entomopathogenic fungus *Beauveria bassiana* alone and in combination with diatomaceous earth (DE) was studied against the larvae of Indian meal moth, *Plodia interpunctella* (*Hübner*) (Lep., Pyralidae). This study clearly showed that simultaneous use of *B. bassiana* and DE against larvae of *P. interpunctella*, not only could reduce the required concentration of fungal conidia or DE, but also could shorten the time need for showing insecticidal effects. The LC_{50} value of fungus at 7 d after treatment was 9.8×10^5 conidia mg^{-1} diet. Larvae showed a dose response to *B. bassiana*, and the addition of diatomaceous earth at 500 and 2000 ppm resulted in a significant increase in mortality. Larval mortality reached to the maximum of 28.3% and 71.7% after 7 d exposure to 500 and 2000 ppm DE concentrations, respectively. The LC_{50} value for *B. bassiana* in the presence of DE 500 ppm was 4.6×10^4 con. mg^{-1} diet and of DE 2000 ppm was 1.65×10^3 con. mg^{-1} diet. According to our results, *B. bassiana* and DE can be considered as two suitable candidates for integration into IPM strategy.

Keywords

Stored Product Pest, *Beauveria bassiana*, Synergism, Diatomaceous Earth, Bioassay

1. Introduction

Entomopathogenic fungi are among the important biological control agents of insect pests, by causing lethal in-

---

*Corresponding author.

fections and regulating insect and mite populations in nature by epizootics [1]-[3]. These biocontrol agents infect a wide range of insect orders including Hemiptera, Coleoptera and Lepidoptera which are of great concern in worldwide agriculture but some strains of these fungi could be host specific with a very low risk of attacking non-target organisms or beneficial insects [4].

The entomopathogenic as comycete, *Beauveria bassiana* (Balsamo) Vuillemin is an important pathogen of insects and it has been developed as a microbial insecticide for use against many major arthropod pests [5] [6]. It has been developed as a microbial insecticide for use against many major pests, including lepidopterans. It is reported to be non-toxic to humans and other vertebrates, so it can be applied on commodities [7].

Plodia interpunctella (Hübner) (Lepidoptera: Pyralidae), is a serious pest of raw and processed food products worldwide [8]-[11]. Some reports have pointed out to the promising control effects of *B. bassiana* on Indian meal moth [12] [13]. However the fungi encountered some limiting factors to show adequate level of pest control.

The use of *B. bassiana* alone for the control of many stored product pests requires a high application rates. The action of fungal penetration to the host integument is dependent to physicochemical properties of cuticle including thickness, sclerotization, and presence of fatty acids as well as cuticle destroying enzymes [14]-[16].

The presence of cuticular lipids in insect integument could protect the insect from pathogenic microorganisms. If any agent could damage this barrier, greater penetration and higher pathogen virulence would be expected.

Diatomaceous Earth (DE), a natural product composed of the fossils of diatoms, can exhibit such role with its special properties like integument scarification and adsorbing wax from cuticle layer [17], causing the release of subcuticular compounds that have a synergistic effect on conidial attachment and virulence of entomopathogenic fungi [18] [19].

Several studies document that DE formulations are very effective against a wide range of stored products pests, persist on the product for a long period and can be easily removed from the grains [20]-[22].

It is generally accepted that these materials act as desiccants on insect cuticle, and that insects exposed to DE particles die from rapid water loss [17] [23] [24]. Low mammalian toxicity is another important characteristic of DE [25], rendering it a potential and safe method for pest control.

Our objective in this study was to determine the effects of *B. bassiana* and DE, alone or in combination, on the Indian meal moth larvae as well as clarify interactions of these two agents.

2. Materials and Methods

2.1. Insect Rearing

Indian meal moth, *P. interpunctella* was acquired from a laboratory population at the Department of Plant Protection, University of Tehran and reared under laboratory conditions ($25°C \pm 1°C$, $65\% \pm 10\%$ RH and 16:8 L:D) in Polyethylene boxes ($30 \times 20 \times 15$ cm) covered with a piece of fine cloth.

Artificial diet including wheat bran, yeast, honey, and glycerol, prepared by the method described by Sait *et al.* [26], was used for insect rearing. No antibiotic or fungicide was added to the diet.

2.2. Fungal Isolate and Cultures

The isolate used in this study was *B. bassiana* ETU105, originally isolated from soil using *Galleria*-bait and preserved in Laboratory of Biological Control at the Department of Plant Protection, University of Tehran. Fungal isolate was grown on Sabouraud dextrose agar plus 1% yeast extract (SDAY) in 9 cm diameter Petri dishes (**Figure 1**) and incubated under conditions of $25°C \pm 1°C$, 75 ± 10 RH and 16:8 L:D for 14 days. Colony was preserved at $4°C$.

Fungal conidia were collected by scraping conidial layer using sterilized scalpel into 0.02% Tween 80. The conidial concentration was estimated using a haemocytometer. Each mg of collected conidia contained 4.5×10^8 spores. Five serial concentrations 10^3 - 10^7 conidia mg^{-1} diet were prepared by mixing adequate conidia with artificial diet. An electric mixer was used for 5 min to prepare an even mixture.

2.3. Diatomaceous Earth

Diatomaceous Earth (DE) was prepared from Kimia Sabz Avar Company located in Iran. It was oven-dried initially at $60°C$ for two hours, and then two concentrations of 500 and 2000 mg/kg were prepared by adding sufficient DE to artificial diet. These two concentrations were selected based on the results of our preliminary test.

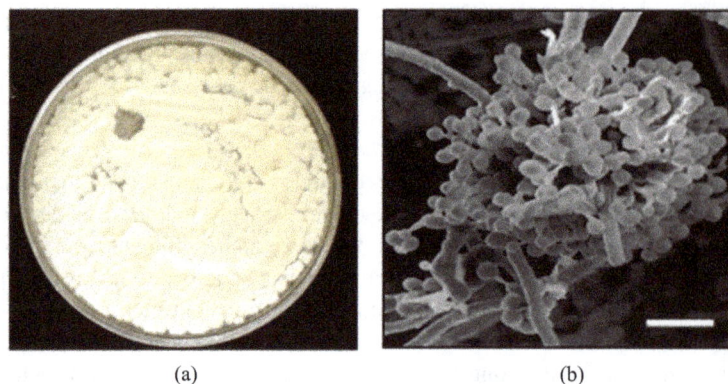

Figure 1. *Beauveria bassiana*, (a) Fully sporulated on SDAY medium; (b) Scanning electron micrograph of conidiophores and conidia clusters (original, bar = 10 μm).

2.4. Bioassay

There were 17 treatments: five fungal dose rates alone, five fungal dose rates in combination with each of the two DE dose rates and the two DE rates alone. Five serial dose rates of fungal conidia were 10^3 - 10^7 con. mg^{-1} diet. Fifteen g of diet containing treatments were placed in series of glass Petri dishes with 9 cm in diameter. An additional series of dishes containing untreated diet served as a control.

After the preparation of the dishes, 15 third instar larvae of *P. interpunctella* were introduced into each dish. The dishes were placed in incubators set up at temperature of 25°C ± 1°C and RH of 65% ± 5% for a week. All the experiments were repeated four times. Larval mortality was assessed daily until 7 days after exposure.

2.5. Statistical Analysis

Mortality counts were corrected by using Abbott's formula [27]. The experiment was designed and conducted in a factorial Complete Randomized Design (CRD). The pooled data were analyzed separately for each treatment category (Fungus alone, DE alone, DE 500+ fungus, DE 2000+ fungus), by submitting the mortality counts to ANOVA's for dose rate. Means were separated using the Tukey test at $P < 0.05$. SAS software version 9.1 was used for statistical analysis. The LC_{50} values were estimated with probit analysis using POLO-Plus [28].

3. Results

3.1. Effect of *B. bassiana* on *P. interpunctella* Larvae

Mortality rates of third instar larvae of *P. interpunctella* treated with *B. bassiana* increased with increasing conidial concentration and time of exposure. Differences among lethal effects established by different conidial concentrations of fungal isolate were significantly different (F = 72.49, df = 5, $P < 0.0001$).

The LC_{50} value of fungus at 7 d after treatment on *P. interpunctella* larvae was 9.8×10^5 con. mg^{-1} diet. Comparative virulence of *B. bassiana* isolate against L_3 instar of *P. interpunctella* indicated that mortality of larvae began at the third day after exposure and reached to 61.7% at the day 7. The LT_{50} value of fungus at 1×10^7 con. mg^{-1} was 6.59 d (6.23 - 7.12 days lower and upper 95% C.I., respectively).

3.2. Effect of Diatomaceous Earth on *P. interpunctella* Larvae

Larval mortality reached to a maximum of 28.3% and 71.7% after 7 d exposure to 500 and 2000 mg/kg DE concentrations, respectively. There were highly significant differences in larval mortality resulting from different concentrations of DE (F = 121.13, df = 2, $P < 0.0001$). Larval mortality ranged from 1.7 to a maximum of 71.7% over the range of tested concentrations.

3.3. Combination of *B. bassiana* and DE

Data on comparative virulence of *B. bassiana* isolate in combination with DE against L_3 instar of *P. interpunctella* is brought in **Table 1**. These results showed that there was a synergistic relationship between fungus and DE targeting larvae of *P. interpunctella*.

The LC_{50} value for *B. bassiana* plus DE 500 mg/kg was 4.6×10^4 con. mg^{-1} diet and with DE 2000 mg/kg was 1.65×10^3 con. mg^{-1} diet in contrast to the LC_{50} of 9.8×10^5 con. mg^{-1} diet for *B. bssiana* alone (**Table 2**). There were significant differences among these treatments ($F = 72.49$, df = 5, $P < 0.0001$).

The combination treatment of fungus with DE caused the highest mortality at 7d after treatment which was significantly greater than all other treatments. With concentration of 500 ppm DE, mortality of larvae began at the second day after exposure to 1×10^7 con. mg^{-1} of fungus and reached to 100% at the day 7.

With DE 500, LT_{50} value for fungus with concentration of 1×10^5 con. mg^{-1}, diminished to 3.9 d which was significantly lower than that of fungus alone with 6.6 d ($F = 11.48$, df = 26, $P < 0.0001$). This value showed a greater reduction when DE 2000 was applied, such that it was reduced to 2.1 d. The log-probit regression lines of LC_{50} had slopes of 0.34 without DE, 0.65 with DE 500 and 0.64 with DE 2000. These parameters for LT_{50s} were 8.21, 6.54 and 5.35, respectively (**Table 3**).

4. Discussion

Results from bioassays on *P. interpunctella* larvae showed that the *B. bassiana* isolate ETU105 was virulent against the pest, however, a relatively high concentration of 9.8×10^5 conidia mg^{-1} diet needs to cause 50% mortality during 7 d. These results stand somewhat in agreement with that of Buda and Pečiulytė [29] who found that after treatment with concentration of 2.6×10^6 conidia mg^{-1} of *B. bassiana*, larval mortality of the species, was reached to 50% during 5 days.

Many of the important pests in grain storage have proven to be susceptible to *B. bassiana*, but its high production costs and high application rates make it economically unreachable [30]-[36].

Table 1. Mean percent mortalities of *P. interpunctella* third instar larvae caused by *B. bassiana* in combination with diatomaceous earth.

DE concentration (ppm)	Fungal concentration (conidia mg^{-1} diet)					
	0	10^3	10^4	10^5	10^6	10^7
0	5	25	28.33	43.33	55	61.66
500	38.33	41.66	60	71.66	80	100
2000	71.66	83.33	91.66	96.66	98.33	100

Table 2. LC_{50} values of *B. bassiana* alone or in combination with two diatomaceous earth doses based on experiments with 3rd instar larvae of *P. interpunctella*.

DE concentration (ppm)	Fungus LC_{50} (con. mg^{-1} diet)	χ^2	Slope	Lower 95% confidence limit	Upper 95% confidence limit
0	0.98×10^6	4.23	0.34	0.28×10^6	0.50×10^7
500	4.06×10^4	8.86	0.65	6.11×10^3	0.12×10^6
2000	1.65×10^3	8.21	0.64	0.26×10^2	1.03×10^4

Number of treated larvae = 360.

Table 3. Values of LT_{50} for *B. bassiana* at 1×10^7 con. mg^{-1} diet alone or in combination with two doses of diatomaceous earth based on experiments with 3rd instar larvae of *P. interpunctella*.

DE concentration (ppm)	Fungus LT_{50} (d)	χ^2	Slope	Lower 95% confidence limit	Upper 95% confidence limit
DE 0	6.59	5.9	8.21	6.23	7.12
DE 500	3.92	9.68	6.54	3.69	4.15
DE 2000	2.09	8.08	5.35	1.90	2.28

Number of treated larvae = 360; df = 26.

The concentration of 500 ppm of DE, caused 28.3% mortality of *P. interpunctella* larvae with increasing the DE concentration to 2000 ppm could enhance the mortality rate to 71.7 percent. This result approved the finding of Sabbour *et al.* [13], however, the LC_{50} of fungus in their study was 1.29×10^9 con. mg^{-1} which was much greater than that of our finding, 1×10^5 con. mg^{-1} diet. This difference may be related to fungus isolate or population of *P. interpunctella*.

Our results also are in conformity with findings of Akbar *et al.* [18] who found that DE increased the efficacy of *B. bassiana* against *Tribolium castaneum* larvae. Also Michalaki *et al.* [19] found that the effectiveness of entomopathogenic fungus, *Metarhizium anisopliae* can be benefitted by the presence of DE against *T. confusum* larvae.

DE does not leave any toxic residue because of its physical effects; however, regarding the possible health issues, lower concentrations should be preferred. Low concentration of DE does not leave any harmful residue on stored commodities. If it could be prepared from good quality and suitable resources, would have very low effect on human breathing system. Also, using suitable breathing mask, such effects could be greatly diminished.

Fast and stable effectiveness, make DE an appropriate choice to substitute chemical compounds for stored product protection. Moisture increase of environment can lead to less effectiveness of DE, so it is more successful in control of stored product pests. Moreover such pests are little in size, have a higher body surface to body volume ratio, and encountered to low water content of commodities. The lack of insect resistance against DE is another important prominence which enhances its capability for use as a part of integrated pest management program.

DE does not have any significant effect on stored products by own itself and can be removed easily with air blowing or washing. This can decrease applying chemical pesticides in grain deposit. Using DE at concentration of 500 ppm, the LC_{50} value of *B. bassiana* was significantly decreased on *P. interpunctella* larvae.

The exact mechanisms by which DE interacts with *B. bassiana* are not clear but may involve a combination of increased availability of water and other nutrients, removal or mitigation of inhibitory materials, alteration of adhesive properties, and physical disruption of the cuticular barrier [18]. Lord [12] [34] proposed that lipid removal may contribute to the synergistic interaction between *B. bassiana* and DE against some stored grain pests.

Akbar *et al.* [18] showed the number of *B. bassiana* conidia attached to the larval cuticle of *Tribolium castaneum* was significantly greater with DE presence than without it. The mean counts of conidia were 212.7 with DE and 90.9 without DE.

Michalaki *et al.* [19] revealed that DE benefits the fungal efficacy only when conidial concentration exceeds a certain "active threshold". Below this "threshold", a considerable amount of conidia may be damaged by the presence of DE, or DE particles may partially lose their desiccant capacity.

5. Conclusion

Finally, this study clearly indicated that simultaneous use of *B. bassiana* and DE against larvae of *P. interpunctella*, not only reduced the required concentration of fungus or DE, but it would also shorten the time need for showing their effects. On the other hand, the greatest factor in the loss of inoculum viability of entomopathogenic fungi under field conditions is inactivation caused by UV light [37] [38], but grain storage environment does not have such disadvantage, thus it could be a suitable environment for integrated application of *B. bassiana* and DE.

References

[1]   Burges, H.D. and Weiser, J. (1973) Occurrence of Pathogens of the Flour Beetle, *Tribolium castaneum*. *Journal of Invertebrate Pathology*, **22**, 464-466. http://dx.doi.org/10.1016/0022-2011(73)90178-X

[2]   Carruthers, R.I. and Soper, R.S. (1987) Fungal Diseases. In: Fuxa, J. and Tanada, Y., Eds., *Epizootiology of Insect Diseases*, John Wiley and Sons Inc., New York, 357-416.

[3]   McCoy, C.W., Samson, R.A. and Boucias, D.G. (1988) Entomogenous Fungi. In: Ignoffo, C.M. and Mandava, N.B., Eds., *Handbook of Natural Pesticides*, *Vol. V, Microbial Insecticides*, *Part A*, CRC Press, Boca Raton, 151-236.

[4]   Roberts, D.W. and Humber, R.A. (1981) Entomogenous Fungi. In: Cole, G.T. and Kendrick, B., Eds., *Biology of Conidial Fungi*, Academic Press, New York, 201-236. http://dx.doi.org/10.1016/b978-0-12-179502-3.50014-5

[5]   Charnley, A.K. and Collins, S.A. (2007) Entomopathogenic Fungi and Their Role in Pest Control. In: Kubicek, C.P. and Druzhinina, I.S., Eds., *Environmental and Microbial Relationships*, Springer, Berlin, 159-187.

[6]   Faria, M.R. and Wraight, S.P. (2007) Mycoinsecticides and Mycoacaricides: A Comprehensive List with Worldwide Coverage and International Classification of Formulation Types. *Biological Control*, **43**, 237-256.

http://dx.doi.org/10.1016/j.biocontrol.2007.08.001

[7]   Mahroof, R. and Subramanyam, B. (2006) Susceptibility of *Plodia interpunctella* (Lepidoptera: Pyralidae) Developmental Stages to High Temperatures Used during Structural Heat Treatments. *Bulletin of Entomological Research*, **96**, 539-545. http://dx.doi.org/10.1017/BER2006454

[8]   Sedlacek, J.D., Weston, P.A. and Barney, R.J. (1996) Lepidoptera and Psocoptera. In: Subramanyam, B. and Hagstrum, D.W., Eds., *Integrated Management of Insects in Stored Products*, Marcel Dekker Inc., New York, 63-66.

[9]   Doud, C.W. and Phillips, T.W. (2000) Activity of *Plodia interpunctella* (Lepidoptera: Pyralidae) in and around Flour Mills. *Journal of Economic Entomology*, **93**, 1842-1847. http://dx.doi.org/10.1603/0022-0493-93.6.1842

[10] Johnson, J.A., Wang, S. and Tang, J. (2003) Thermal Death Kinetics of Fifth-Instar *Plodia interpunctella* (Lepidoptera: Pyralidae). *Journal of Economic Entomology*, **96**, 519-524. http://dx.doi.org/10.1093/jee/96.2.519

[11] Campbell, J.F. and Mullen, M.A. (2004) Distribution and Dispersal Behavior of *Trogoderma variabile* and *Plodia interpunctella* outside a Food Processing Plant. *Journal of Economic Entomology*, **97**, 1455-1464. http://dx.doi.org/10.1093/jee/97.4.1455

[12] Lord, J.C. (2001) Desiccant Dusts Synergize the Effect of *Beauveria bassiana* (Hyphomycetes: Moniliales) on Stored-Grain Beetles. *Journal of Economic Entomology*, **94**, 367-372. http://dx.doi.org/10.1603/0022-0493-94.2.367

[13] Sabbour, M.M., Abd-El-Aziz, S., El-Sayed, A. and Sherief, M. (2012) Efficacy of Three Entomopathogenic Fungi Alone or in Combination with Diatomaceous Earth Modifications for the Control of Three Pyralid Moths in Stored Grains. *Journal of Plant Protection Research*, **52**, 359-364. http://dx.doi.org/10.2478/v10045-012-0059-7

[14] St. Leger, R.J. (1993) Biology and Mechanisms of Insect-Cuticle Invasion by Deuteromycete Fungal Pathogens. In: Beckage, N.E., Thompson, S.N. and Federici, B.A., Eds., *Parasites and Pathogens of Insects*, Volume 2, Academic Press Inc., New York, 211-229.

[15] Gupta, S.C., Leathers, T.D., El-Sayed, G.N. and Ignoffo, C.M. (1994) Relationships among Enzyme Activities and Virulence Parameters in *Beauveria bassiana* Infections of *Galleria mellonella* and *Trichoplusia ni*. *Journal of Invertebrate Pathology*, **64**, 13-17. http://dx.doi.org/10.1006/jipa.1994.1062

[16] Butt, T.M. and Goettel, M.S. (2000) Bioassays of Entomogenous Fungi. In: Navon, A. and Ascher, K.R.S., Eds., *Bioassays of Entomopathogenic Microbes and Nematodes*, CAB International, Wallingford, UK, 141-195. http://dx.doi.org/10.1079/9780851994222.0141

[17] Mewis, I. and Ulrichs, C. (2001) Action of Amorphous Diatomaceous Earth against Different Stages of the Stored Product Pests *Tribolium confusum, Tenebrio molitor, Sitphilus granarius* and *Plodia interpunctella*. *Journal of Stored Products Research*, **37**, 153-164. http://dx.doi.org/10.1016/S0022-474X(00)00016-3

[18] Akbar, W., Lord, J.C., Nechols, J.R. and Howard, R.W. (2004) Diatomaceous Earth Increases the Efficacy of *Beauveria bassiana* against *Tribolium castaneum* Larvae and Increases Conidia Attachment. *Journal of Economic Entomology*, **97**, 273-280. http://dx.doi.org/10.1093/jee/97.2.273

[19] Michalaki, M.P., Athanassiou, C.H., Kavallieratos, N.G., Batta, Y.A. and Balotis, G.N. (2006) Effectiveness of *Metarhizium anisopliae* (Metschnikoff) Sorokin Applied Alone or in Combination with Diatomaceous Earth against *Tribolium confusum* DuVal Larvae: Influence of Temperature, Relative Humidity and Type of Commodity. *Crop Protection*, **25**, 418-425. http://dx.doi.org/10.1016/j.cropro.2005.07.003

[20] Fields, P.G. and Korunic, Z. (2000) The Effect of Grain Moisture Content and Temperature on the Efficacy of Diatomaceous Earths from Different Geographical Locations against Stored Product Beetles. *Journal of Stored Products Research*, **36**, 1-13. http://dx.doi.org/10.1016/S0022-474X(99)00021-1

[21] Athanassiou, C.G. (2006) Influence of Instar and Commodity in Insecticidal Effects of Two Diatomaceous Earth Formulations against Larvae of *Ephestia kuehniella* (Lepidoptera: Pyralidae). *Journal of Economic Entomology*, **99**, 1905-1911. http://dx.doi.org/10.1093/jee/99.5.1905

[22] Vardeman, E.A., Arthur, F.H., Nechols, J.R. and Campbell, J.F. (2006) Effect of Temperature, Exposure Interval, and Depth of Diatomaceous Earth Treatment on Distribution, Mortality, and Progeny Production of Lesser Grain Borer (Coleoptera: Bostrichidae) in Stored Wheat. *Journal of Economic Entomology*, **99**, 1017-1024. http://dx.doi.org/10.1093/jee/99.3.1017

[23] Korunic, Z. (1998) Diatomaceous Earth: A Group of Natural Insecticides. *Journal of Stored Products Research*, **34**, 87-97. http://dx.doi.org/10.1016/S0022-474X(97)00039-8

[24] Subramanyam, B. and Roesli, R. (2000) Inert Dusts. In: Subramanyam, B. and Hagstrum, D.W., Eds., *Alternatives to Pesticides in Stored-Product IPM*, Kluwer Academic Publishers, Boston, 321-380. http://dx.doi.org/10.1007/978-1-4615-4353-4_12

[25] Cox, P.D. and Wilkin, D.R. (1996) The Potential Use of Biological Control of Pests in Stored Grain. Research Review 36, Home-Grown Cereals Authority, London.

[26] Sait, S.M., Begon, M., Thompson, D.J., Harvey, J.A. and Hails, R.S. (1997) Factors Affecting Host Selection in an Insect Host-Parasitoid Interactions. *Ecological Entomology*, **2**, 225-230. http://dx.doi.org/10.1046/j.1365-2311.1997.t01-1-00051.x

[27] Abbott, W.S. (1925) A Method of Computing the Effectiveness of an Insecticides. *Journal of Economic Entomology*, **18**, 265-267. http://dx.doi.org/10.1093/jee/18.2.265a

[28] LeOra Software (2006) POLO-Plus 1.0 Probit and Logit Analysis. LeOra Software, Petaluma.

[29] Buda, V. and Peciulyte, D. (2008) Pathogenicity of Four Fungal Species to Indian Meal Moth, *Plodia interpunctella* (Hubner) (Lepidoptera: Pyralidae). *Ekologija*, **54**, 265-270. http://dx.doi.org/10.2478/v10055-008-0040-y

[30] Hluchy, M. and Samsinakova, A. (1989) Comparative Study on the Susceptibility of Adult *Sitophilus granarius* (L.) (Coleoptera: Curculionidae) and Larval *Galleria mellonella* (L.) (Lepidoptera: Pyralidae) to the Entomogenous Fungus *Beauveria bassiana* (Bals.) Vuill. *Journal of Stored Products Research*, **25**, 61-64. http://dx.doi.org/10.1016/0022-474X(89)90011-8

[31] Adane, K., Moore, D. and Archer, S.A. (1996) Preliminary Studies on the Use of *Beauveria bassiana* to Control *Sitophilus zeamais* (Coleoptera: Curculionidae) in the Laboratory. *Journal of Stored Products Research*, **32**, 105-113. http://dx.doi.org/10.1016/0022-474X(96)00009-4

[32] Rice, W.C. and Cogburn, R.R. (1999) Activity of the Entomopathogenic Fungus *Beauveria bassiana* (Deuteromycota: Hyphomycetes) against Three Coleopteran Pests of Stored Grain. *Journal of Economic Entomology*, **92**, 691-694. http://dx.doi.org/10.1093/jee/92.3.691

[33] Bourassa, C., Vincent, C., Lomer, C.J., Borgemeister, C. and Mauffette, Y. (2001) Effects of Entomopathogenic Hyphomycetes against the Larger Grain Borer, *Prostephanus truncatus* (Horn) (Coleoptera: Bostrichidae), and Its Predator, *Teretriosoma nigrescens* Lewis (Coleoptera: Histeridae). *Journal of Invertebrate Pathology*, **77**, 75-77. http://dx.doi.org/10.1006/jipa.2000.4986

[34] Lord, J.C. (2007) Desiccation Increases the Efficacy of *Beauveria bassiana* for Stored-Grain Pest Insect Control. *Journal of Stored Products Research*, **43**, 535-539. http://dx.doi.org/10.1016/j.jspr.2007.03.002

[35] Meikle, W.G., Cherry, A.J., Holst, N., Hounna, B. and Markham, R.H. (2001) The Effects of an Entomopathogenic Fungus, *Beauveria bassiana* (Balsamo) Vuillemin (Hyphomycetes), on *Prostephanus truncatus* (Horn) (Col.: Bostrichidae), *Sitophilus zeamais* Motschulsky (Col.: Curculionidae), and Grain Losses in Stored Maize in the Benin Republic. *Journal of Invertebrate Pathology*, **77**, 198-205. http://dx.doi.org/10.1006/jipa.2001.5015

[36] Padin, S.B., Dal-Bello, G.M. and Fabrizio, M. (2002) Grain Losses Caused by *Tribolium castaneum, Sitophilus oryzae* and *Acanthoscelides obtectus* in Stored Durum Wheat and Beans Treated with *Beauveria bassiana*. *Journal of Stored Products Research*, **38**, 69-74. http://dx.doi.org/10.1016/S0022-474X(00)00046-1

[37] Ignoffo, C.M. and Garcia, C. (1992) Influence of Conidial Color on Inactivation of Several Entomogenous Fungi (Hyphomycetes) by Simulated Sunlight. *Environmental Entomology*, **21**, 913-917. http://dx.doi.org/10.1093/ee/21.4.913

[38] Braga, G.L., Flint, S.D., Miller, C.D., Anderson, A.J. and Roberts, D.W. (2001) Variability in Response to UV-B among Species and Strains of *Metarhizium* Isolated from Sites at Latitudes from 61°N to 54°S. *Journal of Invertebrate Pathology*, **78**, 98-108. http://dx.doi.org/10.1006/jipa.2001.5048

12

Fish Nutrition Additives in SHK-1 Cells: Protective Effects of Silymarin

Rodrigo Sanchez[1,2]*, Pamela Olivares[1]*, Erico Carmona[3], Allisson Astuya[4], Hector Herrera[2], Jorge Parodi[1]#

[1]Laboratorio Fisiología de la Reproducción, Escuela de Medicina Veterinaria, Núcleo de Investigación en Producción Alimentaria, Facultad de Recursos Naturales, Universidad Católica de Temuco, Temuco, Chile
[2]Empresa Vitapro Chile, Castro, Región de Los Lagos, Chile
[3]Grupo de Genotoxicología, Escuela de Medicina Veterinaria, Núcleo de Estudios Ambientales, Facultad de Recursos Naturales, Universidad Católica de Temuco, Temuco, Chile
[4]Laboratory of Cell Culture and Marine Genomics, Marine Biotechnology Unit, Faculty of Natural and Oceanographic Sciences, University of Concepcion and Sur-Austral COPAS Program, University of Concepcion, Concepción, Chile
Email: #jparodi@uct.cl

Abstract

In nutrition for productive species, additives play an important role in boosting physiological processes. Only in recent years studies include models of the effects on fish cells of these additives. We observed effects of silymarin, a compound highly utilized in aquaculture. The cell line SHK-1 was used derived from the upper liver of the Atlantic salmon as a biological model. Samples were exposed to silymarin in incrementing time and concentrations, to evaluate by MTT and number of cells, the effects on cell viability. Also, oxidative stress models were used to find the protector effects of silymarin against these agents. Our data indicate that a dose of 100 ppm of silymarin is sufficient to stimulate cellular proliferation. Cultures were exposed to high glucose (15 mM) or H_2O_2 (0.1 mM) in presence of or absence of silymarin at 100 ppm. We observed that the toxic effects of both compounds were blocked by the presence of silymarin. Our results indicate that it is important to evaluate additive effects at a cellular level. Also, silymarin does have proliferative effects, and protect against cellular injury in our models. Our study helps to generate more rational applications of additives in the industry and presents new challenges in order to better manipulate the model in the laboratory, allowing us to obtain new evidence regarding the microalgae's biology through *in vitro* studies.

---

*Both authors contributed in an equal form in the manuscript.
#Corresponding author.

Keywords

Fish, Nutrition, Silymarin, Cell Cultures

1. Introduction

Cellular cultures are important for fish research and in vertebrates, albeit in the exploration of additives or their interaction with organs and the appearance of secondary effects. Cell cultures also provide unlimited biological material for the diagnostic of alterations produced by additives [1] [2]. Cell lines are study models for molecular effects' con cell function [3]. In vertebrates, especially mammals, they have been amply utilized for biotechnology development, such as the CHO cells [4]. Although it is novel in aquaculture, the passage from information at a cellular level to complex models common in mammals [3]. The most developed compound search is in frog (*X. laevis*) oocytes, where endogenous receptors are used to understand the pharmacology of nicotine receptors [5]. For this same reason, understanding additive effects at a compound cellular level in fish models is a good option to improve the comprehension of the physiology of them, and avoid toxicity through chronic applications [6] [7]. Silymarin is a complex flavonoligan of *Silybum marianum* (L.), commonly called thistle, and that comprehends a large number of flavonoligans including Silybin (silybin A and B) isosilybin A and B, silychristin A and B, silydianin and other phenolic compounds [8]. It is amply used in the treatment of hepatic diseases in vertebrates [9] which reduces toxic effect on liver [10] and has effect over hepatic system in fish [11]. Literature suggests that the antioxidant properties of silymarin contribute to its pharmacological properties [12]. But this is not its only exclusive mechanism [13] [14]. Apart from antioxidant properties, it has anti-lipid peroxidative, anti-fibrotic, anti-inflammatory, membrane stabilizing, immunomodulation, as well as anti-tumoral properties, and shows anti-arteriosclerotic and anti-diabetic activity [15] [16]. In fish, the use of silymarin as an antioxidant is not widely studied [17], and it is used in diets as an additive that boosts development and growth [18]. Its use in diets is diverse; it can be used in the farming of freshwater fish [19], or in the fattening of saltwater fish [18]. It has been used as an immunomodulator [15] or a protectant against agents such as curcumin [17], and can regulate function in fish glial cells [20]. Its action mechanism in fish is unknown, and is identified as an additive that is absorbed as a flavonoid through the intestinal wall. Once in the bloodstream, it plays a role in improving cellular nutritional conditions. This could be because of undescribed antioxidant properties, or due to being a metal-chelating agent, improving cell function and development in fishes [15] [17].

In this study, SHK-1 derived cells from the head and kidney of Atlantic salmon were treated in increasing time and concentrations of silymarin. Then, these samples were analyzed by MTT for the number of cells in response to proliferation and their viability. Also, models of oxidative stress were used to measure the protective capacity of silymarin against cells exposed to high levels of glucose.

2. Methods

2.1. Cellular Cultures and Cytotoxic Studies

Cell lines derived from leucocytes from Salmo salar SHK-1 (ECACC N°97111106), were maintained in an incubator at 17°C. Cells were cultivated in a Leibowitz L-15 medium and supplemented with 10% fetal bovine serum, glutamine 1%, penicillin/streptomycin 1% and 2-mercaptoetanol 72 µl/ml. The cellular expansion procedures were carried out in a biosafety cabinet and then are seed in plates of 24 wells for parallel experiment. Cultures were exposed to temporal curves (one to seven days) and concentrations (0.1 to 1000 ppm) of generic silymarin to evaluate its toxicity, staining and morphology were analyzed to explore the reaction to the additive. Modifying [21], the effect was added of high glucose 30 mM at for 24 hours to generate free radicals, and for seven days to generate an oxidative state. Also, 100 µM de H_2O_2 was used for 24 hours and 7 days to observe the direct effect of free radicals and the antioxidant action of the additives.

2.2. Staining

Cells grown in culture plates were exposed to temporal curves and concentrations of the additives. Cells were then washed of the culture medium and incubated with eosin 0.3%, glacial acetic acid and distilled water for one minute. Then, then solution was removed and a phosphate buffer was added. One hundred cells were counted randomly, and those that presented a reddish-orange tone were considered dead, and this percentage was recorded. The methodology was based on [22] and modified.

2.3. Cellular Proliferation

Cells were washed with PBS and removed with a trypsin solution 1%, later they were placed in complete culture medium to inhibit the trypsin. Samples were collected in sterile 15 ml tubes and were centrifuged at 1200 rpm for 10 minutes. Cells in the bottom were suspended in 1 ml of base medium and a sample was taken to be recounted in a haemocytometer diluted 1/100. With this value, samples were seeded in at a density of 100,000 cells per ml, in culture with 1.6 mm diameter, with 250 ul of medium. This culture was exposed for 24 hours to the additives and the number of cells was recounted by area. The density of cells was recorded as an indicator of the change in cellular proliferation or growth rate, modified from [23].

2.4. May-Grunwald and Giemsa Staining

Cells grown in culture plates were exposed to temporal curves and concentration of additives were washed of culture medium and incubated with a PanOptic kit for May-Grunwald and Giemsa staining (Quimica Clinica Aplicada, Spain). Samples were left for 5 seconds in reactive 1, 10 seconds in May-Grunwald reactive and 10 seconds in Giemsa reactive. The plates were washed with ultrapure water and left to be microphotographed with a NIKON Labophot 2 optic microscope and a 519CU 5.OM CMOS camera, modified from [24].

2.5. Statistical Analysis

Unless specified, all results including image analysis are presented using the average $\pm$ SEM, we used prism 5 software for the analysis. Statistical comparisons were carried out using ANOVA two way, indicated in the figure legend and Bonferroni test, are used after the analysis (post-test). The sample are exposed to normality test, Shapiro-Wilk, for see how for the distribution is from Gaussian. A probability p less than 0.05 is considered statistically significant.

3. Results

3.1. Silymarin Promotes Cell Proliferation in a Concentration-Response Manner

The cells are seeded and the beginning of the experitmen observed at time = 0 (initial condtion) that the number of cells per area upon adding 150 ppm of silymarin. **Figure 1(a)** shows an example image of the control condition and the cultures treated with 150 ppm silymarin from time = 0 and at 5 days of exposure, showing an increased number of cells per area in the exposed sample. **Figure 1(b)** represents a quantification of cell numbers per area at different times, and an increase in cell number can be observed in the treated sample. **Figure 1(c)** shows the quantified effect of 150 ppm of silymarin. **Figure 1(d)** shows that there was no deleterious effect on the culture upon being acutely exposed. These data suggest that a dose of 150 ppm does not acutely affect the culture, but it increases chronic cell proliferation.

3.2. Concentration Dependent Effect and Toxic Model

We evaluated the concentration-dependent effect for silymarin. For this, we made proliferation curves with different doses of silymarin (0.1 - 1000 ppm) and we observed the number of cells at the end of the proliferative curve. We observed at 5 days had the maximum proliferation effect and we used the number of the cells at these time for see the effect of the different concentration of the silymarin in the proliferation. In **Figure 2(a)**, a sigmoidal curve is presented of the effect, where 1000 ppm of silymarin is the concentration to generate the median cellular proliferation effect. In mammals, it is described that an increase in metabolism generates oxidative stress conditions, and a manner to induce these states is by using glucose. We did not know what the effect of glucose was on the SHK-1 model, being a culture that is developed at a lower temperature. For this purpose we measured the viability of the culture using MTT at different concentrations of glucose on the external medium for 24 hours. **Figure 2(b)** shows the development of a curve, indicating that concentrations of 30 mM of glucose generates a significant decrease in the viability of the culture, and can bused as a cellular injury model for oxidative stress.

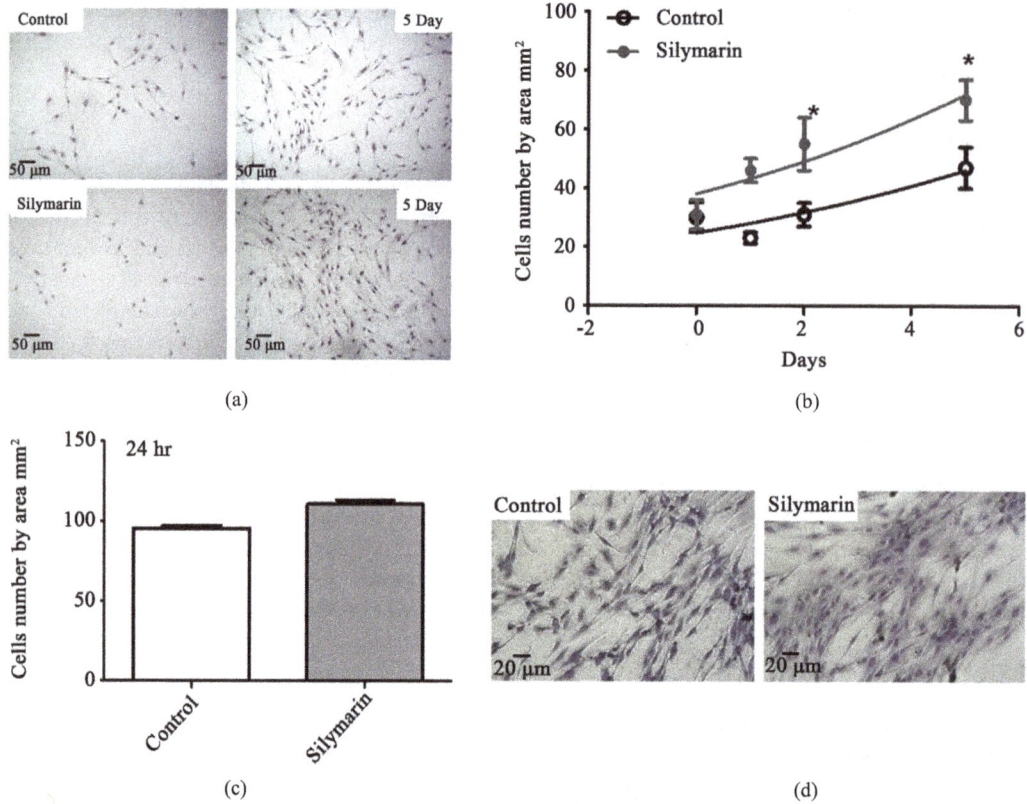

(a)

(b)

(c)

(d)

Figure 1. Effect on cellular proliferation. In (a), example of microphotographs in control, and with silymarin 150 ppm at times 0 and 5 days; (b) quantification of the number of cells after 5 days of culture in absence or presence of silymarin; (c) shows the quantification of the number of cells of the culture exposed to 150 ppm of silymarin after 24 hours; (d) is representative of the quantification at 24 hours of treatment with silymarin 150 ppm. The microphotographs are representative of the 5 independent observations. Each bar or point represents (mean ± SEM), la measurements of at least 5 independent culture and experiments. The asterisk indicates $p < 0.05$ (ANOVA).

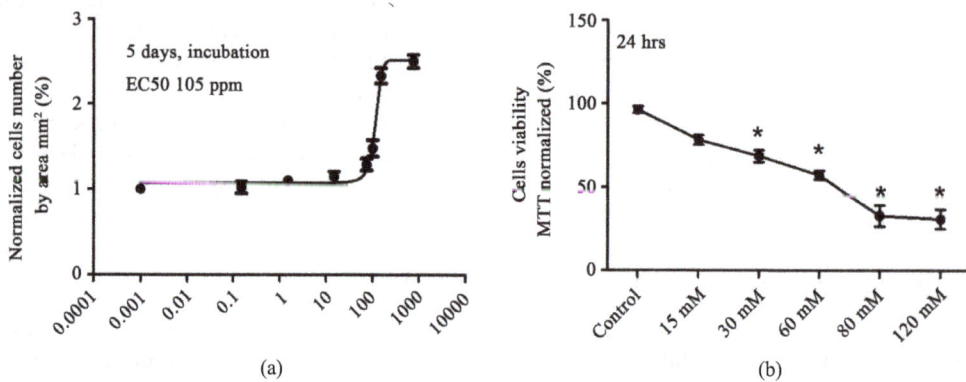

(a)

(b)

Figure 2. Doses concentration response. In (a), doses concentration curve, proliferation at 5 days of silymarin 0.1 to 1000 ppm. An EC of 105 ppm is observed for proliferation; (b) shows cellular viability of cell cultures exposed to increasing concentrations of glucose, 15 mM to 120 mM, for doses of toxicity at 24 hours of treatment. Each point represents (mean ± SEM), the measurement of at least 3 independent culture and experiments. The asterisk indicates $p < 0.05$ (ANOVA).

3.3. Silymarin Reduces the Effects of High Glucose

Observing that glucose at 30 mM alters the viability of the cultures, we tested its effect when it was maintained in the presence of 100 ppm silymarin. **Figure 3(a)** shows samples that were exposed for seven days to high glucose, in the absence and presence of silymarin. **Figure 3(b)** shows the quantification of the number of cells ex-

(a)

(b)

(c)

(d)

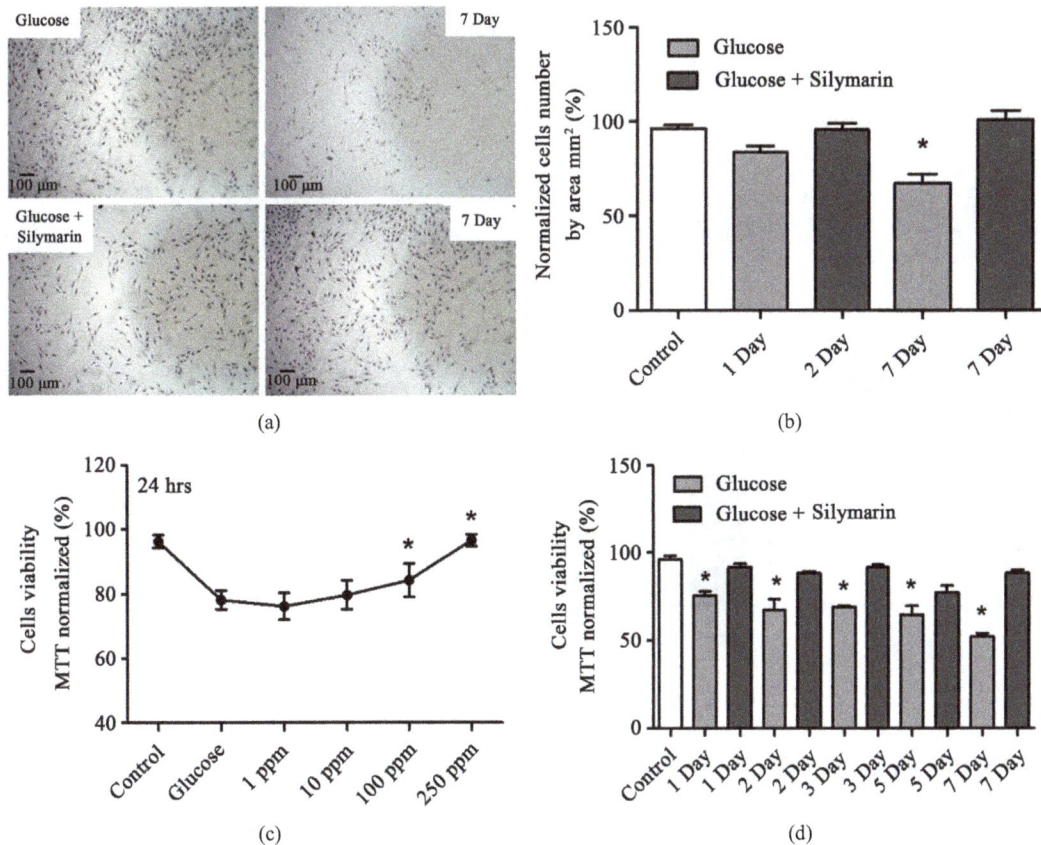

Figure 3. Cellular protect effect. In (a), example of microphotography H_2O_2 0.1 mM in the presence or absence of silymarin 100 ppm, 7 days of incubation; (b) demonstrates the quantification of the cells after 7 days of culture in H_2O_2 0.1 mM in the absence or presence of silymarin 100 ppm; (c) gives the effect of H_2O_2 0.1 mM after 24 hours on cell viability in the absence or presence of increasing concentrations of silymarin. Microphotography is representative of 6 independent observations made in 6 different cultures. Each bar or point represents (mean ± SEM), the measurement of at least 6 independent culture and experiments. The asterisk indicates $p < 0.05$ (ANOVA).

posed to glucose in absence of precense of silymarin. We observed a reduction on the toxic glucose effect, when exposed to 100 ppm of silymarin. With the objective of determining if the effect is concentration dependent, cultures exposed to 30 mM of glucose for 24 hours, were incubated in incrementing concentrations of silymarin (1 to 250 ppm). As is shown in **Figure 3(c)**, we observed that very low concentrations of silymarin did not revert the deleterious effects of the glucose, and that values over 100 ppm protect the culture. Upon using the concentration in a chronic form, and evaluating viability by MTT, as is shown in **Figure 3(d)**, we observed that 100 ppm is sufficient to significantly reduce the effect of 30 mM of glucose on the cells. These data suggest that silymarin is capable of reducing an environment of cellular stress generated by high glucose.

3.4. Silymarin Reduces the Effects of Hydrogen Peroxide

The metabolic effect, associated to a culture with high glucose, is possible to associate to generation of free radicals that alter the REDOX of the cell. To confirm that silymarin can participate in reduces this effect, we used H_2O_2 directly on the culture, to generate an intense toxic effect, suggesting an oxidative stress state. **Figure 4(a)** shows images of the cultures exposed to H_2O_2 0.1 mM in the absence and presence of silymarin 100 ppm. **Figure 4(b)** shows a quantification of the chronic effect of the H_2O_2 on the culture in the absence and presence of silymarin, observing that after 7 days of co-incubation, the number of cells increased with respect to the culture exposed to H_2O_2 in a solitary form, although still less than the control. Upon evaluating the viability of the culture by MTT in acute applications, we observed that silymarin reduces the effects of H_2O_2 in a concentration dependent form (**Figure 4(c)**). Concentrations less than 100 ppm did not significantly reduce the mortality of the

(a)

(b)

(c)

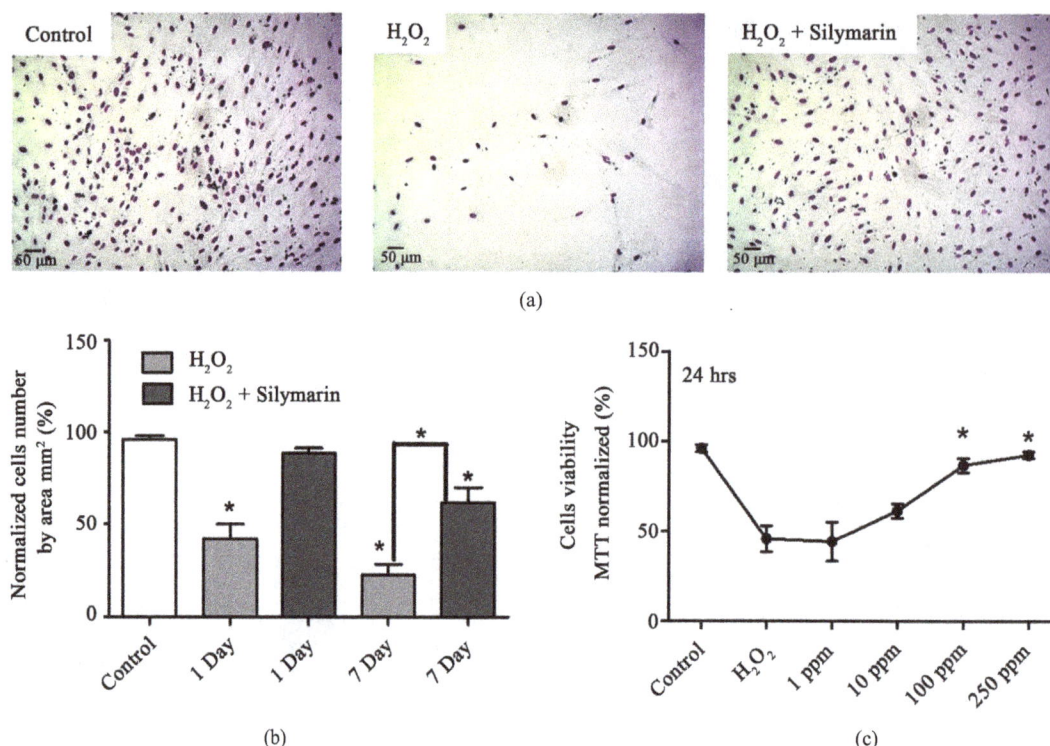

Figure 4. Cellular protect effect. In (a) example of microphotography H_2O_2 0.1 mM in the presence or absence of silymarin 100 ppm, 7 days of incubation; (b) demonstrates the quantification of the cells after 7 days of culture in H_2O_2 0.1 mM in the absence or presence of silymarin 100 ppm; (c) gives the effect of H_2O_2 0.1 mM after 24 hours on cell viability in the absence or presence of increasing concentrations of silymarin. Microphotography is representative of 6 independent observations made in 6 different cultures. Each bar or point represents (mean ± SEM), the measurement of at least 6 independent culture and experiments. The asterisk indicates $p < 0.05$ (ANOVA).

culture. These date suggest that silymarin acts as protect, reducing the effect of H_2O_2 on the cultures in concentrations superior to 100 ppm.

4. Discussion

The use of cells for the study of additives is a widely used tool in mammalian studies [3]; recently, this type of testing has started to be used in fish, as a way to evaluate procedures [6] and obtain functional information about additives [15]. Our results were obtained using SHK-1 cells exposed to silymarin, an additive used in the salmon industry.

We found that silymarin did not alter cellular viability, and promoted proliferation in this cell line (**Figure 1**), as well as proliferation in a dose response manner with a value of EC50 of 105 ppm (**Figure 2(a)**). Silymarin is described as being a good antioxidant in other cellular model [12] and can complete this function in fish [17]. We exposed the effects of silymarin on cultures exposed to high glucose, for made simulation of high oxidative stress (**Figure 2(b)**), which has been previous described in other cell model [25]. We found that the effects of glucose were reduced when the culture was exposed to 100 ppm silymarin (**Figure 3(b)**), and this effect in dependent concentration shows that values of 100 ppm of silymarin protect the culture.

Our objective was to determine the direct effect of free radicals on SHK-1 cultures, and if silymarin can reduce their toxicity. So, we exposed the cultures to 0.1 nM of H_2O_2 in the presence and absence of 100 ppm silymarin (**Figure 4(b)**). We observed that the silymarin reduced the mortality of the culture, suggesting a protective effect on oxidative stress conditions. We found that this effect is also concentration dependent (**Figure 4(c)**), as observed with glucose; values less than 100 ppm did not reduce the effect of H_2O_2, increasing mortality in these conditions.

Our findings indicate that concentrations of 100 ppm of silymarin are sufficient to promote cellular division in SHK-1 cells and significantly reduce the toxic effects in cellular injury model. This suggests that in these conditions, silymarin is an additive which has functions as a cell protector. These results are applicable to the formulation of fish diets, and further studies should focus on the bioavailability of silymarin and the effect it could have on aquaculture.

Acknowledgements

Financed by the Convenio de Asistencia Técnicas UCT 278-2450 Tonalli 2013. Jorge Parodi received the grant MECESUP UCT 0804. We thank the business SalmoFood-VitaPro, for donating the additives for testing. Rodrigo Sánchez, postgraduate student at the UCT received financing as part of his thesis from SalmoFood-VitaPro.

References

[1] Wuest, D.M., Harcum, S.W. and Lee, K.H. (2012) Genomics in Mammalian Cell Culture Bioprocessing. *Biotechnology Advances*, **30**, 629-638. http://dx.doi.org/10.1016/j.biotechadv.2011.10.010

[2] Blagodatski, A. and Katanaev, V.L. (2011) Technologies of Directed Protein Evolution *in Vivo*. *Cellular and Molecular Life Sciences*, **68**, 1207-1214. http://dx.doi.org/10.1007/s00018-010-0610-5

[3] Majors, B.S., Chiang, G.G. and Betenbaugh, M.J. (2009) Protein and Genome Evolution in Mammalian Cells for Biotechnology Applications. *Molecular Biotechnology*, **42**, 216-223. http://dx.doi.org/10.1007/s12033-009-9156-x

[4] Kim, J.Y., Kim, Y.G. and Lee, G.M. (2012) CHO Cells in Biotechnology for Production of Recombinant Proteins: Current State and Further Potential. *Applied Microbiology and Biotechnology*, **93**, 917-930. http://dx.doi.org/10.1007/s00253-011-3758-5

[5] Garcia-Colunga, J. and Miledi, R. (1994) Serotoninergic Agents Block Neuronal Nicotinic Receptors Expressed in Oocytes. *FASEB Journal*, **8**, A107-A107.

[6] Rolland, J.B., Bouchard, D., Coll, J. and Winton, J.R. (2005) Combined Use of the ASK and SHK-1 Cell Lines to Enhance the Detection of Infectious Salmon Anemia Virus. *Journal of Veterinary Diagnostic Investigation*, **17**, 151-157. http://dx.doi.org/10.1177/104063870501700209

[7] Yamaha, E., Saito, T., Goto-Kazeto, R. and Arai, K. (2007) Developmental Biotechnology for Aquaculture, with Special Reference to Surrogate Production in Teleost Fishes. *Journal of Sea Research*, **58**, 8-22. http://dx.doi.org/10.1016/j.seares.2007.02.003

[8] Wu, J.W., Lin, L.C. and Tsai, T.H. (2009) Drug-Drug Interactions of Silymarin on the Perspective of Pharmacokinetics. *Journal of Ethnopharmacology*, **121**, 185-193. http://dx.doi.org/10.1016/j.jep.2008.10.036

[9] Abenavoli, L., Capasso, R., Milic, N. and Capasso, F. (2010) Milk Thistle in Liver Diseases: Past, Present, Future. *Phytotherapy Research*, **24**, 1423-1432. http://dx.doi.org/10.1002/ptr.3207

[10] Yurtcu, E., Iseri, O.D. and Sahin, F.I. (2014) Genotoxic and Cytotoxic Effects of Doxorubicin and Silymarin on Human Hepatocellular Carcinoma Cells. *Human and Experimental Toxicology*, **33**, 1269-1276. http://dx.doi.org/10.1177/0960327114529453

[11] Suh, H.J., Cho, S.Y., Kim, E.Y. and Choi, H.S. (2015) Blockade of Lipid Accumulation by Silibinin in Adipocytes and Zebrafish. *Chemico-Biological Interactions*, **227**, 53-62. http://dx.doi.org/10.1016/j.cbi.2014.12.027

[12] Sherif, I.O. and Al-Gayyar, M.M.H. (2013) Antioxidant, Anti-Inflammatory and Hepatoprotective Effects of Silymarin on Hepatic Dysfunction Induced by Sodium Nitrite. *European Cytokine Network*, **24**, 114-121.

[13] Khazanov, V.A. and Vengerovsky, A.I. (2007) Effect of Silimarin, Succinic Acid, and Their Combination on Bioenergetics of the Brain in Experimental Encephalopathy. *Bulletin of Experimental Biology and Medicine*, **144**, 806-809. http://dx.doi.org/10.1007/s10517-007-0436-9

[14] Razavizadeh, M. and Arj, A. (2012) Comparison of the Therapeutic Effects of Vitamin E and Silymarin in Nonalcoholic Steatohepatitis. *Journal of Gastroenterology and Hepatology*, **27**, 270.

[15] Ahmadi, K., Banaee, M., Vosoghei, A.R., Mirvaghefi, A.R. and Ataeimehr, B. (2012) Evaluation of the Immunomodulatory Effects of Silymarin Extract (*Silybum marianum*) on Some Immune Parameters of Rainbow Trout, *Oncorhynchus mykiss* (Actinopterygii: Salmoniformes: Salmonidae). *Acta Ichthyologica et Piscatoria*, **42**, 113-120. http://dx.doi.org/10.3750/AIP2011.42.2.04

[16] Skottova, N., Vecera, R., Urbanek, K., Vana, P., Walterova, D. and Cvak, L. (2003) Effects of Polyphenolic Fraction of Silymarin on Lipoprotein Profile in Rats Fed Cholesterol-Rich Diets. *Pharmacological Research*, **47**, 17-26. http://dx.doi.org/10.1016/S1043-6618(02)00252-9

[17] Shiau, R.J., Shih, P.C. and Wen, Y.D. (2011) Effect of Silymarin on Curcumin-Induced Mortality in Zebrafish (*Danio rerio*) Embryos and Larvae. *Indian Journal of Experimental Biology*, **49**, 491-497.

[18] Banaee, M., Sureda, A., Mirvaghefi, A.R. and Rafei, G.R. (2011) Effects of Long-Term Silymarin Oral Supplementation on the Blood Biochemical Profile of Rainbow Trout (*Oncorhynchus mykiss*). *Fish Physiology and Biochemistry*, **37**, 885-896. http://dx.doi.org/10.1007/s10695-011-9486-z

[19] Jia, R., Cao, L.P., Du, J.L., Xu, P., Jeney, G. and Yin, G.J. (2013) The Protective Effect of Silymarin on the Carbon Tetrachloride (CCl₄)-Induced Liver Injury in Common Carp (*Cyprinus carpio*). *In Vitro Cellular & Developmental Biology—Animal*, **49**, 155-161.

[20] Malekinejad, H., Taheri-Brujerdi, M., Janbaz-Acyabar, H. and Amniattalab, A. (2012) Silymarin Regulates HIF-1 Alpha and iNOS Expression in the Brain and Gills of Hypoxic-Reoxygenated Rainbow Trout *Oncorhynchus mykiss*. *Aquatic Biology*, **15**, 261-273. http://dx.doi.org/10.3354/ab00427

[21] Jofre, I., Gomez, P.N., Parodi, J., Romero, F. and Salazar, R. (2013) Chilean Crude Extract of *Ruta graveolens* Generates Vasodilatation in Rat Aorta at Subtoxic Cellular Concentrations. *Advances in Bioscience and Biotechnology*, **4**, 8.

[22] Navarrete, G.P., Alvarez, J.G., Parodi, J., Romero, F. and Sánchez, R. (2012) Effect of Aracnotoxin from *Latrodectus mactans* on Bovine Sperm Function: Modulatory Action of Bovine Oviduct Cells and Their Secretions. *Andrologia*, **44**, 764-771.

[23] Parodi, J., Flores, C., Aguayo, C., Rudolph, M.I., Casanello, P. and Sobrevia, L. (2002) Inhibition of Nitrobenzylthioinosine-Sensitive Adenosine Transport by Elevated D-Glucose Involves Activation of P2Y2 Purinoceptors in Human Umbilical Vein Endothelial Cells. *Circulation Research*, **90**, 570-577. http://dx.doi.org/10.1161/01.RES.0000012582.11979.8B

[24] Flores, C., Rojas, S., Aguayo, C., Parodi, J., Mann, G., Pearson, J.D., Casanello, P. and Sobrevia, L. (2003) Rapid Stimulation of L-Arginine Transport by D-Glucose Involves p42/44(Mapk) and Nitric Oxide in Human Umbilical Vein Endothelium. *Circulation Research*, **92**, 64-72. http://dx.doi.org/10.1161/01.RES.0000048197.78764.D6

[25] Robertson, R.P., Harmon, J., Tran, P.O., Tanaka, Y. and Takahashi, H. (2003) Glucose Toxicity in Beta-Cells: Type 2 Diabetes, Good Radicals Gone Bad, and the Glutathione Connection. *Diabetes*, **52**, 581-587. http://dx.doi.org/10.2337/diabetes.52.3.581

13

Molecular Cloning of a Chitinase Gene from the Ovotestis of Kuroda's Sea Hare *Aplysia kurodai*

Gaku Matsunaga, Syuuji Karasuda, Ryo Nishino, Hideto Fukushima, Masahiro Matsumiya

Department of Marine Science and Resources, College of Bioresource Sciences, Nihon University, Kanagawa, Japan
Email: matsumiya@brs.nihon-u.ac.jp

Abstract

In this study, we report that we successfully cloned and sequenced a chitinase gene from the ovotestis of Kuroda's sea hare *Aplysia kurodai*. By using reverse transcription-polymerase chain reaction (RT-PCR) and a system for the 5' and 3' rapid amplification of cDNA ends, we obtained a 1352 bp chitinase gene (*AkChi*) from the ovotestis of *A. kurodai*. *AkChi* contains a 1263 bp open reading frame that encodes 421 amino acids. The domain structure predicted from the deduced amino acid sequence was an N-terminal signal peptide and a catalytic domain of glycoside hydrolase (GH) family 18 chitinase. A comparative analysis of the deduced amino acid sequences of *AkChi* with those of the acidic mammalian chitinase of the California sea hare *Aplysia californica* revealed the highest homology at 83%. The purified chitinase from the ovotestis was digested by trypsin, and 119 residues of digested peptides were consistent with the deduced amino acid sequence of *AkChi*. We used RT-PCR to evaluate the expression of *AkChi* in various tissues of *A. kurodai*, and we observed that *AkChi* was expressed only in the ovotestis. A phylogenetic tree analysis, performed using the amino acid sequences of *AkChi* and known GH family 18 chitinases, showed that *AkChi* was separated from the molluscan chitinases with a chitin binding domain. To our knowledge, this is the first study demonstrating the cDNA cloning of an ovotestis chitinase from a sea hare.

Keywords

Chitinase, Molecular Cloning, Kuroda's Sea Hare *Aplysia kurodai*, Mollusc, Ovotestis, Phylogenetic Tree Analysis

1. Introduction

Chitin, a major molecular constituent of the exoskeleton of insects and crustaceans, is a straight-chain homopolymer of β-1,4-linked N-acetyl-D-glucosamine units [1]-[3]. Chitinases (EC 3.2.1.14) are enzymes that randomly hydrolyze the β-1,4 glycosidic bonds of chitin [4]. They have been found in various organisms, and they play important physiological roles in functions such as attack, defense, morphological changes, and digestion [5] [6].

The characterization and cDNA cloning of chitinases from several fishes have been reported [7]-[9]. The stomach chitinases of fish have been identified and are classified into two groups, acidic fish chitinase-1 (AF-Case-1) and acidic fish chitinase-2 (AFCase-2) based on the differences in their primary structure and the activity toward short substrates [8]. Chitinases from molluscs play important physiological roles in the digestion of food [10] [11], attacking crustaceans [12], and shell formation [13] [14]. However, reports on the distribution, characterization, and cDNA cloning of molluscan chitinases are limited [10]-[16]. In this study, we were using the Kuroda's sea hare, *Aplysia kurodai*. *A. kurodai* is a kind of herbivorous gastropoda seen in the vicinity of the coast from April to June. In addition, this creature was allowed to degenerate shells despite the shellfish. In a previous study, we detected chitinase activity in the ovotestis and egg of *A. kurodai* [16], whereas lysozyme activity (antibacterial enzyme activity) was not detected in all of the organs [16]. We also reported the purification and properties of a chitinase from the ovotestis of *A. kurodai* [16]. Together the results indicated that the physiological role of this chitinase was as a defense against nematodes and fungus which had chitin in the body wall as a structural component [16].

In the present study, we cloned the cDNA encoding chitinase from the ovotestis of *A. kurodai* and determined the primary structure of the chitinase.

2. Materials and Methods

2.1. Materials

Kuroda's sea hare *Aplysia kurodai* and laid egg were captured from the tide pools of Shimoda Bay (Shizuoka, Japan) in June.

2.2. Cloning of the Chitinase cDNA from *A. Kurodai*

The sequences of all primers are presented in **Table 1**. Total RNA was extracted from the ovotestis of *A. kurodai* using ISOGEN II reagent (Nippon Gene, Tokyo) according to the manufacturer's instructions. First-strand cDNA was synthesized using 500 ng of total RNA and oligo dT primers with Prime Script Reverse Transcriptase (Takara Bio, Shiga, Japan) according to the manufacturer's instructions. Six degenerate primers were designed for the reverse transcriptase-polymerase chain reaction (RT-PCR) from conserved sequences of molluscan chitinase, including those from California sea hare (*Aplysia californica*; GenBank: XM_005112601), triangle sail mussel (*Hyriopsis cumingii*; GenBank: JN582038), Pacific oyster (*Crassostrea gigas*; GenBank: AJ971239), Hawaiian bobtail squid (*Euprymna scolopes*; GenBank: KF015222), and golden cuttlefish (*Sepia esculenta*; GenBank: AB986212).

The first PCR was performed using *A. kurodai* cDNA as a template and P1 and P2 as primers (**Figure 1**). The PCR parameters were as follows: 94°C for 2 min, followed by 30 cycles of 94°C for 30 s, 55°C for 30 s, and 72°C for 30 s. Nested PCR was performed using the products of the first PCR as templates and P3, P4, P5, and P6 as primers, with the same PCR parameters as described above. The nucleotide sequence analysis of the RT-PCR amplified chitinase cDNA fragments from the ovotestis of *A. kurodai* detected one nucleotide sequence (*AkChi*).

For the 3' rapid amplification of cDNA ends (RACE), we designed primers specific to *AkChi* (*i.e.*, P7, P8, and P9, respectively; **Table 1**) based on the detected sequences. We amplified cDNA fragments encoding the 3' region of *AkChi* using *A. kurodai* cDNA as the template and the primer pairs P7 and 3R, P8 and 3R, and P9 and 3R (**Figure 1**). The PCR parameters were as follows: 94°C for 2 min, followed by 30 cycles of 94°C for 30 s, 56°C for 30 s, and 72°C for 30 s. For 5' RACE, specific primers (P10, P11, and P12 for *AkChi*; **Table 1**) were designed based on the nucleotide sequences obtained from RT-PCR. cDNA fragments encoding the 5' regions of *AkChi* were amplified using PCR. The first PCR was performed using the newly synthesized first-strand cDNA as a template and the primer pairs P10 and P11 for *AkChi*. Nested PCR was performed using the first PCR products as templates and the primer pairs P10 and P12 for *AkChi*. The PCR parameters were as follows:

Table 1. Primers used for PCR, RACE, and tissue-specific expression.

Primer	Sequence (5' → 3')	Purpose
P1*	TNGCNGCNTTYGARTGGAAYGA	Primary PCR
P2*	CATNCCNSWRAARTCRTCRTTRTC	Primary PCR
P3*	GGNGGNTGGAAYATGGG	Primary PCR
P4*	ACCCAYTGRTTNCCNARNACNA	Primary PCR
P5*	GNAAYTTYGAYGGNYTNGA	Primary PCR
P6*	TTDATCATYTCRCANACYTCRTARTA	Primary PCR
P7	GCCGGATACGAAGTGGAC	3' RACE
P8	GGAACTTAACGAGTACTT	3' RACE
P9	GACAGACGAGAGCGACTCTGGTCG	3' RACE
3R	CTGTGAATGCTGCGACTACGAT	3' RACE
P10	CACAATGACGTTGCAAG	5' RACE, Full-length PCR
P11	ATGGCCTGGGCTCATTTT	5' RACE
P12	TTATCCTCTGGAGGGCT	5' RACE
P13	CACGTTATGATTGCGAC	Full-length PCR
P14	TCTGCTGCTGTGAGTGCTGGCAAGG	tissue-specific expression
P15	GCATTTCGCACACCTCGTAGTAAGA	tissue-specific expression
β-actin-a*	GAYAAYGGNWSNGGNATGTG	tissue-specific expression
β-actin-b*	TCRAACATDATYTGNGTCAT	tissue-specific expression

Note: *Degenerate primers.

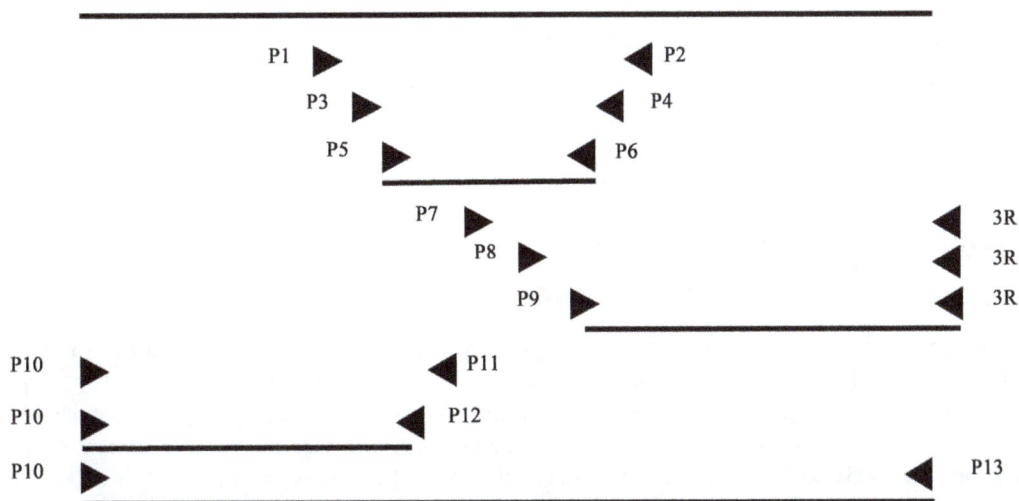

Figure 1. Schematic representation of the cDNA structure of *AkChi* and location of the primers. Arrowheads indicate the primers, and lines between the arrowheads indicate the amplified cDNA fragments.

94°C for 1 min, followed by 30 cycles of 94°C for 30 s, 49°C for 30 s, and 72°C for 30 s.

The nucleotide sequences of cDNA fragments containing a full-length open reading frame (ORF) were confirmed by PCR using specific primers (P10 and P13 for *AkChi*; **Table 1**) and Platinum *Pfx* DNA Polymerase (Invitrogen, Carlsbad, CA).

2.3. Nucleotide Sequence Analysis

The RT-PCR, 3' RACE, and 5' RACE amplification products, and the full-length amplification products were subcloned into pGEM-T Easy Vector (Promega, Madison, WI), according to the manufacturer's instructions.

Sequences were determined on an ABI PRISM 3130 genetic analyzer (Applied Biosystems, Foster City, CA) using a Big Dye Terminator v3.1 cycle sequencing kit (Applied Biosystems).

2.4. Amino Acid Sequence of the Peptide of the Purified Chitinase from the Ovotestis of *A. kurodai*

A chitinase from the ovotestis of *A. kurodai* was purified as described [16]. The purified chitinase was subjected to sodium dodecyl sulfate-polyacrylamide gel electrophoresis (SDS-PAGE) and stained with AE-1360 EzStain Silver (ATTO, Tokyo). A gel slice was cut into small pieces and destained by destaining solution (15 mM $K_3[Fe(CN)_6]$, 50 mM $Na_2S_2O_3$). Destained gel pieces were trypsinized as described in the manual of In-Gel Tryptic Digestion Kit manual (Thermo Scientific, Waltham, MA). The peptide mixtures thus obtained were subjected to a nano-scale liquid chromatography-electrospray ionization-tandem mass spectrometry (nanoLC-ESI-MS/MS) analysis using a Q Exactive mass spectrometer (Thermo Scientific) equipped with a captive spray ionization source (Michrom Bioresources, Auburn, CA) and an Advance UHPLC System (Michrom Bioresources).

2.5. Tissue-Specific Expression of *AkChi*

Total RNA was prepared from the ovotestis, egg, skin, gill, crop, anterior gizzard, and posterior gizzard as described in the cloning methods section (2.2) above. First-strand cDNA was pre-cloned from the RNA isolated from each tissue and egg as described in the RT-PCR section (2.2) above. For tissue-specific expression, we designed primers specific to *AkChi* (P14 and P15, respectively; **Table 1**) based on the detected sequences. *AkChi* was amplified using the first-strand cDNA as template and the primer pairs P14 and P15 (**Table 1**). The PCR parameters were as follows: 94°C for 1 min, followed by 35 cycles of 94°C for 30 s, 62°C for 30 s, and 72°C for 30 s. To determine the amount of total RNA in each tissue, we amplified β-actin mRNA fragments using specific primer pairs (**Table 1**).

2.6. Phylogenetic Tree Analysis of *AkChi*

In order to classify the chitinase from the ovotestis of *A. kurodai* among the GH family 18 chitinases, we constructed a phylogenetic tree based on the enzyme precursor sequences by the neighbor-joining method, using the ClustalW program (http://www.genome.jp/tools/clustalw/). A bacterial chitinase (GenBank: X03657) was used as the out group.

3. Results and Discussion

3.1. Cloning of *A. kurodai* Chitinase cDNA

The structure of *AkChi* and the location of primer sequences are schematically represented in **Figure 1**. The internal sequence of the cDNA of *A. kurodai* ovotestis chitinase was amplified by RT-PCR using degenerate primers (from P1 to P6, respectively; **Table 1**); an amplified product of approx. 400 bp was obtained. The product was sequenced, and 86% homology with the acidic mammalian chitinase of *A. californica* was confirmed (accession no. XM_005112601). Because the sequence was part of ovotestis chitinase cDNA from *A. kurodai*, we used it to design gene-specific primers for 3' and 5' RACE (from P7 to P12; **Table 1**). An amplified product of approx. 430 bp was obtained by 3' RACE, and its sequence contained a stop codon. An amplified product of approx. 520 bp was also obtained by 5' RACE; its sequence contained a start codon. Based on these results, we designed full-length primers (P10 and P13; **Table 1**) to incorporate these start and stop codons. cDNA was amplified using the primers and the amplified product was sequenced.

The full-length cDNA of *A. kurodai* ovotestis chitinase (*AkChi*) was 1352 bp in length and contained an ORF of 1263 bp encoding 421 amino acids (**Figure 2**). The size of ORF of *AkChi* was smaller than it from *H. cumingii* [14], 1962 bp encoding 653 amino acids. A poly-A sequence in eukaryotes was detected at the 3' end of *AkChi*. *AkChi*, which encodes *A. kurodai* ovotestis chitinase, has been registered in the database of the DNA Data Bank of Japan (DDBJ) (accession no. LC085435). We compared the deduced amino acid sequence of *AkChi* with that of other organisms using BLAST, and the highest homology, 83%, was confirmed with the acidic mammalian chitinase of *A. californica* (accession no. XM_005112601). **Figure 3** compares amino acid

```
CACA                                                                                                              4

ATGACGTTGCAAGTTTCAGCTCTCTGCGTTCTGCTTGGGATTGCTCTAGCGGTGTGTCAAGGAGGTGCGTTCCATGGGTGTGCACAAAACCGACTCCAAACCCGTAACGGGCGCAGG  121

M  T  L  Q  V  S  A  L  C  V  L  L  G  I  A  L  A  V  C  Q  A  G  A  F  H  G  C  A  Q  N  R  L  Q  T  R  N  G  R  R

GGTTCTGATGATGAGAAAAACCAGCTTGTGTGCTACTACACGAACTGGGCCCAATACCGGCCGGGCAAAGGCGGCGTTTTTCCCCGAGGACATAGACGCGCAACTTGTGCACTCACATT  238

G  S  D  D  E  K  N  Q  L  V  C  Y  Y  T  N  W  A  Q  Y  R  P  G  K  G  A  F  F  P  E  D  I  D  A  N  L  C  T  H  I

CATTACGCCTTTGCCATTCTAGTGGACGGTCTTCTGGCTCCCTTCGAGTGGAATGATGACGACACGGAGTGGTCGGAGGGAATGTACACTCGTGTGAACAAACTAAAGGAGGATAAC  355

H  Y  A  F  A  I  L  V  D  G  L  L  A  P  F  E  W  N  D  D  D  T  E  W  S  E  G  M  Y  T  R  V  N  K  L  K  E  D  N

CCTGCTCTAAAGACTATGCTGTCTCTGGGAGGCTGGAACATGGGAACCCAAAACTGGACTCTGATGGTGAAAGATGAGTCGTCCAGACAGAAATTCATTCAGAACGCCATCCCCTTT  472

P  A  L  K  T  M  L  S  L  G  G  W  N  M  G  T  Q  N  W  T  L  M  V  K  D  E  S  S  R  Q  K  F̲ ̲I̲ ̲Q̲ ̲N̲ ̲A̲ ̲I̲ ̲P̲ ̲F̲

CTGCGACAAAGGAACTTTGATGGTCTGGATCTGGACTGGGAGTACCCAGGCTCAAGAGGCAGCCCTCCAGAGGATAAGCAAAAGTTCACCACGCTGATTCAGGAACTGTTGATCGCCT  589

L̲ ̲R̲ ̲Q̲ ̲R̲ ̲N̲ ̲F̲ ̲D̲ ̲G̲ ̲L̲ ̲D̲ ̲L̲ ̲D̲ ̲W̲ ̲E̲ ̲Y̲ ̲P̲ ̲G̲ ̲S̲ ̲R̲  G  S  P  P  E  D  K  Q  K  F  T  T  L  I  Q  E  L  L  I  A

TCGAAAGTGAGCCCAGGCCATCGGGCACACCACGGCCTTCTCCTGTCTGCTGCTGTGAGTGCTGGCAAGGACACTATTGACGGCCGGATACGAAGTGGACCTTATTTCTGAGAATCTGG  706

F  E  S  E  P  R  P  S  G  T  P  R̲ ̲L̲ ̲L̲ ̲L̲ ̲S̲ ̲A̲ ̲A̲ ̲V̲ ̲S̲ ̲A̲ ̲G̲ ̲K̲  D  T  I  D  A  G  Y  E  V  D  L  I  S  E  N  L

ACTACCTCGTCCTAATGACCTACGACTTCTTCGGTGCCTGGGACCCTGTAACGGGACATAATAGTCCTCTCTACAAGGCTGATGATCAAACCTCGGAACTTAACGAGTACTTTAATG  823

D  Y  L  V  L  M  T  Y  D  F  F  G  A  W  D  P  V  T  G  H  N  S  P  L  Y  K  A̲ ̲D̲ ̲D̲ ̲Q̲ ̲T̲ ̲S̲ ̲E̲ ̲L̲ ̲N̲ ̲E̲ ̲Y̲ ̲F̲ ̲N̲

TGGACTATGCATCCAACTACTGGGTGGAATTGGGCTGTCCCAAAGACAAACTGTACATCGGACTGGCTACGTACGGACGGTCGTTCACTCTGACAGACGAGAGCGACTCCGGTCGTG  940

V̲ ̲D̲ ̲Y̲ ̲A̲ ̲S̲ ̲N̲ ̲Y̲ ̲W̲ ̲V̲ ̲E̲ ̲L̲ ̲G̲ ̲C̲ ̲P̲ ̲K̲  D  K  L̲ ̲Y̲ ̲I̲ ̲G̲ ̲L̲ ̲A̲ ̲T̲ ̲Y̲ ̲G̲ ̲R̲ ̲S̲ ̲F̲ ̲T̲ ̲L̲ ̲T̲ ̲D̲ ̲E̲ ̲S̲ ̲D̲ ̲S̲ ̲G̲ ̲R̲

GTGCTCCGGCCAGTGGTGCCGGAAATGCTGGCGAGTTCACCCGGGGAGGCTGGCTTCTTGTCTTACTACGAGGTGTGTGAAATGCTCCAAGCCGGCGGCGCCAAGAGAACGTTTCTGGATG  1057

G̲ ̲A̲ ̲P̲ ̲A̲ ̲S̲ ̲G̲ ̲A̲ ̲G̲ ̲N̲ ̲A̲ ̲G̲ ̲E̲ ̲F̲ ̲T̲ ̲R̲ ̲E̲ ̲A̲ ̲G̲ ̲F̲ ̲L̲ ̲S̲ ̲Y̲ ̲Y̲ ̲E̲ ̲V̲ ̲C̲ ̲E̲ ̲M̲ ̲L̲ ̲Q̲ ̲A̲ ̲G̲ ̲A̲ ̲K̲  R  T  F  L  D

ACCAGAAAGTTCCTTACCTGGTGCTGGGGAACCAGTGGGTGGGCTACGAGGACGAGGACAGTATTGCGGAAAAGATTCTATACATTCAAAACCATGCATTTGCTGGTGGCATGGTAT  1174

D  Q  K  V  P  Y  L  V  L  G  N  Q  W  V  G  Y  E  D  E  D  S  I  A  E  K  I  L  Y  I  Q  N  H  A  F  A  G  G  M  V

GGGACTACGACTTGGATGATTTCGGTGGAGAATTTTGCGGGCAAGGGAACTACCCGCTGATAAGTTTGATTAGCCAGTATTTGTCGCAATCATAACGTGTTACCGTTACCCGTATGC  1291

W  D  Y  D  L  D  D  F  G  G  E  F  C  G  Q  G  N  Y  P  L  I  S  L  I  S  Q  Y  L  S  Q  S  *

TGATTTATTCCCAACAAATGAAGTTATTGCAAAACTGAAAAAAAAAAAAAAAAAAAAAA//                                                      1352
```

Figure 2. cDNA and deduced amino acid sequences of *AkChi*. Underlined sequences show matching with the peptide fragments of the purified and tripsinized enzyme (coverage: 35.39%, 119 residues).

sequences from *AkChi* and some other known molluscan chitinases (*A. californica*, *H. cumingii*, *C. gigas*, *E. scolopes*, and *S. esculenta*). The deduced amino acid sequence of *AkChi* was shown to have a structure of the GH family 18 chitinase, with an N-terminal signal peptide and a GH 18 catalytic domain. The catalytic domain also contained an active site that is a conserved sequence of GH family 18 chitinases (**Figure 3**). Though the chitinase of *H. cumingii* [14] and *E. scolopes* [15] had two chitin binding domains (CBDs) and the chitinase of *S. esculenta* had one CBD, *AkChi* lacked a CBD. It was reported that fish chitinases have one CBD [8]. This result suggests that the structure of molluscan chitinase is diverse compared to the fish chitinases.

3.2. Amino Acid Sequence of the Chitinase

We analyzed the sequences of the peptide fragments obtained by the tryptic treatment of the purified chitinase from the ovotestis of *A. kurodai* [16] were analyzed and compared them to the deduced amino acid sequence of *AkChi*. The obtained sequences from peptide fragments were consistent with the deduced amino acid sequence of *AkChi* (coverage: 35.39%, 119 residues) (**Figure 2**). This result suggests that *AkChi* is a gene coding the purified enzyme. In addition, trypsin is cut the C-terminal side of lysine and arginine. In this result, it was confirmed that the trypsin is working properly in the all of cleavage site.

3.3. Tissue-Specific Expression of *AkChi*

We investigated the tissue-specific expression of *AkChi* in *A. kurodai* by RT-PCR using the housekeeping β-actin gene as a control (**Figure 4**). It is reported that fish express chitinase to the digestive organs for digestion of chitin from food [17]. The expression profile results indicated that *AkChi* was present only in the ovotestis. We previously detected chitinase activity in the ovotestis and egg from *A. kurodai* [16], whereas lysozyme activity (antibacterial enzyme activity) was not detected in any of the organs [16]. *A. kurodai* has to prey on seaweed.

Signal peptide   Glycoside hydrolase family 18 catalytic domain

```
AkChi     --MTLQVSALCVLLGIALAVCQAGAFHGCAQNRLQTRNGRRGSDDEKNQLVGYYTNWADYRPGKGAFFEDIDANLGTIHYGAILVDGILAPFEWNDDDTEWSEGYYTRVNKLAEDNP
AcAMCase  --MTLQALSLCILLGIALVVCQAGCAR-----NHRVSNGHGGSDEKNKLVGYYTNWADYRPGKGAFFEDIDADLGTIHYGAIIVDGILAPFEWNDDDTPWSEGYYTRVNKLQQKNP
HcChi-3   MEISPRVFLLTAICVLYLQVRVHAYNR--------------------VGYYTNWADYRPGQGKFVIEDIDPNLCSIIYAGAKLNGNQIQAFEWNDETTTWMKNFDRFNAVGSKNP
CgChi3    ---MSRLNLPSLLVFLVILKVSHSYMR---------------------VGYYTNWAIYRPNNGKYVRENLGSLIFAGMKMNGNFIVAFEWNDESTDWMRGNYAKFNDIGLKNP
EsChito   -----MLAVSLLFLLAIGGVSSAGYRR--------------------VGYHTNWSIYRPAPGKYFIESIDPHLCTLCYAGAKLNGNHITAFEWNDESEPWMKGVYDRTMALGKKN
SeChi     ---MSMKCFFSLLLFLFIASRIEASRR--------------------WIEYTNWAIYRKGGARFLIKDIIARFITHISYAGVTLKNGBIAAIEWNDDDTPYAEIGYKQVNNVGKQLI
```

Active site

```
AkChi     ALRTMLSLGGWNMGTQNWTLMVKDESSRQKSIQNAIPFLRQRNFDGLDLDWEYPGSRGSLPEDRQKBTTIIQELLIABESEPRPSGTPRILLSAAISAGKDTIDAGYEVDLISENLDYLV
AcAMCase  ALRTMLSVGGWNMGTKNWTLMVQDTSSRQKSIQNAIPFLRQRNFDGLDLDWEYPGSRGSFPEBKFKGTYMVQELLIABDNEPRPSGTPRILLSAAIAAGKDTIDAGYQVASISKELDYIV
HcChi-3   SIKTLLAVGGWNMGSEPFTSVWSTAASSREGAVTSAKFLRDRNFDGLDLDWEYPANRGSPBVPKQRFTELMVKELKRVGDEDARQTGKPPILLTAAIAAGKSKIDTAIDVPDICRYLDFIS
CgChi3    TVKTLLAVGGWNMGSKPFTQKVKTPESSAEFTKSTIKFLRERNFDGLDLDWEYPANRGSEKEBKDRFGTKMVIQLRSAINQEAAMTGKPPILLTAAIAAGKDKVDTGGDVATIAKHMDFIN
EsChito   ALRILISVGGWNMGSPPPTAIWSSAANSKDSIKHGIKWMRDRGFDGLDLDWEYPANRGSEPEBKNRFSDGIRETRLABNAEAKETGNERILLATAISAGKDKIDTGGDIPEVSKYFDFIT
SeChi     GLRTLLAIGGWNMGSNLFSDIWATKQTRQKSITSTISFLSSIREDGLDICBEYGTKRGSNPQLKBERBGILMLKELRTAIDENAKK-GLSKIIIGIVIGTDENLIENAIDIDAIKSSVDAVS
```

```
AkChi     LMTMDFFGAWDPVTGHNSPLYKADDQTSELNEYFIVDYASNYMVELSCPKDKLYIGLATYGRSFTITDESDSGRGAPASGAGNAGEFIREAGFISMYIVGEMLQABAKRTFLDDQKVEYL
AcAMCase  LMTMDFFGAWDPVTGHNSPLYKADDSTSELNQYFIVDYASHYMVDESCPKDKLYIGLATYGRSFTGRDESDSGRGAPASGAGIMGTYBREAGFISMYIVGEMIIABARMEFLEDQRVEYL
HcChi-3   VMTMDLHGSWEDRTGHNSPLFPHAGETGDG-RYLILAWASNYMHQTSCPKHKLNIGLGLYGRSFTISNPSDNDVMASARGTGEBAGQFIREGGFISMYIIQMKASBGQARYIRDQQAFYM
CgChi3    VMTMDLHGPWEKVTGHNSPLHSRREEMGPD-TQLMIAWAAEYGYKLSLPKQKLNVGLALYGRSFTIASDPYNNGVGAPAKEKGNAGNYBREAGFISMYIIDMIKKBGTTHWIKEQEVEYV
EsChito   IMTMDLHGAWEKFTGHNSPLYARSDESGAQ-KNLIMKWASEYMVSKSAPKSILNIGMALYGRSFTISNKAKTQPGDTTKGPCHBGRYIREKGFISMYISDMIKKBGTTHWIKEQEVEYV
SeChi     LLSMAFYSAMSTDSAVHTSALYASNITKGSDGKKSVEYVAKSIVKNSIIPRNLINIGIALYGHSYRIKDTNAKGEGALISBPGAGRYINTFBGIMAIMIVGEMINNBGIVTFIKGRGVEYL
```

```
AkChi     VLGNBDWVGYEDEDSIAEKILYIQNHAFAGGMVWDYDLDDFGGEFCGQGNYPLISLISQYLSQS-------------------------------------------
AcAMCase  VLGNBDWVGYENKTBIGEKISYIQDNDFAGGMVWDYDLDDFAGGMVWDYDLDDFGGEFCGQGRYPLISLISQSLL---------------------------------
HcChi-3   VKGDBDWVZYDDTDSLQIKIDWIKQNDFGGIMVWALDLDDFT-NVCGQGMYPLLTAINNALGHSTMHIITPPPTAHIVTPPPTRTQPNHNQAQSTRAPIAHEVKTTQAPPNIPSIAKDFTC
CgChi3    TYADBDWVGYDDADSLAVKVSDCRAEVTQSEIIKAAVN----------------------------------------------------------------------
EsChito   VKGDBDWVGYDDQKSLTIKTNWVKSNGYGGIAVWALPLDDFG-GMCGGEKYPLLKSIVRTLGDSVVPSEKP--------------------------VVVTKKPVTLPSGNEDTLC
SeChi     VLGNBDWVAFENEESVTLKTKFALNEGYGGVMIWSFDNDDFSGMCQGGKIYPLFKAFYNAMQMP-----------------------------QTTPDPNWPKKFC
```

Chitin binding domain type2

```
AkChi     ------------------------------------------------
AcAMCase  ------------------------------------------------
HcChi-3   SQAGNGYHSDPTSCMQYFICAGGTAFKFKCAQGLAWNSANNFCDWPDKVTCPPGTSIIGNQRPAEVVRPPPVTQRPTPPPTKPPTTAFSANLNAFDTNYPVATSPPTPSWMDWSRWATDW
CgChi3    ------------------------------------------------
EsChito   SGKADGTYAHPKSCTDYVLCQNGQTYVDHCTAGMWN--------DEIKDCDPTPGFECRRGNKVVTNPSVVTRKPITHP-----------------------------
SeChi     LKHGNGFFG--LDCKRFMICTNGNGFVSQCTQGQLWDKKLNTCVNAKLTTCT-------------------------------
```

Chitin binding domain type2

```
AkChi     ------------------------------------------------
AcAMCase  ------------------------------------------------
HcChi-3   LNSVATTNSYWNQPPSAGGTVNFCTGKSDGIHANPTSCRKYYDCSNGYVYEYTCPAGTGFSAIYKICDYIDNIPGCRP
CgChi3    ------------------------------------------------
EsChito   --------------SGRKNDQFCSGKADGLYADPNDCGAYFNCAAGLTFAEKCGPGTGFDPKIKSCNFKSSIPGCS-
SeChi     ------------------------------------------------
```

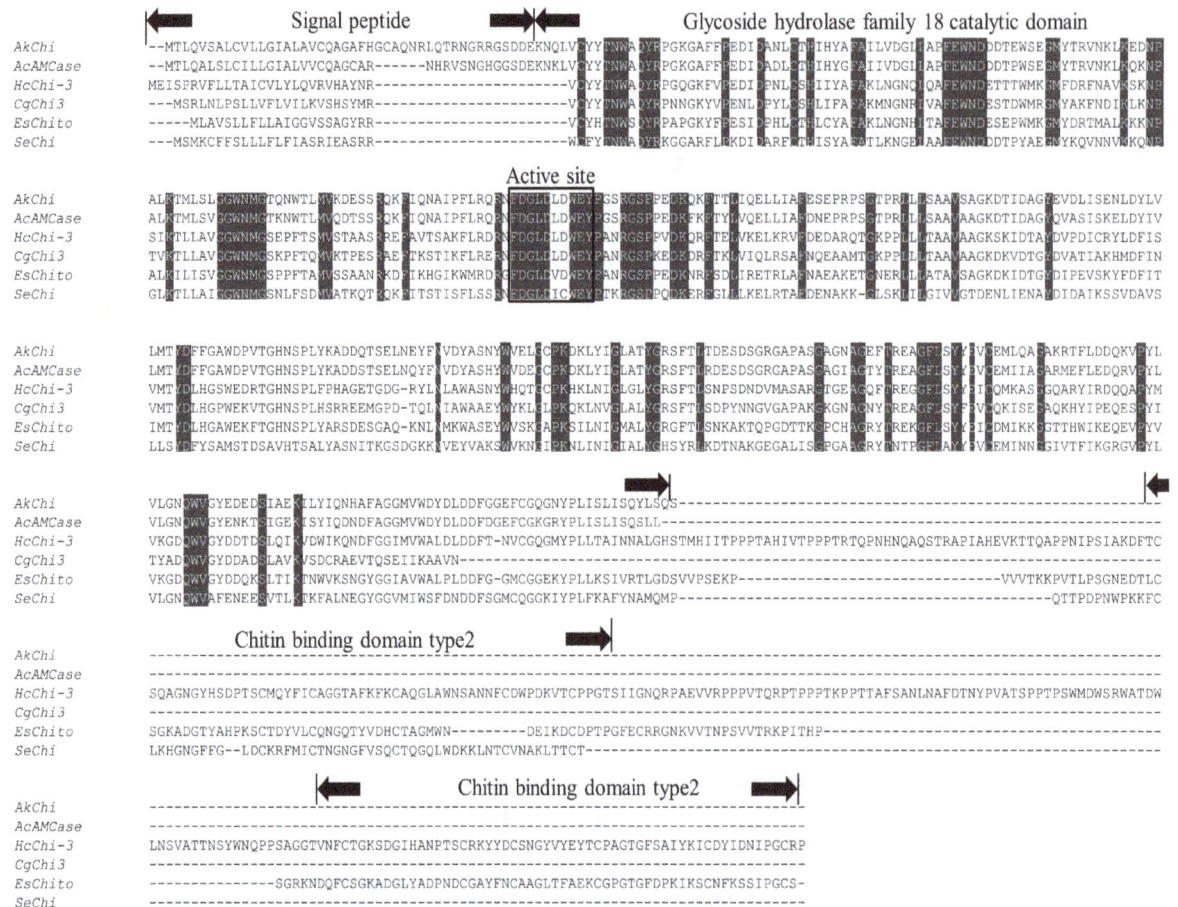

Figure 3. Multiple alignment of duduced amino acid sequences of *A. kurodai* chitinase (*AkChi*) with *Aplysia californica* acidic mammalian chitinase (*AcAMCase*), *Hyriopsis cumingii* chitinase-3 (*HcChi-3*), *Crassostrea gigas* Chit3 protein A (*CgChi3*), *Euprymna scolopes* chitotriosidase (*EsChito*), and *Sepia esculenta* chitinase (*SeChi*). GenBank accession nos.: *AcAMCase*, XM_005112601; *HcChi-3*, JN582038; *CgChi3*, AJ971239; *EsChito*, KF015222; *SeChi*, AB986212. Matched sequences are shown in black.

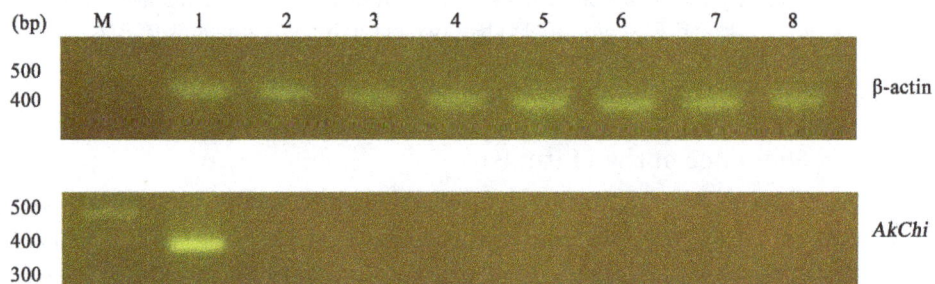

Figure 4. Expression profiles of *AkChi* and *β*-actin mRNA in tissue using RT-PCR. M, markers; 1, ovotestis; 2, egg; 3, skin; 4, gill; 5, buccal mass; 6, crop; 7, anterior gizzard; 8, posterior gizzard.

Thus, *A. kurodai* is not necessary chitinase in digestion and attack of food as squid [10] [11] and octopus [12], respectively. In addition, there is not necessary to shell formation because it does not even have shells. These results suggest that the role of this chitinase is as a defense against nematodes and fungus which have chitin in the body wall as a structural component.

3.4. Phylogenetic Tree Analysis of *AkChi*

We performed a phylogenetic tree analysis of GH family 18 chitinases and *AkChi* (**Figure 5**). Acidic mamma-

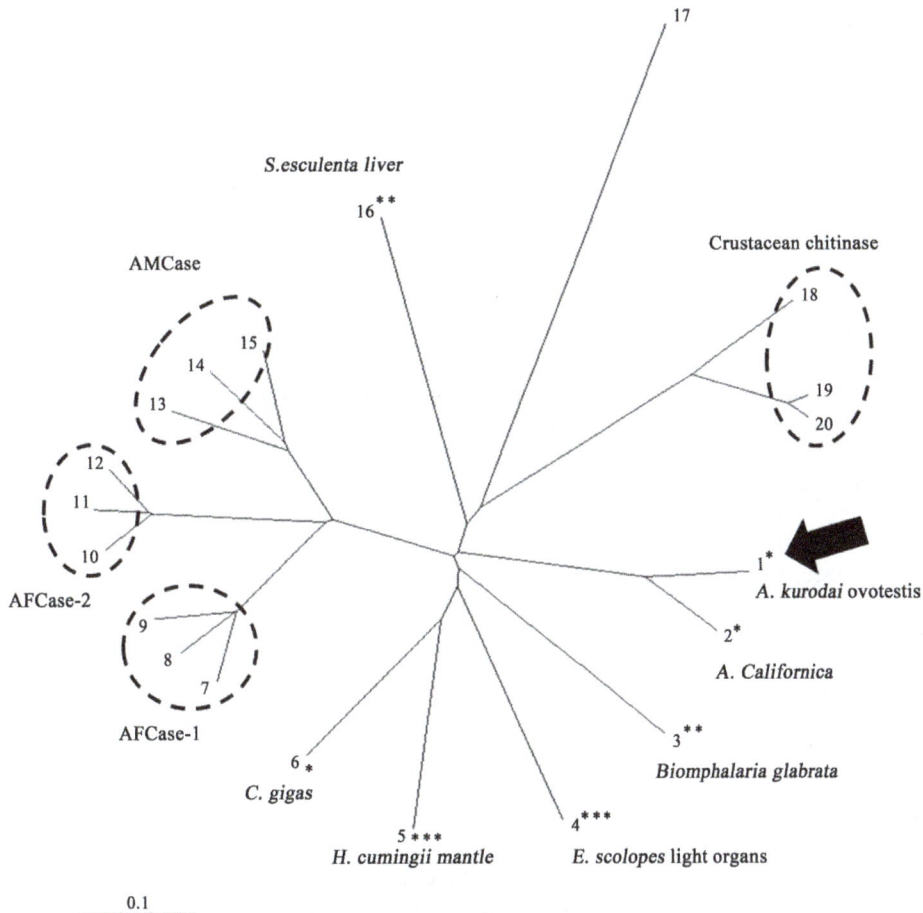

No.	Species	Genbank accession number
1	*Aplysia kurodai* (chitinase)	LC085435
2	*Aplysia californica* (acidic mammalian chitinase)	XM_005112601
3	*Biomphalaria glabrata* (chitinase-3-like protein 1)	XP_013090777
4	*Euprymna scolopes* (chitotriosidase)	KF015222
5	*Hyriopsis cumingii* (chitinase-3)	JN582038
6	*Crassostrea gigas* (Chit3 protein)	AJ971239
7	*Epinephelus coioides* (chitinase1)	AB686658
8	*Sebastiscus marmoratus* (chitinase1)	FJ169895
9	*Parapristipoma trilineatum* (chitinase1)	AB642677
10	*Epinephelus coioides* (chitinase2)	FJ169894
11	*Parapristipoma trilineatum* (chitinase2)	AB642678
12	*Sebastiscus marmoratus* (chitinase2)	AB686659
13	*Bos Taurus* (chitin binding protein b04)	AB051629
14	*Mus musculus* (acidic chitinase)	EF094027
15	*Homo sapiens* (acidic mammalian chitinase)	AF290004
16	*Sepia esculenta* (chitinase)	AB986212
17	*Serratia marcescens* (chiA protein precursor)	X03657
18	*Portunus trituberculatus* (chitinase1)	AB874469
19	*Portunus trituberculatus* (chitinase2)	AB890123
20	*Scylla serrata* (chitinase)	EU402970

Figure 5. Phylogenetic tree analysis of chitinase amino acid sequence by the neighbor-joining method of the program Clustal W. A bacterial chitinase, *Serratia marcescens* chitinase, was used as the out group. The scale bar indicates the substitution rate per residue. The arrow shows *AkChi* obtained in the present study. * Molluscan chitinase without a CBD; ** Molluscan chitinase with one CBD; *** Molluscan chitinase with two CBDs.

lian chitinases (AMCases) have been found in the stomach of mammals. Two chitinase groups with different structures and activity toward short substrates, AFCase-1 and AFCase-2, have been found in the stomach of fish [8]. Crustacean showed a chitinase group [18]. In contrast, molluscan chitinases did not show clear chitinase groups. The reason for this might be the differences in the chitinase domain structure that are due to the presence or absence of a CBD and the number of CBDs. We previously detected chitinase activity in the ovotestis and oviduct from the Walking sea hare *Aplysia juliana* [16]. If the success in cloning the chitinase from *A. juliana*, it will be conceivable to form a group of sea hare chitinase.

4. Conclusion

The cDNA of the ovotestis chitinase obtained from *A. kurodai* contained a 1263 bp open reading frame with a coding potential for 421 amino acid peptides. *AkChi* had the structural motifs of GH family 18 chitinase, but it did not have chitin binding domain. This study is the first report of the cloning of chitinase from the ovotestis of a sea hare.

Acknowledgements

This work was supported in part by a Grant-in-Aid for Scientific Research (C) (no. 25450309) and a College of Bioresource Science, Nihon-University Grant (2015).

References

[1]   Khandeparker, L., Gaonkar, C.C. and Desai, D.V. (2013) Degradation of Barnacle Nauplii: Implications to Chitin Regulation in the Marine Environment. *Biologia*, **68**, 696-706. http://dx.doi.org/10.2478/s11756-013-0202-6

[2]   Arbia, W., Arbia, L., Adour, L. and Amrane, A. (2013) Chitin Extraction from Crustacean Shells Using Biological Methods—A Review. *Food Technology and Biotechnology*, **51**, 12-25.

[3]   Kramer, K.J. and Koga, D. (1986) Insect Chitin: Physical State, Synthesis, Degradation and Metabolic Regulation. *Insect Biochemistry*, **16**, 851-877. http://dx.doi.org/10.1016/0020-1790(86)90059-4

[4]   Umemoto, N., Ohnuma, T., Mizuhara, M., Sato, H., Skriver, K. and Fukamizo, T. (2013) Introduction of a Tryptophan Side Chain into Subsite +1 Enhances Transglycosylation Activity of a GH-18 Chitinase from *Arabidopsis thaliana*, AtChiC. *Glycobiology*, **23**, 81-90. http://dx.doi.org/10.1093/glycob/cws125

[5]   Gooday, G.W. (1999) Aggressive and Defensive Roles for Chitinases, Chitin and Chitinases. *Cellular and Molecular Life Sciences*, **87**, 157-169.

[6]   Henrissat, B. (1991) A Classification of Glycosyl Hydrolases Based on Amino Acid Sequence Similarities. *Biochemical Journal*, **280**, 309-316. http://dx.doi.org/10.1042/bj2800309

[7]   Ikeda, M., Miyauchi, K. and Matsumiya M. (2012) Purification and Characterization of a 56 kDa Chitinase Isozyme (PaChiB) from the Stomach of the Silver Croaker, *Pennahia argentatus*. *Bioscience, Biotechnology, and Biochemistry*, **76**, 971-979. http://dx.doi.org/10.1271/bbb.110989

[8]   Ikeda, M., Kondo, Y. and Matsumiya, M. (2013) Purification, Characterization, and Molecular Cloning of Chitinases from the Stomach of the Threeline Grunt *Parapristipoma trilineatum*. *Process Biochemistry*, **48**, 1324-1334. http://dx.doi.org/10.1016/j.procbio.2013.06.016

[9]   Laribi-Habchi, H., Dziril, M., Badis, A., Mouhoub, S. and Mameri, N. (2012) Purification and Characterization of a Highly Thermostable Chitinase from the Stomach of the Red Scorpionfish *Scorpaena scrofa* with Bioinsecticidal Activity toward Cowpea Weevil *Callosobruchus maculates* (Coleoptera: bruchidae). *Bioscience, Biotechnology, and Biochemistry*, **76**, 1733-1740. http://dx.doi.org/10.1271/bbb.120344

[10]   Nishino, R., Suyama, A., Ikeda, M., Kakizaki, H. and Matsumiya, M. (2014) Purification and Characterization of a Liver Chitinase from Golden Cuttlefish, *Sepia esculenta*. *Journal of Chitin and Chitosan Science*, **2**, 238-243. http://dx.doi.org/10.1166/jcc.2014.1065

[11]   Matsumiya, M., Miyauchi, K. and Mochizuki, A. (2002) Characterization of 38 kDa and 42 kDa Chitinase Isozymes from the Liver of Japanese Common Squid *Todarodes pacificus*. *Fisheries Science*, **68**, 603-609. http://dx.doi.org/10.1046/j.1444-2906.2002.00467.x

[12]   Ogino, T., Tabata, T., Ikeda, M., Kakizaki, H. and Matsumiya, M. (2014) Purification of a Chitinase from the Posterior Salivary Gland of Common Octopus *Octopus vulgaris* and Its Properties. *Journal of Chitin and Chitosan Science*, **2**, 135-142. http://dx.doi.org/10.1166/jcc.2014.1049

[13]   Zhang, G., Fang, X., Guo, X., Li, L., Luo, R., Xu, F., Yang, P., Zhang, L., Wang, X., Qi, H., Xiong, Z., Que, H., Xie,

Y., Holland, P.W.H., Paps, J., Zhu, Y., Wu, F., Chen, Y., Wang, J., Peng, C., Meng, J., Yang, L., Liu, J., Wen, B., Zhang, N., Huang, Z., Zhu, Q., Feng, Y., Mount, A., Hedgecock, D., Xu, Z., Liu, Y., Domazet-Lošo, T., Du, Y., Sun, X., Zhang, S., Liu, B., Cheng, P., Jiang, X., Li, J., Fan, D., Wang, W., Fu, W., Wang, T., Wang, B., Zhang, J., Peng, Z., Li, Y., Li, N., Wang, J., Chen, M., He, Y., Tan, F., Song, X., Zheng, Q., Huang, R., Yang, H., Du, X., Chen, L., Yang, M., Gaffney, P.M., Wang, S., Luo, L., She, Z., Ming, Y., Huang, W., Zhang, S., Huang, B., Zhang, Y., Qu, T., Ni, P., Miao, G., Wang, J., Wang, Q., Steinberg, C.E.W., Wang, H., Li, N., Qian, L., Zhang, G., Li, Y., Yang, H., Liu, X., Wang, J., Yin, Y. and Wang, J. (2012) The Oyster Genome Reveals Stress Adaptation and Complexity of Shell Formation. *Nature*, **490**, 49-54. http://dx.doi.org/10.1038/nature11413

[14] Wang, G.-L., Xu, B., Bai, Z.-Y. and Li, J.-L. (2012) Two Chitin Metabolic Enzyme Genes from *Hyriopsis cumingii*: Cloning, Characterization, and Potential Functions. *Genetics and Molecular Research*, **11**, 4539-4551. http://dx.doi.org/10.4238/2012.October.15.4

[15] Kremer, N., Philipp, E.E.R., Carpentier, MC., Brennan, C.A., Kraemer, L., Altura, M.A., Augustin, R., Häsler, R., Heath-Heckman, E.A.C., Peyer, S.M., Schwartzman, J., Rader, B., Ruby, E.G., Rosenstiel, P. and McFall-Ngai, M.J. (2013) Initial Symbiont Contact Orchestrates Host-Organ-Wide Transcriptional Changes that Prime Tissue Colonization. *Cell Host & Microbe*, **14**, 183-194. http://dx.doi.org/10.1016/j.chom.2013.07.006

[16] Karasuda, S., Ikeda, M., Miyauchi, K. and Matsumiya, M. (2011) Existence and Physiological Role of Chitinase in the Gonad of Two Species of Sea Hare, Kuroda's Sea Hare *Aplysia kurodai* and Walking Sea Hare *Aplysia Juliana*. *Proceedings of the* 9th *Asia-Pacific Chitin and Chitosan Symposium*, Vietnum, 3-6 August 2011, 169-172.

[17] Kakizaki, H., Ikeda, M., Fukushima, H. and Masahiro, M. (2015) Distribution of Chitinolytic Enzymes in the Organs and cDNA Cloning of Chitinase Isozymes from the Stomach of Two Species of Fish, Chub Mackerel (*Scomber japonicus*) and Silver Croaker (*Pennahia argentata*). *Open Journal of Marine Sciences*, **5**, 398-411. http://dx.doi.org/10.4236/ojms.2015.54032

[18] Fujitani, N., Hasegawa, H., Kakizaki, H., Ikeda, M. and Masahiro, M. (2014) Molecular Cloning of Multiple Chitinase Genes in Swimming Crab *Portunus trituberculatus*. *Journal of Chitin and Chitosan Science*, **2**, 149-156. http://dx.doi.org/10.1166/jcc.2014.1046

Abbreviations

RT-PCR: reverse transcription-polymerase chain reaction;
RACE: rapid amplification of cDNA ends;
GH: glycoside hydrolase;
AFCase-1: acidic fish chitinase-1;
AFCase-2: acidic fish chitinase-2;
CBD: chitin binding domain;
$K_3[Fe(CN)_6]$: potassium ferricyanide;
$Na_2S_2O_3$: sodium thiosulfate.

14

Effect of Interleukin-1Beta (IL-1β) on the Cortical Neurons Survival and Neurites Outgrowth

Ebtesam M. Abd-El-Basset

Department of Anatomy, Faculty of Medicine, Kuwait University, Kuwait city, Kuwait
Email: ebtesam@hsc.edu.kw

Abstract

Insults to the brain are known to cause a myriad of downstream effects, including the release of cytokines by astrocytes and resultant reactive gliosis. The author has examined effect of cytokine IL-1β on the survival of cortical neurons using mouse astrocyte-neuron co-culture. Five groups were used. These were neurons alone (Group 1), neurons with added IL-1β (Group 2), neurons co-cultured with astrocytes (Group 3), neurons co-cultured with astrocytes that was pre-treated with IL-1β before co-culture (Group 4) and neurons co-cultured with astrocytes and IL-1β added (post-treated) (Group 5). In Group 1 only a few neurons grew and survived only for 5-6 days. In Group 2, it was observed that more neurons survived up to 11 days. Moreover, in Group 3, more neurons grew and survived up to 16-18 days. They had large cell bodies and many long neurites that formed anastomosing networks. In Group 4, few neurons survived up to 13 days, whereas in Group 5, the growth of neurons were affected but to a much lesser extent than Group 4 and survived up to 15 days. In addition, it was found that IL-1β stimulated the expression of glial fibrillary acidic protein (GFAP) by astrocytes. This study indicates that IL-1β affects the survival of cortical neurons and modulates the astrocytic support to neuronal survival and neurites outgrowth by acting directly on the astrocytes.

Keywords

Astrocytes, IL-1β, Cell Culture, Neuronal Survival, Cytokines, Gliosis

1. Introduction

Astrocytes play an important role in neuronal migration and survival during the development of the central nervous system (CNS) [1] [2]. Astrocytes act also as important regulators of brain inflammation [3]. In addition, numerous studies have demonstrated the importance of astrocytes in a variety of neurophysiological processes such as the supply of energy metabolites, defense against oxidative stress or neurotransmitter reuptake and recycling [4]-[6].

Insults to the brain are known to cause a myriad of downstream effects, including the release of cytokines by astrocytes and resultant reactive gliosis [7] [8]. Cytokines are thought to be a major mediator of reactive gliosis [9]. IL-1β is a pro-inflammatory cytokine that exerts an important role in developing brains as a maintenance and growth-promoting factor [10] [11] and in migration of cortical neurons [12]. IL-1β is secreted by activated astroglia and microglia in brain, where it exerts a diverse range of activities on immune function and coordination of many aspects of the acute phase response to trauma and infection [8] [10]. Inflammation is the key host-defense response to infection and injury but is also thought to be a major contributor to a diverse range of diseases. It is generally believed that the inflammatory processes stimulated by IL-1β is detrimental and can aggravate the primary damage caused by infections of the CNS [13]. The mechanisms by which IL-1β exerts its activities are not totally understood. The hypothesis of this study is that pro-inflammatory cytokine IL-1β acts on astrocytes to alter their chemical and physical properties, which in turn affects the survival of neurons. In this study the author have observed the direct effect of IL-1β on survival of cortical neurons and how IL-1β-treated astrocytes modulate the survival of cortical neurons.

2. Materials and Methods

Animal care followed the recommendations of NIH Guidelines for Care and Use of Laboratory Animals and the Guide for the Care and Use of Laboratory Animals (Kuwait University-Faculty of Health Publication). All efforts were made to minimize animal suffering, to reduce the number of animals used, and to utilize alternatives to *in vivo* techniques.

2.1. Astroglia Culture

New-born Balb/c mice (from Health and Science Animal Laboratory, Kuwait University) were used. The brain was dissected and cells were disaggregated by gently forcing the neopallia through a nitex mesh of 75 μm. Cells were cultured both in culture flasks and on cover slips. A concentration of 5×10^4 and 5×10^6 nigrosine excluding cells were plated on 11×22 mm glass cover slips in Petri dishes and on 75 mm flasks, respectively. The Petri dishes and the flasks were incubated with 3 ml and 13 ml of growth medium (MEM) (GibcoBRL) containing 5% horse serum (GibcoBRL) respectively. Some flasks were treated with 200 international units (U)/ml of IL-1β (Sigma) for 4 days. The cells were then incubated at 37°C in a humidified atmosphere with 5% CO_2 in air, the medium was changed every 2 days.

2.2. Neuronal Culture

The cerebral neuronal cultures were prepared according to the method reported in [14]. Balb/c mouse fetuses of embryonic day 15 were used. The brain was dissected to obtain the neopallia, and then disaggregated in 0.25% trypsin (GibcoBRL) in Pucks solution for 5 min at room temperature. Horse serum was added to stop the action of trypsin. The suspension was then centrifuged at 1000 rpm for 5 min. The pellet of cells was suspended in MEM containing 30 mM glucose and further disaggregated by passing through a 75-μm nitex mesh. These cells were then plated in 60-mm Petri dish containing 11×22 mm cover slips coated with Poly-L-lysine(Sigma) (Group 1), then incubated in freshly prepared MEM containing 5% horse serum and 30 mM glucose(Sigma). Some neuronal cultures were treated with 200 U/ml of IL-1β for 4 days (Group 2).

2.3. Neuron-Astroglia Co-Culture

Astroglia cultures were prepared as described above and then maintained for 10 days till a monolayer of cells were obtained. Neuronal cells obtained by the above procedure were seeded over this monolayer of astrocytes [15]. After 15 min of incubation the culture was rinsed and incubated with MEM containing 5% horse serum and 30 mM glucose (Group 3). Some astroglia were treated with 200 U/ml IL-1β for 4 days before adding the neurons (pre-treated) (Group 4), whereas, some co-cultures were treated with 200 U/ml IL-1β for 4 days (post-treated) (Group 5).

2.4. Immunostaining for Neurons

Cells grown on cover slips were fixed in methanol at $-20°C$ for 4 min, and then treated with PBS containing 3% hydrogen peroxide. The cells were incubated with mouse momoclonalanti-neurofilament (NF) (Sigma) diluted 1:50 for one hour. This was then incubated with biotinylated-anti-mouse IgG (1:200) for 45 min followed by incubation of extra-avidin peroxidase (1:20) (Sigma) for 45 min. The color was developed using DAB solution for 5 - 10 min, then the cells on coverslips were washed with distilled water followed by dehydration in ascending grades of alcohol and cleared in the xylene and mounted in DPX.

2.5. Morphological Studies and Neurite Outgrowth Essay

The morphological changes in neurons after 3, 5, 7, 13, 17, 20 days of cultures were examined by phase contrast microscopy. To assess the number of survived neurons, the number of neurons extending neurites longer than two cell body diameters was counted in 8 randomly chosen microscopic fields using a 20x objective lens. The fields were examined from the center of each coverslip inside the tissue culture dish. Three dishes were examined from each experiment. Three separate independent experiments were used. The length of the neurites from the above neurons was measured using an imaging analysis program (Image J, version 1.04b, Wayne Rasband, NIH). In each experiment, 50 - 60 neurites were measured.

2.6. Polyacrylamide Gel Electrophoresis (PAGE) and Immunoblotting

Flasks of IL-1β treated astrocytes for 4 days and control of non-treated astrocytes were washed three times in PBS, scraped off and collected in PBS. The cells were immediately sonicated and the protein concentration was determined according to the method reported in [16]. The sonicated cells were dissolved in 10% SDS-containing sample buffer and boiled for 3 min. The same amount of protein (20 μg/well) from each sample was analyzed electrophoretically in 10% SDS-PAGE gel according to the method reported in [17]. The proteins in the gel were transferred to nitrocellulose sheets by a modification of the method reported in [18]. After transfer, the nitrocellulose sheets were incubated for 1 h with 3% bovine serum albumin (BSA) in PBS. The nitrocellulose sheets were then incubated overnight at room temperature with rabbit polyclonal antibody to GFAP, diluted 1:500 in PBS containing 3% BSA, then followed by affinity-purified goat anti-rabbit IgG conjugated to horse-radish peroxidase. The color was developed using freshly prepared 0.05% 4-chlor-1-naphthol and 0.015% H_2O_2 in PBS. The reaction was stopped by washing in tap water. All the chemicals and materials for electrophoresis and immunobloting were purchased from BioRad (Missisauga, Ont, Canada). Immunoblots of five independent experiments were scanned using the Snapescan 1212 scanner and Adobe Photoshop 5.0 program. The relative levels of GFAP expression were determined by analyzing the pixel intensity of the bands using an imaging analysis program (Image J, version 1.04b, Wayne Rasband, NIH). The percentage of the protein expression was calculated in the following manner. The average background protein levels in the lane, excluding the bands, were first subtracted from both the control and treatment bands. The percentage of the increase in intensity (I) of IL-1β treated cells (T) compared to the control (C) was calculated as follows:

$$\text{Percentage of increase in the intensity} = 100 \times \frac{\text{Mean } I \text{ of } T - \text{Mean } I \text{ of } C}{\text{Mean } I \text{ of } C}$$

2.7. Statistical analysis

Experiments were performed at least three times on different cell preparations. The data were analyzed with one-way Analysis of Variance (ANOVA) which rejected the null hypothesis. P-values < 0.05 were considered statistically significant. A post-hoc Bonferroni's test was used to compare the five groups individually

3. Results

3.1. The Effect of IL-1β on Neuronal Survival

When neurons were cultured alone on Poly-L-lysine coated coverslips (Group 1), few of them survived up to 5 - 6 days (**Figure 1(a)** and **Figure 1(b)**). These neurons had small cell bodies and few short neurites. When these neurons were treated with IL-1β (Group 2) they survived up to 10 days (**Figure 1(c)**).

3.2. The Effect of Astrocytes on the Neuronal Survival

When the same number of neurons was cultured on top of normal astrocytes (Group 3), more neurons grew and they made anastomosing networks. The neurons had larger cell bodies and extended long neurites (**Figure 2(a)**). These neurons were grouped together and their neurites formed anastomosing networks in culture in 13-days culture (**Figure 2(b)**). The neurons survived up to 16 - 18 days (**Figure 2(c)**).

3.3. Effect of IL-1β-Pre-Treated Astrocytes on the Neuronal Survival

When the same number of neurons were cultured on top of IL-1β pre-treated astrocytes (Group 4), few neurons attached (**Figure 3(a)**), and survived only for 13 days and had small cell bodies and short neurites (**Figure 3(b)**). The neurites did not form bundles as compare to the neurons grew on the top of non-treated astrocytes in (**Figure 2(b)**).

3.4. Effect of IL-1β on the Neuronal Survival in Astrocyte-Neuron Co-Culture (Post-Treated)

When astrocyte-neuron co-culture were treated with IL-1β for 12 days (Group 5), the survival of the neurons

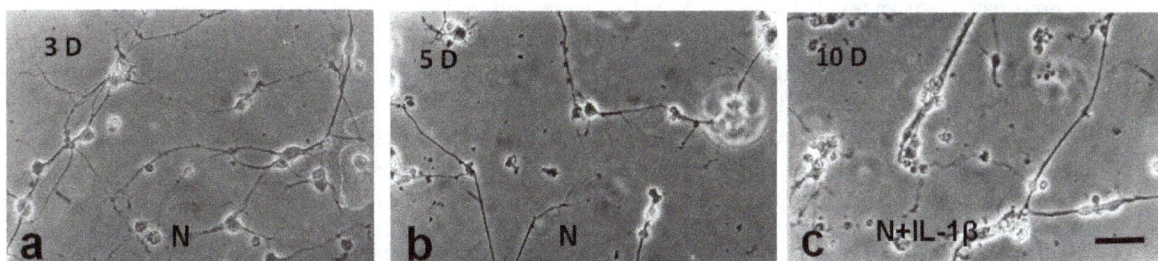

Figure 1. Phase contrast image of 3-day (a) and 5-day (b) old neuronal cultures growing alone without astrocytes (Group 1). Images show few numbers of neurons with small cell bodies and short neuritis. Phase contrast image of 10-days old neuronal cultures growing alone without astrocytes and treated with IL-1β (Group 2), the neurons survived up to 10 days (c). Scale bar is 50 μm.

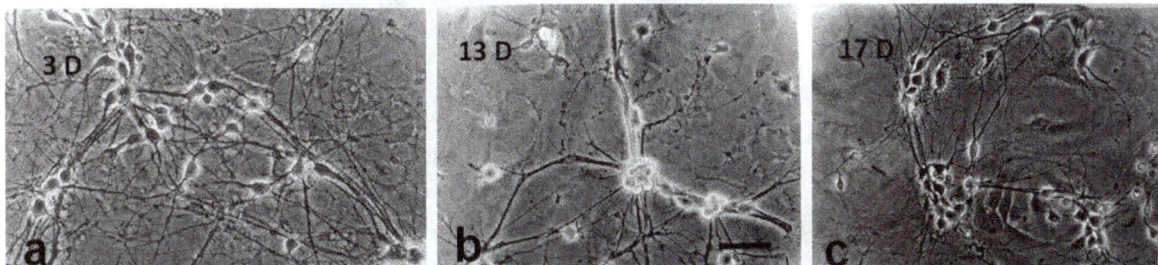

Figure 2. Phase contrast images of 3-day (a), 13-day (b), and 17-day (c) old neuronal cultures on the top of normal astrocytes (Group 3) showing many neurons grow and extend long neurites. Neuronal cultures on top of normal astrocytes of 13-days (b) showing many neurons form group together and the neurites form bundles. Scale bar is 50 μm.

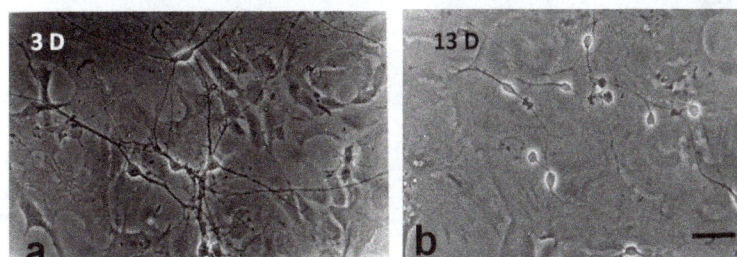

Figure 3. Phase contrast images of neurons on top of IL-1β pre-treated astrocytes of 3-day (a), and 13-day (b) old culture (Group 4) showing less survived neurons than when they grow on non-treated astrocytes. Scale bar is 50 μm.

were affected but to a much lesser extent than the neuron grown on IL-1β pre-treated-astrocytes. The neurons survived up to 15 days (**Figure 4**).

3.5. Immunostaining for Neurons

When neurons and their neurites were stained for neurofilaments, many neurons and neurites were stained in 7-days old cultures grew on top of normal astrocyte (Group 3) compared to the neurons that grew on top of IL-1β pre-treated-astrocytes (Group 4) (**Figure 5**).

3.6. The Neuronal Survival Essay

In 3-day old culture, when neurons were cultured alone without astrocytes (Group 1) only few of them grew (10 ± 1) and survived up to 5 - 6 days, while when they were treated with IL-1β (Group 2) they survived up to 10 days. When the same number of neurons was cultured on the top of astrocytes (Group 3), many neurons attached and survived (75 ± 4). However, when the same number of neurons were cultured on the top of IL-1β pre-treated astrocytes (Group 4) the number of survived neurons decreased to about (40 ± 3) neurons. In IL-1β post-treated co-culture (Group 5) the number of neurons attached is similar to the non-treated culture. In the older cultures (15, 17-days old) no neurons survived when cultured alone as well as when cultured on top of IL-1β pre-treated astrocytes. However, when they were cultured on top of normal astrocytes, about (35 ± 5) neurons survived in the 13-days old culture, about (30 ± 2) neurons survived in the 15-days old culture, and (25 ± 2) neurons in the 17-days old culture (**Figure 6**). In post-treated co-culture about (10 ± 1) neurons survived in 13-days and no neurons survived in 15-days old culture. There was a statistically significant increase in the number of neurons growing on the top of astrocytes (Group 3) (P-values < 0.001) and longer period of survival (17 days) compared to the neurons growing alone (Group 1) or on the top of IL-1β pre-treated astrocytes (Group 4). There was no significant increase in the number of neurons alone (Group 1) or when they were treated with

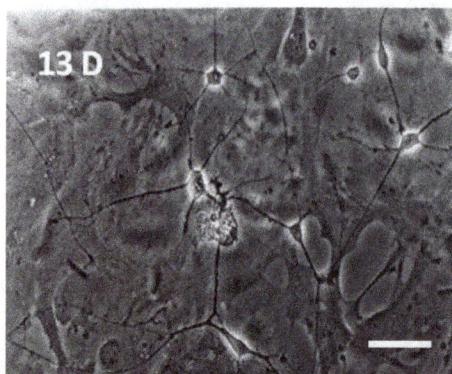

Figure 4. Phase contrast image of 13-day-old neuronal co-cultures (Group 5) showing only few neurons survived. Scale bar is 20 μm.

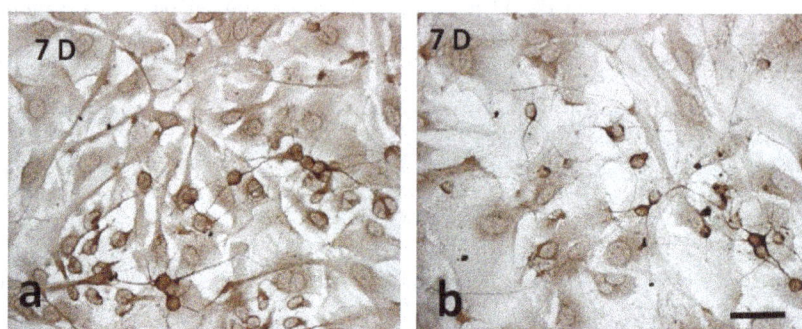

Figure 5. Immunocytochemistry staining with monoclonal antibody to neurofilaments of 7-day-old neuronal cultures on the top of normal astrocytes (a) (Group 3) and on pre-treated astrocytes with IL-1β (b) (Group 4) showing that many neurons with long neurites in figure (a) compare to figure (b). Scale bar is 50 μm.

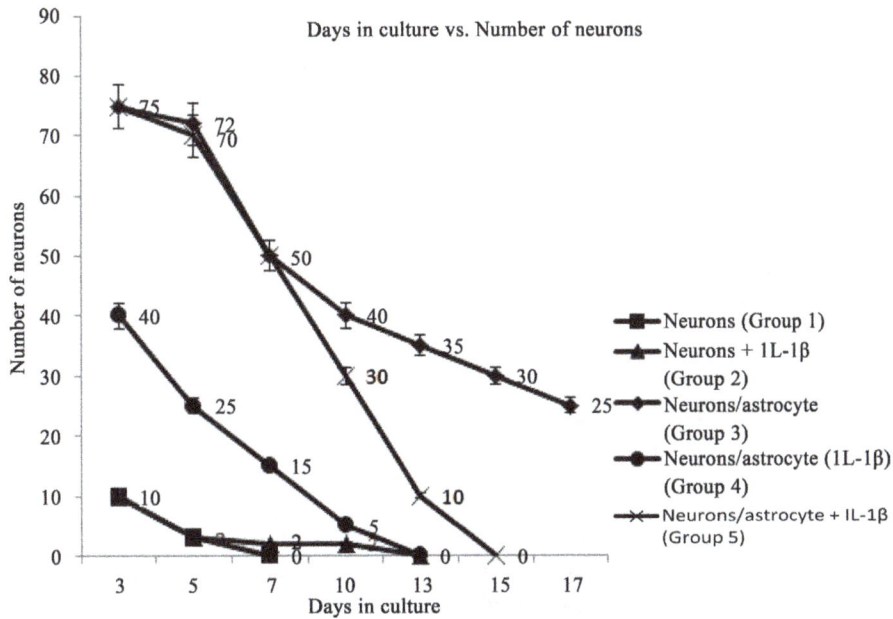

Figure 6. Graph showing number of neurons in five different conditions: Groups 1-5. Note the significant increase in the number of neurons growing on the top of astrocytes (Group 3) (P-values < 0.001) and longer period of survival (17 days) compared to the neuron growing alone (Group 1) or on the top of IL-1β pre-treated astrocytes (Group 4). There is no significant increase in the number of neurons alone when they were treated with IL-1β (Group2). Also there is no significant changes when the neuron on the top of normal astrocytes was post-treated with IL-1β (Group 5) compare to pre-treated astrocytes (Group 4).

IL-1β (Group 2). Also there were no significant changes when the neuron on the top of normal astrocytes (Group 3) compared to post-treated astrocytes (Group 5). A post-hoc Bonferroni's test was used to compare the three groups individually (Groups 1, 3 and 4), and showed that there was a significant difference between each of the three groups with a P-value of <0.001 for all groups.

3.7. The Neurites Outgrowth Essay

Measurement of the length of the neurites using the Image J analysis program showed that neurons cultured on the top of normal astrocytes (Group 3) have the longest neurites with progressive increases in the length with the days in culture compared to Group 1, Group 2, Group 4 or Group 5 (**Figure 7**). There was a significant difference between three groups, Group 1, Group 3 and Group 4 (P-values < 0.001). A post-hoc Bonferroni's test was used to compare the three groups individually and showed that there was a significant difference between each of the three groups with a P-value of <0.001 for all groups. However, there was not a statistically significant difference in the length of neutites when comparing (Group 1) with (Group 2).

3.8. Immunoblotting

To determine the effect of IL-1β on the expression of GFAP in astrocytes, protein from 4 day IL-1β treated astrocytes and control astrocytes (non-treated) were separated by 10% SDS-PAGE, transferred to nitrocellulose sheets and treated with antibody to GFAP (**Figure 8**). Quantification of the western blots of five independent experiments showed that the amount of GFAP in treated astrocytes increased to about 125% of control.

4. Discussion

This study shows that neuronal survival and neuritis outgrowth depends upon the presence of astrocytes. IL-1β induces gliosis in astrocytes which in turn inhibits neuronal survival. Although the direct effect of IL-1β on promoting neuronal survival is not significant, it can be explained by the cytokines ability to promote the use of growth factors by neurons [19]. The major effect of IL-1β is its effect on astrocytes. It inhibits the astrocytic support of neuronal survival and neuritis outgrowth. This modulation can be explained by astrogliosis induced in

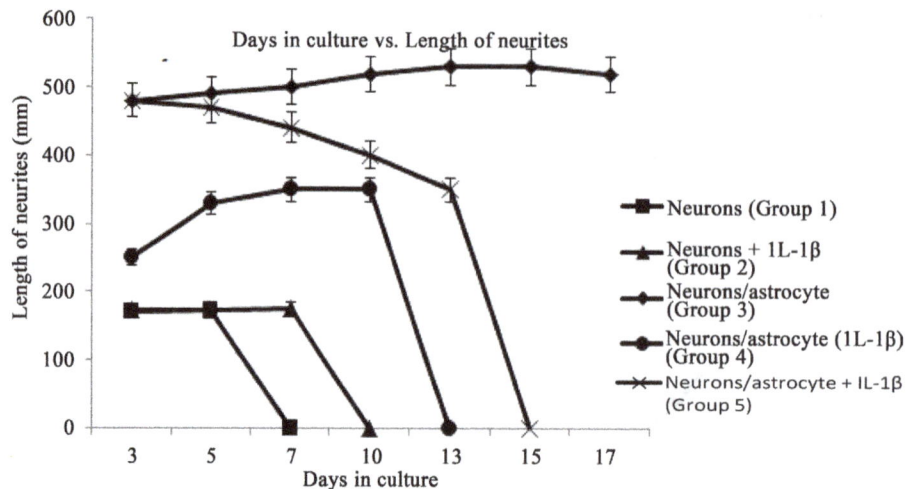

Figure 7. Graph showing the lengths of the neurites in 5 different conditions: (Groups 1-5) Note the significant increase lengths of the neurites of the neurons growing on top of astrocytes (Group 3) (P-values < 0.001), and significant decrease lengths of neurites of neurons growing on the top of IL-1β pre-treated astrocytes (Group 4) (P-values < 0.001). There is no significant increase in length of neutites of neurons cultured alone and treated with IL-1β (Group 2) or in co-culture treated with IL-1β (Group 5). It should be notes that the sudden drop (to zero) in the length of the neurites in (Group 1) at day 7 or in (Group 2) at day 10 or in (Group 4) at day 13 and in (Group 5) at day 15 was due to the absence of the neurons and therefore the absence of neurites These lengths were magnified by 1220 x.

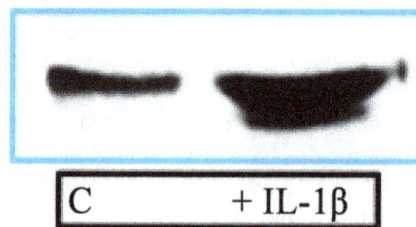

Figure 8. Immunoblotting of the transferred total protein extract of control astrocytes and 4 days IL-1β-treated astrocytes stained with antibodies to GFAP. Note the increase level of GFAP in IL-1β treated astrocytes

IL-1β pre-treated astrocytes which may affect the chemical and physical properties of astrocytes so few neurons attached and survived. Previous studies have shown the role of IL-1β in induction of gliosis and neurodegeneration [20] [21]. Whether IL-1β regulates extracellular matrix (ECM) and adhesion molecules in astrogliosis is completely unclear, however, recent studies indicate the role of ECM in neuronal function and survival [22] [23].

The modifications of the astrocytic metabolism induced by IL-1β may modulate neuronal susceptibility to an exotoxic insult in neuron-astrocyte co-cultures. Together, these results suggest that IL-1β affects the metabolic profile of astrocytes, and that these changes have functional consequences for surrounding neurons. A previous study showed that IL-1β inhibited astroglia glutamic transport, which led to neurotoxicity [24].

Brain immune response is an essential process for protection against potentially deleterious threats such as infection, injury, cellular debris or abnormal protein aggregates. Although it is usually transient, persisting insults may give rise to chronic neuroinflammation which has been implicated in several neuropathology including Alzheimer's disease, Parkinson's disease, amyotrophic lateral sclerosis, multiple sclerosis, and stroke [25]. Neuroinflammation is a complex process, its overall effect is the result of a fine balance between a wide array of cytokines and growth factors, which may exert either neuroprotective or neurotoxic effects [26] [27]. Cytokines are major effectors in this fine balance as they have a dual function, potentially promoting or repressing neuroinflammation, hence their traditional classification as pro- or anti-inflammatory. Despite this classic denomination of cytokines, understanding their exact individual effect is far more complex as many of them interact with each other either synergistically or antagonistically and may additionally have pleiotropic effects [10]. Adding another level of complexity, different sets of cytokine receptors are expressed by the various cell types present in brain. Research over last 2 decades has revealed that IL-1β is an important mediator of neuronal injury and blocking the actions of IL-1β is beneficial in a number of experimental models of brain injury [28]. Mechanisms underlying the actions of IL-1β in brain injury remain unclear though increasing evidence indicates the cerebrovascular as a key target [29] [30]. The therapeutic use of anti-inflammatory agents may reduce CNS pathology induced by inflammatory responses. However, it would appear that the key to developing feasible anti-inflammatory drugs is to minimize the neurotoxic effects while promoting the beneficial and neurotrophic effects [31]-[33].

Many factors protect neurons through the inhibition of pro-inflammatory cytokines IL-1β, such as erythropotein [34], fullerenols and glucosamine fluerences [35], omega-3 polyunsaturated fatty acids [36], and curcumin [37] [38]. In addition, recent evidence using a MAPKs inhibitor [39] and an IL-1β receptor antagonist [28] suggest that these may act as anti-inflammatory drugs.

In conclusion, this study indicates that IL-1β modulates the astrocytic support to neuronal survival and neurites outgrowth by acting directly on the astrocytes.

Acknowledgements

This work is supported by Kuwait University, Grant MA01/14. The author would like to thank Mrs. Amna Najem and Josily Joy for technical assistance.

References

[1] Furukawa, S., Furukawa, Y. and Satoyoshi, E. (1986) Synthesis and Secretion of Nerve Growth Factor by Mouse Astroglial Cells in Culture. *Biochemical and Biophysical Research Communications*, **136**, 57-63. http://dx.doi.org/10.1016/0006-291X(86)90876-4

[2] Clarke, L.E. and Barres, B.A. (2013) Emerging Roles of Astrocytes in Neural Circuit Development. *Nature Reviews Neuroscience*, **14**, 311-321. http://dx.doi.org/10.1038/nrn3484

[3] Farina, C., Aloisi, F. and Meinl, E. (2007) Astrocytes Are Active Players in Cerebral Innate Immunity. *Trends in Immunology*, **28**, 138-145. http://dx.doi.org/10.1016/j.it.2007.01.005

[4] Belanger, M. and Magistretti, P.J. (2009) The Role of Astroglia in Neuroprotection. *Dialogues in Clinical Neuroscience*, **11**, 281-295.

[5] Liddell, J.R., Robinson, S.R., Dringen, R. and Bishop, G.M. (2010) Astrocytes Retain Theirantioxidant Capacity into Advanced Old Age. *Glia*, **58**, 1500-1509.

[6] Walls, A.B., Waagepetersen, H.S. and Bak, L.K. (2015) The Glutamine-Glutamate/GABA Cycle: function, Regional Differences in Glutamate and GABA Production and Effects of Interference with GABA Metabolism. *Neurochemical Research*, **40**, 402-409. http://dx.doi.org/10.1007/s11064-014-1473-1

[7] Laping, N.J., Teter, B., Nichols, N.R., Rozovsky, I. and Finch, C.E. (1994) Glial Fibrillary Acidic Protein: Regulation by Hormones, Cytokines, and Growth Factors. *Brain Pathology*, **1**, 259-275. http://dx.doi.org/10.1111/j.1750-3639.1994.tb00841.x

[8] Murray, K.N., Parry-Jones, A.R. and Allan, S.M. (2015) Interleukin-1 and Acute Brain Injury. *Frontiers in Cellular Neuroscience*, **9**, 1-17. http://dx.doi.org/10.3389/fncel.2015.00018

[9] Little, A.R. and O'Callagha, J.P. (2001) Astrogliosis in Adult and Developing CNS: Is There a Role for Poinflammatory Cytokines? *NeuroToxicology*, **22**, 607-618. http://dx.doi.org/10.1016/S0161-813X(01)00032-8

[10] John, G.R., Lee, S.C., Song, X., Rivieccio, M. and Brosnan, C.F. (2005) IL-1-Regulated Responses in Astrocytes: Relevance to Injury and Recovery. *Glia*, **49**, 161-176. http://dx.doi.org/10.1002/glia.20109

[11] Boato, F., Hechler, D., Rosenberge, K., Lüdecke, D., Peters, E.M., Nitsch, R. and Hendrix, S. (2011) Interleukin-1 Beta and Neurotrophin-3 Synergistically Promote Neurite Growth *in Vitro*. *Journal of Neuroinflammation*, **8**, 183. http://dx.doi.org/10.1186/1742-2094-8-183

[12] Ma, L., Li, X.W., Zhang, S.J., Yang, F., Zhu, G.-M., Yuan, X.-B. and Jiang, W. (2014) Interleukin-1 Beta Guides the Migration of Cortical Neurons. *Journal of Neuroinflammation*, **11**, 114. http://dx.doi.org/10.1186/1742-2094-11-114

[13] Medel-Matus, J.S., Álvarez-Croda, D.M. and Martínez-Quiroz, J. (2014) IL-1β Increases Necrotic Neuronal Cell Death in the Developing Rat Hippocampus after Status Epilepticus by Activating Type I IL-1 Receptor (IL-1RI). *International al Journal of Developmental Neuroscience*, **38**, 232-240.

[14] Zhang, S. and Fedoroff, S. (1996) Neuron-Microglia Interactions *in Vitro*. *Acta Neuropathologica*, **91**, 385-395. http://dx.doi.org/10.1007/s004010050440

[15] Abd-El-Basset, E.M. (2013) Proinflammatory Tumor-Necrosis Factor-Alpha (TNF-α) Inhibits Astrocytic Support of Neuronal Survival and Neurites Outgrowth. *Advances in Bioscience and Biotechnology*, **4**, 73-80. http://dx.doi.org/10.4236/abb.2013.48A2010

[16] Bradford, M.M. (1976) A Rapid and Sensitive Method for Quantification of Microgram Quantities of Protein Utilizing the Principles of Protein-Dyebinding. *Analytical Biochemistry*, **72**, 248-254. http://dx.doi.org/10.1016/0003-2697(76)90527-3

[17] Laemmli, U.K. (1970) Cleavage of Structural Proteins during Assembly of Head of the Bacteriophage T4. *Nature*, **227**, 680-685. http://dx.doi.org/10.1038/227680a0

[18] Towbin, H., Staehlin, T. and Gordon, J. (1979) Electrophoretic Transfer of Proteins from Polyacrylamide Gels to Nitrocellulose Sheets: Procedure and Some Applications. *Proceedings of the National Academy of Sciences of the United States of America*, **76**, 4350-4354. http://dx.doi.org/10.1073/pnas.76.9.4350

[19] Ho, A. and Blum, M. (1997) Regulation of Astroglial-Derived Dopaminergic Neurotrophic Factors by Interleukin-1 Beta in the Striatum of Young and Middle-Aged Mice. *Experimental Neurology*, **148**, 348-359. http://dx.doi.org/10.1006/exnr.1997.6659

[20] Boutin, H., LeFeuvre, R.A. and Horai, R. (2001) Role of IL-1α and IL-1β in Ischemic Brain Damage. *The Journal of Neuroscience*, **21**, 5528-5534.

[21] Abd-El-Basset, E.M. and Abd-El-Barr, M.M. (2011) Effect of Interleukin-1β on the Expression of Actin Isoforms in Cultured Mouse Astroglia. *The Anatomical Record*, **294**, 16-23. http://dx.doi.org/10.1002/ar.21303

[22] Summers, L., Kangwantas, K., Nguyen, L., Kielty, C. and Pinteaux, E. (2010) Adhesion to the Extracellular Matrix Is Required for Interleukin-1 Beta Actions Leading to Reactive Phenotype in Rat Astrocytes. *Molecular and Cellular Neuroscience*, **44**, 272-281. http://dx.doi.org/10.1016/j.mcn.2010.03.013

[23] Summers, L., Kangwantas, K., Rodriguez-Grande, B., Denes, A., Penny, J., Kielty, C. and Pinteaux, E. (2013) Activation of Brain Endothelial Cells by Interleukin-1 Is Regulated by the Extracellular Matrix after Acute Brain Injury. *Molecular and Cellular Neuroscience*, **57**, 93-103. http://dx.doi.org/10.1016/j.mcn.2013.10.007

[24] Prow, N.A. and Irani, D.N. (2008) The Inflammatory Cytokine, Interleukin-1 Beta, Mediates Loss of Astroglial Glutamate Transport and Drives Excitotoxic Motor Neuron Injury in the Spinal Cord during Acute Viral Encephalomyelitis. *Journal of Neurochemistry*, **105**, 1276-1286. http://dx.doi.org/10.1111/j.1471-4159.2008.05230.x

[25] Allan, S.M. and Rothwell, N.J. (2003) Inflammation in Central Nervous System Injury. *Philosophical Transactions of the Royal Society B: Biological Sciences*, **358**, 1669-1677. http://dx.doi.org/10.1098/rstb.2003.1358

[26] Ross, F.M., Allan, S.M., Rothwell, N.J. and Verkhratsky, A. (2003) A Dual Role for Interleukin-1 in LTP in Mouse Hippocampal Slices. *Journal of Neuroimmunology*, **144**, 61-67. http://dx.doi.org/10.1016/j.jneuroim.2003.08.030

[27] Song, C., Zhang, Y. and Dong, Y. (2012) Acute and Subacute IL-1β Administrations Differentially Modulate Neuroimmune and Neurotrophic Systems: Possible Implications for Neuroprotection and Neurodegeneration. *Journal of Neuroinflammation*, **10**, 59.

[28] Schizas, N., Andersson, B., Hilborn, J. and Hailer, N.P. (2014) Interleukin-1 Receptor Antagonist Promotes Survival of Ventral Horn Neurons and Suppresses Microglial Activation in Mouse Spinal Cord Slice Cultures. *Journal of Neuroscience Research*, **92**, 1457-1465. http://dx.doi.org/10.1002/jnr.23429

[29] Murray, K.N., Girad, S., Holemes, W.M., Parkes, L.M., Williams, S.R., Parry-Jones, A.R. and Allan, S.M. (2014) Systemic Inflammation Impairs Tissue Reperfusion through Endothelin-Dependent Mechanisms in Cerebral Ischemia. *Stroke*, **45**, 3412-3419. http://dx.doi.org/10.1161/STROKEAHA.114.006613

[30] Chapouly, C., Tadesse Argaw, A., Horng, S., Castro, K., Zhang, J., Asp, L., *et al.* (2015) Astrocytic TYMP and VEGFA Drive Blood-Brain Barrier Opening in Inflammatory Central Nervous System Lesions. *Brain*, **138**, 1548-1567. http://dx.doi.org/10.1093/brain/awv077

[31] Kumar, A. and Loane, D.J. (2012) Neuroinflammation after Traumatic Brain Injury: Opportunities for Therapeutic Intervention. *Brain, Behavior, and Immunity*, **26**, 1191-1201. http://dx.doi.org/10.1016/j.bbi.2012.06.008

[32] Finnie, J.W. (2013) Neuroinflammation: Beneficial and Detrimental Effects after Traumatic Brain Injury. *Inflammopharmacology*, **21**, 309-320. http://dx.doi.org/10.1007/s10787-012-0164-2

[33] Ransohoff, R.M., Schafer, D., Vincent, A., Blachère, N.E. and Bar-Or, A. (2015) Neuroinflammation: Ways in Which the Immune System Affects the Brain. *Neurotherapeutics*, **12**, 896-909. http://dx.doi.org/10.1007/s13311-015-0385-3

[34] Noh, M.Y., Cho, K.A., Kim, H., Kim, S.M. and Kim, S.H. (2014) Erythropoietin Modulates the Immune-Inflammatory Response of a SOD1[G93A] Transgenic Mouse Model of Amyotrophic Lateral Sclerosis (ALS). *Neuroscience Letters*, **574**, 53-58. http://dx.doi.org/10.1016/j.neulet.2014.05.001

[35] Fluri, F., Grünstein, D., Cam, E., Ungethuem, U., Hatz, F., Schäfer, J., *et al.* (2015) Fullerenols and Glucosamine Fullerenes Reduce Infarct Volume and Cerebral Inflammation after Ischemic Stroke in Normotensive and Hypertensive

Rats. *Experimental Neurology*, **265**, 142-151. http://dx.doi.org/10.1016/j.expneurol.2015.01.005

[36] Zendedel, A., Habib, P., Dang, J., Lammerding, L., Hoffmann, S., Beyer, C. and Slowik, A. (2015) Omega-3 Polyunsaturated Fatty Acids Ameliorate Neuroinflammation and Mitigate Ischemic Stroke Damage through Interactions with Astrocytes and Microglia. *Journal of Neuroimmunology*, **278**, 200-211. http://dx.doi.org/10.1016/j.jneuroim.2014.11.007

[37] Li, Y., Li, J., Li, S., Li, Y., Wang, X., Liu, B., *et al.* (2015) Curcumin Attenuates Glutamate Neurotoxicity in the Hippocampus by Suppression of ER Stress-Associated TXNIP/NLRP3 Inflammasome Activation in a Manner Dependent on AMPK. *Toxicology and Applied Pharmacology*, **286**, 53-63. http://dx.doi.org/10.1016/j.taap.2015.03.010

[38] Yuan, J., Zou, M., Xiang, X., Zhu, H., Chu, W., Liu, W., *et al.* (2015) Curcumin Improves Neural Function after Spinal Cord Injury by the Joint Inhibition of the Intracellular and Extracellular Components of Glial Scar. *Journal of Surgical Research*, **195**, 235-245. http://dx.doi.org/10.1016/j.jss.2014.12.055

[39] Liu, X.W., Ji, E.F., He, P., Xing, R.X., Tian, B.X., Li, X.D., *et al.* (2014) Protective Effects of the p38 MAPK Inhibitor on NMDA-Induced Injury in Primary Cerebral Cortical Neurons. *Molecular Medicine Reports*, **10**, 1942-1948. http://dx.doi.org/10.3892/mmr.2014.2402

15

Improving Oxygen Binding of Desiccated Human Red Blood Cells

Steingrimur Stefansson[1], David S. Chung[2], Jamie Yoon[3], Won Seok Yoo[4],
Young Wook Park[5], George Kim[6], David Hahn[6], Huyen Le[7], Sung-Jae Chung[6,8],
Stephen P. Bruttig[1], David H. Ho[1*]

[1]HeMemics Biotechnologies Inc., Rockville, MD, USA
[2]Division of Biology & Medicine, Brown University, Providence, RI, USA
[3]Earl Warren College, UC San Diego, CA, USA
[4]Department of Chemistry, Michigan University, Ann Arbor, MI, USA
[5]Seoul International School, Seongnam, South Korea
[6]Fuzbien Technology Institute, Rockville, MD, USA
[7]Nauah Solutions, Mclean, VA, USA
[8]Marymount University School of Arts & Science, Arlington, VA, USA
Email: [*]dho@hememics.com

Abstract

Desiccating human red blood cells (RBCs) to increase their storage life has been the subject of intense research for a number of years. However, drying RBCs invariably compromises their integrity and has detrimental effects on hemoglobin function due to autoxidation. We have previously demonstrated an RBC desiccation and rehydration process that preserves RBC antigenic epitopes better than frozen RBCs. This study expands on those observations by examining what effects this desiccation process has on RBC hemoglobin function with respect to oxygen binding properties. In this paper, we examined RBCs from normal donors which were desiccated to 25% moisture content and stored dry for 2 weeks at room temperature prior to rehydration with plasma followed by structural and functional studies. Our data showed that approximately 98% of the RBCs were intact upon rehydration based on hemolysis assays. Oxygen dissociation curves for the desiccated/rehydrated RBCs showed a left shift compared to fresh RBCs (pO$_2$ = 17 mmHg vs. 26 mmHg, respectively). The desiccated/rehydrated RBCs also showed an increase in methemoglobin compared to fresh RBCs (4.5% vs 0.9%, respectively). 2,3-Diphosphoglycerate concentration of the desiccated/rehydrated RBCs was reduced by 20%. In conclusion, although this RBC dehydration process preserves RBC integrity and hemoglobin oxygen binding properties better than most other dehydration techniques described so far, further optimization and long-term studies are needed to make this procedure acceptable for human transfusion.

---

[*]Corresponding author.

Keywords

Desiccation, Red Blood Cells, Oxygen Binding, Hemolysis, Methemoglobin, 2,3-Diphosphoglycerate

1. Introduction

Stabilization of RBCs that allows for dried storage at room temperature, without causing prohibitive hemolysis and hemoglobin oxidation, could potentially change the future of blood banking because it offers numerous advantages over 4°C and frozen storage. Dry storage that could extend the product's shelf life, will ease logistical problems, facilitate efficient transportation and save energy by eliminating the need for refrigeration [1].

Currently, the only viable alternative to storing RBCs at 4°C is freezing. However, freezing cells in the absence of cryoprotective agents cause irreparable cellular damage because ice crystals puncture membranes and organelles [2]. Apart from the physical damage, ice crystals inflict on cellular structures; the freeze-drying process itself can damage and/or oxidize proteins, including hemoglobin (Hb) [1]. Hb inside RBCs is also vulnerable to oxidation in the freeze-drying process. Oxidized Hb (met-hemoglobin) does not bind oxygen [3]. The oxidation of Hb within the RBCs can also lead to formation of intracellular reactive oxygen species that can damage lipids and structural proteins [4]. Attempts at freeze-drying RBCs have mostly resulted in poor cellular recovery [1] [2] [4].

Many factors can influence recovery of functional RBCs after freeze-drying. This includes addition of cryoprotectants (e.g. glycerol and DMSO), lyoprotectants (e.g. trehalose and polymers), manipulating the operating condition during the drying process (primary and secondary drying) and the rehydration process. As of yet, recovery of intact RBCs after drying and rehydration has been reported to be between 50% and 85% [5]-[8], but because of elevated Hb oxidation, no RBC desiccation procedures meet the standards for transfusion at present [1] [4].

Here, we present a storage process for RBCs that utilizes desiccation. The desiccation process does not involve freezing; thus ice-crystal damage and hemolysis is avoided. The overall desiccation/rehydration process involves several steps including incubation in HemSol™ which stabilizes the RBC morphology prior to desiccation. HemSol™ contains a proprietary mixture of non-toxic high and low molecular weight carbohydrates, including trehalose, mannose and dextrans that protect RBC integrity and protein function during the desiccation and rehydration process. Previously we demonstrated that RBCs desiccated and stored in HemSol™ for over 3 weeks at room temperature can be used, in lieu of fresh RBCs, as standards for hemagglutination assays [8].

This RBC desiccation process did not involve freezing. HemSol™ treated RBCs were dried in a vacuum oven at low temperature (42°C) until approximately 25% of the water was remaining. That study showed that our HemSol™ desiccation process preserves RBC morphology and blood typing epitopes better than RBCs that had been frozen. This opens up the possibility of using desiccated RBCs to archive serological phenotypes for use in reference laboratories.

In this study we expand on these previous results by examining whether HemSol™ treatment alters the main physiological property of RBCs, namely the oxygen carrying capacity of Hb. RBCs used in this study were stored desiccated in HemSol™ at room temperature for 2 weeks prior to being rehydrated. Our results demonstrate that the desiccated/rehydrated cells have functional Hb. Compared to fresh RBCs, desiccated and rehydrated RBCs were 98% ± 1% intact based on hemolysis and could be used for these assays directly without a washing step. The rehydrated RBCs demonstrated a left shift in the pO_2 (p50) oxygen saturation value compared to fresh control RBCs (17.10 mmHg vs. 26.48 mmHg, respectively). The reasons for this left shift could be due to increased methemoglobin levels (4.5%) and/or decreased 2,3-diphosphoglycerate (DPG) levels (20%).

To our knowledge, this is the first time these types of experiments have been performed on Hb function in RBC that have been desiccated for 2 weeks at RT. This is a promising start for developing a procedure for producing desiccated RBCs suitable for human transfusion, but further studies are needed to optimize this process.

2. Materials and Methods

Red Blood Cells and Morphology: In-dated human blood, no older than 1 - 3 day was obtained from the Red Cross (American Red Cross, Rockville, MD) as packed red cells. Cells were checked for count per unit volume, morphology, hemolysis and crenated cells. Evaluations were performed according to standard manual methods for counting (hemocytometer), microscopy and digital photo-microscopy. For morphology experiments, the RBCs were divided into 2 equal parts, one part to be used for desiccation process and room temperature storage and the other part was kept refrigerated at 4°C as a control.

Aliquots of the reconstituted RBCs were diluted 1:10,000 in saline and examined for morphology under a compound microscope at 400× magnification. For each sample, 5 independent fields each containing at least 100 RBCs were examined.

Desiccation and Rehydration of Red Blood Cells: RBCs were mixed with HeMemics' desiccation buffer called HemSol™ as described previously [8]. Briefly, a HEPES buffer pH 7.4 contained a proprietary ratio of low molecular weight sugars (Trehalose, Mannitol and Glucose) and high molecular weight sugar Dextran. For these experiments, RBCs were washed with saline 3 times using low speed centrifugation (100 ×g for 10 minutes) to remove residual plasma. The packed RBCs were then mixed with HemSol™ at 1:1 ration and allowed to equilibrate at room temperature for 30 minutes. Then, 0.5 mL volumes of the red blood cells mixed with HemSol™ were dried in 20 mL screw top glass vials. Specifically, aliquots of the blood-desiccation buffer mix which were dispensed into glass vials were loaded into a vacuum drying oven and dehydrated under mild heat (42°C) until visibly dry. The percent moisture of the resulting RBC preparations was determined by *gravimetric* drying *method* using the Sartorius MA45 Gravimetric Moisture Balance (PWB, Bradford, MA). The samples were then capped and stored at room temperature (23°C - 25°C) until use.

For the rehydration process, the desiccated cells were rehydrated in human plasma. Plasma (1 mL) was added to the cells in the vial and allowed to incubate for 30 minutes at 37°C with gentle swirling or vial rotation to insure uniform rehydration. To ensure sufficient volume for testing once rehydration was completed, contents of several vials were combined and adjusted to 45% hematocrit with human plasma for testing.

Post-Rehydration Evaluation for Hb-Oxygen Dissociation Curve Analysis: After 2 weeks of storage in Hem-Sol™ at RT, the desiccated blood was reconstituted as described above and oxygen dissociation curve testing was conducted. The Hemox-Analyzer (TSC Scientific Corporation, New Hope, PA) was used to determine the oxygen dissociation curve (ODC) by exposing 50 µL of blood to an increasing partial pressure of oxygen and deoxygenating it with nitrogen gas. A Clark oxygen electrode was used to detect the change in oxygen tension, which was recorded with on the x-axis of an x-y recorder. The resulting increase in oxyhemoglobin fraction was monitored simultaneously by dual wavelength spectrophotometry at 560 nm and 576 nm and displayed on the y-axis. For the assays, reconstituted RBC (45% HCT) prepared in plasma with heparin anticoagulant was kept on wet ice until the assay. Fifty µL of RBC was diluted in 5 µL of Hemox buffer (7.4 ± 0.01). The sample-buffer was drawn into a cuvette and the temperature of the mixture was equilibrated and brought to 37°C; the sample was oxygenated to 100% with air. After adjustment of the pO_2 value, the sample was deoxygenated with nitrogen; during the deoxygenation process the curve was recorded on graph paper. The pO_2 (p50) value was extrapolated on the x-axis as the point at which O_2 saturation was 50%. Two different lots of desiccated RBCs were tested and from each lot, samples were tested in triplicate.

Determination of Hemolysis: A sample of control and reconstituted RBCs were centrifuged at 3000 ×g for 10 min to pellet the cells. Absorbance of the supernatant was measured at 450 nm and compared to the absorbance of the lysed pellet to give % hemolysis.

Met-Hb Determination: To lyse RBC, 0.5 mL of desiccated/reconstituted RBC at 45% hematocrit was washed in saline 3 times via centrifugation at 500 g for 15 minutes. To the RBC pellet, 5.5 mL of distilled water was added and allowed to stand for 5 min at RT. The samples was diluted with 4 mL of 0.1 M PBS, pH 6.8, mixed well and centrifuged at 3000 ×g for 10 min. The supernatant was removed for Met Hb analysis. Briefly, 0.5 mL of 4% (w/vol) $K_3Fe(CN)_6$ was added to 3 mL RBC supernatant and allowed to sit for 10 min, and A630 nm measured using a spectrophotometer. To calculate baseline, 0.5 mL water was added to a fresh 3 mL sample of RBC supernatant and absorbance was measured at 630 nm. The reaction was then neutralized by the addition of 0.050 mL 5% (w/vol) KCN and after 5 min. absorbance 630 nm was measured again. Met-Hb was calculated as the ratio of the absorbance change at 630 nm after adding KCN to the RBC supernatant. Control RBC sample was treated in the same manner to extract Met-Hb for measurements. All samples were tested in triplicate.

Measurement of 2,3-DPG (DPG): The assay was performed on 2 week old RBC samples stored desiccated in HemSol™ and fresh RBCs using a commercially available human DPG ELISA Kit (antibodiesonline.com, Atlanta, GA). The microtiter plate provided in this kit had been pre-coated with an antibody specific to DPG. Procedures were well described in the protocol provided with the kit. Briefly, the RBC supernatant prepared as described in the Met-Hb determination section was added to the appropriate microtiter plate wells with a bio-tin-conjugated polyclonal antibody preparation specific for DPG. Avidin conjugated to Horseradish Peroxidase (HRP) was added to each well and incubated, followed by addition of TMB substrate solution. The peroxidase reaction was terminated by the addition of a sulphuric acid solution, and A450 was measured in a spectropho-tometer. The concentration of DPG in the samples was calculated from a DPG standard curve. Control RBC sample was treated in the same manner to extract Met-Hb for measurements. All samples were tested in tripli-cate.

3. Results

Red blood cells were desiccated to different levels of dryness and then the percentage of overall hemolysis was determined. Samples were reconstituted with plasma and assayed for hemolysis as described. **Table 1** shows that removal of more than 25% of residual moisture resulted in an increase in cell fragility upon reconstitution with increased time of storage. Thus, based on these observations, the subsequent data were performed on RBC at 25% residual moisture at 2 week storage cut off.

Changes in RBC morphology after the desiccation/rehydration process were evaluated by gross microscopic examination. Typical, healthy red blood cells are round, smooth bi-concave disks with a pink or red color. Other forms of RBCs, such as crenated (partially shrunken) and spherocytes can be present [9]. Crenated RBCs are associated with storage lesion, but also occur in response to high osmotic conditions or drying. Crenation is of-ten reversible and while exposure to the high osmotic load of HemSol™ does cause some crenation, with proper rehydration, most of the crenated cells revert to smooth bi-concave disks after 30 min in human plasma (**Figure 1(a)**). **Figure 1(b)** shows control RBCs kept for 2 weeks at 4°C.

The critical physiologic parameter for RBC preparations is their ability to properly bind, carry and release oxygen. The determinations of this ability are conducted by exposing the RBCs to oxygen to assure complete loading, and then exposure to complete nitrogen to determine the continuous unloading of the oxygen (oxyhe-moglobin dissociation curve). Typical oxyhemoglobin dissociation curve determinations are presented in **Figure 2** and demonstrate a sigmoidal curve for both the fresh control RBCs and the desiccated/rehydrated RBCs, indi-cating cooperative oxygen binding.

Met-Hb is a measure of hemoglobin oxidation and is expressed as a proportion of total hemoglobin. Met-Hb is a naturally occurring oxidized metabolite of hemoglobin; under normal physiologic conditions, or in fresh RBCs, around 1% of total Hb is Met-Hb [10]. In desiccated/reconstituted RBCs, Met-Hb levels were 4.56% ± 0.15% upon reconstitution. Fresh control RBCs showed 0.9% ± 0.12% Met-Hb (**Figure 3(a)**).

Table 1. Correlation of RBC hemolysis to moisture content.

% Residual Moisture	% Hemolysis at Day 1	% Hemolysis at Day 7	% Hemolysis at Day 14	% Hemolysis at Day 21
10 ± 0.8	9.5 ± 1.1	11.4 ± 1.4	11.8 ± 2.2	15.5 ± 2.4
15 ± 0.7	5.5 ± 1.3	6.1 ± 1.1	7.3 ± 1.8	10.1 ± 2.1
20 ± 0.8	2.0 ± 0.7	2.5 ± 0.6	3.0 ± 0.8	5.5 ± 1.2
25 ± 1.0	0.4 ± 0.2	0.5 ± 0.1	0.6 ± 0.3	1.5 ± 0.5
30 ± 1.2	0.3 ± 0.2	0.4 ± 0.1	0.7 ± 0.2	1.8 ± 0.5
35 ± 1.3	0.5 ± 0.1	0.4 ± 0.3	0.4 ± 0.2	0.9 ± 0.3
40 ± 1.3	0.3 ± 0.2	0.5 ± 0.4	0.7 ± 0.3	0.8 ± 0.2
45 ± 1.2	0.5 ± 0.3	0.6 ± 0.2	0.5 ± 0.3	1.0 ± 0.2
50 ± 1.1	0.2 ± 0.1	0.5 ± 0.2	0.7 ± 0.3	0.8 ± 0.2
55 ± 1.3	0.5 ± 0.3	0.4 ± 0.1	0.6 ± 0.2	0.8 ± 0.4
Control RBC	0.4 ± 0.2	0.3 ± 0.1	0.4 ± 0.1	0.6 ± 0.1

Data from 5 lots of RBC purchased commercially. Each samples were tested in triplicate ± Standard deviation. Samples were pooled from storage and tested at time intervals indicated. Control RBC samples were stored refrigerated in accordance with FDA requirements.

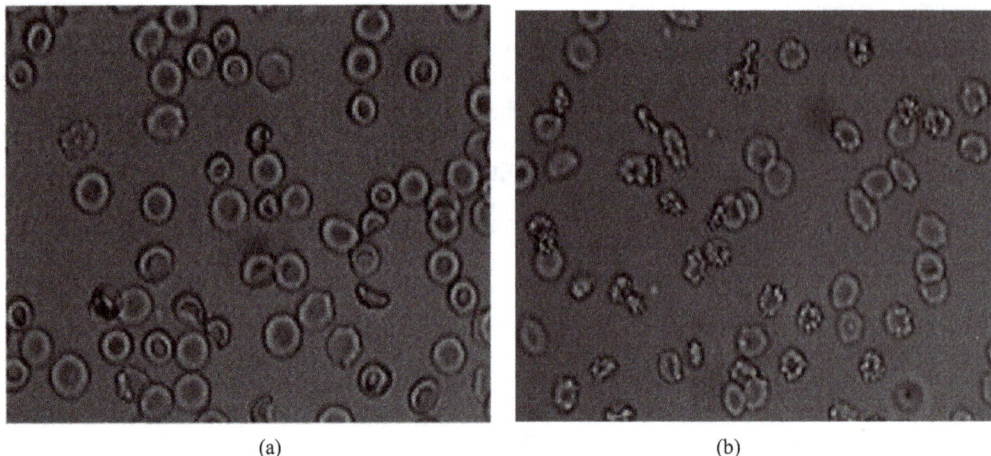

(a)   (b)

Figure 1. Light field images of RBCs. (a) RBCs stored desiccated for 2 weeks and reconstituted in human plasma and (b) RBCs stored for 2 weeks at 4˚C (400× magnification).

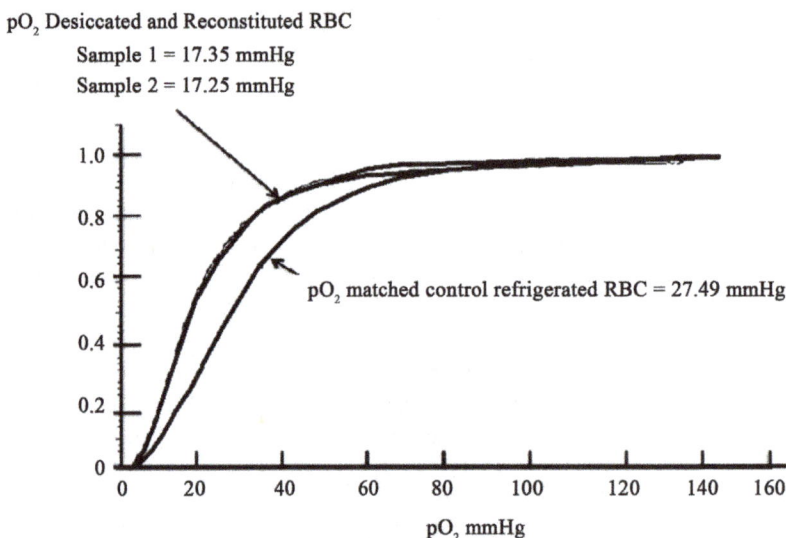

Figure 2. Oxygen binding and release profile for reconstituted RBCs. Dried RBCs were prepared in two separate sets from two different donors and kept for 2 weeks at room temperature before testing. The fresh and rehydrated RBCs (from both sets) showed similar oxygen saturation kinetics (99.4%). The P50 of the rehydrated RBCs (17.35 for set 1 and 17.25 for set 2) were somewhat reduced compared to the fresh RBC (P50 = 27.49). These assays were performed in triplicate.

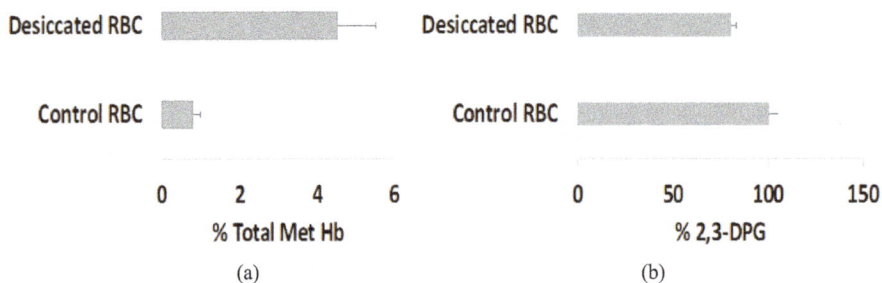

Figure 3. Measurements of Met Hb and DPG levels in RBC. (a) Met-Hb content increased in desiccated RBC compared to fresh RBC; (b) The assay for DPG content showed a decrease of approximately 20% for desiccated RBCs upon reconstitution compared to fresh RBCs. These assays were performed in triplicate.

2,3-Diphosphoglycerate (DPG) is an important regulator of Hb oxygen affinity and we measured DPG concentrations in desiccated/rehydrated RBCs stored for 2 weeks at room temperature in HemSol™. Using fresh RBCs as control, we found that the concentrations of DPG in the desiccated/rehydrated RBCs were reduced by 20 % (**Figure 3(b)**).

4. Discussion

RBCs destined for transfusion must be stored refrigerated at 4°C with a maximum allowable storage time of 42 days [9] [11]. Freezing RBCs and storing them in frozen state is an alternative to cold storage. Subzero temperatures decrease most biochemical reactions and frozen RBCs have been shown to have a longer shelf life [12]. However, the cryopreservation technique has disadvantages because of the need to introduce sometimes toxic antifreezing agents (e.g. DMSO, high concentrations of glycerol) into the blood before the freezing process and removing them before transfusion, all the while maintaining sterility. Because of the labor-intensive and technically demanding nature of processing coupled with the high cost, frozen RBCs are not in routine clinical use. Furthermore, freezing RBCs, even in the presence of cryoprotective agents does not eliminate hemolysis or hemoglobin oxidation. The damage caused by freezing and thawing also leaves behind cellular debris and materials unacceptable for transfusion [1] [13].

Drying RBCs in the absence of freezing step and keeping them functional is even more challenging than freeze-drying [13]. Most biochemistry occurs in an aqueous environment but a few organisms can survive extensive dehydration, a process termed anhydrobiosis. Drying cells without destroying them requires the presence of lyoprotectants. The most common lyoprotectant found in nature is the disaccharide trehalose, although trehalose alone is not sufficient to ensure full integrity upon rehydration [14].

The HemSol™ formulation used in this study contains trehalose along with other carbohydrates that have been shown to be non-toxic in humans. This RBC drying process does not involve freezing and has been shown to preserve RBC integrity and antigenicity better than freezing [8]. For this study we used an improved HemSol™ formulation that recovered 98% ± 1% of the RBCs after reconstition compared to 50% as we reported previously [8]. In contrast, most studies on drying and reconstituting RBCs have reported recoveries between 20% - 50% [5]-[8].

Before assessing the oxygen binding properties of RBC, we performed experiments to determine the optimal moisture content and storage time for the desiccated RBC. Our data showed that the 25% residual moisture was the optimal level of dryness for RBC using our process. This data also suggested that removal of more than 75% of cell water content may compromise cell membrane structure that could cause higher breakage upon reconstitution. Furthermore, we also showed that beyond 2 weeks of storage, the cells also became more fragile upon reconstitution as shown in **Table 1**. It is possible that a longer storage time at room temperature may further decrease the moisture content of the cells. As such, in this study, we confined our functional studies to 2 weeks in storage of cells at 25% residual moisture.

We examined the morphology of the reconstituted RBCs after a 2 week storage in HemSol™ at room temperature. The RBCs were then hydrated with plasma as described in the materials and methods and imaged under a microscope. **Figure 1(a)** show representative fields of the rehydrated RBCs stored in HemSol™ at room temperature for 2 weeks and control RBCs (**Figure 1(b)**) stored in Adsol™ at 4°C for 2 weeks, respectively. As seen in this figure, the majority of HemSol™ stored RBCs show typical smooth biconcave structures, while the majority of Adsol stored RBCs showed some crenation. These results demonstrate that the RBCs treated with HemSol™ show less ultrastructural damages than RBCs stored at 4°C in Adsol™ for the same time period.

The main function of RBCs is to carry oxygen to tissues. The ability of RBCs to bind and release oxygen is influenced by many factors including Hb oxidative state and regulators of Hb oxygen binding [9]. The standard way to measure this property is to perform oxygen saturation studies. **Figure 2** shows oxygen dissociation curves of RBCs stored desiccated in HemSol™ for 2 weeks. The reconstituted RBCs were compared to fresh RBCs and there is a left shift in the HemSol™ treated RBCs compared to fresh RBCs (pO_2 of 26.48 mmHg vs 17.10 mmHg respectively). Both RBC preparations demonstrate a sigmoidal curve, indicating cooperative oxygen binding of hemoglobin. Additionally, both RBC preparations have similar O_2 saturation concentrations, which indicates that most of the Hb in the desiccated/reconstituted RBCs is functional. Interestingly, the pO_2 value of 17 mmHg is comparable to RBCs stored in acid-citrate-dextrose-adenine for approximately 5 days at 4°C [15].

To further examine Hb oxidation levels in desiccated cells, we measured the levels of methemoglobin (Met-Hb), which is an indication of oxidative stress associated with freeze-drying and desiccation [1] [13]. The heme group in Met-Hb has been oxidized from the oxygen carrying ferrous (Fe^{2+}) state to ferric (Fe^{3+}). Autoxidation of Hb to Met-Hb occurs naturally under normal physiologic conditions, accounting for around 1% of total Hb [3], but prolonged exposure of stored blood to atmospheric oxygen levels can increase Met-Hb levels. Desiccation of RBCs has been shown to increase the Met-Hb levels to 15% - 50% [1] [13]. **Figure 3(b)** shows that the Met-Hb was 4.5% in desiccated/reconstituted RBCs, compared to 0.9% in control RBCs. This data supports the results of **Figure 2**, which indicates that the majority of Hb in the desiccated/resuspended RBCs is functional. Although 4.5% Met-Hb is lower than most desiccation procedures have reported, it is still higher than physiological concentrations (0% - 3%) [3].

Since the data shown in **Figure 2** and **Figure 3(a)** suggest that most of the Hb in desiccated/resuspended is capable of binding oxygen, we examined DPG levels in the HemSol™ treated RBCs to account for the left shift in the oxygen saturation experiment (**Figure 2**). This left shift is consistent with lower concentrations of 2,3-diphosphoglycerate [16].

DPG is an important regulator of Hb oxygen binding. It binds deoxygenated Hb, enhancing release of remaining bound oxygen. We measured the concentrations of DPG in the desiccated/reconstituted RBCs and those results shown in **Figure 3(b)** demonstrate there is a 20% reduction in DPG concentrations in the HemSol™ treated cells. This is consistent with studies of RBC storage lesions, in which the steady depletion of DPG leads to a left shift in Hb oxygen binding properties.

5. Conclusion

Here, we describe a storage process for red cells that utilizes desiccation rather than refrigeration or freezing. The drying process does not involve freezing, thus ice-crystal damage and subsequent debris formation. It is clear that more work is needed to show HemSol™ RBC desiccation as an alternative to regular RBC storage; however, to our knowledge, the present study shows that this process is superior to most other RBC desiccation procedures with respect to preserving Hb oxygen binding properties during desiccation and storage at room temperature for 2 weeks.

Acknowledgements

The authors wish to acknowledge the financial support of the Fort Detrick Technology Transfer Initiative, Ft. Detrick, MD and the award, administrative and contract support of TEDCO (the Maryland Technology Development Corporation, Columbia, MD). In addition, HeMemics appreciates the enthusiastic support thru a Cooperative Research and Development Agreement with US Army Institute of Surgical Research Fort Sam Houston, Texas.

References

[1]   Kanias, T. and Acker, J.P. (2010) Biopreservation of Red Blood Cells—The Struggle with Hemoglobin Oxidation. *FEBS Journal*, **277**, 343-356. http://dx.doi.org/10.1111/j.1742-4658.2009.07472.x

[2]   Meryman, H.T. (1971) Osmotic Stress as a Mechanism of Freezing Injury. *Cryobiology*, **8**, 489-500. http://dx.doi.org/10.1016/0011-2240(71)90040-X

[3]   Mansouri, A. and Lurie, A.A. (1993) Methemoglobinemia. *American Journal of Hematology*, **42**, 7-12. http://dx.doi.org/10.1002/ajh.2830420104

[4]   Lynch, A.L., Dury, B.A.P., Guyader, C.P.E. and Slater, N.K.H. (2011) Sugars Comparable to Glutathione as Hemoglobin Oxidation Protectants During Vacuum Drying. *Biopreservation and Biobanking*, **9**, 411-413. http://dx.doi.org/10.1089/bio.2011.0021

[5]   Török, Z., Satpathy, G.R., Banerjee, M., Bali, R., Little, E., Novaes, R., Ly, H.V., Dwyre, D.M., Kheirolomoom, A., Tablin, F., Crowe, J.H. and Tsvetkova, N.M. (2005) Preservation of Trehalose-Loaded Red Blood Cells by Lyophilization. *Cell Preservation Technology*, **2**, 96-111. http://dx.doi.org/10.1089/cpt.2005.3.96

[6]   Weinstein, R., Sowemimo-Coker, S.O. and Goodrich, R.P. (1995) Survival of Lyophilized and Reconstituted Human Red Blood Cells *in Vivo*. *Transfusion Clinique et Biologique*, **2**, 427-432. http://dx.doi.org/10.1016/S1246-7820(05)80067-9

[7]   Han, Y., Quan, G.B., Liu, X.Z., Ma, E.P., Liu, A., Jin, P. and Cao, W. (2005) Improved Preservation of Human Red

Blood Cells by Lyophilization. *Cryobiology*, **51**, 152-164. http://dx.doi.org/10.1016/j.cryobiol.2005.06.002

[8]   Ho, D., Schierts, J., Zimmerman, Z., Gadsden, I. and Bruttig, S. (2009) Comparison of Frozen versus Desiccated Reference Human Red Blood Cells for Hemagglutination Assays. *Transfusion*, **49**, 2173-2180. http://dx.doi.org/10.1111/j.1537-2995.2009.02270.x

[9]   Klein, H.G., Spahn, D.R. and Carson, J.L. (2007) Red Blood Cell Transfusion in Clinical Practice. *The Lancet*, **370**, 415-426. http://dx.doi.org/10.1016/S0140-6736(07)61197-0

[10] Faivre, B., Menu, P., Labrude, P. and Vigneron, C. (1998) Hemoglobin Autoxidation/Oxidation Mechanisms and Methemoglobin Prevention or Reduction Processes in the Bloodstream Literature Review and Outline of Autoxidation Reaction. *Artificial Cells, Blood Substitutes and Biotechnology*, **26**, 17-26. http://dx.doi.org/10.3109/10731199809118943

[11] Holme, S., Elfath, M.D. and Whitley, P. (1998) Evaluation of *in Vivo* and *in Vitro* Quality of Apheresis-Collected RBC Stored for 42 Days. *Vox Sanguinis*, **75**, 212-217. http://dx.doi.org/10.1046/j.1423-0410.1998.7530212.x

[12] Hess, J.R. (2004) Red Cell Freezing and Its Impact on the Supply Chain. *Transfusion Medicine*, **14**, 1-8. http://dx.doi.org/10.1111/j.0958-7578.2004.00472.x

[13] Kanias, T. and Acker, J.P. (2010) Mechanism of Hemoglobin-Induced Cellular Injury in Desiccated Red Blood Cells. *Free Radical Biology and Medicine*, **49**, 539-547. http://dx.doi.org/10.1016/j.freeradbiomed.2010.04.024

[14] Loi, P., Iuso, D., Czernik, M., Zacchini, F. and Ptak, G. (2013) Towards Storage of Cells and Gametes in Dry Form. *Trends in Biotechnology*, **31**, 688-695. http://dx.doi.org/10.1016/j.tibtech.2013.09.004

[15] Franklin Bunn, H., May, M.H., Kocholaty, W.F. and Shields, C.E. (1969) Hemoglobin Function in Stored Blood. *Journal of Clinical Investigation*, **48**, 311-321. http://dx.doi.org/10.1172/JCI105987

[16] Benesch, R., Benesch, R.E. and Yu, C.I. (1968) Reciprocal Binding of Oxygen and Diphosphoglycerate by Human Hemoglobin. *Proceedings of the National Academy of Sciences of the United States of America*, **59**, 526. http://dx.doi.org/10.1073/pnas.59.2.526

16

Centrosome Functions as a Molecular Dynamo in the Living Cell

Yue Zhao

Department of Radiation Oncology, University of Texas Southwestern Medical Center, Dallas, USA
Email: alexanderyz@gmail.com

Abstract

Recent development in the field of quantum biology highlights that the intracellular electromagnetic field (EMF) of microtubules plays an important role in many fundamental cellular processes such as mitosis. Here I propose an intriguing hypothesis that centrosome functions as molecular dynamo to generate electric flow over the microtubules, leading to the electric excitation of microtubule EMF that is required for spindle body microtubule self-assembly. With the help of motors proteins within the centrosome, centrosome transforms the energy from ATP into intracellular EMF in the living cell that shapes the functions of microtubules. There will be a general impact for the cell biology field to understand the mechanistic function of centrosome for the first time in correlation with its structural features. This hypothesis can be tested with technics such as super resolution live cell microscope.

Keywords

Centrosome, Dynamo, Electro Magnetic, Microtubule, Electron

1. Introduction

Centrosome was first discovered by Theodor Boveri in the 1880's [1], it is the key organelle that is responsible for mitosis and meiosis in metazoan lineage of eukaryotic cells [2]. In animal cells, centrosome regulates the nucleation and spatial organization of microtubules, functioning as the primary microtubule-organizing center (MTOC) [3]. The centrosome is comprised of two centrioles that are surrounded by pericentriolar material (PCM). The two centrioles are perpendicularly arranged, one centriole has additional appendages at the end farthest from the other centriole (distal) and is called the mother or maternal centriole, the subdistal appendages of the maternal centriole also act like microtubule-anchoring sites [4] [5]. Pelletier *et al.* reported that subdiffraction imaging of centrosomes revealed pericentriolar material which had higher-order organizational features. Centrosome components adopt a toroidal pattern with progressively larger, overlapping diameters around the proximal end of the mother centriole in interphase cells. On one side, the toroid is slightly opened (gap) in the area where the daughter centriole is positioned [6], this higher order structural feature of centrosome may help the mother and daughter centriole to form an orthogonal configuration. In most of cases, each centriole is composed with 9 MT triplets and is ~0.5 μm in length and 0.2 μm in diameter [7] [8].

Recent studies in the field of quantum biology point to the possibility that electric magnetic interactions may involve in many fundamental cellular processes [9] [10]. In particular, the electromagnetic property of microtubule has been reported with both computation modelling and experimental evidences, Cifra *et al.* used computation model to simulate the electric pulse moving along microtubules, Bandyopadhyay *et al.* reported that nano sized electric pump was required for the self-assembly of microtubules in live cell simulator[11]-[13]. Medical treatment of cancer with cancer cell specific interfering EMF has been developed to disrupt the mitotic spindle microtubules of cancer cells [14]. Centriole produces an electromagnetic field apparently due to the longitudinal oscillation of its microtubules (MTs). Centrosome clustering is a hallmark of cancer cells. A cluster of centrioles is therefore presumed to produce an enhanced electromagnetic field. It is possible to target cancer cells using nano particles based on the enhanced electromagnetic property of cancer cells [15]. However, some important questions remain unanswered, what is the energy source of the intracellular electric field and what is the molecular mechanism that leads to the excitation of intracellular electric field? ATP is the most common cellular energy source. To transform the chemical energy in ATP into electric magnetic field within the living cell, cell needs to have a molecular dynamo to transform the mechanistic movement of protein complexes to directional movements of intracellular electrons, leading to the electric excitation of the spindle body microtubules as well as the M phase chromosomes, which is essential for mitosis [9].

2. Hypothesis

Here I present a novel hypothesis that centrosome functions as a molecular dynamo in the living cell to generate electric current from the cytosol electrolyte to the spindle body and M phase chromosome, leads to the electric excitation of the spindle body and chromosome during mitosis. Based upon the structure of the centrosome, there is one microtubule in the center, and 9 microtubule triplicate outside, connected by motor proteins such as dynein and kinesin [3] (https://vimeo.com/58347006). The mechanistic movement of these motor proteins will trigger the rotation of the microtubules triplets forming the barrel structure of the centriole to rotate around the center microtubule. The rotation and electric oscillation of each centriole will generate a dynamic electromagnetic field that mimic the physical structure of the centriole, and the orthogonal arrangement of centrioles of each centrosome will result in the microtubules of the barrel structure of each centriole to cut the electromagnetic field generated by the other centriole when rotating (**Figure 1**). Such a natural design makes centrosome to function as a molecular dynamo, generating directional electron flow through the dipolar structure of each individual microtubule in the centrosome, transforming the energy from ATP to electric current. During mitosis, centrosome is known to locate at the microtubule organizing center (MTOC), and only the mother centriole contains the sub-distal appendages that connect with spindle body microtubules, which allow the electrons to move from centrosome to the spindle body microtubules.

During mitosis chromosomes are connected with spindle body microtubules with K-fibres, which are microtubule bundles that join kinetochores to the spindle poles. Pericentriolar material (PCM) is composed primarily of hyaluronic acid (HA) and has a similar negative charge density as DNA [16]. The electrons flow over the spindle body leads to the electric excitation of the spindle body and chromosome, generating an enhanced intracellular electromagnetic field during mitosis, which is consistent with the observation of yeast cell at M phase [17] [18]. Bandyopadhyay *et al.* reported nano-sized electromagnetic pumping is required for microtubule self-assembly [13], the electric excitation of spindle body microtubules is required for the self-assembly and growth of the spindle body microtubules during M phase. The basal body of cilium is a centrosome like cellular organelle [19], similar molecular mechanism is applied for the basal body centriole to generate electric flow over the cilium microtubules, forming the nano electromagnetic field that is required for the self-assembly of cilium microtubule. Collectively, my hypothesis proposes centrosome is at the center of the electric network that continuously drawn the cellular chemical energy from ATP to feed the intracellular electromagnetic field.

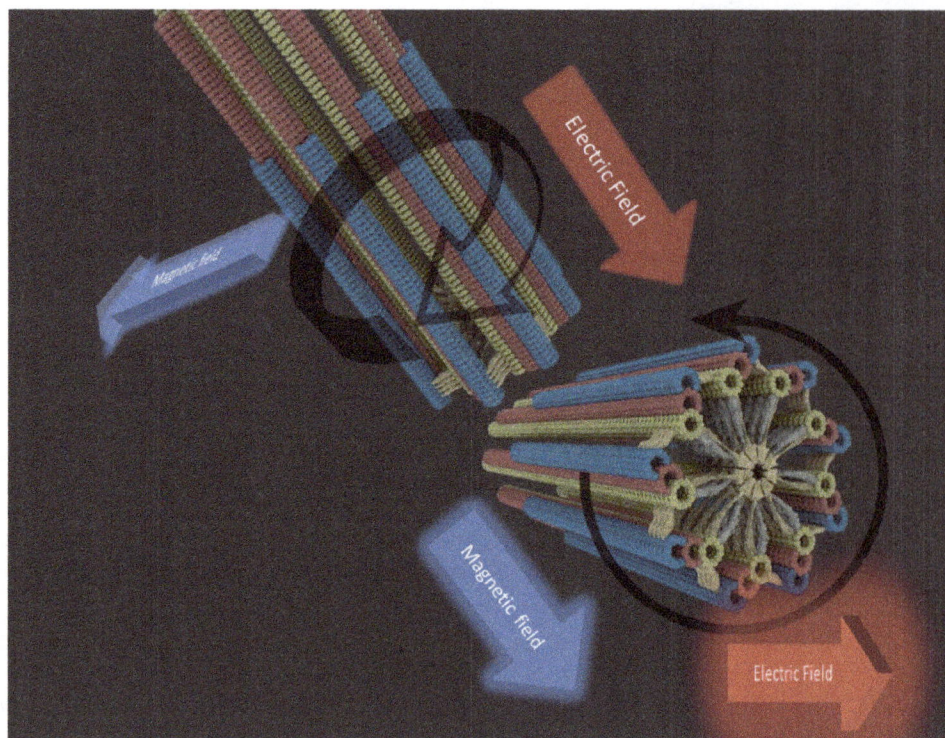

Figure 1. Cartoon illustration of how centrosome functions as a molecular dynamo within the living cell.

3. Testing of the Hypothesis

I am currently searching collaborators to do some experiment with the super resolution microscopy live cell imaging technique, hopefully to observe the rotation of barrel structure of the centrosome during mitosis in the living cell. Despite of current technical limitations, we still need to develop new methods for both direct measurements or experimental simulations of the intracellular electric field with advanced photonic and nano technologies.

This hypothesis is at the intersection of structural biology and quantum biology, which highlights a new avenue for basic biology researches. The hypothesis is supported by multiple lines of evidences and clearly elucidates the function-structure relationship of centrosome during mitosis for the first time, not only advance our knowledge to the function of centrosome but also have great implications for mitosis, one of the fundamental cellular events in biology.

Conflict of Interests

The author declares no conflict of interests for this hypothesis.

Acknowledgements

I thank Dr. Anirban Bandyopadhyay for his discussion with me about the hypothesis, and Dr. Laurent Schwartz for his interesting questions.

References

[1]   Doxsey, S. (2001) Re-Evaluating Centrosome Function. *Nature Review Molecular Cellular Biology*, **2**, 688-698. http://dx.doi.org/10.1038/35089575

[2]   Bornens, M. and Azimzadeh, J. (2007) Origin and Evolution of the Centrosome. *Advances in Experimental Medicine and Biology*, **607**, 119-129. http://dx.doi.org/10.1007/978-0-387-74021-8_10

[3]   Bettencourt-Dias, M. and Glover, D. (2007) Centrosome Biogenesis and Function: Centrosomics Brings New Understanding. *Nature Review Molecular Cellular Biology*, **8**, 451-463. http://dx.doi.org/10.1038/nrm2180

[4]   Mogensen, M.M. (1999) Microtubule Release and Capture in Epithelial Cells. *Biology of the Cell*, **91**, 331-341. http://dx.doi.org/10.1111/j.1768-322X.1999.tb01091.x

[5]   Chretien, D., Buendia, B., Fuller, S.D. and Karsenti, E. (1997) Reconstruction of the Centrosome Cycle from Cryoe-lectron Micrographs. *Journal of Structural Biology*, **120**, 117-133. http://dx.doi.org/10.1006/jsbi.1997.3928

[6]   Lawo, S., Hasegan, M., Gupta, G.D. and Pelletier, L. (2012) Subdiffraction Imaging of Centrosomes Reveals High-er-Order Organizational Features of Pericentriolar Material. *Nature Cell Biology*, **14**, 1148-1158. http://dx.doi.org/10.1038/ncb2591

[7]   Paintrand, M., Moudjou, M., Delacroix, H. and Bornens, M.J. (1992) Centrosome Organization and Centriole Archi-tecture: Their Sensitivity to Divalent Cations. *Journal of Structural Biology*, **108**, 107-128. http://dx.doi.org/10.1016/1047-8477(92)90011-X

[8]   Bornens, M. (2002) Centrosome Composition and Microtubule Anchoring Mechanisms. *Current Opinion in Cell Bi-ology*, **14**, 25-34. http://dx.doi.org/10.1016/S0955-0674(01)00290-3

[9]   Zhao, Y. and Zhan, Q. (2012) Electric Fields Generated by Synchronized Oscillations of Microtubules, Centrosomes and Chromosomes Regulate the Dynamics of Mitosis and Meiosis. *Theoretical Biology and Medical Modelling*, **9**, 26. http://dx.doi.org/10.1186/1742-4682-9-26

[10]   Zhao, Y. and Zhan, Q. (2012) Electric Oscillation and Coupling of Chromatin Regulate Chromosome Packaging and Transcription In Eukaryotic Cells. *Theoretical Biology and Medical Modelling*, **9**, 27. http://dx.doi.org/10.1186/1742-4682-9-27

[11]   Havelka, D., Kučera O., Deriu M.A. and Cifra M. (2014) Electro-Acoustic Behavior of the Mitotic Spindle: A Semi-Classical Coarse-Grained Model. *PLoS ONE*, **9**, Article ID: e86501. http://dx.doi.org/10.1371/journal.pone.0086501

[12]   Havelka, D., Cifra, M. and Kučera, O. (2014) Multi-Mode Electro-Mechanical Vibrations of a Microtubule: *In Silico* Demonstration of Electric Pulse Moving along a Microtubule. *Applied Physics Letters*, **104**, Article ID: 243702. http://dx.doi.org/10.1063/1.4884118

[13]   Sahu, S., Ghosh, S., Fujita, D. and Bandyopadhyay, A. (2014) Live Visualizations of Single Isolated Tubulin Protein Self-Assembly via Tunneling Current: Effect of Electromagnetic Pumping during Spontaneous Growth of Microtubule. *Scientific Reports*, **4**, 7303. http://dx.doi.org/10.1038/srep07303

[14]   Zimmerman, J.W., Pennison, M.J., Brezovich, I., *et al.* (2012) Cancer Cell Proliferation Is Inhibited by Specific Mod-ulation Frequencies. *British Journal of Cancer*, **106**, 307-313. http://dx.doi.org/10.1038/bjc.2011.523

[15]   Hutson, R.L. (2015) Using the Electromagnetics of Cancer's Centrosome Clusters to Attract Therapeutic Nanoparticles. *Advances in Bioscience and Biotechnology*, **6**, 172-181. http://dx.doi.org/10.4236/abb.2015.63017

[16]   Van Oosten, A.S. and Janmey, P.A. (2013) Extremely Charged and Incredibly Soft: Physical Characterization of the Pericellular Matrix. *Biophysical Journal*, **104**, 961-963. http://dx.doi.org/10.1016/j.bpj.2013.01.035

[17]   Jelínek, F., Cifra, M., Pokorný, J., *et al.* (2009) Measurement of Electrical Oscillations and Mechanical Vibrations of Yeast Cells Membrane around 1 kHz. *Electromagnetic Biology and Medicine*, **28**, 223-232. http://dx.doi.org/10.1080/15368370802710807

[18]   Pohl, H.A., Braden, T., Robinson, S., *et al.* (1981) Life Cycle Alternations of the Micro-Dielectrophoretic Effects in Cells. *Journal of Biological Physics*, **9**, 133-154. http://dx.doi.org/10.1007/BF01988247

[19]   Adams, M. (2010) The Primary Cilium: An Orphan Organelle Finds a Home. *Nature Education*, **3**, 54.

Abbreviations

EMF   electromagnetic field
MTOC   microtubule organizing center
PCM   Pericentriolar material

17

Isolation, Partial Purification and Characterization of Texas Live Oak (*Quercus fusiformis*) Lectin

Ruby A. Ynalvez[1]*, Carmen G. Cruz[1], Marcus A. Ynalvez[2]

[1]Department of Biology and Chemistry, Texas A & M International University, Laredo, Texas, USA
[2]Department of Social Sciences, Texas A & M International University, Laredo, Texas, USA
Email: *rynalvez@tamiu.edu

Abstract

Lectins are carbohydrate-binding proteins with agglutination properties. There is a continuous interest in lectins due to their biological properties that can be exploited for medicinal and therapeutic purposes. The objective of this study was to isolate and characterize lectin activity in Texas Live Oak (*Quercus fusiformis*). More specifically, the study aimed to determine the lectin's blood group specificity and pH stability, determine effects of seasonal variation, soil moisture and soil pH on lectin activity. The study also aimed to determine the presence of antifungal activity in *Q. fusiformis* extracts. Lectin activity was detected and compared via agglutination and protein assays. Protein partial purification was accomplished using diethylaminoethyl ion-exchange chromatography matrix. High Performance Liquid Chromatography (HPLC) was used to assess purity of the lectin. Results showed that *Q. fusiformis* extracts' lectin activities are stable at a pH range of 5.2 - 9.2 but with a significant decrease in activity above pH 9.2. The lectin activity was significantly higher when assayed against sheep red blood cells as compared to other blood groups tested. *Quercus fusiformis* extract is devoid of antifungal activity against *Aspergillus niger* and *Rhizopus stolonifer*. The effects of seasonal variation, soil moisture and soil pH do not significantly correlate with lectin activity. Results from HPLC showed presence of three peaks indicating a partial purification of the *Q. fusiformis* lectin.

Keywords

Lectin, *Quercus fusiformis*, Protein Purification, HPLC, Blood Group Specificity, pH Stability

---

*Corresponding author.

1. Introduction

Lectin is a protein of non-immunological origin and contains at least one non-catalytic carbohydrate-binding site. Lectins' ability to recognize and specifically bind reversibly to carbohydrates distinguishes them among other proteins [1] [2]. Their capability of specifically recognizing and binding to sugars of erythrocytes *in vitro* merits the term agglutinin [2] [3], the process is termed agglutination. Lectins' specificities such as binding to glyco-protein receptors on cellular membranes allow the study of the physiological means in cellular communication such as cell-to-cell recognition [4] [5]. Thus, lectins play a pivotal role in cellular communication and pro-tein-carbohydrate interactions [6] [7]. Lectin's sugar-binding property readily distinguishes them as biotechno-logical tools to define carbohydrate structure and physiological dynamics [8]-[12]. This characteristic is ex-ploited for biomedical research for wide applications [13]-[20]. Sources of lectins within an organism vary from organism to organism. Lectins will also vary in terms of their function, structures, biological activity, concentra-tion, as well as in organ and cellular location [21]-[23].

Isolation and purification of lectins may be done through a variety of protein purification methods [24]-[33]. Methods for purifying lectins vary due to lectin sources (*i.e.* plant or animal). Methods also will depend on lec-tins' structure, specificity, physiochemical properties and biological activity [34]. The combinations of methods for lectin purification will vary. For example, in studying seeds of the pepper plant, *Capsicum annum* lectin was isolated and purified by a four-step procedure [34]. The four-step protein purification procedure for *Capsicum annum* included lyophilization of samples followed by purification using two column chromatography tech-niques. The *C. annum* samples were partially purified with two separate columns, DEAE-cellulose and with a second column, QAE-Sephadex. Lastly, partially purified *C. annum* samples were subjected to affinity column chromatography, using Sephadex G-100 column [34]. Another example was lectin isolation and purification from *Artocarpus camansi* Blanco seeds, which was a three-step procedure. Blanco seed lectin was isolated with the ammonium sulfate precipitation (salting out) technique. Isolated lectin was then dialyzed and applied to gel-filtration column chromatography using a Sephadex G-200 column [35]. Another method was the extraction of lectin from fresh leaves of *Kalanochoe crenata*. The lectin was isolated and purified in a two-step procedure, one was ion-exchange column chromatography, DEAE-cellulose and second was gel filtration column chroma-tography, using Sephadex G-100 [36].

Texas Live Oak (*Quercus fusiformis*) is a thicket-forming shrub (or a large spreading tree). It is a drought-to-lerant and cold-hardy plant compared with its look-a-like *Quercus virginiana*. The Texas Live Oak is native to Oklahoma and in north central, central and southern Texas [37] [38]. *Q. fusiformis* belongs to the family Faga-ceae and may also be known as Escarpment Live Oak, Plateau Live Oak, Scrub Live Oak and West Texas Live Oak [37]. Several native plant species of Texas have yet to be examined as potential lectin sources, including Texas Live Oak. Previous study reported lectin to be present in the leaf, stem and fruit of *Q. fusiformis* with the highest lectin activity expressed in the leaves [22]. In this regard, the objective of this study was to isolate, par-tially purify and characterize leaf lectin activity from *Q. fusiformis*. This study focused on conventional methods of protein purification, which included centrifugation and column chromatographic techniques. High Perfor-mance Liquid Chromatography (HPLC) assessed the purity of the lectin. The lectin was characterized in terms of its animal blood group specificity, pH stability, seasonal variation's effect on lectin activity, and soil moisture and soil pH effect on lectin activity. The antifungal property of the *Q. fusiformis* extract was also determined. The research study was significant since it contributed to the inventory of lectins found in plants. It also estab-lished an efficient and effective protocol to isolate and partially purify lectin from *Q. fusiformis*.

2. Methodology

2.1. Sample Collection

Leaf samples were collected from Texas Live Oak trees found on the Texas A & M International University campus. Five trees were selected to sample and represent the Texas Live Oak tree population (N27°34'24.3" W99°26'00.3"; N27°34'23.4" W99°26'12.3"; N27°34'22.3", W99°26'17.3"; N27°34'34.5" W99°25'49.0"; N27°34'23.3" W99°25'59.1"). Leaf samples collected were mature sized leaves that ranged from 8 centimeter to 12 centimeter in size. Leaf samples were stored in assigned labeled Ziploc™ bags placed in a −40°C freezer or samples were immediately homogenized.

2.2. Crude Extraction

Leaves were washed, ground and homogenized using a Waring® Laboratory variable-speed blender. Cold 0.01 M Tris-Cl buffer (0.15 M NaCl), pH 9.4 (1:8 w/v) was added and homogenized for one minute for each respective sample. Samples were stirred for one hour using a VWR® Dyla-Dual™ Hot Plate Stirrer at 1000 rpm in a cold room (4°C) to optimize homogenization. Then, the homogenate was filtered using cheesecloth and centrifuged at 8000 rpm at 10°C for 30 minutes using an Avanti® JE Centrifuge JA-20 Rotor. The crude extract was stored (at −40°C) for later use or immediately used for agglutination assays.

2.3. Preparation of Red Blood Cells

Two-hundred microliter aliquot of blood samples including human (collected from Laredo Medical Center Laboratory), horse, rabbit and sheep (purchased from Biomérieux® company) were mixed with 10 mL 0.01 M, phosphate buffer saline (PBS), pH 7.2 (0.15 M NaCl) in a 15 mL tube. The blood suspension was centrifuged at room temperature using a Hamilton Bell® VanGuard V6500 Biohazard centrifuge with fixed speed of 3500 rpm for five minutes. At the end of centrifugation, the supernatant was discarded. A second wash or until the supernatant was clear to remove lysed red blood cells was done using 10 mL of 0.01 M, PBS. The pellet was dissolved with PBS to obtain a 2% blood suspension.

2.4. Agglutination Assays

The crude extract and purified fractions were assayed for the presence of lectin activities using Corningware™ 96-well microtiter U-plates, 0.01 M PBS and 2% blood suspension. The sample was diluted by a serial two-fold dilution in PBS (50 μL) and incubated with a 2% suspension of RBCs (50 μL) at room temperature for an hour or until the negative control showed a red button formation. Agglutination activity was detected based on the RBCs appearance on the well; a positive result appears as a red-carpet layer, while negative results, appear as a red button in the bottom of the well.

2.5. Protein Content Determination

The protein content of crude extract samples was determined with the Bradford method [39] using a Quick-Start™ Bradford Protein Assay Kit.

2.6. Characterization of Lectin Activities

2.6.1. Blood Group Specificity
The blood group specificity was tested in four different blood groups; horse, human, rabbit and sheep. The animal blood was purchased as defibrinated blood from Biomérieux®. Human red blood cells were collected from the Laredo Medical Center Laboratory Department, as postpartum blood samples and screened negative for both HIV and blood-borne communicable diseases.

2.6.2. Effect of pH on *Q. fusiformis* Lectin Activity
Crude extracts of *Q. fusiformis* were studied for the effect of pH by using buffers of different pH levels: 5.2, acetate buffer; 6.2 and 7.2 phosphate buffer; 8.2 and 9.2, Tris-Cl buffer.

2.6.3. Effect of Seasonal Variation on *Q. fusiformis* Lectin Activity
Leaf samples were collected in three seasons (fall, winter and summer) to determine if seasonal variation had an effect on *Q. fusiformis* lectin activity. Summer season consisted of hot temperatures with an average of 104°F, with mostly clear skies, and long days. The fall season, had an average temperature of 93°F, with partly cloudy skies and with the most rain due to hurricane season. The winter season consisted of an average temperature of 50°F, with cloudy skies and short days. Two fall seasons and two winter seasons were observed while only one summer season was observed. Crude extracts of *Q. fusiformis* were prepared with 0.01 M phosphate buffer, 0.15 M NaCl, pH 7.2.

2.6.4. Effect of Soil Moisture on *Q. fusiformis* Lectin Activity
The soil samples were collected by the Auger Method. Soil was sampled and subsampled by depth. Soil ex-

amined was under the *Q. fusiformis* crown and collected from three *Q. fusiformis* locations. *Q. fusiformis* roots are at least 50 centimeters deep [37] [38], thus depth focused on the 20 - 30 centimeter depth, which was the best estimation for soil analysis. Samples were stored in a cold room (4°C) or used immediately. Soil moisture was determined by examining moisture loss of soil subsamples on aluminum weight trays. Samples were weighed using an analytical balance. A moist, 2.00 gram subsample was used for each triplicate subsample. Soils were oven-baked for 24 hours at 105°C to dry and weighed once more to record dry soil mass.

2.6.5. Effect of Soil pH on *Q. fusiformis* Lectin Activity

Soil samples collected for soil moisture determination were also sampled to determine soil pH. Soil samples were put in paper bags to air-dry for three to seven days, followed by sieving with a two-millimeter (mm) mesh screen #10. Five g of air-dry soil sample was placed into a 50 mL labeled centrifuge tube. A total of three replicates were performed. Ten mL of deionized water was added to the 50 mL centrifuge tube for a soil water ratio of 1:2. Solutions were shaken for one hour at 120 revolutions per minute (rpm); subsequently, solutions were centrifuged at 15,000-×g for 5 minutes. The resulting supernatant was used to record pH using the VWR® symphony pH meter.

2.6.6. Determination of the Antifungal Activity of *Q. fusiformis* Extracts

Two antifungal assays, an anti-*Aspergillus niger* and an anti-*Rhizopus stolonifer* assays were done. Potato Dextrose Agar (PDA) plates were marked with a circular circumference on the bottom of the plate with a 60 millimeter (mm) perimeter. Fungal cultures were grown at 28°C until they reached the drawn boundaries which took approximately 24 hours. After the mycelial colony had grown, sterile blank paper discs (6 mm in diameter, Grade AA) were saturated with 20 µL aliquot solutions of *Q. fusiformis* freeze-dried crude extracts that were diluted with sterile water. The concentrations, 125 µg/µl, 250 µg/µl and 500 µg/µl were tested. The discs were placed at the outer rim of the mycelial colony (which was within the new circumference of 60 mm away from the original circumference of the petri plate). The plates were incubated at 28°C for 24 hours. After incubation, crescents or zones of inhibition were measured using a vernier caliper. A negative control was used (sterile water) which allowed mycelial growth. On the other hand, a positive control (nystatin) was used to demonstrate the formation of crescents or zones of inhibition around the disc (**Figure 1**).

2.7. Purification of *Q. fusiformis* Lectin

The lectin from *Q. fusiforms* crude extracts was partially purified by one step chromatography using an ion-exchange column chromatography with a Bio-Rad® BioLogic™ low-pressure chromatography system. All absorbance readings were at 280 nm. The flow rate was 1 ml/min and collections of eluate with 3 mL/fraction. Twenty-six mL of *Q. fusiformis* crude extract was applied to DEAE-cellulose column, which was equilibrated and washed with Buffer A, 0.01 M Trizma-Cl with 0.15 M NaCl, pH 9.2. After washing, Buffer B, 0.01 M Trizma-Cl with 0.5 M NaCl , pH 9.2, was run into the column until absorbance was below < 0.01, then elution Buffer C, 0.01 M Trizma-Cl with 1.0 M NaCl, pH 9.2, was applied into the column. A Bio-Rad® fraction collector

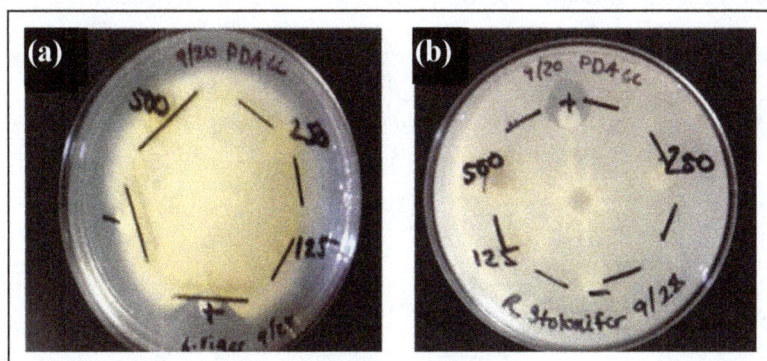

Figure 1. Antifungal activity assay against (a) *A. niger* and (b) *R. stolonifer*. The negative control was sterile water and the positive (+) control was the antifungal agent, nystatin showing crescent and zone of inhibition for *A. niger and R. stolonifer*, respectively.

(#731 - 8304) was used to collect fractioned proteins. Fractioned proteins were pooled and dialyzed to remove the salts. Dialysis was performed with Thermo Scientific SnakeSkin® dialysis tubing. The dialysate used was deionized water. The dialyzed samples were then freeze-dried.

High Performance Liquid Chromatography (HPLC) separation was performed on Prostar PS 220 system which has a UV detector and Rheodyne injector with 20 µl loop volume. Galaxie© Software was applied for data collecting and processing. Purity assessment of the dialyzed and freeze-dried sample (2 µL of freeze-dried sample diluted in 100 µL deionized H_2O) was analyzed by HPLC. A Phenomenex RP-C18 (250 × 4.60 mm) column was used as the stationary phase. A gradient of 100% water to 100% methanol was used as mobile phase. The mobile phase was pumped at 1 ml/min and the eluents were monitored at 215 nm and 280 nm. The injection volumes of samples were 20 µl.

2.8. Statistical Analysis

Data for blood group specificity and pH stability were obtained from five replicates. An analysis of variance (ANOVA) associated with a 5 × 4 factorial experiment in randomized complete blocks design was performed using the Statistical Analysis System (SAS 9.3)® software. Mean comparisons were performed by way of a Bonferroni test using a type I error rate of 0.05. Data analysis for seasonal variation was carried out using a one-way ANOVA. Bivariate correlations among seasonal variation, soil moisture and pH were estimated and tested using a Pearson correlation analysis with level of significance for a two-tailed test set at 0.01.

3. Results and Discussion

Lectin activity from *Q. fusiformis* leaves was investigated, using agglutination assays. In the presence of lectin, sugars on the surface of red blood cells form an interaction with the lectin resulting in agglutination. This is evident by the formation of a carpet layer on the bottom of a microtiter plate wells. On the other hand, in the absence of lectin, a distinctive red button is formed on the bottom of the microtiter plate well. The reciprocal of dilution is calculated as titer value, which reflects lectin activity. The higher the titer value the higher is the lectin activity. Different factors *i.e.* blood group specificity; pH and seasonal variation are known to affect lectin activities [40]-[43]. In this study, how these factors affect *Q. fusiformis* lectin activity were investigated. Results of the study will aid in the characterization of *Q. fusiformis* lectin.

3.1. Blood Group Specificity Study of *Q. fusiformis* Lectin

The blood group specificity of *Q. fusiformis* lectin activity was investigated in four blood groups (horse, human, rabbit and sheep). The extracts of *Q. fusiformis* agglutinated to all blood groups tested, making *Q. fusiformis* a non-blood group specific lectin. Lectin specific activity (SA) was expressed as titer over milligrams of protein. For analysis, lectin specific activity was transformed to natural logarithm of specific activity (ln SA). ANOVA results reported a difference among the four blood groups tested, which was subjected to a Bonferroni test using a Type I error rate (α of 5%) for mean comparisons to determine significant differences (**Table 1**, **Figure 2**). The Bonferroni test showed a significant difference in the ln specific activity values of *Q. fusiformis* lectin between

Table 1. Determinations of blood group specificity of *Q. fusiformis* crude extract lectin.

Blood group	Titer[a]	Protein Content(mg/mL)[b]	HA[c]	SA (titer/mg)[d]	lnSA[e]
Horse	6.08	0.743	304	959.59	6.29[f]
Human	18.16	0.743	908	2094.11	7.21[f]
Rabbit	15.2	0.743	760	2542.26	7.02[f]
Sheep	49.44	0.743	2472	10,654.62	8.3[g]

[a] Titer is the reciprocal of the lowest dilution that was positive for lectin activity.
[b] Protein content was determined using Bradford assay.
[c] Hemagglutination Activity (HA), is titer multiplied with sample volume (50 uL).
[d] SA, Specific Activity is HA divided by the protein content.
[e] ln Specific activity (SA), is the natural logarithm of specific activity. Values of ln SA are mean values from Bonferroni test using a type I error rate of 5%.
[f] Blood groups horse, human and rabbit show no significant difference from each other.
[g] Sheep blood group is significantly different from horse, human and rabbit.

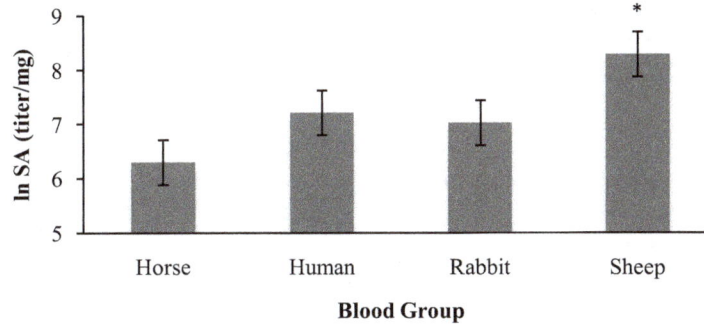

*Significant (p<0.05) compared to horse, human and rabbit blood groups.

Figure 2. Blood group specificity of *Q. fusiformis* lectin based on ln SA of mean comparisons analyzed with a Bonferroni test using a type I error rate of 5%.

sheep blood (ln SA = 8.30) and the three blood groups (horse, ln SA = 6.29), human (ln SA = 7.21) and rabbit (ln SA = 7.02). As shown in **Figure 2**, ln specific activity using sheep blood showed a significantly higher lectin activity as compared to the other three blood groups. However, lectinln specific activities of *Q. fusiformis* using horse, human and rabbit blood groups were not significantly different from each other.

Specificity is influenced by the limited number of contacts with carbohydrates and depth of the sugar binding sites [44]. In addition, any modification or substitution to a binding site can influence binding specificity [45]. The observed differences in *Q. fusiformis* lectin activity with different blood groups may be due to differences in carbohydrate-lectin binding interactions which can be attributed to differences in carbohydrates presented on the cell surfaces of the different blood groups.

The carbohydrates on the erythrocyte cellular surfaces are distinguishable among the four different blood groups [46]. There are reported monosaccharide determinants in the different blood groups, fucose in horse [47], galactose in human and [47]-[49] and mannose in rabbit erythrocytes [50]. Horse, human and rabbit red blood cells may contain carbohydrate components on the cellular surface binding sites that are relatively less recognized by the *Q. fusiformis* lectin binding site. On the other hand, the carbohydrates found on the cellular surface of sheep red blood cells may contain carbohydrate units in a structure and position more specific and with higher affinity for the binding of *Q. fusiformis* lectin, subsequently increasing sheep erythrocyte agglutination. Moreover, lectin characteristics such as, multivalency may determine cross-linking interaction in binding recognition. Spatial distribution of multivalency among lectin structures may produce a higher level of specificity [51] [52].

Quercus fusiformis lectin activity is similar to other plant lectins that are also non-blood group specific. *Erythrina speciosa* lectin was characterized as a non-blood group specific. Its lectin activity was examined in the human blood ABO system and animal blood groups, rabbit, mouse, sheep and horse [53]. Likewise, the blood group specificity for leaf lectin in *Kalanochoe crenata* was characterized as a non-blood group specific lectin, agglutinating to the different types of human blood red cells of the ABO system [36]. *Artocarpus incisa* seeds were also examined in a wide range of blood groups including human ABO system and animal blood groups, cow, goat, rabbit, pig and sheep. *Artocarpus incisa* seed lectin resulted in non-blood group specificity in humans ABO system, while rabbit blood group activity was not different from human ABO and the other four blood groups were significantly different from human and rabbit agglutination [54]. Similarly, *Bryopsis plumosa* lectin from a green marine alga agglutinated sheep and horse erythrocytes [55]. The blood group specificity of *Q. fusiformis* lectin activity suggests *Q. fusiformis* lectin is non-blood group specific with higher specificity directed towards sheep erythrocytes. The present study is the first to report the non-blood group specificity of the *Q. fusiformis* lectin.

3.2. Effect of pH on *Q. fusiformis* Lectin Activity

Crude extracts of *Q. fusiformis* lectin were prepared and incubated in wide range of pH values with stirring for 24 hours at 4°C. A Bonferroni test using a Type I error rate (α of 5%) was used to compare mean natural logarithm-transformed specific activity (ln SA) across pH levels (5.2, 6.2, 7.2, 8.2 and 9.2). Results indicated that *Q. fusiformis* lectin was stable at a pH range 5.2 to 8.2 while that for pH 9.2 was significantly lower than at pH 5.2 - 8.2 (**Table 2**, **Figure 3**). Different pH conditions were found to have profound effects on the tertiary and qua-

Table 2. Effect of pH on *Q. fusiformis* crude extract lectin.

pH	Titer[a]	Protein Content (ug/uL)[b]	HA[c]	SA(titer/mg)[d]	lnSA[e]
5.2	18.6	0.532	930	2912.27	7.33[f]
6.2	19.8	0.697	990	3053.03	7.25[f]
7.2	36.1	0.856	1805	8147.53	7.73[f]
8.2	24.5	0.78	1225	4944.42	7.32[f]
9.2	12.5	0.847	607	1255.98	6.39[g]

[a] Titer is the reciprocal of the lowest dilution that was positive for lectin activity.
[b] Protein content was determined using Bradford assay.
[c] Hemagglutination Activity (HA), is titer multiplied with sample volume (50 μL).
[d] SA, Specific Activity is HA divided by the protein content.
[e] ln Specific activity (SA), is the natural logarithm of specific activity. Values of ln SA are mean values from Bonferroni test using a type I error rate of 5%.
[f] pH levels not significant difference from each other, lectin activity was stable.
[g] pH level significantly different from 5.2 - 9.2.

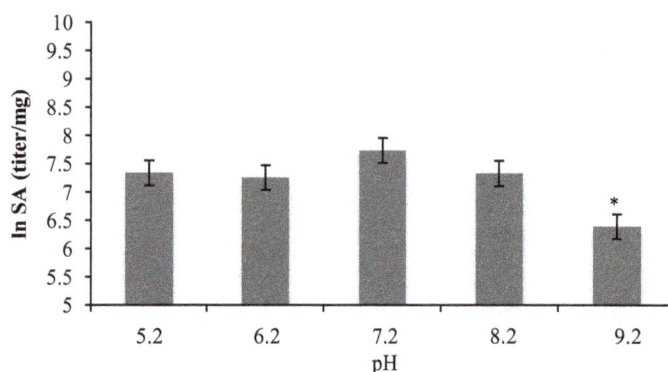

*Significant, ($p < 0.05$) compared to pH range of 5.2 - 8.2.

Figure 3. The effect of pH on *Q. fusiformis* lectin activity based on mean comparisons of ln SA analyzed with a Bonferroni test using a 5% type-I error rate.

ternary structure of proteins and can perturb protein conformational stability [36] [56]. Lectins have different optimum pH to maintain their stabilities [57].

Literature reported that pH stabilities of plant lectins vary. For example, *Morus rubra* [58] and *Ipomoea asarifolia* [59] leaf lectin crude extracts were stable at pH 7.5, while the Korean mistletoe lectin crude extract was stable at pH 8.0 [60]. On the other hand, the pH stability of *Kalanochoe crenata* leaf lectin was at pH values 2 to 12 [36]. *Kalanochoe crenata* leaf lectin activity was stable from pH 2 to 7.5, and the activity was lost at a pH higher than 9.2 [36]. The effect of pH on lectin activity was studied in the leaves of *Chorchorus olitours* also in a broad range of pH values from pH 2.0 to 10.5. *Chorchorus olitours* exhibited a high agglutination activity from pH 7.2 to 8.0 with a dramatic decrease of lectin activity below acidic conditions of pH 6.5 and above basic pH of 9.0 [46]. The pH stability of *Q. fusiformis* lectin activity at a wide range of pH is quite similar to other leaf lectins.

3.3. Effect of Seasonal Variation on *Q. fusiformis* Lectin Activity

Previous studies have reported lectin sources that display a seasonal variation in lectin activity [61]. For example, Elderberry (*Sambucus nigra*) and Black locust (*Robinia pseudoacacia*) bark lectins demonstrated a seasonal variation in content where lectin accumulation was higher during summer to winter compared to spring [42]. On the other hand, lectin activities in cultivated red alga *(Kappaphycus alvarezii)* varied with environmental characteristics such as heavy rain, solar radiation and low seawater temperatures [61]. It was hypothesized that lectin fluctuation found in both Elderberry and Black locust barks as contributed by seasonal variation resembles the behavior of plant storage proteins [42]. Although, many plant lectins mimic the behavior of plant storage proteins, these lectins should not be classified as storage proteins [2]. Costa *et al.* [43] studied *Phthirusa pyrilfo-*

lia leaf lectin activity with an activity higher during sunny weather. This was attributed to mistletoe's slower rate of photosynthesis promoting high lectin production.

In this study, *Q. fusiformis* leaf samples were collected during the fall, winter and summer seasons and lectin activities were compared. The mean ln SA (specific activity) was 7.6, 7.8 and 8.2 for fall, winter and summer, respectively (**Table 3**). Results of the correlation analysis determined that there was no significant correlation in mean ln SA for *Q. fusiformis* lectin during the different seasons, fall, winter and summer (**Figure 4**). Compared to previous studies, this study has shown that *Q. fusiformis* lectin activity was not affected by seasonal variations.

3.4. Effect of Soil Moisture and Soil pH on *Q. fusiformis* Lectin Activity

Since the productivity of plants including biomolecule synthesis is affected by environmental factors, soil properties, moisture and pH were examined to determine if any of these factors affect lectin activity in *Q. fusiformis*. To date, the literature with regards to soil properties namely, moisture and pH affecting lectin activity are scarce. The lectin activity of common beans (*Phaseolus vulgaris*) was examined in three different soil types in a semiarid region. The result of this study reported differences in lectin concentration and suggested environmental factors to contribute to lectin difference [62]. Results of the present study showed the average soil pH reading for 20 - 30 centimeter depth in fall was pH 8.36 and in winter was pH 8.31. On the other hand, the average moisture loss for the 20 - 30 centimeter depth in fall was 3% and for winter 5.6%.

The soil moisture and pH were subjected to a correlation analysis with level of significance for a two-tailed test at level 0.01 as shown in **Table 4**. Soil moisture and pH reflect the chemical status of soil conditions for the *Q. fusiformis*. Laredo soil is usually alkaline throughout [63]. Since the pH results of soil are basic, this may indicate that there may be some nutrients unavailable to *Q. fusiformis*, the production and expression of lectin may be affected. On the other hand, *Q. fusiformis* is a drought-tolerant tree and only requires water every three to

Table 3. Effect of seasonal variation of *Q. fusiformis* crude extract lectin.

Seasons	Titer[a]	Protein Content[b]	HA[c]	SA (titer/mg)[d]	lnSA[e]
Fall	5	0.15683	200	2540.06	7.6676
Winter	9	0.165167	450	2792.97	7.8715
Summer	4	0.112667	200	3930.9	8.2296

[a] Titer is the reciprocal of the lowest dilution that was positive for lectin activity.
[b] Protein content was determined using Bradford assay.
[c] Hemagglutination Activity (HA), is titer multiplied with sample volume (50 uL).
[d] SA, Specific Activity is HA divided by the protein content.
[e] ln Specific activity (SA), is the natural logarithm of specific activity. Values of ln SA are mean values from one-way ANOVA, t-test.

Table 4. Pearson correlation analysis for the effect of soil moisture and soil pH on lectin activity of *Q. fusiformis*.

		ln SA	SA	soil moisture 20 - 30 cm	soil pH 20 - 30 cm
ln SA	Correlation	1	0.972	0.454**	0.479**
	Sig. (2-tailed)		0.000	0.089	0.071
	N	15	15	15	15
SA	Correlation	0.972	1	0.432**	0.442**
	Sig. (2-tailed)	0.000		0.108	0.099
	N	15	15	15	15
soil moisture 20 - 30 cm	Correlation	0.454**	0.432**	1	0.844
	Sig. (2-tailed)	0.089	0.108		0.000
	N	15	15	15	15
soil pH 20 - 30 cm	Correlation	0.479**	0.442**	0.844	1
	Sig. (2-tailed)	0.071	0.099	0.000	
	N	15	15	15	15

**Correlation is significant at the level 0.01 (2-tailed).

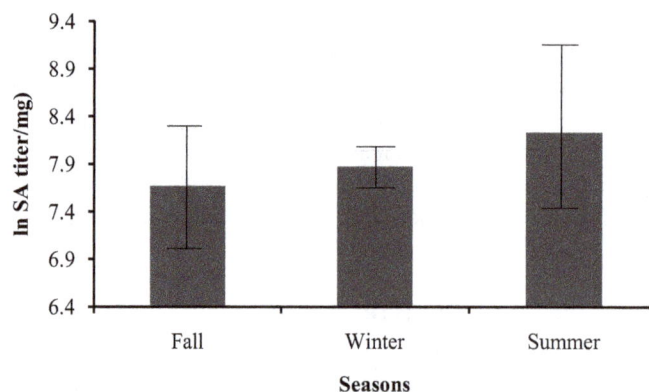

Figure 4. Seasonal variation results from correlation analysis of lectin mean ln SA.

four weeks. Analysis of the results for soil moisture and soil pH showed no significant correlation with ln SA for *Q. fusiformis* lectin.

3.5. Determination of the Antifungal Activity of *Q. fusiformis* Extracts

There are two physiological proposed roles for plant lectins; the first is in plant defense mechanisms and the second, symbiotic nitrogen-fixing bacteria association [7] [64] [65]. Plant lectins have been investigated for their resistance to insects [66], bacteria [1] [27] [67] [68] and fungi [1] [2] [69] [70]. The study of *Q. fusiformis* lectin is relevant since new compounds and especially plant lectins with antifungal activity are of high interest due to rising fungal resistance to common fungicides. There are only a few plant lectins reported to have antifungal activities.

The antifungal activity of *Q. fusiformis* lectin was tested against *Aspergillus niger* and *Rhizopus stolonifer*. These two fungi are common pathogens in fruits and vegetables and cause rot disease [71]. Freeze-dried crude extracts of 125 µg/µl, 250 µg/µl and 500 µg/µl concentrations did not show any antifungal activity against *A. niger* and *R. stolonifer*. Similarly, the knife bean (*Canavalia gladiata*) lectin, was devoid of antifungal activity when tested against three fungi *Botrytis cinerea*, *Myocospharella arachidcola* and *Fusarium oxysporum* [72]. In contrast, plant mistletoe *(Phthirusapyrfolia)* lectin was tested against eleven fungi (*Aspergillus niger*, *A. flavus*, *A. fumigatus*, *Rhizopus arrhizue*, *Paeciloyces variottie*, *Fusarium moniliforme*, *F. laterituium*, *Candida albicans*, *C. burneses*, *C. tropicalis*, *C. parapsilosis*, *Saccharomyces cerevisiae* and *Rhizoctnia solani*) and demonstrated antifungal activity against two fungi *F. lateritium and R. solani* [43]. Red kidney bean *(Phaseolus vulgaris)*lectin reported by Ye *et al.* [73], also expressed antifungal activity but towards three fungal species, *Fusarium oxysporum*, *Rhizoctonia solani* and *Coprinus comatus*.

3.6. Purification of *Q. fusiformis* Lectin

The crude extraction of lectin from *Q. fusiformis* was achieved by homogenization in Tris-Cl, pH 9.2 (0.15 M NaCl) buffer and stirred at 4°C for one hour. Partial purification of *Q. fusiformis* crude extract was carried out by a one-step column chromatography technique using the ion-exchange chromatography column. The weak anionic DEAE-cellulose column (Pall Life Sciences AcroSep™, Product No. 20067-C001) consisted of a matrix with positive charges that will bind negatively charged proteins. Crude extracts of *Q. fusiformis* at pH 9.2 ran through the DEAE column chromatography resulted in the binding of proteins. The chromatography profile showed three peaks, peaks are labeled with abbreviation letter D to indicate DEAE. As shown in **Figure 5**, the *Q. fusiformis* crude extract resulted in the separation of proteins that corresponded to 3 peaks. The first peak (D1), the major peak, contained unbound proteins. The second peak (D2) and third peak (D3) corresponded to bound proteins that were eluted from the DEAE column using half stepwise salt concentration gradient 0.5 and 1.0M, respectively.

Quercus fusiformis DEAE fractions, D1, D2 and D3 were tested for agglutination activities. These fractions did not exhibit any agglutinating activity. The absence of lectin activity in DEAE fractions may be for two reasons. First, fractions may not contain enough lectin concentration to have an activity detectable by the sensitivity of the agglutination assay due to dilution effect of column chromatography. Secondly, the high salt concentration used to elute proteins in peaks D2 and D3 may indicate an effect on conformational structure and stability of the lectin. Thus, fractions from each peak were pooled, dialyzed and lyophilized. Dialyzed samples were measured for pH with litmus paper and demonstrated a neutral pH.

Lectin activity is affected when pH conditions are modified for basic or acidic conditions [57]. This is true for *Q. fusiformis* lectin activity, and provides support for the effect of pH results when lectin activity decrease at pH level 9.2. The lyophilized samples were diluted with 200 µl of millipore water and subjected to agglutination assays to detect presence of lectin activity. **Table 5** shows results of the agglutination assays. Results showed that all peaks contained lectin activity. The binding of *Q. fusiformis* lectin to the DEAE column could be attributed to the lectin amino acid composition. It can be assumed that *Q. fusiformis* lectin has a predominance of negatively charged aspartic acid and/or glutamic acid residues. These amino acids would give the lectin a net negative charge at the basic pH due to the R-carboxylic groups (COO⁻) present. Thus, the negatively charged *Q. fusiformis* lectins bind to the positive charged matrix of DEAE. For example, red marine alga (*Vidalia obtusiloba* C.) lectin was purified through two chromatographic steps, with its first step being an ion exchange DEAE-cellulose column. This lectin was reported to be rich in aspartic acid, glutamic acid and leucine [74].

3.7. Determination of *Q. fusiformis* Purity via HPLC

The first peak, D1 contained all unbound proteins including unbound lectins, thus D1 was not subjected to purity determination. Between D2 and D3, D2 was chosen for purity determination owing to the relatively higher protein concentration as well as higher lectin activity observed in peak 2, D2. The purity of sample D2 was assessed using high performance liquid chromatography (HPLC). Based on the analytical assessment of D2, *Q. fusiformis*

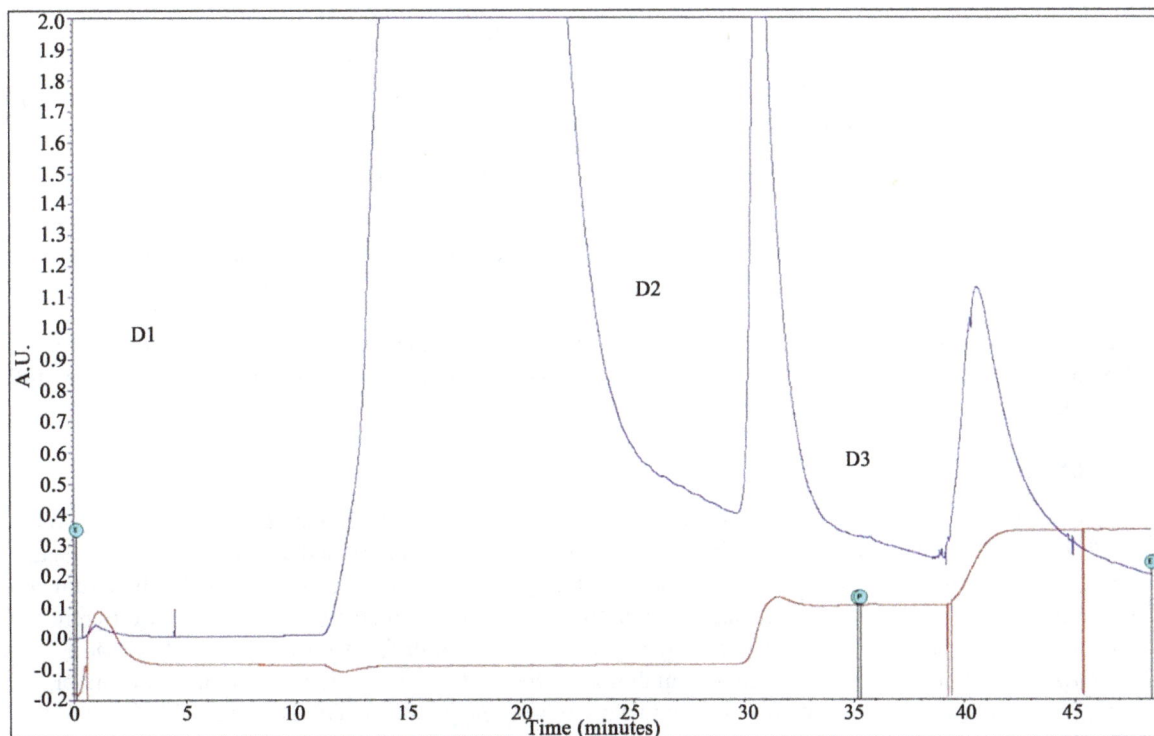

Figure 5. Purification of *Q. fusiformis* lectin by DEAE ion-exchange column chromatography.

Table 5. Lectin activity of the dialyzed and lyophilized fractions from D1, D2 and D3.

Peak	Titer	HA
D1	144	7200
D2	20	1000
D3	10	500

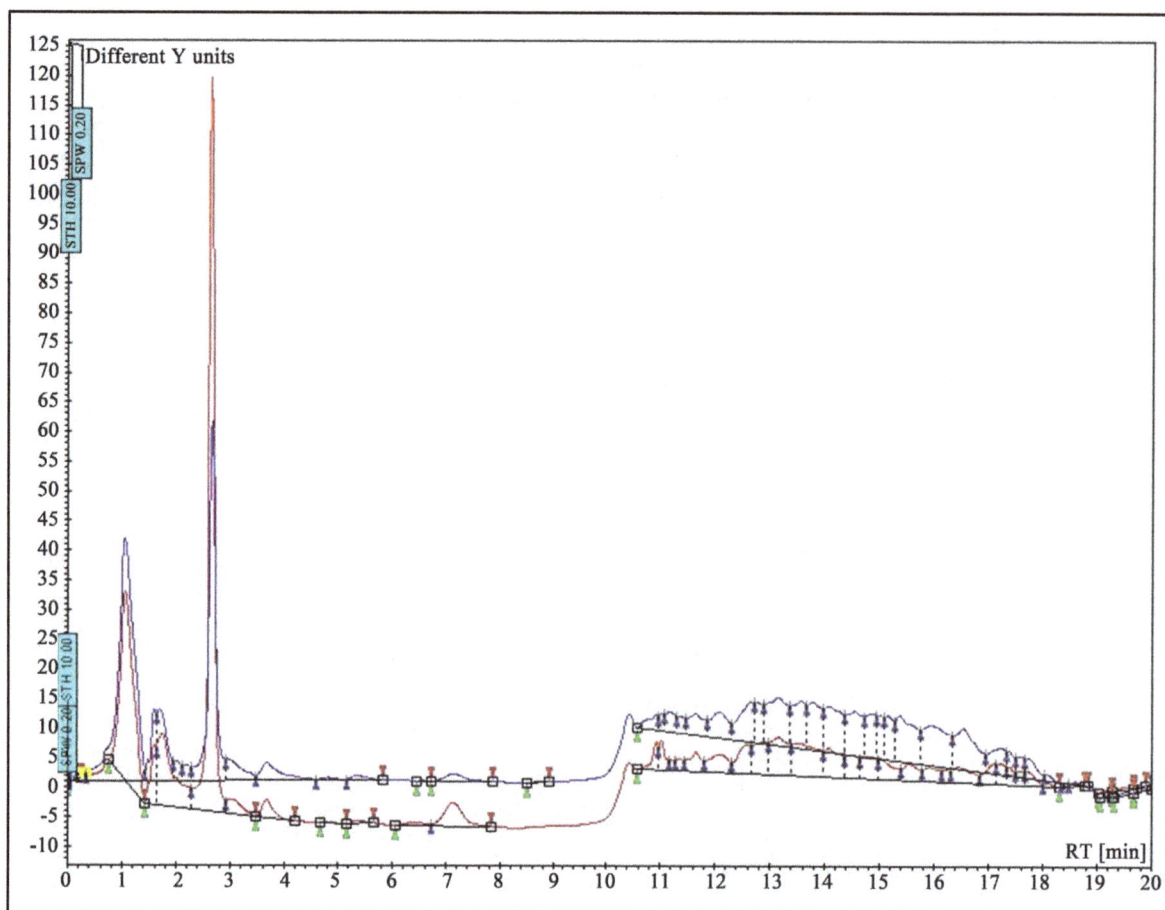

Figure 6. HPLC elution profile of peak 2, D2 from DEAE ion exchange column chromatography of *Q. fusiformis* extract showed three peaks.

lectin is either only partially purified or D2 contains at least three isolectins since 3 peaks (retention times, 1.0, 1.5 and 2.5 min) were obtained after HPLC (**Figure 6**). Isolectins are defined as closely related lectins difficult to separate since they have a similar amino acid composition.

4. Conclusions

Quercus fusiformis' lectin was non-blood group specific, with significantly higher affinity for sheep blood. Also, *Q. fusiformis* lectin was stable within a pH range of pH 5.2 to 8.2, with optimal activity at 7.2 and a significant decrease in activity at pH 9.2. The pH sensitivity of *Q. fusiformis* lectin indicates that the three-dimensional conformational structure and its binding sites are disrupted by ionic interactions occurring in different buffers and pH values. Seasonal variation was examined and no significant difference was found in lectin activity in *Q. fusiformis*. Environmental factors examined in this study focused on soil properties, soil moisture and pH. These soil properties had no significant effect on lectin activity in *Q. fusiformis*. Lectin is hypothesized to play a role in plant defense and exhibit antifungal properties. Crude extracts from *Q. fusiformis* were examined for antifungal activity for potential inhibition and was devoid of fungal activity in the two fungi tested, *A. niger* and *R. stolonifer*.

The one step ion-exchange column chromatography technique with DEAE column resulted in two adsorbed peaks, with lectin activities. HPLC results indicated a partially purified lectin. Results of the present study provide a characterization and the partial purification of *Q. fusiformis* lectin. Future efforts are needed that include upscale purification and further characterization of the lectin under study. The sugar inhibition test may reveal sugar specificity of *Q. fusiformis* lectin. Highly concentrated lectin and purified lectin may be examined for other biological activities including antifungal activities. In addition, determination of antifungal activities against other fungal species is recommended. The DEAE isolate can be further purified using a galactose affinity column owing to *Q. fusiformis* observed affinity to sheep's blood.

Acknowledgements

The Avanti JE Centrifuge was acquired through a U.S. National Science Foundation Major Research Instrument grant award (DBI 0959395). This research was funded by a Texas A & M International University Research Development Award to R. A. Ynalvez. The authors would like to thank Dr. A. Addo-Mensah for the technical advice on the use of the HPLC.

References

[1] Charungchitrak, S., Petsom, A., Sangvanich, P. and Karnchanatat, A. (2011) Antifungal and Antibacterial Activities of Lectinfrom the Seeds of *Archidendron jiringa* Nielsen. *Food Chemistry*, **126**, 1025-1032. http://dx.doi.org/10.1016/j.foodchem.2010.11.114

[2] Peumans, W.J. and Van Damme, E.J.M. (1995) Lectins as Plant Defense Proteins. *Plant Physiology*, **109**, 347-352. http://dx.doi.org/10.1104/pp.109.2.347

[3] Hajtó, T., Hostanska, K., Berki, T., Pálinkas, L., Boldizsár, F. and Németh, P. (2005) Oncopharmacological Perspectives of a Plant Lectin (*Viscum* Album Agglutinin-I): Overview of Recent Results from *in-Vitro* Experiments and *in-Vivo* Animal Models, and Their Possible Relevance for Clinical Applications. *Evidence-Based Complementary and Alternative Medicine*, **2**, 59-67. http://dx.doi.org/10.1093/ecam/neh058

[4] Jiang,Q. L., Zhang, S., Tian, M., Zhang, S.Y., Xie, T., Chen, D.Y., Chen, Y.J., He, J., Liu, J., Ouyang, L. and Jiang, X. (2015) Plant Lectins, from Ancient Sugar-Binding Proteins to Emerging Anti-Cancer Drugs in Apoptosis and Autophagy. *Cell Proliferation*, **48**, 17-28. http://dx.doi.org/10.1111/cpr.12155

[5] Gonzalez de Mejía, E. and Prisecaru, V.I. (2005) Lectins as Bioactive Plant Proteins: A Potential in Cancer Treatment. *Critical Reviews in Food Science and Nutrition*, **45**, 425-445. http://dx.doi.org/10.1080/10408390591034445

[6] Nagre, N.N., Chachadi, V.B., Eligar, S.M., Shubhada, C., Pujari, R., Shastry, P., Swamy, B.M. and Inamdar, S.R. (2010) Purification and Characterization of a Mitogenic Lectin from *Cephalosporium*, a Pathogenic Fungus Causing Mycotic Keratitis. *Biochemistry Research International*, **2010**, Article ID: 854656. http://dx.doi.org/10.1155/2010/854656

[7] Sharon, N. and Lis, H. (2004) History of Lectins: From Hemagglutinins to Biological Recognition Molecules. *Glycobiology*, **14**, 53R-62R. http://dx.doi.org/10.1093/glycob/cwh122

[8] Christian, B., Thomas, J.B., Melissa, W., de Graaf, A.M.A., van Delft, F.L. den Brok, M.H. and Adema, G.J. (2013) Targeting Aberrant Sialylation in Cancer Cells Using a Fluorinated Sialic Acid Analog Impairs Adhesion, Migration, and *in Vivo* Tumor Growth. *Molecular Cancer Therapeutics*, **12**, 1935-1946. http://dx.doi.org/10.1158/1535-7163.MCT-13-0279

[9] Lam, S.K. and Ng, T.B. (2011) Lectins: Production and Practical Applications. Applied *Microbiology Biotechnology*, **89**, 45-55. http://dx.doi.org/10.1007/s00253-010-2892-9

[10] Oliveira, C., Nicolau, A., Teixeira, J.A. and Domingues, L. (2011) Cytotoxic Effects of Native and Recombinant Frutalin, a Plant Galactose-Binding Lectin, on Hela Cervical Cancer Cells. *Journal of Biomedicine and Biotechnology*, **2011**, Article ID: 568932. http://dx.doi.org/10.1155/2011/568932

[11] Poiroux G., Pitié, M., Culerrier, R., Ségui, B., Van Damme, E.J.M., Peumans, W.J., Bernadou, J., Levade, T., Rougé, P., Barre, A. and Benoist, H. (2011) Morniga G: A Plant Lectin as an Endocytic Ligand for Photosensitizer Molecule Targeting Toward Tumor-Associated T/Tn Antigens. *Photochemistry and Photobiology*, **87**, 370-377. http://dx.doi.org/10.1111/j.1751-1097.2010.00858.x

[12] Mahalingam, A., Geonnotti, A.R., Balzarini, J. and Kiser, P.F. (2011) Activity and Safety of Synthetic Lectins Based on Benbenzoboroxole-Functionalized Polymers for Inhibition of HIV Entry. *Molecular Pharmaceutics*, **8**, 2465-2475. http://dx.doi.org/10.1021/mp2002957

[13] Koharudin, L.M. and Gronenborn, A.M. (2014) Antiviral Lectins as Potential HIV Microbicides. *Current Opinion in Virology*, **7**, 95-100. http://dx.doi.org/10.1016/j.coviro.2014.05.006

[14] Liu, Z., Luo, Y., Zhou, T.T. and Zhang, W.Z. (2013) Could Plant Lectins Become Promising Anti-Tumor Drugs for Causing Autophagic Cell Death? *Cell Proliferation*, **46**, 509-515.

[15] Randy, C.F., Cheung, H., Leung, H., Pan, W.L. and Ng, T.B. (2013) A Calcium Ion-Dependent Dimeric Bean Lectin with Antiproliferative Activity toward Human Breast Cancer MCF-7 Cells. *Protein Journal*, **32**, 208-215. http://dx.doi.org/10.1007/s10930-013-9477-2

[16] Li, W.W., Yu, J.Y., Xu, H.L. and Bao, K.K. (2011) Concanavalin A: A Potential Anti-Neoplastic Agent Targeting

Apoptosis, Autophagy and Anti-Angiogenesis for Cancer Therapeutics. *Biochemical and Biophysical Research Communications*, **414**, 282-286. http://dx.doi.org/10.1016/j.bbrc.2011.09.072

[17] Fang, E.F., Lin, P., Wong, J.H., Tsao, S.W. and Ng, T.B. (2010) A Lectin with Anti-HIV-1 Reverse Transcriptase, Antitumor And Nitric Oxide Inducing Activities From Seeds of *Phaseolus vulgaris* Cv. Extralong Autumn Purple Bean. *Journal Agricultural Food Chemistry*, **58**, 2221-2229. http://dx.doi.org/10.1021/jf903964u

[18] Lima, R.F., Criddle, D.N., Soares, P.M.G., Ribeiro, S.P., Cavada, B.S., Nascimento, K.S., Sampaio, A.H. and Assreuy, A.M.S. (2010) *Bryothamnion seaforthii* Lectin Relaxes Vascular Smooth Muscle: Involvement of Endothelium and NO Synthase. *Protein and Peptide Letters*, **17**, 305-310. http://dx.doi.org/10.2174/092986610790780332

[19] Liu, B., Bian, H.J. and Bao, J.K. (2010) Plant Lectins: Potential Antineoplastic Drugs from Bench to Clinic. *Cancer Letters*, **287**, 1-12. http://dx.doi.org/10.1016/j.canlet.2009.05.013

[20] Swanson, M.D., Winter, H.C., Goldstein, I.J. and Markovitz, D.M. (2010) A Lectin Isolated from Bananas Is a Potent Inhibitor of HIV Replication. *Journal of Biological Chemistry*, **285**, 8646-55. http://dx.doi.org/10.1074/jbc.M109.034926

[21] Raja, S.B., Murali, M.R., Kumar, N.K. and Devaraj, S.N. (2011) Isolation and Partial Characterization of a Novel Lectin from *Aegle marmelos* Fruit and Its Effect on Adherence and Invasion of Shigellae to HT29 Cells. *PLoS ONE*, **6**, e16231-e16231. http://dx.doi.org/10.1371/journal.pone.0016231

[22] Ynalvez, R.A., Fuentes, L.M. and Sanchez, C.V. (2011) Comparison and Temperature Study of Lectin Activities in Texas Live Oak (*Quercus fusiformis*) Crude Extracts. *Journal of Plant Science*, **6**, 124-134. http://dx.doi.org/10.3923/jps.2011.124.134

[23] Chrispeels, M.J. and Raikhel, N.V. (1991) Lectins, Lectin Genes, and Their Role in Plant Defense. *Plant Cell*, **3**, 1-9. http://dx.doi.org/10.1105/tpc.3.1.1

[24] Xiao, X., He, H., Ding, X., Yang,Q., Liu, X., Liu, S., Rang, J., Wang, T., Zuo, M. and Xia, L. (2015) Purification and Cloning of Lectin that Induce Cell Apoptosis from *Allium chinense*. *Phytomedicine*, **22**, 238-244. http://dx.doi.org/10.1016/j.phymed.2014.12.004

[25] Silvia, M.C.C., Santana, L.A., Mentele, R., Ferreiraa, R.S., de Miranda, A., Silva-Luccac, R.A., Sampaioa, M.U., Correiad, M.T.S. and Oliva, M.L.V. (2012) Purification, Primary Structure and Potential Functions of a Novel Lectin from *Bauhinia forficata* Seeds. *Process Biochemistry*, **47**, 1049-1059. http://dx.doi.org/10.1016/j.procbio.2012.03.008

[26] Bashir, H., Khan, T., Masood, A. and Hamid, R. (2010) Isolation, Purification and Characterization of a Lectin from a Local Kashmiri Variety of Soybean (*Glycine Max*). *Asian Journal Biochemistry*, **5**, 145-153. http://dx.doi.org/10.3923/ajb.2010.145.153

[27] Gomes, F.S., Procópio, T.F., Lima, T.A., Napoleão, T.H., Coelho, L.C.B. and Paiva, P.M.G. (2010) Isolation and Antimicrobial Activity of Lectin from *Schinus terebinthifolius* Leaves. *Journal of Biotechnology*, **150**, 453. http://dx.doi.org/10.1016/j.jbiotec.2010.09.661

[28] Laija, S.N., Mahesh, S., Smitha, L.S. and Remani, P. (2010) Isolation and Partial Characterization of Two Plant Lectins. *Current Research Journal of Biological Sciences*, **2**, 232-237. http://www.maxwellsci.com/print/crjbs/v2-232-237.pdf

[29] Silva, M.C., Corrêa, A.D., Donizete dos Santos, C., Marcos, F.C.A. and Patto de Abreu, C.M. (2010) Partial Purification of Leaf Lectin from *Manihot esculenta* and Screening Its Fungicidal Activity. *Journal of Agricultural Biotechnology and Sustainable Development*, **2**, 136-141. http://www.isabb.academicjournals.org/article/article1379332438_Silva%20et%20al.pdf

[30] Lin, P., Ye, X. and Ng, T. (2008) Purification of Melibiose-Binding Lectins from Two Cultivars of Chinese Black Soybeans. *Acta Biochimica et Biophysica Sinica*, **40**, 1029-1038. http://dx.doi.org/10.1111/j.1745-7270.2008.00488.x

[31] Chen, H.P. and Xu, L.L. (2005) Isolation and Characterization of a Novel Chitosan-Binding Protein from Non-Heading Chinese Cabbage Leaves. *Journal of Integrative Plant Biology*, **47**, 452-456. http://dx.doi.org/10.1111/j.1744-7909.2005.00022.x

[32] Kaur, A., Singh, J., Kamboj, S.S., Sexana, A.K., Pandita, R.M. and Shamnugavel, M. (2005) Isolation of an N-Acetyl-D-Glucosamine Specific Lectin from the Rhizomes of *Arundo donax* with Antiproliferative Activity. *Phytochemistry*, **66**, 1933-1940. http://dx.doi.org/10.1016/j.phytochem.2005.06.026

[33] Almanza, M., Vega, N. and Pérez, G. (2004) Isolating and Characterizing a Lectin from *Galactialindenii* Seeds That Recognizes Blood Group H Determinants. *Archives of Biochemistry and Biophysics*, **429**, 180-190. http://dx.doi.org/10.1016/j.abb.2004.06.010

[34] Kuku, A., Odekanyin, O., Adreniran, K., Adewusi, M. and Olonade, T. (2009) Purification of a Mannose/Glucose-Specific Lectin with Antifungal Activity from Pepper Seeds (*Capsicum annuum*). *African Journal of Biochemistry Research*, **3**, 272-278. http://www.researchgate.net/publication/228472330_Purification_of_a_mannoseglucose-specific_lectin_with_antifung

al_activity_from_pepper_seeds_%28Capsicum_annuum%29

[35] Occeña, I.V., Mojica, E.R.E. and Merca, F.E. (2007) Isolation and Partial Characterization of a Lectin from the Seeds of *Artocarpus camansi* Blanco. *Asian Journal of Plant Science*, **6**, 757-764. http://dx.doi.org/10.3923/ajps.2007.757.764

[36] Adenike, K. and Eretan, O.B. (2004) Purification and Partial Characterization of a Lectin from the Fresh Leaves of *Kalanchoe crenata* (Andr.) Haw. *Journal of Biochemistry and Molecular Biology*, **37**, 229-233. http://dx.doi.org/10.5483/BMBRep.2004.37.2.229

[37] Simpson, B.J. (1999) A Field Guide to Texas Trees. Lone Star Books, Houston.

[38] Tull, D. and Miller, G.O. (1999) A Field Guide to Wildflowers, Trees and Shrubs of Texas. Lone Star Books, Houston.

[39] Bradford, M.M. (1976) A Rapid and Sensitive Method for the Quantitation of Microgram Quantities of Protein Utilizing the Principle of Protein-Dye Binding. *Analytical Biochemistry*, **72**, 248-254. http://dx.doi.org/10.1016/0003-2697(76)90527-3

[40] Naeem, A., Saleemuddin, M. and Hasan Khan, R. (2007) Glycoprotein Targeting and Other Applications of Lectins in Biotechnology. *Current Protein and Peptide Science*, **8**, 261-271. http://dx.doi.org/10.2174/138920307780831811

[41] Muramoto, K., Kado, R., Takei, Y. and Kamiya, H. (1991) Seasonal Changes in the Multiple Lectin Compositions of the Acorn Barnacle *Megabalanus rosa* as Related to Ovarian Development. *Comparative Biochemistry and Physiology*, **98**, 603-607.

[42] Nsimba-Lubaki, M. and Peumans, W.J. (1986) Seasonal Fluctuations of Lectins in Barks of Elderberry (*Sambucus nigra*) and Black Locust (*Robinia pseudoacacia*). *Plant Physiology*, **80**, 747-751. http://dx.doi.org/10.1104/pp.80.3.747

[43] Costa, R.M.P., Vaz, A.F.M., Oliva, M.L.V., Coelho, L.C.B., Correia, M.T.S. and Carneiro-da-Cunha, M.G. (2010) A New Mistletoe *Phthirusa pyrifolia* Leaf Lectin with Antimicrobial Properties. *Process Biochemistry*, **45**, 526-533. http://dx.doi.org/10.1016/j.procbio.2009.11.013

[44] Drickamer, K. (1995) Multiplicity of Lectin-Carbohydrate Interactions. *Nature Structural Biology*, **2**, 437-439. http://dx.doi.org/10.1038/nsb0695-437

[45] Drickamer, K. and Taylor, E. (2002) Glycan Arrays for Functional Glycomics. *Genome Biology*, **3**, Reviews1034.1-Reviews1034.4. http://genomebiology.com/2002/3/12/reviews/1034

[46] Khan, M.M.H., Rahman, A.T.M., Uddin, M.S., Khatun, S., Pervin, F. and Absar, N. (2008) Purification and Characterization of Lectins from Jute (*Chorchorus olitorius*) Leaves. *Journal of the Chinese Chemical Society*, **55**, 1171-1177. http://dx.doi.org/10.1002/jccs.200800173

[47] Wu, A.M., Lisowska, E., Duk, M. and Yang, Z. (2009) Lectins as Tools in Glycoconjugate Research. *Glycoconjugate Journal*, **26**, 899-913. http://dx.doi.org/10.1007/s10719-008-9119-7

[48] Kusui, K. and Takasaki, S. (1998) Structural Study of N-Linked Sugar Chains of Sheep Erythrocyte Membrane Glycoproteins. *Glycoconjugate Journal*, **15**, 3-10. http://dx.doi.org/10.1023/A:1006967614009

[49] Shibuya, N., Goldstein, I.J., Van Damme, E.J.M. and Peumans, W.J. (1988) Binding Properties of a Mannose-Specific Lectin from the Snowdrop (*Galanthus nivalis*) Bulb. *Journal of Biological Chemistry*, **263**, 728-734. http://www.jbc.org/content/263/2/728

[50] Yagi, F., Iwaya, T., Haraguchi, T. and Goldstein, I.J. (2002) The Lectin from Leaves of Japanese Cycad, *Cycas revoluta* Thunb. (Gymnosperm) Is a Member of the Jacalin-Related Family? *European Journal of Biochemistry*, **269**, 4335-4341. http://dx.doi.org/10.1046/j.1432-1033.2002.03127.x

[51] Loris, R. (2002) Principles of Structures of Animal and Plant Lectins. *Biochimica et Biophysica Acta*, **1572**, 198-208. http://dx.doi.org/10.1016/S0304-4165(02)00309-4

[52] Sacchettini, J.C., Baum, L.G. and Brewer, C.F. (2001) Multivalent Protein-Carbohydrate Interactions. A New Paradigm for Supermolecular Assembly and Signal Transduction. *Biochemistry*, **40**, 3009-3015. http://dx.doi.org/10.1021/bi002544j

[53] Konozy, E.H.E., Mulay, R., Faca, V., Ward, R.J., Greene, L.J., Roque-Barriera, M.C., Sabharwal, S. and Bhide, S.V. (2002) Purification, Some Properties of a D-Galactose-Binding Leaf Lectin from *Erythrina indica* and Further Characterization of Seed Lectin. *Biochimie*, **84**, 1035-1043. http://dx.doi.org/10.1016/S0300-9084(02)00003-2

[54] Moreira, R.A., Castelo-Branco, C.C., Monteiro, A.C.O., Tavares, R.O. and Beltramini, L.M. (1998) Isolation and Partial Characterization of a Lectin from *Artocarpus incisa* L. Seeds. *Phytochemistry*, **47**, 1183-1188. http://dx.doi.org/10.1016/S0031-9422(97)00753-X

[55] Han, J.W., Jung, M.G., Kim, M.J., Yoon, K.S., Lee, K.P. and Kim, G.H. (2010) Purification and Characterization of a D-Mannose Specific Lectin from the Green Marine Alga, *Bryopsis plumosa*. *Phycology Research*, **58**, 143-150. http://dx.doi.org/10.1111/j.1440-1835.2010.00572.x

[56] Ugwu, S.O. and Apte, S.P. (2004) The Effect of Buffers on Protein Conformational Stability. *Pharmaceutical Tech-*

nology, **28**, 86-108.

[57] Utarabhand, P. and Akkayanont, P. (1995) Purification of a Lectin from *Parkiajavanica* Beans. *Phytochemistry*, **38**, 281-285. http://dx.doi.org/10.1016/0031-9422(94)00550-D

[58] Sureshkumar, T. and Priya, S. (2012) Purification of a Lectin from *M. rubra* Leaves Using Immobilized Metal Ion Affinity Chromatography and Its Characterization. *Applied Biochemistry and Biotechnology*, **168**, 2257-2267. http://dx.doi.org/10.1007/s12010-012-9934-y

[59] Salles, H.O., Vasconcelos, I.M., Santos, L.F.L., Oliveira, H.D., Costa, P.P., Nascimento, N.R.F., Santos, C.F., Sousa, D.F., Jorge, A.R.C., Menezes, D.B., Monteiro, H.S.A., Gondim, D.M.F. and Oliveira, J.T.A. (2011) Towards a Better Understanding of *Ipomoea asarifolia* Toxicity: Evidence of the Involvement of a Leaf Lectin. *Toxicon*, **58**, 502-508. http://dx.doi.org/10.1016/j.toxicon.2011.08.011

[60] Park, W.B., Han, S.K., Lee, M.Y. and Han, K.H. (1997) Isolation and Characterization of Lectins from Stem and Leaves of Korean Mistletoe (*Viscum album* var. *coloratum*) by Affinity Chromatography. *Archives of Pharmacal Research*, **20**, 306-312. http://dx.doi.org/10.1007/BF02976191

[61] Hung, L.D., Hori, K., Nang, H.Q., Kha, T. and Hoa, L.T. (2009) Seasonal Changes in Growth Rate, Carrageenan Yield and Lectin Content in the Red Alga *Kappaphycus alvarezii* Cultivated in Camranh Bay, Vietnam. *Journal of Applied Phycology*, **21**, 265-272. http://dx.doi.org/10.1007/s10811-008-9360-2

[62] Gonzalez de Mejía, E., Maldonado, S.H.G., Gallegos, J.A.A., Camacho, R.R., Rodriguez, E.R., Hernandez, J.L.P., Chavira, M.M.G., Castellanos, J.Z. and Kelly, J.D. (2003) Effect of Cultivar and Growing Location on the Trypsin Inhibitors, Tannins, and Lectins of Common Beans (*Phaseolus vulgaris* L.) Grown in the Semiarid Highlands of Mexico. *Journal of Agricultural Food Chemistry*, **51**, 5962-5966. http://dx.doi.org/10.1021/jf030046m

[63] United States Department of Agriculture (1985) Natural Resources Conservation Service,. Texas Online Soil Survey Manuscript, Webb County.

[64] De Souza Cândido, E., Pinto, M.F.S., Pelegrini, P.B., Lima, T.B., Silva, O.N., Pogue, R., Grossi-de-Sá, M.F. and Franco, O.L. (2011) Plant Storage Proteins with Antimicrobial Activity: Novel Insights into Plant Defense Mechanisms. *Federation of American Societies for Experimental Biology Journal*, **25**, 3290-3305. http://dx.doi.org/10.1096/fj.11-184291

[65] Lerouge, P., Roche, P., Faucher, C., Malliet, F., Truchet, G., Promé, J.C. and Denarié, J. (1990) Symbiotic Host Specificity of *Rhizobium meliloti* Is Determined by a Sulphated and Acylated Glucosamine Oligosaccharide Signal. *Nature*, **344**, 781-784. http://dx.doi.org/10.1038/344781a0

[66] Vandenborre, G., Smagghe, G. and Van Damme, E.J.M. (2011) Plant Lectins as Defense Proteins against Phytophagous Insects. *Phytochemistry*, **72**, 1538-1550. http://dx.doi.org/10.1016/j.phytochem.2011.02.024

[67] Ratanapo, S., Ngamjunyaporn, W. and Chulavatnatol, M. (2001) Interaction of a Mulberry Leaf Lectin with a Phytopathogenic Bacterium, *P. syringaepvmori*. *Plant Science*, **160**, 739-744. http://dx.doi.org/10.1016/S0168-9452(00)00454-4

[68] Ayouba, A., Causse, H., Van Damme, E.J.M. , Peumans, W.J., Bourne, Y., Cambillau, C. and Rougé, P. (1994) Interactions of Plant Lectins with the Components of the Bacterial Cell Wall Peptidoglycan. *Biochemical Systematics and Ecology*, **22**, 153-159. http://dx.doi.org/10.1016/0305-1978(94)90005-1

[69] Albuquerque, L.P., Santana, G.M.S., Melo, A.M.A., Coelho, L.C.B., Silva, M.V. and Paiva, P.M.G. (2010) Deleterious Effects of *Microgramma vaccinifolia* Rhizome Lectin on *Fusarium* Species and *Artemiasalina*. *Journal of Biotechnology*, **150**, S453. http://dx.doi.org/10.1016/j.jbiotec.2010.09.662

[70] Hossain, M.A., Maiti, M.K., Basu, A., Sen, S., Ghosh, A.K. and Sen, S.K. (2006) Transgenic Expression of Onion Leaf Lectin Gene in Indian Mustard Offers Protection against Aphid Colonization. *Crop Science*, **46**, 2022-2032. http://dx.doi.org/10.2135/cropsci2005.11.0418

[71] Yildirim, I., Turhan, H. and Özgen, B. (2010) The Effects of Head Rot Disease (*Rhizopus stolonifer*) on Sunflower Genotypes at Two Different Growth Stages. *Turkish Journal of Field Crops*, **15**, 94-98. http://www.field-crops.org/assets/pdf/product513202a2be6f4.pdf

[72] Wong, J.H. and Ng, T.B. (2005) Isolation and Characterization of a Glucose/Mannose/Rhamnose-Specific Lectin from the Knife Bean *Canavalia gladiata*. *Archives of Biochemistry and Biophysics*, **439**, 91-98. http://dx.doi.org/10.1016/j.abb.2005.05.004

[73] Ye, X.Y., Ng, T.B., Tsang, P.W.K. and Wang, J. (2001) Isolation of a Homodimeric Lectin with Antifungal and Antiviral Activities from Red Kidney Bean (*Phaseolus vulgaris*) Seeds. *Journal of Protein Chemistry*, **20**, 367-375. http://dx.doi.org/10.1023/A:1012276619686

[74] Melo, F.R., Benevides, N.M.B., Pereira, M.G., Holanda, M.K., Mendes, F.N.P., Oliveira, S.R.M., Freitas, A.L.P. and Silva, L.M.C. (2004) Purification and Partial Characterization of a Lectin from the Red Marine Alga *Vidalia obtusiloba* C. Agardh. *Brazilian Journal of Botany*, **27**, 263-269. http://dx.doi.org/10.1590/S0100-84042004000200006

18

Inter-Specific Biochemical Diversity between *Echis pyramidum* and *Eryx colubrinus* Inhabiting El-Faiyum, Egypt

Mohamed A. M. Kadry[1*], Eman M. E. Mohallal[2], Doha M. M. Sleem[3], Mohamed A. S. Marie[1]

[1]Zoology Department, Faculty of Science, Cairo University, Cairo, Egypt
[2]Ecology Unit of Desert Animals, Desert Research Center, Cairo, Egypt
[3]Zoology Department, Faculty of Science, Ain Shams University, Cairo, Egypt
Email: [*]ecokadry@yahoo.com

Abstract

Discontinuous polyacrylamide gel electrophoreses for lactate dehydrogenase (*Ldh*) and alfa-esterase (*α-Est*) isoenzymes were conducted for biochemical differentiation between *Echis pyramidum* and *Eryx colubrinus* inhabiting El-Faiyum, Egypt. Total lipids and total protein of liver and muscle tissues in both species were also analyzed. Three *Ldh* isoforms were recorded in both species and the activity and relative front (RF) of *Ldh*-1 seemed to be higher in *E. pyramidum* than in *E. colubrinus*. This high activity could be supported by the significant increase in the total lipids and total protein in liver and muscle tissues of this species. It is thus possibly reasonable to suppose that *E. pyramidum* is more active, energetic and adaptable in its habitat than *E. colubrinus*. *α-Est* showed five isozymic forms fractions in *E. pyramidum*, while it showed only four isoforms in *E. colubrinus*. *α-Est*-1 was the first clear, dense and thick isoform in *E. pyramidum*, but it was completely absent in *E. colubrinus*. High activity of the esterase isoform; *α-Est*-1 only in the heart tissue of *E. pyramidum* may reflect the high ability of *E. pyramidum* to be more resistant to the accumulated toxic residues in its body tissues than *E. colubrinus*.

Keywords

Echis pyramidum (Viperidae), *Eryx colubrinus* (Boidae), *Ldh* Isoenzyme, *Est* Isoenzyme, Total Lipids

1. Introduction

The squamates are the most diverse group containing the lizards and snakes [1]. Several investigations have

---

[*]Corresponding author.

been recorded on the fauna of Egyptian reptiles [2] [3]. The suborder Serpentes is distributed all over the world [4] [5]. Previous descriptions of the external and taxonomical features of some snakes have been ambiguous and unreliable. Therefore, several authors used the biochemical electrophoresis [6] and molecular sequence analysis [3] [7]-[13] to resolve the cladistic relationships among snakes and to clarify their phylogeny.

The family, viperidae comprises approximately 270 species of venomous snakes. The saw-scaled viper, *Echis pyramidum* (Geoffroy 1827) belongs to the family viperidae and is found in Asia, Africa, India, Iraq, Iran and Afghanistan. It occurs on a range of different substrates, including sand, rock, and soft soil and in scrublands, often found hiding under loose rocks [14]. Boidae, is a family of non-venomous snakes found in America, Africa, Europe, Asia, and some pacific islands. It includes relatively primitive snakes and comprises eight genera and 43 species. The kenyan sand boa, *Eryx colubrinus* (Linnaeus 1758) is found in northern Africa from Egypt as far west as Niger including Somalia, Ethiopia, Sudan, Kenya, and northern Tanzania [15]. It occurs in semi-desert, scrub savannahs and rock outcroppings. Also it prefers sandy and friable soil [16]. Extensive molecular genetic diversity has been discovered within and among populations and species in total protein [17], isoenzymes/alloenzymes [18] and DNA [19].

Isoenzymes are multiple forms of a single enzyme, which often have different isoelectric points and therefore can be separated by electrophoresis. Lactate dehydrogenases (*Ldhs*) isoenzymes are very suitable systems for studying several metabolic, genetic, ecological features, and are very useful in systematic studies [20]. *Ldhs* are a hydrogen transfer enzyme that catalyzes the oxidation of L-lactate to pyruvate with nicotinamide-adenine dinucleotide $(NAD)^+$ as hydrogen acceptor, the final step in the metabolic chain of anaerobic glycolysis. Esterase isoenzymes (*Est*) are one of the lipid-hydrolyzing enzymes, possessing high significance in genetics and toxicology [21].

Our present study aimed to investigate the patterns of inter-specific biochemical and genetic diversity between *Echis pyramidum* and *Eryx colubrinus* inhabiting two different sites in El-Faiyum desert of Egypt using the electrophoretic analyses of two isoenzymes.

2. Materials and Methods

2.1. Taxon Sampling and Study Area

A total of 10 individuals of two common Egyptian snake species were collected from two different sites in El-Faiyum desert (5 samples for each species). The first species was the saw-scaled viper, *Echis pyramidum* (Geoffroy 1827) (Reptilia: Viperidae). It is a medium sized snake with a short stocky body, moderate eyes separated from supralabials by scales and gray dorsum with a mid-dorsal series of dark-edged with whitish narrow-like marks [22]. It was collected from Gabal El-Nagar (Mahatet El-Rafa) [29°22'N 30°37'E]; a rocky area near Water drainage and cultivated area planted with wheat (**Figure 1**).

The other one was the Kenyan sand boa, *Eryx colubrinus* (Linnaeus 1758) (Reptilia: Boidae). It is a short thick snake with a relatively short tail end with a conical scale. It has also scales around the eyes, sandy dorsum with large irregularly shaped dark-brown blotches and yellowish ventral plain [22]. It was collected from Ab-showy Kahek [29°24'N 30°38'E]; an area planted near a sandy road (**Figure 1**).

Figure 1. Photos of *Echis pyramidum* (a) and *Eryx colubrinus* (b) inhabiting El-Faiyum desert, Egypt.

2.2. Sample Preparation and Isoenzyme Assay

Tissue samples of liver and heart were removed and immediately taken to the lab and stored at −80°C for further laboratory use. For isoenzyme extraction, approximately 0.5 g of tissue was homogenized in 10 mL saline solution (PBS, pH = 6.8) using a manual Homogenizer. The homogenates were centrifuged at 5000 rpm for 10 minutes and the supernatants were kept at −20°C until use. The enzymes; alfa-esterase (α-Est) in heart and Lactate dehydrogenase (Ldh) in liver supernatants were separated by discontinuous polyacrylamide gel electrophoresis [23] [24].

Electrophoresis was carried out conveniently in discontinuous polyacrylamide gels. An amount of 50 μl of the clear supernatant of the liver and muscle homogenate of each sample was mixed with 20 μl of protein dye (1% bromophenol blue) and 20 μl of 2% sucrose. 30 μl of the mixture per gel slot were used to be applied per each sample for isoenzymes electrophoresis. After electrophoresis, the gel was transferred into a staining solution (50 - 70 ml) according to [25] which was then replaced by a destaining mixture of methanol, acetic acid and water (5:1:5 v/v/v). A potential gradient of high voltage electrode [(20 v/cm), anode] across the gel was applied for 4 h at 8°C for separation of the enzymes.

2.2.1. Ldh Isoenzyme

For Ldh and after electrophoresis, the gel was soaked in 100 mL of 0.2 M Tris-HCl (pH 8.0) containing 30 mg NBT, 25 mg EDTA, 50 mg NAD, 10 mg L-lactic acid and 2 mg PMS. 0.05 M Tris-HCl pH 8.5 was prepared by dissolving 0.605 g Tris in 50 mL distilled water. The pH was adjusted to 8.5 by HCl. Then the solution was completed to 100 ml by distilled water [26].

2.2.2. α-EST Isoenzyme

Regarding α-Est, after electrophoresis, the gel was soaked in 0.5 M borate buffer (pH 4.1) for 90 minutes at 4°C. This procedure lowers the pH of the gel from 8.8 to about 7 at which the reaction proceeds readily. The low temperature minimizes diffusion of the protein within the gel. The gel then was rinsed rapidly in two changes of double distilled water. The gel was stained for esterase activity by incubation at 37°C in a substrate solution of 100 mg α-naphthyl acetate and 100 mg fast blue RR salt in 200 ml of 0.1 M phosphate buffer pH 6.5 [27].

After the appearance of the enzyme bands, the reaction was stopped by washing the gel two or three times with tap water. This was followed by adding the fixative solution, which consists of ethanol and 20% glacial acetic acid (9:11 v/v). The gel was kept in the fixative solution for 24 hours and then was photographed.

All gels were scanned using Gel Doc-2001 Bio-Rad system. For isoenzymes, the bands of enzyme activity were designated according to the system nomenclature proposed by [28]. An abbreviation which corresponds to the name of the isoenzyme designated each locus. When multiple loci were involved, the fastest anodal protein band was designated as locus one, the next as locus two and so on.

2.3. Metabolic Study

Immediately after collection, snakes were dissected. Pieces of liver and thigh muscles were removed and immediately weighed in grams (g) to the nearest 0.01 g. They were stored frozen at −20°C till use. Livers and thigh muscles were processed for estimation of total lipids and total protein according to the method of [29] and [30], respectively using a kit of Biodiagnostics Company.

Our Institutional Animal Care and Use Committee (IACUC) at Zoology Department, Faculty of Science, Cairo University has approved this study protocol from the ethical point of view and according to Animal welfare Act of the Ministry of Agriculture in Egypt that enforces the humane treatment of animals.

3. Statistical Analysis

Student t-test in the PASW package v. 20 was used to calculate the significance difference of total lipids and total proteins within and between the snake species.

4. Results and Discussion

Ldhs isoenzymes are very suitable systems for studying several metabolic, genetic, ecological features, and are very useful in systematic studies [20] [31]. Ldhs are a hydrogen transfer enzymes that catalyze the oxidation of

L-lactate to pyruvate with nicotinamide-adenine dinucleotide (NAD)$^+$ as hydrogen acceptor, the final step in the metabolic chain of anaerobic glycolysis. The reaction is reversible and the reaction equilibrium strongly favours the reverse reaction, namely the reduction of pyruvate (P) to lactate (L):

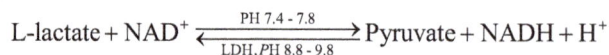

$$\text{L-lactate} + \text{NAD}^+ \xrightleftharpoons[\text{LDH, PH 8.8 - 9.8}]{\text{PH 7.4 - 7.8}} \text{Pyruvate} + \text{NADH} + \text{H}^+$$

Three *Ldhs* isozymic forms were revealed in the liver tissues of both *Echis pyramidum* and *Eryx colubrinus*. The activity of *Ldh*-1 isoform seemed to be higher in *E. pyramidum* than in *E. colubrinus*. Such higher activity was reflected in the thicker and denser bands as well as their higher relative front (RF) in *E. pyramidum* (**Figure 2**).

The apparent increase in the activity of *Ldh* in liver tissues of *E. pyramidum*, in the present study, could be supported by the highly significant increase in the total protein and the highly significant decrease in the total lipids in liver and muscle tissues of this species (**Table 1**) [31]. It is thus possibly reasonable to suppose that *E. pyramidum* is more active, energetic and adaptable in the desert habitat than *E. colubrinus*.

α-Est showed five isozymic forms fractions in *E. pyramidum*, while it showed only four isoforms in *E. colubrinus*. *α-Est*-1 was the first clear, dense and thick isoform in *E. pyramidum* but it was completely absent in *E. colubrinus*. While the second isoform was revealed in *E. colubrinus*, it recorded only in three samples of the other species. The other isoforms; *α-Est*-3, *α-Est*-4 and *α-Est*-5 were dense and thick by the same extent in both *E. pyramidum* and *E. colubrinus* and also their relative front was nearly close to each other (**Figure 3**). The present results revealed the higher activity of *alfa-esterases* in the examined heart tissues of *E. pyramidum* than *E. colubrinus*. Esterases are used as bio-indicators to measure the toxic potency of pesticide and heavy metal residues usually applied in the field [20] [21]. The presence of the highly active esterase isoform; *α-Est*-1 only in the heart tissue of *E. pyramidum* may reflect to some extent, the unsafety of the diet applied to this species of snake and its high ability to resist and accumulate the toxic residues in its body tissues than in *E. colubrinus*.

Table 1 recorded the mean and standard error values of total lipids and total protein in liver and muscle tissues of both *E. pyramidum* and *E. colubrinus*. While the total lipids showed a very highly significant increases

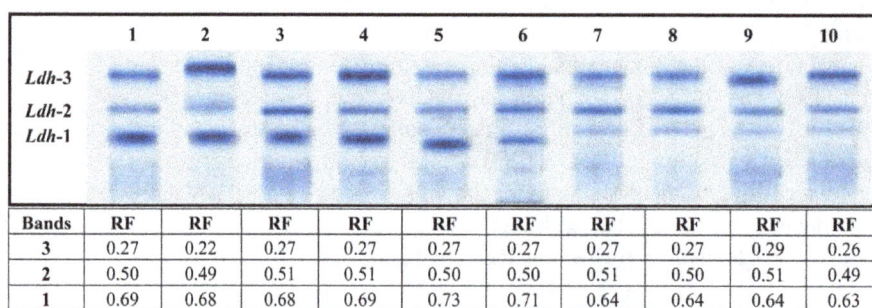

Bands	RF	RF	RF	RF	RF	RF	RF	RF	RF	RF
3	0.27	0.22	0.27	0.27	0.27	0.27	0.27	0.27	0.29	0.26
2	0.50	0.49	0.51	0.51	0.50	0.50	0.51	0.50	0.51	0.49
1	0.69	0.68	0.68	0.69	0.73	0.71	0.64	0.64	0.64	0.63

Figure 2. The electrophoretic profile of *Ldh* isoenzymes in liver tissues. Lanes are as follow: 1 - 5 (*E. pyramidum*), 6 - 10 (*E. colubrinus*).

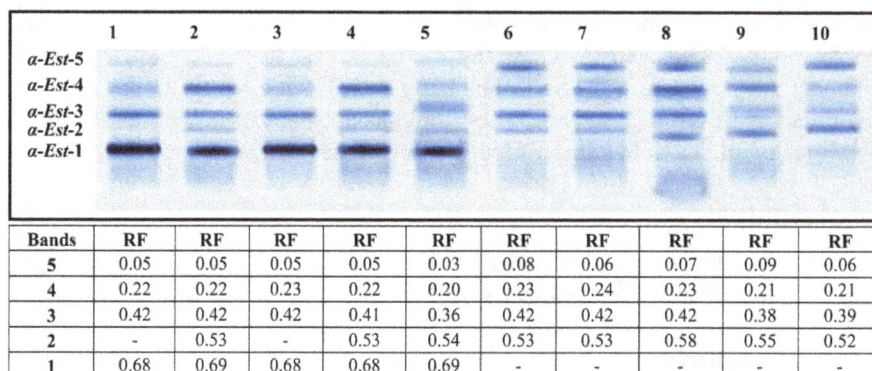

Bands	RF	RF	RF	RF	RF	RF	RF	RF	RF	RF
5	0.05	0.05	0.05	0.05	0.03	0.08	0.06	0.07	0.09	0.06
4	0.22	0.22	0.23	0.22	0.20	0.23	0.24	0.23	0.21	0.21
3	0.42	0.42	0.42	0.41	0.36	0.42	0.42	0.42	0.38	0.39
2	-	0.53	-	0.53	0.54	0.53	0.53	0.58	0.55	0.52
1	0.68	0.69	0.68	0.68	0.69	-	-	-	-	-

Figure 3. The electrophoretic profile of *α-Est* isoenzymes in the studied heart tissues. Lanes are as follow: 1 - 5 (*E. pyramidum*), 6 - 10 (*E. colubrinus*).

Table 1. Comparison of total lipids and total proteins in liver and muscle tissues of *E. pyramidum* and *E. colubrinus*. Data are expressed as mean ± standard error. Number of individuals between parentheses.

Parameters	E. pyramidum	E. colubrinus	t-test
Liver total lipids (mg/100mg)	61.66 ± 9.66 (5)	154.58 ± 14.87 (5)	6.14***
Thigh muscle total lipids (mg/100mg)	19.17 ± 4.03 (5)	18.33 ± 6.09 (5)	5.44***
t-test	4.68**	3.61***	---
Liver total proteins (mg/100mg)	45.93 ± 10.28 (5)	28.58 ± 2.72 (5)	6.44***
Thigh muscle total proteins (mg/100mg)	31.69 ± 4.23 (5)	18.95 ± 1.44 (5)	8.46***
t-test	6.75***	10.98***	

**Highly significant at $P < 0.01$, ***Very highly significant at $P < 0.001$.

in liver ($P < 0.001$) and muscle ($P < 0.001$) tissues of *E. colubrinus* than in *E. pyramidum*, *E. pyramidum* recorded a very highly significant increases ($P < 0.001$) in the total protein in liver and muscle tissues than in *E. colubrinus*. Within each species, while total lipids and total protein were significantly higher in liver than in muscle tissues ($P < 0.01$, $P < 0.001$ respectively). In *E. pyramidum*, they revealed a very highly significant increase ($P < 0.001$) in liver than in muscle tissues in *E. colubrinus*.

5. Conclusion

In conclusion, *E. pyramidum* acquired high physiological performance and activity than *E. colubrinus*, where *Ldh* isoenzyme expression in the first species was higher than in the second. The accumulations of total lipids and total protein were also significantly higher in the first species than in the second. The present data also revealed a high activity of the esterase isoform, *α-Est*-1 only in the heart tissue of *E. pyramidum* which may reflect the high ability of *E. pyramidum* to resist and accumulate the toxic residues in its body tissues than in *E. colubrinus*.

Acknowledgements

We are grateful to Prof. Dr. Hany Hassan, professor at Animal Reproduction Research Institute for his technical support in conducting the practical part of isoenzyme assay in this work.

References

[1]   Vidal, N. and Hedges, S.B. (2009) The Molecular Evolutionary Tree of Lizards, Snakes, and Amphisbaenians. *Comptes Rendus Biologies*, **332**, 129-139. http://dx.doi.org/10.1016/j.crvi.2008.07.010

[2]   Goodman, S.M. and Hobbs, J.J. (1994) The Distribution an Ethnozoology of Reptiles of the Northern Portion of the Egyptian Eastern Desert. *Journal of Ethnobiology*, **14**, 75-100.

[3]   Sayed, N.H.M. (2012) Genetic Diversity among Eight Egyptian Snakes (Squamata-Serpentes: Colubridae) Using RAPD-PCR. *Life Science Journal*, **9**, 423-430.

[4]   Zug, E.R., Vitt, L.J. and Caldwell, J.P. (2001) Herpetology: An Introductory Biology of Amphibians and Reptiles. Academic Press, New York.

[5]   Sayed, N.H.M. (2011) Genetic Diversity among Five Egyptian Non-Poisonous Snakes Using Protein and Isoenzymes Electrophoresis. *Life Science Journal*, **8**, 1034-1042.

[6]   Dowling, H.E., Hass, C.A., Hedges, S.B. and Highton, R. (1996) Snake Relationships Revealed by Slow-Evolving Total Protein: A Preliminary Survey. *Journal of Zoology*, **240**, 1-28. http://dx.doi.org/10.1111/j.1469-7998.1996.tb05482.x

[7]   Lawson, R., Slowinski, J.B., Crother, B.I. and Burbrink, F.T. (2005) Phylogeny of the Colubroidea (Serpentes): New Evidence from Mitochondrial and Nuclear genes. *Molecular Phylogenetics and Evolution*, **37**, 581-601. http://dx.doi.org/10.1016/j.ympev.2005.07.016

[8]   Burbrink, F.T. and Pyron, R.A. (2008) The Taming of the Skew: Estimating Proper Confidence Intervals for Divergence Dates. *Systematic Biology*, **57**, 317-328. http://dx.doi.org/10.1080/10635150802040605

[9]   Wiens, J.J., Kuczynski, C.A., Smith, S.A., Mulcahy, D.E., Sites Jr., J.W., Townsend, T.M. and Reeder, T.W. (2008) Branch Lengths, Support, and Congruence: Testing the Phylogenomic Approach with 20 Nuclear Loci in Snakes. *Systematic Biology*, **57**, 420-431. http://dx.doi.org/10.1080/10635150802166053

[10] Kelly, C.M.R., Barker, N.P., Villet, M.H. and Broadley, D.E. (2009) Phylogeny, Biogeography and Classification of the Snake Superfamily Elapoidea: A Rapid Radiation in the Late Eocene. *Cladistics*, **25**, 38-63. http://dx.doi.org/10.1111/j.1096-0031.2008.00237.x

[11] Vidal, N., Rage, J.C., Couloux, A. and Hedges, S.B. (2009) Snakes (Serpentes). In: Hedges, S.B. and Kumar, S., Eds., *The Time Tree of Life*, Oxford University Press, New York, 390-397.

[12] Zaher, H., Grazziotin, F.E., Cadle, J.E., Murphy, R.W., Moura-Leite, J.C. and Bonatto, S.L. (2009) Molecular Phylogeny of Advanced Snakes (Serpentes, Caenophidia) with an Emphasis on South American Xenodontines: A Revised Classification and Descriptions of New Taxa. *Papéis Avulsos de Zoologia*, **49**, 115-153. http://dx.doi.org/10.1590/S0031-10492009001100001

[13] Pyron, R.A., Burbrink, F.T., Colli, E.R., de Oca, A.N.M., Vitt, L.J., Kuczynski, C.A. and Wiens, J.J. (2011) The Phylogeny of Advanced Snakes (Colubroidea), with Discovery of a New Subfamily and Comparison of Support Methods for Likelihood Trees. *Molecular Phylogenetics and Evolution*, **58**, 329-342. http://dx.doi.org/10.1016/j.ympev.2010.11.006

[14] Mallow, D., Ludwig, D. and Nilson, E. (2003) True Vipers: Natural History and Toxinology of Old World Vipers. Krieger Publishing Company, Malabar, 359 p.

[15] McDiarmid, R.W., Campbell, J.A. and Touré, T. (1999) Snake Species of the World: A Taxonomic and Geographic Reference, Volume 1. Herpetologists' League, Washington, District of Columbia, 511 p.

[16] Mehrtens, J.M. (1987) Living Snakes of the World in Color. Sterling Publishers, New York, 480 p.

[17] Zuckerkandl, E. and Pauling, L. (1965) Evolutionary Divergence and Convergence in Total Protein. In: Bryson, V. and Vogel, H.J., Eds., *Evolving Genes and Total Protein*, Academic Press, New York, 97-166.

[18] Lewontin, R.C. (1974) The Genetic Basis of Evolutionary Change. Columbia University Press, New York.

[19] Kimura, M. (1983) The Neutral Theory of Molecular Evolution. Cambridge University Press, Cambridge. http://dx.doi.org/10.1017/CBO9780511623486

[20] Al-Harbi, M.S. and Amer, S.A.M. (2012) Comparison of Energy-Related Isoenzymes between Production and Racing Arabian Camels. *Advances in Bioscience and Biotechnology*, **3**, 1124-1128. http://dx.doi.org/10.4236/abb.2012.38138

[21] Shahjahan, R.M., Karim, A., Begum, R.A., Alam, M.S. and Begum, A. (2008) Tissue Specific Esterase Isozyme Banding Pattern in Nile Tilapia (*Oreochromis niloticus*). *Universal Journal of Zoology* (*Rajshahi University*), **27**, 1-5.

[22] El Din, M.B. (2006) A Guide to Reptiles and Amphibians of Egypt. American University in Cairo Press, Cairo, 359 p.

[23] Maurer, R. (1968) Disk Electrophorese. W. de Gruyter and Co., Berlin, 222 p.

[24] Shaw, C.R. and Prasad, R. (1970) Starch Gel Electrophoresis of Enzymes: A Compilation of Recipes. *Biochemistry and Genetics*, **4**, 297-329. http://dx.doi.org/10.1007/BF00485780

[25] Mulvey, M. and Vrijenhoek, R.C. (1981) Genetic Variation among Laboratory Strains of the Planorbid Snail *Biomphalaria glabrata*. *Biochemical Genetics*, **19**, 1169-1182. http://dx.doi.org/10.1007/BF00484572

[26] Jonathan, F.W. and Wendel, N.F. (1990) Visualization and Interpretation of Plant Isozymes. In: Soltis, D.E. and Soltis, P.S., Eds., *Isozymes in Plant Biology*, Champan and Hall, London, 5-45.

[27] Scandaliojs, G. (1964) Tissue-Specific Isozyme Variations in Maize. *Journal of Heredity*, **55**, 281-285.

[28] Shaklee, J.B., Allendorf, F.W., Morizot, D.C. and Whitt, G.S. (1990) Gene Nomenclature for Protein Coding Loci in Fish. *Transactions of the American Fisheries Society*, **119**, 2-15. http://dx.doi.org/10.1577/1548-8659(1990)119<0002:GNFPLI>2.3.CO;2

[29] Zöllner, N. and Kirsch, K. (1962) Colorimetric Method for Determination of Total Lipids. *Journal of Experimental Medicine*, **135**, 545-550. http://dx.doi.org/10.1007/BF02045455

[30] Gornall, A.G., Bardawill, C.J. and David, M.M. (1949) Determination of Serum Proteins by Means of the Biuret Reaction. *Journal of Biological Chemistry*, **177**, 751-766.

[31] Kadry, M.A.M. and Mohamad, H.R.H. (2014) Genetic and Metabolic Variability between Two Subspecies of *Chamaeleo chamaeleon* (Reptilia: Chamaeleonidae) in Egypt. *Advances in Bioscience and Biotechnology*, **5**, 699-703. http://dx.doi.org/10.4236/abb.2014.58083

19

The Participation of Wall Monolignols in Leaf Tolerance to Nature Flooding of Hydrophytes

Olena M. Nedukha

Department of Cell Biology and Anatomy, Institute of Botany, National Academy of Sciences of Ukraine, Kiev, Ukraine
Email: o.nedukha@hotmail.com

Abstract

The comparative analysis of the monolignols localization in epidermis, photosynthesizing parenchyma and vessels walls of *Myriophyllum spicatum*, *Potamogeton pectinatus* and *P. perfoliatus* submerged leaves carried out on the basis of cytochemical method and laser confocal microscopy. The images of quantitative distribution of syringyl and guaiacyl in the cell walls were obtained at cellular level depending on the type of leaf tissues and plant species. The increase of relative content of monolignols was established in walls of vessels and in the corners of parenchyma cells. Cytochemical analysis indicates that ratio of syringyl/guaiacyl in leaf tissues changes depending on species. The role of of syringyl and guaiacyl monolignols in the cellular mechanisms of adaptation to nature flooding is discussed.

Keywords

Flooding, Monolignols, Hydrophytes, Leaves, Laser Scanning Microscopy

1. Introduction

Soil water logging and flooding are the one of the major abiotic stresses for plants. The appearance of serious problem in agriculture as well as in plant natural populations is provoked by the stresses. The stress effects plant growth, because during flood occurs the shortage of oxygen and CO_2 due to the change diffusion rates of gasses in water [1]-[3]. The outer cell walls of submerged leaves and shoots are the one and the main barriers between plant and water environment. Epidermis and mesophyll of submerged stem and leaves of hydrophytes adapt to water environment by change of wall structure and absent of stomata and wax. The recent data showed the presence of cuticle pores in the outer cell walls of the submerged leaves of some hydrophytes and rice seedlings

[4] [5]. Besides, the effect of submergence on lignin and cellulose content in the submerged shoots of *Ludwigia repens* [6], *Trapa natans* and *Sagittaria sagittifolia* [7] was established.

Lignin is the biopolymer of aromatic alcohols, that is synthesized in secondary cellular walls that made off growth by enlargement, and participates in adaptation of plant to the stress, changing structure of cellular wall matrix, providing impassability of water and water solutions through xylem vessel walls of leading bunches and forming in an epidermis a barrier to the pathogens. A lignin is characterized by hydrophobic, this polymer substitute for a water in a wall, forming to hydrogen and covalently association between polysaccharides and hemicelluloses [8]. Content of lignin in the walls is big (from 27% to 81.7%), and depends on plant age and cell differentiation [9]. Lignification of walls strengthens the protective function of cells at infecting, reduces speed of growth cell, strengthens mechanical durability of cells, and also reduces permeability to water through apoplast [10].

Lignin is highly branched and composed of cross-linked units, monolignols that derive from three hydroxy-cinnamyl alcohol monomers differing in their degree of methoxylation, *p*-coumaryl, coniferyl, and sinapyl alcohols [11] [12]. These monolignols produce, respectively, *p*-hydroxyphenyl (H), guaiacyl (G), and syringyl (S) phenylpropanoid units when incorporated into the lignin polymer. It is known that a syringyl monolignol mainly enters in the composition of lignin of floral plants and spore plants, while guaiacyl, syringyl and *p*-hydroxyphenyl monolignols enter in the composition of lignin of many vascular plants [13].

Real hydrophytes and marsh plants grow in condition of soil flooding or complete submergence. These plants adapt to reduction of free oxygen in soil and to constant water environment. Such plants are generated the specified adaptative mechanisms at different levels: tissular, cellular and subcellular [14]-[16]. Submerged leaves become thinner, structure and type photosynthesizing and epidermal cells change; stomata and wax no form, ultrastructure cells and cellulose content of epidermal tissue change significantly [5]. So much the lignin and monolignols of cell walls of hydrophytes species that adapt to complete submergence are not studied enough. Therefore, in this research we used submerged leaves of three species of hydrophytes (*Myriophyllum spicatum* L., *Potamogeton pectinatus* L. and *P. perfoliatus* L.). Selected aquatic plant species are typical for the Ukrainian flora. *M. spicatum* L., *P. pectinatus* and *P. perfoliatus* are known to be aquatic plants with submerged leaves, which grow and reproduce (from June to September, in Ukraine). These plants grow at the riverside, in water, at differential depth (from soil flooding to water surface), which successfully grow and reproduce in the Dnipro River in Ukraine. This paper presents a current picture of localization, distribution and relative content of monolignols in the leaf cell walls of some hydrophytes.

2. Material and Methods

2.1. Plant Material

Research objects were leaves of three hydrophytes: *Myriophyllum spicatum* (Haloraceae), *Potamogeton pectinatus* (Potamogetonaceae) and *P. perfoliatus* (Potamogetonaceae), which were identified according to Manual on High Plants of Ukraine [17]. The leaves of plants were collected at the beginning of June at the stage of vegetative growth. We used upper leaves that finished growth by enlargement. Plants were grown in water along-shore of the Rusanivskiy canal (left Shore of Dnipro River, in Kyiv) on the depth of 80 - 100 cm. The sun illumination [photosynthetic photon fluency rate (PPFR)] on water surface was 1200 - 1250 μmol quantum·m^{-2}·sec^{-1}, and on surface of the upper leaves of studied plants (about 8 - 10 cetimetres below the water surface) was 12 - 15 μmol quantum·m^{-2}·sec^{-1}. PPFR was measure by the means of Light Meter Li-250 (USA, LI-COR). The leaves situated above 8 - 10 cm under water surface were used for cytochemical analysis. The temperature of water was +15°C and the temperature of air was +23°C. The leaves of three plants from each species: submerged dissected leaves of *M. spicatum* (**Figure 1(a)**); submerged entire leaves of *P. perfoliatus* (**Figure 1(b)**), and submerged dissected leaves of *P. pectinatus* (**Figure 1(c)**) have been used for the cytological investigations.

The information concerning the physic-chemical property of water in Dnipro river is given by the Management of Dniepr-water resources (2015, http://www.dbuwr.com.ua/sendmail.php) and the data of ecologists [18]. Water of Dnipro River is the second class. Salts composition of water in Dnipro River (near Kiev) are carried out over last several years shown that water is clean and alkalescent, characterized by the middle salts value, notable: mineralization (sum of ions) average from 250 to 500 mg/dm^3; chlorides—16 - 24 mg/dm^3; sulphates—13 - 53 mg/dm^3; ammonium ions—0.29 - 1.08 mg/dm^3; and dissolved oxygen—8.7 - 13.5 mg/dm^3 (Management of Dniepr-water resources, 2015, http://www.dbuwr.com.ua/sendmail.php; [18]).

(a)   (b)   (c)

Figure 1. General view of *Myriophyllum spicatum* (a); *Potamogeton perfoliatus* (b) and *P. pectinatus* (c) leaves.

2.2. Microscopy Analysis

The cytochemical method accordingly to [19] was used for the study of both distribution monolignols (syringyl and guaiacyl) and relative content of these monolignols in walls. Leaf sections were hand cut from fresh leaf samples. For the detection of phenylpropanoids, sections were stained within 2 min with saturated (0.25%, w/v) diphenylboric acid-2-aminoethyl ester (DPBA) (Sigma) in MilliQ water containing 0.02% (v/v) Triton-X-100, then samples intensive washed with H_2O. The sections were visualized immediately with an laser scanning microscope LSM 5 Pascal (Carl Zeiss, Germany). For detection of complex DPBA-syringyl fluorescence a laser was excited at 340 - 380 nm, and the fluorescence emission detected at 430 nm wave length; and for detection of complex DPBA-guaiacyl fluorescence a laser was excited at 450 - 490 nm, and the fluorescence emission detected at 520 nm wave length, using an ×10, ×20 and ×40 objectives the PASCAL program. Chlorophyll auto fluorescence was excited at 440 nm and fluorescent emission detected at 660 nm. Fluorescence intensity of monolignols was measured in the cell walls as a function of emissions wave length using the PASCAL program.

For statistical treatment took for two/three leaves from each plant, in every leaf took at least 30 different cells (epidermis, photosynthesizing parenchyma, and xylem vessels). Values of cytochemical results were expressed as the mean and standard errors. Three replications of cytological results were expressed as the mean and standard arrows. Statistical significance of relative content of monolignols in cell wall was determined using the BIO software (Institute of Botany, Kiev, Ukraine) and a Student's test ($p < 0.05$).

3. Results and Discussion

Submerged leaves of *Myriophyllum spicatum*, *Potamogeton pectinatus* and *P. perfoliatus* had various form and size (**Figure 1**). Leaves of *Myriophyllum spicatum* were feather-like, dissected (**Figure 1(a)**), average size of leaf dissected particle of upper leaf was 18 ± 1.7—at long axis and 1.25 ± 0.5 mm—at short axis. Clasping leaves of *Potamogeton perfoliatus* had solid leaflet (**Figure 1(b)**), size of that was 51 ± 5.3 mm at long axis and 23 ± 2.7 mm—at short axis (in the middle of leaf). Leaves of *P. pectinatus* were needle-like, linear, highly dissected (**Figure 1(c)**), average size of leaf dissected particle was 57 ± 5.9 mm at long axis and 1.3 ± 0.5 mm—at short axis.

3.1. *Myriophyllum spicatum*

Cytochemical analysis of monolignols in the dissected leaves of *M. spicatum* are shown as blue fluorescence for syringyl and as green fluorescence for guaiacyl in the walls of epidermis, photosynthesizing parenchyma and also in xylem vessels (**Figures 2(a)-(d)**). But the fluorescence intensity of DPKK-syringyl and DPKK-guaiacyl complex was different in the tissues (**Table 1**; **Figure 2(e)** and **Figure 2(f)**). Wall of vessels and anticlinal wall of epidermis had the greatest fluorescence intensity of DPKK-syringyl and DPKK-guaiacyl complex in comparison with other walls of epidermis and parenchyma (1st - 4th layers). It should be noted that in the corners of parenchyma cells fluorescence of monolignols was considerably higher, than along walls. Except it, attitude of content of syringyl toward guaiacyl in the walls of photosynthesizing parenchyma was high enough, especially in the first layers to the parenchyma. The size of S/G ratio in cells takes place in the next order: anticlinal walls of epidermis > walls in the corners of parenchyma > vessel walls > parenchyma walls > periclinal walls of epidermis. It is revealed that maximum frequency for syringyl in the epidermal cell was 225787 (pixels, blue graph), the maximum frequency for guaiacyl was 286463 (pixels, green graph), and the maximum frequency for auto fluorescence of chlorophyll was 358132 (pixels, red graph) (**Figure 2(g)**).

(a)   (b)   (c)

(d)   (e)   (f)

(g)

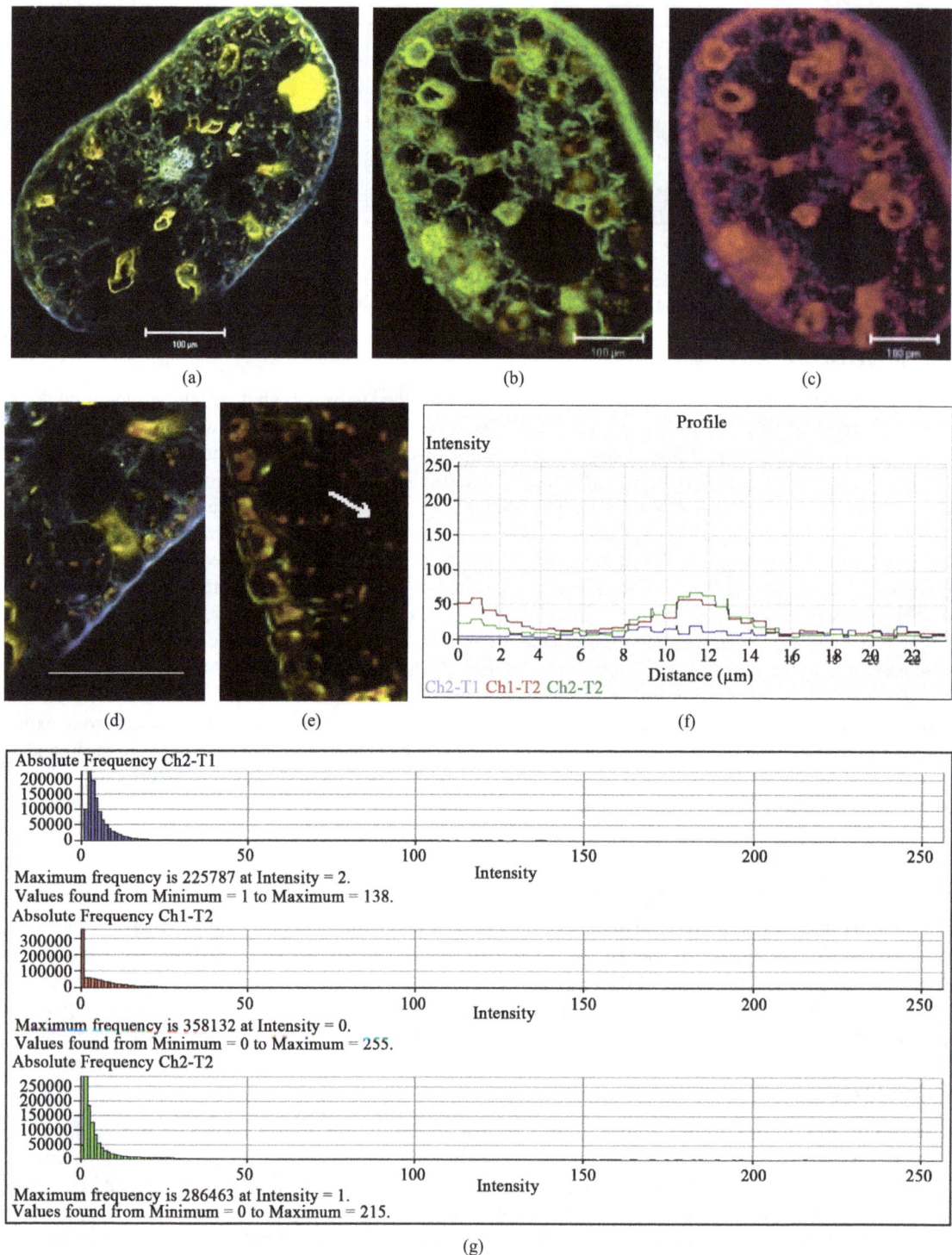

Figure 2. Micrographs of cytochemical fluorescence of monolignols in the leaf cells of *Myriophyllum spicatum.* Localization of syringyl has blue fluorescence and guaiacyl—green fluorescence; chlorophyll—red auto fluorescence. Monolignols are shown: on figures (a), (d) syringyl + guaiacyl in walls; on (b) guaiacyl, on (c) syringyl. On the (e), (f) histogram of fluorescence intensity of syrigyl (blue line), guaiacyl (green line) and chlorophyll auto fluorescence intensity (red line). Ordinate—Fluorescence intensity, relative units (pixels). Abscissa—distance (μm), which was scanned on the (e). This distance is shown as white line on the (e). On the (g) is shown absolute frequency of pixels for syringyl (blue graph), for guaiacyl (green graph) and for auto fluorescence of chlorophyll (red graph). Bars = 100 μm.

Table 1. The intensity of complex DPBA-syringyl and DPBA-guaiacyl fluorescence in leaves of *Myriophyllum spicatum*. (data are means ± SD of 3 replicates. 30 - 35 cells of epidermis, 50 - 60 cells of parenchyma and 15 - 17 cells of vessels used in each replicate).

Tissue/cell wall	Intensity of monolignols fluorescence, relative units			S/G
	Syringyl	Guaiacyl	S + G	
Epidermis:				
Outer periclinal wall	19 ± 0.7	48 ± 2.7*	67 ± 1.7	0.396
Inner periclinal wall	62 ± 5.3	49 ± 3.7	111 ± 5	1.26
Anticlinal wall	140 ± 17	80 ± 7.9*	220 ± 13	1.75
Photosynthesizing parenchyma:				
1st layer	65 ± 5.7	50 ± 3.9	115 ± 5	1.3
2nd - 4th layers	30 ± 1.5	46 ± 2.7*	76 ± 2.1	0.65
Walls in corners of cell junction	120 ± 11	70 ± 6.9*	190 ± 9	1.7
Vessels	175 ± 18	125 ± 19 *	300 ± 18	1.4

*Values of guaiacyl is significantly different ($p \leq 0.05$) from syringyl.

3.2. *Potamogeton perfoliatus*

Like fluorescence of monolignols observed in *P. perfoliatus* leaves. Cytochemical analysis of the complex DPKK-syringyl and DPKK-guaiacyl in leaves shown blue fluorescence of syringyl and green fluorescence of guaiacyl in cell walls of epidermis, photosynthesizing parenchyma and vessels cells (**Figures 3(a)-(c)**). There were some differences in fluorescence intensity of monolignols in cell walls of epidermis, in the corner of parenchyma cells and cell of vessels. The level of fluorescence intensity of monolignols is presented in the **Table 2** and **Figure 3(d)** and **Figure 3(e)** (diagram). It is necessary to note that syringyl was not exposed in cellular walls of 2nd, 3rd and 4th layers of parenchyma (of three plants) and in plans four-intensity of fluorescent was 7 ± 0.6 relative units; notably, minute intensity. On the size of the relation of S/G cells in *P. perfoliatus* are situated in such order: walls in the corners of parenchyma > vessels > periclinal walls of epidermis > anticlinal walls of epidermis > parenchyma. Luminescence intensity of monolignols in *P. perfoliatus* leaves differ from *M. spicatum* leaves: 2.1 - 3.2 times guaiacyl increased in epidermis walls and 5 times—in walls of 1st layer of parenchyma. It is necessary to note that independently from leaf species the relative content of monolignols in vessel walls was more than other tissue walls. It is revealed that maximum frequency for syringyl in the epidermal cell was 314166 (pixels, blue graph), the maximum frequency for guaiacyl was 209942 (pixels, green graph), and the maximum frequency for auto fluorescence of chlorophyll was 400695 (pixels, red graph) (**Figure 3(f)**).

3.3. *Potamogeton pectinatus*

Cytochemical analysis of monolignols in leaves of *P. pectinatus* is showed blue fluorescence of syringyl and green fluorescence of guaiacyl in the cell wall of epidermis, parenchyma and vessels cells, similar to that in leaves of *M. spicatum* leaves and *P. perfoliatus* (**Figures 4(a)-(c)**). The level of luminescence intensity is presented in **Table 3** and **Figure 4(d)** and **Figure 4(e)** (diagram). The analysis of monolignols luminescence intensity showed that relative content of guaiacyl in walls of periclinal walls of epidermis, in corners of parenchyma cells and in vessels was more than that in cell walls of anticlinal walls of epidermis and parenchyma cell walls. The size of S/G was lowest in the corners of parenchyma cells. The size of S/G ratio in cells takes place in the next order: parenchyma > epidermis > vessels > walls in the corners of parenchyma. It is revealed that maximum frequency for syringyl in the epidermal cell was 104477 (pixels, blue graph), the maximum frequency for guaiacyl was 24534 (pixels, green graph), and the maximum frequency for auto fluorescence of chlorophyll was 61963 (pixels, red graph) (**Figure 4(f)**).

Thus, we showed the presence of syringyl and guaiacyl monolignols in submerged leaves of three species real hydrophytes (*Myriophyllum spicatum*, *Potamogeton pectinatus* and *P. perfoliatus*) and also the increased content

Table 2. The intensity of complex DPBA-syringyl and DPBA-guaiacyl fluorescence in leaves of *Potamogeton perfoliatus*. (data are means ± SD of 3 replicates. 30 - 35 cells of epidermis, 50 - 60 cells of parenchyma and 15 - 17 cells of vessels used in each replicate).

Tissue/cell wall	Intensity of monolignols fluorescence, relative units			S/G
	Syringyl	Guaiacyl	S + G	
Epidermis adaxial:				
Outer periclinal wall	32 ± 2.9	103 ± 13*	135 ± 8	0.31
Inner periclinal wall	31 ± 3.0	200 ± 17*	231 ± 12	0.15
Anticlinal wall	30 ± 3.7	176 ± 19*	206 ± 12	0.17
Epidermis abaxial:				
Outer periclinal wall	47 ± 3.9	120 ± 13*	167 ± 8	0.39
Inner periclinal wall	45 ± 3.3	205 ± 19*	250 ± 12	0.22
Anticlinal wall	37 ± 4.5	255 ± 19*	292 ± 12	0.14
Photosynthesizing parenchyma:				
1st layer	30 ± 3.3	250 ± 13*	280 ± 9	0.12
2nd - 4th layers	No expose**	25 ± 3.1	25 ± 3.1	-
Walls in corners of cell junction	147 ± 13	59 ± 7.2*	206 ± 10	2.49
Vessels	215 ± 22	255 ± 21	470 ± 21	0.84

*Values of guaiacyl is significantly different ($p \leq 0.05$) from syringyl; **In leaf cells of three plants syringyl was not exposed, and in plans four-intensity of fluorescent was 7 ± 0.6 relative units; notably, minute intensity.

Table 3. The intensity of complex DPBA-syringyl and DPBA-guaiacyl fluorescence in leaves of *Potamogeton pectinatus* (data are means ± SD of 3 replicates. 30 - 35 cells of epidermis, 50 - 60 cells of parenchyma and 15 - 17 cells of vessels used in each replicate).

Tissue/cell wall	Intensity of monolignols fluorescence, relative units			S/G
	Syringyl	Guaiacyl	S + G	
Epidermis:				
Outer periclinal wall	61 ± 8.3	126 ± 12*	187 ± 11	0.48
Inner periclinal wall	22 ± 2.9	25 ± 3.1	47 ± 3.1	0.88
Anticlinal wall	15 ± 2.8	23 ± 3.5	38 ± 3.2	0.65
Photosynthesizing parenchyma:				
1st layer cells	20 ± 3.9	23 ± 3.5	43 ± 3.7	0.87
2nd - 4th layers cells	27 ± 3.1	52 ± 7.3*	59 ± 5.2	0.52
Walls in corners of cell junction	40 ± 3.9	220 ± 21*	260 ± 13	0.18
Vessels	88 ± 7.1	170 ± 13*	258 ± 11	0.52

*Values of guaiacyl is significantly different ($p \leq 0.05$) from syringyl.

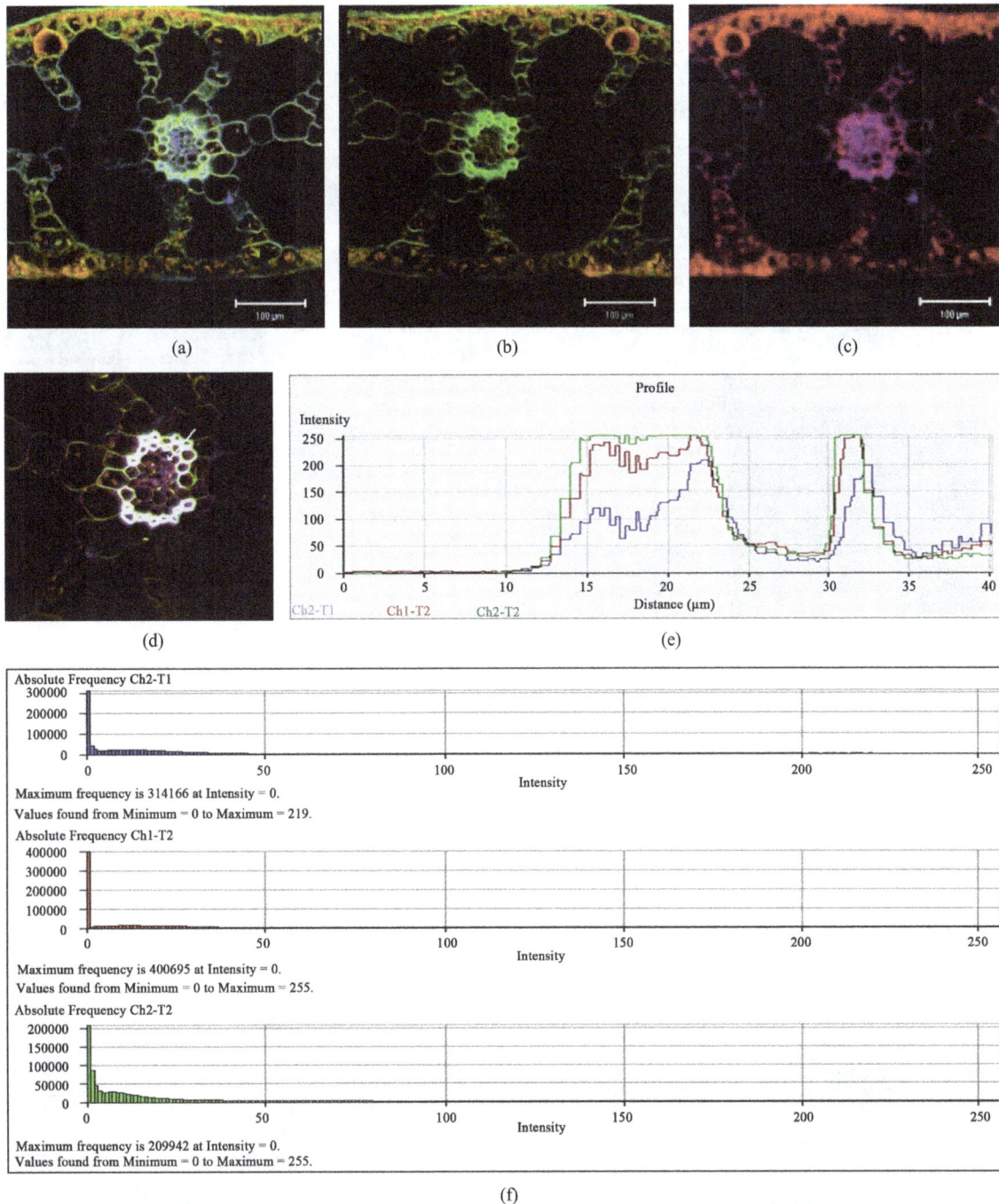

Figure 3. Micrographs of cytochemical fluorescence of monolignols in the leaf cells of *Potamogeton perfoliatus*. Localization of syringyl has blue fluorescence and guaiacy—green fluorescence; and chlorophyll—red auto fluorescence. Monolignols are shown: on (a), (d) syringyl + guaiacyl in walls; on (b) guaiacyl: on (c) syringyl. On (d), (e) histograms of fluorescence intensity of syrigyl (blue line), guaiacyl (green line) and chlorophyll auto fluorescence intensity (red line). Ordinate—Fluorescence intensity, relative units (pixels). Abscissa—Distance (µm), which was scanned on the (d). This distance is shown as white line on the (e); On (f) is shown absolute frequency of pixels for syringyl (blue graph), for guaiacyl (green graph) and for auto fluorescence of chlorophyll (red graph). Bars = 100 µm.

of guaiacyl in walls only of two species of *Potamogeton*. There were the general and differentia signs concerning these monolignols in hydrophytes leaves. General signs were: 1) presence of syringyl and guaiacyl at the investigated species regardless of conditions of growth of leaves; 2) almost identical (sufficiently great) content of S/G

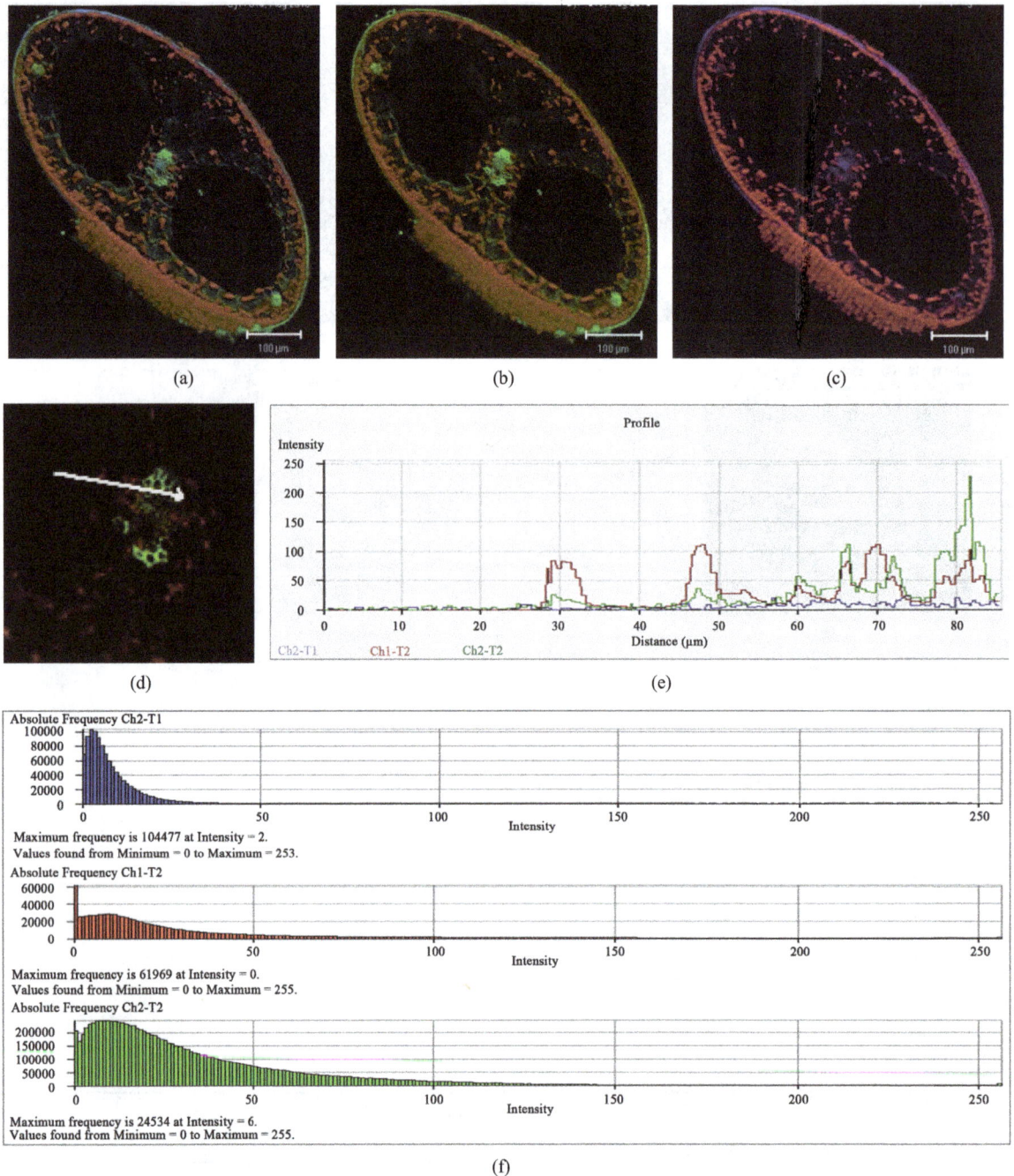

Figure 4. Micrographs of cytochemical fluorescence of monolignols in the leaf cells of *Potamogeton pectinatus*. Localization of syringyl has blue fluorescence and guaiacyl—green fluorescence; and chlorophyll—red auto fluorescence. Monolignols are shown: on (a) syringyl + guaiacyl in walls; on (b) guaiacyl; on (c) syringyl. On (d), (e) histogram of fluorescence intensity of syrigyl (blue line), guaiacyl (green line) and chlorophyll auto fluorescence intensity (red line). Ordinate—Fluorescence intensity, relative units (pixels). Abscissa—Distance (μm), which was scanned on the (d). This distance is shown as white line on the (d). On (f) is shown absolute frequency of pixels for syringyl (blue graph), for guaiacyl (green graph) and for auto fluorescence of chlorophyll (red graph). Bars = 100 μm.

in the vessels cells of all hydrophytes, and in the corners of parenchyma cells of *M. spicatum* and *P. perfoliatus*; and 3) certain polarity of S/G ratio, that characteristic for every species. Differentia cytochemical signs were: relative content of syringyl and guaiacyl in different tissues of hydrophytes, and also different quantity of S/G in epidermis, parenchyma and vessels walls of all studied hydrophytes.

In consideration of presence of magnified content of syringyl and guaiacyl in the corners of parenchyma cells of the investigated samples of hydrophytes and the well-known data that the middle lamella and the cell corners are the so-called nucleation sites of lignin and also they rich in Ca^{2+} pectate [8] [20] and are the first sites to be lignified, we can to draw a conclusion that cell corners of parenchyma cells of hydrophytes is also leading spots for lignin synthesis.

The cytochemical analysis of monolignols in different tissues of submerged leaves of *M. Spicatum*, *P. pectinatus* and *P. perfoliatus*) revealed the increase content of syringyl and guaiacyl in epidermis cell walls in comparison with walls of inner layers of photosynthesizing parenchyma. It is known that submerged leaves of hydrophytes are losing trichomes, stomata and thicker cuticle [5]; therefore the surface of epidermal cells in submerged leaves become more accessible for an invasion of pathogens. Besides, it is known that in cells of leaves and shoots are occurs an increase in lignifications in responses to attack by pathogens (bacteria, fungi) [21]. Besides, it is known that epidermis is the one barrier and protection of leaf from action of ultraviolet, which is provoke both the intensified synthesis of lignin [22]. Taking into account the above and the results of our experiments the greatest fluorescence of monolignols the epidermis cell of leaves that situated under water surface, it is suggested that walls epidermis of studied leaves of hydrophytes protects of leaves from pathogens and even from little UV radiation in water.

Consider that with the increase of quantity of S/G as a chemical barrier increases to protecting of cells from penetration of water and invasion of pathogens [23]. Except it, it is known that this sign (S/G) testifies to the increase of mechanical durability of cells [24]. Taking into account, that the flooded leaves constantly are in a contact with a surrounding water micro flora and numerous alga, and also that flooded leaves undergo permanent pressure of water and action of waves [5] [25], it is possible to assume that periclinal and anticlinal walls of hydrophytes epidermis protect the surface of cells from exogenous influence of water surroundings therefore occurs defined synthesis of components of lignin. Notably, permanent water surroundings are one of basic exogenous factors of an increase synthesis of syringyl and guaiacyl in the leaves of the investigated hydrophytes. We see in addition, that a presence of these monolignols, and also their ratio in the cell walls of epidermis, parenchyma and vessels correlates with the data that were received of angiosperms dicotyledons upland plants [13] [26]. We do not eliminate, that the leaf cell walls of investigated hydrophytes contains and the third monolignol-*p*-hydroxyphenyl, that also plays a substantial role in lignin structure, and it is able determine with using other method, UV spectroscopy [27].

Thus, the obtained results revealed the particularities of syringyl and guaiacyl monolignols of cell walls of *Myriophyllum spicatum*. *Potamogeton pectinatus* and *P. perfoliatus* submerged leaves, which can be used as a basis for future study of both adaptation mechanisms of agricultural plants during flood, and possible for gene-engineering of plant stability in flooded area.

4. Conclusion

Localization and relative content of syringyl and guaiacyl monolignols in *Myriophyllum spicatum*, *Potamogeton pectinatus* and *P. perfoliatus* submerged leaves were studied by the laser scanning microscopy. It was found that natural flooding caused the allocation of monolignols in cell walls dependence on leaf tissue and on species. The most content of monolignols was in walls of vessels, epidermis and in the interspaces (corners of cell junction) of photosynthetic parenchyma. We suggest that cell walls of leaf epidermis, as one main and first barrier between photosynthesizing parenchyma cells and water environment, protects of submerged plant from water pressure and from waves by the redistribution of syringyl and guaiacyl monolignols.

References

[1]   Armstrong, W. and Drew, M.C. (2002) Root Growth and Metabolism under Oxygen Deficiency. In: Wasel, Y., *et al.*, *Plant Roots: The Hidden Yalf*, 3rd Edition, New York and Basel, 729-761.

[2]   Jackson, M.B. and Colmer, T.D. (2005) Response and Adaptation by Plants to Flooding Stress. *Annals of Botany*, **96**, 501-505. http://dx.doi.org/10.1093/aob/mci205

[3]   Parent, C., Capelli, N., Berger, A., Crevecoeur, M. and Dat, J. (2008) An Overview of Plant Responses to Soil Waterlogging. *Plant Stress*, **20**, 20-27.

[4]   Rose-John, S. and Kende, H. (1984) Effect of Submergence on the Cell Wall Composition of Deep-Water Rice Internodes. *Plant Physiology*, **76**, 106-111. http://dx.doi.org/10.1104/pp.76.1.106

[5]   Nedukha, O.M. (2011) Heterophylly in Plants. Alt Press, Kiev, 1-191. (In Ukrainian)

[6]   Little, S.T. (2003) Adaptation and Acclimaton of Populatons of Ludwigia Repens to Growth in High- and Lower-CO$_2$ Springs. A Dissertation Presented to the Graduate School of the University of Florida in Partial Fulfillment of the Requirements for the Degree of Doctor of Philosophy. University of Florida, USA, 1-157.

[7]   Nedukha, O. and Kordyum, E. (2013) The Participation of Cell Wall Polysaccharides in Cellular Mechanisms of Leaf Tolerance to Nature Flooding of Plant. In: *Plant Functioning under Environmental Stress*, 9th Intern Conf., Cracow, 12-15 September 2012, 137-152.

[8]   Boerjan, W., Ralph, J. and Baucher, M. (2003) Lignin Biosynthesis. *Annual Review of Plant Biology*, **54**, 519-546. http://dx.doi.org/10.1146/annurev.arplant.54.031902.134938

[9]   Fengel, D. and Wegener, G. (1984) Wood Chemistry, Ultrastructure, Reaction. Walter de Gruyter, Berlin, 611.

[10]   Monties, B. (1998) Novel Structures and Properties of Lignins in Relation to Their Natural and Induced Variability in Ecotypes, Mutants and Transgenic Plants. *Polymer Degradation and Stability*, **59**, 53-64. http://dx.doi.org/10.1016/S0141-3910(97)00166-3

[11]   Adler, E. (1977) Lignin Chemistry—Past, Present and Future. *Wood Science and Technology*, **11**, 169-218.

[12]   Leisola, M., Pastinen, O. and Axe, D. (2012) Lignin—Designed Randomness. *Bio-Complexity*, **3**, 1-11.

[13]   Weng, J.K., Takuya, A., Nichlolas, D., *et al.* (2010) Convergent Evolution of Syringyl Lignin Biosynthesis via Distinct Pathways in the Lycophyte *Selaginella* and Flowering Plants. *The Plant Cell*, **22**, 1033-1045.

[14]   Vartapetian, B., Andreeva, I., Maslova, I. and Davtian, N. (1970) The Oxygen and Ultrastructure of Root Cells. *Agrochimica*, **15**, 1-19.

[15]   Vartapetian, B. and Jackson, M. (1997) Plant Adaptation to Anaerobic Stress. *Annals of Botany*, **3**, 3-20.

[16]   Nekrasova, G., Ronzhina, D. and Korobizyna, E. (1998) Formation of the Photosynthetic Apparatus during Growth of Submerged, Floating and Above-Water Leaves in Hydrophytes. *Russian Journal of Plant Physiology*, **45**, 539-548. (In Russian)

[17]   Prokudin, J. (1987) Manual on High Plants of Ukraine. Naukova Dumka, Kiev.

[18]   Dudar, T. and Zosimovich, A. (2011) Ecological Rating of Water Quality in Kiev Region. *National Academy of Sciences of Ukraine*, **2**, 125-130.

[19]   Wuyts, N., Lognay, G., Swennen, R. and De Waele, D. (2003) Secondary Metabolites in Roots and Implications for Nematode Resistance in Banana (Musa spp.) *Proceedings of an International Symposium "Banana Root System: Towards a Better Understanding for Its Productive Management"*, San José, 3-5 November 2003, 238-246.

[20]   Carpita, N.C. and Gibeaut, D.M. (1993) Structural Models of Primary Cell Walls in Flowering Plants: Consistency of Molecular Structure with the Physical Properties of the Walls during Growth. *The Plant*, **3**, 1-30.

[21]   Moura, J.C., Bonine, C.A., Viana, J., Dornelas, M.C. and Mazzafera, P. (2010) Abiotic and Bitioc Stresses and Changes in the Lignin Content and Composition in Plants. *Journal of Integrative Plant Biology*, **52**, 360-376. http://dx.doi.org/10.1111/j.1744-7909.2010.00892.x

[22]   Hilal, M., Parrado, M., Rosa, M., Gallardo, M., Orce, L., Massa, M., Gonzabel, J. and Prado, F. (2004) Epidermal Lignin Deposition in Quinoa Cotyledons in Response to UV-B Radiation. *Photochemistry and Photobiology*, **79**, 205-210. http://dx.doi.org/10.1562/0031-8655(2004)079<0205:ELDIQC>2.0.CO;2

[23]   Menden, B., Kohlhoff, M. and Moerschbacher, B.M. (2007) Wheat Cell Accumulate a Syringil-Rich Lignin during the Hypersensitive Resistance Response. *Phytochemistry*, **68**, 513-529.

[24]   Christiernin, M. (2006) Composition of Lignin in Outer Cell-Wall Layers. PhD Thesis, Division of Wood Chemistry and Pulp Technology, Royal Institute of Technology, Stockholm, 1-53.

[25]   Tyree, M.T. and Cheung, Y.N.S. (1997) Resistance to Water Flow in *Fagus grandifolia* Leaves. *Canadian Journal of Botany*, **55**, 2591-2599.

[26]   Baucher, M., Monties, B., Van Montagu, M. and Boerjan, W. (1998) Biosynthesis and Genetic Engineering of Lignin. *Critical Reviews in Plant Sciences*, **17**, 125-197. http://dx.doi.org/10.1016/S0735-2689(98)00360-8

[27]   Koch, G. and Schmitt, U. (2013) Topochemical and Electron Microscopic Analyses on the Lignification of Individual Cell Wall Layers during Wood Formation and Secondary Changes. In: Fromm, J., Ed., *Cellular Aspects of Wood Formation*, Springer-Verlag, Berlin, 41-69.

20

Purification of the *Drosophila melanogaster* Proteins Inscuteable and Staufen Expressed in *Escherichia coli*

Xristo Zárate[1*], Megan M. McEvoy[2], Teresa Vargas-Cortez[1], Jéssica J. Gómez-Lugo[1], Claudia J. Barahona[1], Elena Cantú-Cárdenas[1], Alberto Gómez-Treviño[1]

[1]Universidad Autónoma de Nuevo León, Facultad de Ciencias Químicas, Cd. Universitaria, San Nicolás de los Garza, México
[2]Department of Chemistry and Biochemistry, University of Arizona, Tucson, USA
Email: *xristo.zaratekl@uanl.edu.mx

Abstract

The proteins Inscuteable and Staufen are key components during asymmetric cell division of neuroblasts for the development of *Drosophila melanogaster*. Expression and purification of both proteins has been a difficult task for structure-function studies. Based on codon optimization for protein expression in *Escherichia coli*, we have been able to produce, in soluble form, the C-terminal domains of Inscuteable and Staufen as chimeras with N-terminal maltose binding protein tag that contains a rigid linker between them for feasible crystallization. In addition, using an optimized synthetic gene, corresponding to the amino acid region 250 - 623 of Inscuteable fused to glutathione-S-transferase, low-scale expression experiments showed production of soluble protein. Finally, eukaryotic expression of Inscuteable in the methylothropic yeast *Pichia pastoris* failed to produce the *Drosophila* protein at detectable amounts, reinforcing the fact that *E. coli* still was the microorganism of choice for high-yield protein expression.

Keywords

Inscuteable, Staufen, Protein Expression and Purification, Maltose-Binding Protein, *Escherichia coli*

---

*Corresponding author.

1. Introduction

Asymmetric cell division is of one nature's processes to generate cellular diversity; in *Drosophila melanogaster* neuroblasts are considered the stem cells of the central nervous system [1]. Neuroblasts undergo repeated asymmetric cell divisions to produce a lineage of ganglion mother cells (GMCs) and neurons [2].

The cell fate determinants are biomolecules that regulate differentiation between neuroblasts and GMCs, and the asymmetric localization of these determinants is the key for the proper development and lineage of GMCs [3]. Prospero and Numb are examples of these cell fate determinants in *Drosophila* central nervous system [1]. Prospero is a multidomain protein with DNA-binding motifs that is required for the expression of several genes [4]. Prospero localization during neuroblast-GMC cell division requires the function of Inscuteable and Miranda, while Inscuteable and Staufen localize *prospero*RNA [5] [6].

Inscuteable is the most upstream component for the asymmetric division of neuroblasts and GMCs [7]. It has been demonstrated that the C-terminal domain of Inscuteable (751 - 859 aa) interacts with the mRNA-binding protein Staufen (in its 769 - 1026 aa region) for the proper asymmetric localization of *prospero*RNA [6]. The Inscuteable 252 - 615 amino acid region forms a complex with Miranda and Prospero for the appropriate localization of the latter [5]; also this region is sufficient for most of Inscuteable function [8].

The understanding of asymmetric localization of cell determinants during cell division at the molecular level would be extremely useful in order to further comprehend development and cellular diversity from stem cells. Here we describe the expression and purification of the two domains of Inscuteable mentioned above and the C-terminal of Staufen for protein crystallization and structure-function studies. The task was undertaken with different methodologies in order to obtain several milligrams of protein for structural studies: from eukaryotic protein expression in the eukaryote *Pichia pastoris* to gene synthesis for codon and DNA secondary structure optimization for expression in *Escherichia coli*. We found that with the right construct *E. coli* is still the most suitable host microorganism for high-yield production of recombinant proteins.

2. Materials and Methods

2.1. DNA Constructs

Table 1 shows all the constructs made for this study. The DNA encoding proteins for expression in *Escherichia coli* with a glutathione-S-transferase (GST) tag were cloned into pGEX-4T-2 plasmid (GE Life Sciences). For those proteins expressed as chimeras with maltose binding protein (MBP), the DNA was cloned in a special pET plasmid containing the gene for MBP and a multiple cloning site to produce a rigid linker between MBP and the protein of interest. Cloning was done in pPIC3.5K (Life Technologies) for protein expression in *Pichia pastoris*. The 50 μL PCR reactions consisted of 10 ng of template DNA (DNA library or synthesized genes), 60 pmoles of each primer, 2 μL of 10 mM dNTPs mix, and 2.5 units of Pfu DNA polymerase (Stratagene). The thermocycler conditions were 95°C for 2 minutes; 30 cycles of 95°C-45 seconds, 56°C-45 seconds, and 72°C 1 minute per kb; and a final extension for 10 minutes at 72°C. Plasmids were linearized with BamHI and EcoRI restriction enzymes (New England Biolabs) and Inscuteable or Staufen DNA ligated into them after digestion with the same enzymes; other restriction enzymes were used as necessary. All DNA constructs were confirmed by sequencing.

2.2. Gene Synthesis

The primary sequences of Staufen and Inscuteable were entered into the Primo Optimum 3.4 software (Chang Bioscience), which produced DNA sequences containing the codons most commonly used by *E. coli*. The program also designed the alternating oligonucleotides needed for the PCR synthesis of the gene [9]. A two-step PCR methodology, gene assembly and gene amplification, was followed in order to obtain the full-length double-stranded DNA constructs for the Staufen C-terminal (aa 761 - 1026) and two domains of Inscuteable (aa 719 - 859 and aa 250 - 623). For the gene assembly, 10 μL of each oligo solution, at 25 mM, were mixed. 5 μL of this solution were added to a 50 μL PCR solution with the components mentioned above (except primers). The thermocycler conditions consisted of one step at 94°C; 25 cycles of 94°C-30 seconds, 52°C-30 seconds and 72°C-2 minutes. For the gene amplification, 5 μL from the gene assembly reaction were added to a second PCR mixture with 60 pmoles each of the outer primers (the two external ones used during the assembly). The thermocycler program was 94°C-60 seconds, 25 cycles of 94°C-45 seconds, 68°C-45 seconds, 72°C for 3 minutes and a final extension for 10 minutes at 72°C. All genes were sequenced for authentication.

Table 1. DNA constructs for expression and purification of Inscuteable and Staufen.

NAME	PROTEIN	AA REGION SIZE§	AFFINITY TAG	HOST CELL
Insc1	Inscuteable	719 - 859 15.3 kDa	GST* N-terminal	E. coli
Insc2	Inscuteable	719 - 857 15.0 kDa	GST N-terminal	E. coli
Insc3	Inscuteable	1 - 859 96.4 kDa	His-tag C-terminal	P. pastoris
Insc4	Inscuteable	719 - 857 15.0 kDa	His-tag C-terminal	P. pastoris
Insc5	Inscuteable	719 - 857 15.0 kDa	GST N-terminal	P. pastoris
Insc6	Inscuteable	250 - 601 37.1 kDa	GST N-terminal	P. pastoris
Stau1	Staufen	761 - 1026 27.7 kDa	GST N-terminal	E. coli
Insc7	Inscuteable	719 - 857 15.0 kDa	GST N-terminal	E. coli
Insc8	Inscuteable	719 - 857 15.0 kDa	His-tag N-terminal	E. coli
Insc9	Inscuteable	250 - 623 41.6 kDa	GST N-terminal	E. coli
Insc10	MBP⁺/Inscuteable chimera	719 - 857 55.9 kDa	MBP N-terminal	E. coli
Insc11	MBP/Inscuteable chimera	719 - 859 57.2 kDa	MBP N-terminal His-tag C-terminal	E. coli
Stau2	MBP/Staufen chimera	761 - 1026 68.6 kDa	MBP N-terminal	E. coli

§Indicates the molecular weight without considering the removable affinity tag. For the chimeras, it includes the affinity tag(s). *Glutathione-S-transferase. ⁺Maltose binding protein.

2.3. Protein Expression

For *E. coli* expression, the BL21(DE3) or Rosetta strains were used. For pilot protein expression experiments, 2 mL of LB medium (containing 100 µg/mL ampicillin) were inoculated with a single *E. coli* colony and incubated at 37°C in a shaking incubator. During log-phase growth, expression was induced with 0.5 mM isopropyl-D-1-thiogalactopyranoside (IPTG). After induction, the temperature was lowered to 25°C and the cells were incubated overnight. Cells were harvested and resuspended in 100 µL of SDS-PAGE sample buffer and boiled for 10 minutes. After centrifugation the supernatant was analyzed by electrophoresis. Large-scale protein expression was performed in baffled fernbach flasks using LB-amp medium, cells were incubated at 37°C until O.D.$_{600}$ reached 0.6. Then the temperature was lowered to 25°C and IPTG was added to a final concentration of 0.5 mM. Cells were further incubated for 16 hrs. and harvested by centrifugation. For *P. pastoris* expression, the GS115 strain (Life Technologies) was transformed by electroporation with 10 µg of linear DNA and selection of transformants was performed using the manufacturer's suggested protocol. For *P. pastoris* protein expression induction, cells were grown in minimal methanol medium in baffled fernbach flasks at 30°C.

2.4. Protein Purification

After cell disruption (sonication for *E. coli* and treatment with YeastBuster (Novagen) for *P. pastoris*) and clarification of lysates by centrifugation, proteins were purified by affinity column chromatography. For GST-tagged proteins, lysates were incubated with glutathione sepharose 4B in PBS buffer at room temperature for one hour. The resin was loaded into a column and washed several times with PBS buffer; proteins were eluted in three column-volumes using elution buffer: 50 mM TRIS-HCl, 10 mM reduced glutathione, pH 8.0. Purification of proteins with a His-tag was performed using the His-Pur Ni-NTA resin from Thermo Scientific following manufacturer's instructions. Insc10 and Insc11 were further purified by anion exchange and size exclusion chromatography using the ÄKTA PrimePlus FPLC system from GE Life Sciences. After affinity chromatography, proteins were loaded into a MonoQ anionic exchange column in 50 mM TRIS-HCl buffer at pH 7.8. The column was washed extensively with the same buffer; a 0 - 250 mM NaCl gradient was applied in 40 column-volumes, the fractions that showed the higher protein concentration were analyzed by SDS-PAGE. The purest fractions were pooled and concentrated up to 240 µL, then loaded in a Superose 12 column for size-exclusion separation. Final protein purity was determined by SDS-PAGE.

3. Results and Discussion

3.1. Prokaryotic Expression of Inscuteable

The C-terminal domain of Inscuteable that interacts with Staufen for *prospero* RNA localization is reported to be the amino acid region 751 to 859 [6]. Two C-terminal Inscuteable DNA constructs that include this region, Insc1 and Insc2 (see **Table 1**), were tested using the glutathione-S-transferase expression system. Both constructs were transformed for protein expression in *E. coli* strains BL21(DE3) and Rosetta. **Figure 1** shows a typical result after expression, in any of the two strains, and purification of Insc1 and Insc2; mostly just GST protein (approximately 26 kDa) was obtained after affinity purification. An expected band for GST-Insc2 at approximately 41 kDa was not observed. Purification was also done under denaturing conditions to see if the protein was being expressed as inclusion bodies but no protein was detected.

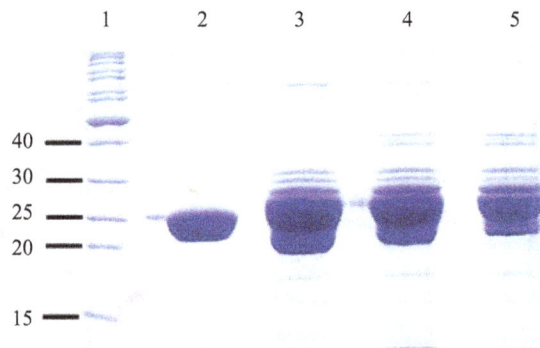

Figure 1. 12% SDS-PAGE showing expression of Insc2 in the *E. coli* strain Rosetta. Lanes 1: protein marker, 2: GST, 3. elution fraction 1, 4: elution fraction 2, 5: elution fraction 3. Sizes of the protein marker bands are shown in the left (in kDa).

This kind of result is not uncommon for eukaryotic proteins expressed in *E. coli* [10], as each unique protein has certain features that allow its proper expression in bacteria. Even though the Rosetta strain, which contains all codons for translation of eukaryotic proteins, was used, *E. coli* failed to express the C-terminal of Inscuteable.

3.2. Eukaryotic Expression of Inscuteable

Several DNA constructs were made for expression of Inscuteable in the yeast *Pichia pastoris*. Full-length Inscuteable DNA Insc3 and its C-terminal region Insc4, both with a C-terminal his-tag, were cloned into pPIC3.5K plasmid vector. **Figure 2** shows the expression and purification of Insc3. After high-imidazole elution the stronger band indicates a molecular weight of approximately 100 kDa, near the calculated size of 96.4 kDa, although this band also appears in lane 5, during the first wash step with low concentration of imidazole. Further experiments will confirm the expression of Insc3; larger protein preparations and characterization, like mass spectrometry, need to be performed. **Figure 3** shows the same experiment but for Insc4, in which an expected protein band of approximately 15 kDa was not observed. Expression of proteins in *P. pastoris* is not as consistent as in *E. coli*; each colony can produce different amounts of protein, and that is the reason why *Pichia* needs an extensive screening for heterologous protein expression.

3.3. Gene Synthesis of Staufen and Inscuteable

Because of unsuccessful attempts to amplify Staufen DNA from a cDNA library, a different approach was needed

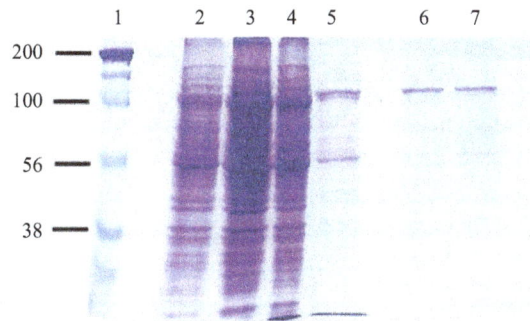

Figure 2. 10% SDS-PAGE from small-scale preparation of Insc3 using the Ni-NTA chromatography columns. Lanes 1: protein marker, 2: 5 μL lysate, 3: 10 μL lysate, 4: lysate after spin column, 5: first wash, 6: first elution step, 7: second elution step.

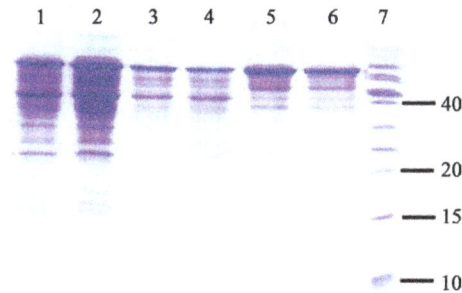

Figure 3. 15% SDS-PAGE showing the expression and purification from cultures of two different colonies of *P. pastoris* expressing Insc4. Lanes 1: lysate 1, 2: lysate 2, 3: first wash 1, 4: first wash 2, 5: elution 1, 6: elution 2, 7: protein marker.

to make the constructs for protein expression. PCR-based DNA synthesis is a method that allows rapid production of a nucleotide sequence optimized for expression in a certain system [9]. **Figure 4** and **Figure 5** show the results of the two-step PCR method for the gene synthesis of the C-terminal domain of Staufen, and two domains of Inscuteable, aa 250 - 623 and aa 719 - 859. After gel-purification of the amplification reactions, genes were cloned and then sequenced. The sequencing indicated some mistakes, though everyone was fixed by site-directed mutagenesis until the desired gene sequence was obtained for protein expression.

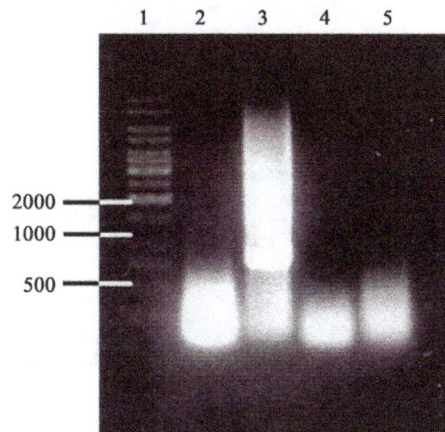

Figure 4. 0.7% agarose gel electrophoresis showing the 798 bp gene synthesis of C-terminal Staufen. Lanes 1: GeneRuler 1kb DNA ladder, 2: gene assembly reaction, 3: gene amplification reaction, 4: no-template negative control for the amplification reaction, 5: no-enzyme negative control. Length, in base pairs, of some of the DNA marker bands is shown.

3.4. *E. coli* Expression and Purification of Inscuteable and Staufen with Optimized DNA

The newly synthesized Staufen DNA was cloned into pGEX-4T-2 vector for the Stau1 construct. After GST af-

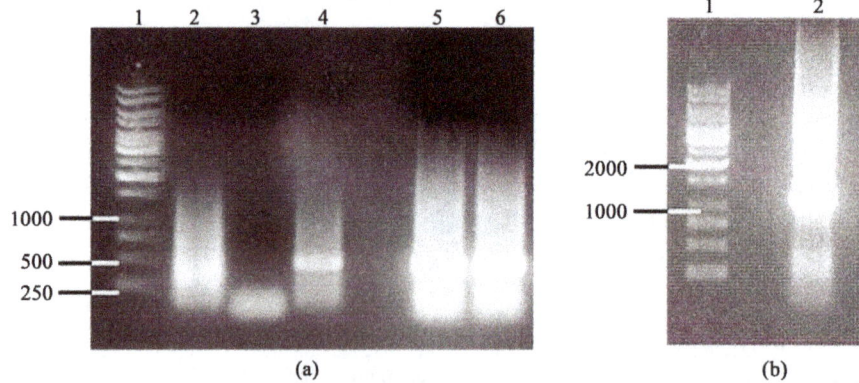

Figure 5. 0.7% agarose gel electrophoresis of synthetic gene products of two domains of Inscuteable. (a) C-terminal domain, 423 bp, Lanes 1: GeneRuler 1kb ladder, Lane 2: gene assembly mixture, Lane 3: no-template control for the amplification reaction, Lane 4: no-enzyme control, Lanes 5 and 6: amplification reaction showing the 423 bp product. (b) Main Inscuteable domain, 1122 bp, Lanes 1: GeneRuler 1kb, Lane 2: amplification reaction.

finity chromatography the construct yielded several milligrams of protein. It was observed that the band appeared as a higher molecular weight protein than expected, so the protein was further characterized by mass spectrometry and sequencing. The mass spectrum showed a value of 54,287.06 m/z that is in good agreement with the calculated value of 54,051.90 Daltons. **Figure 6** shows the results for Stau1 purification; after affinity chromatography the protein was treated with thrombin to remove the GST tag. Stau1 was transferred on to a PVDF membrane, stained with Coomassie blue and N-terminal sequenced. The first ten amino acids obtained from the sequencing were: GSGSNSKKLAK. The first two amino acids, Gly and Ser, are remnants after the thrombin digestion (from its recognition site LVPR/GS), the rest of the amino acids correspond to the C-terminal domain of Staufen. The C-terminal Inscuteable DNA was cloned into pGEX-4T-2 as well, Insc7, but even with the optimized DNA we found the same expression pattern as Insc1 or Insc2: just GST was present. The DNA was also cloned into pET28b vector for a N-terminal his-tag (Insc8) but pilot expression experiments did not show presence of the protein. The main Inscuteable domain DNA was cloned for GST-based expression, Insc9. Pilot expression experiments showed a new protein band appearing at approximately 68 kDa corresponding to GST-Insc9 (**Figure 7**).

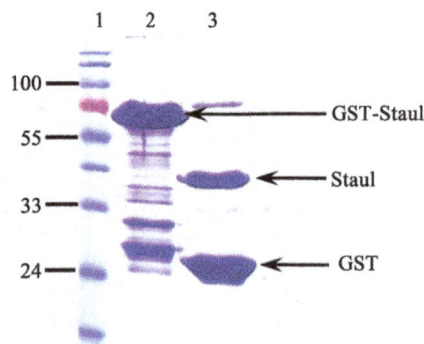

Figure 6. Expression and purification results for Stau1.12% SDS-PAGE showing thrombin digestion for GST removal for Stau1 purification. Lanes 1: protein marker, 2: elution from GST affinity chromatography, 3: protein mixture after 3 hour-treatment with thrombin.

3.5. Expression and Purification of Inscuteable and Staufen as Chimeras with Maltose Binding Protein

Maltose binding protein is a well-known fusion protein that produces high quantity of protein in a soluble form in *E. coli*. A different study has shown that MBP can be a powerful tool for protein crystallization and structure determination [11]. Based on these previous studies, construct Insc10 was design to create a chimera of MBP and the C-terminal domain of Inscuteable. A rigid linker (composed of amino acids AAAEF) was engineered in

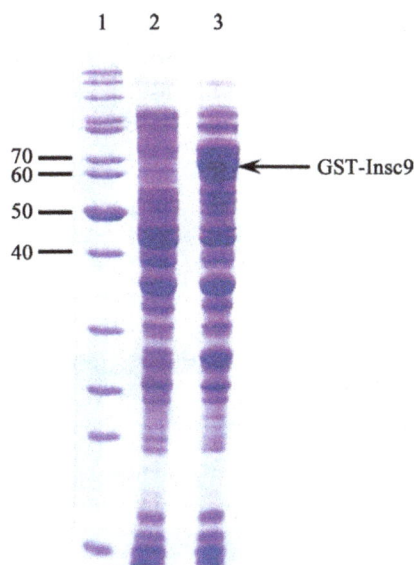

Figure 7. 12% SDS-PAGE showing pilot expression of Insc9 in BL21(DE3). Lanes 1: protein marker, 2: lysate of non-induced cells, 3: lysate of IPTG-induced cells.

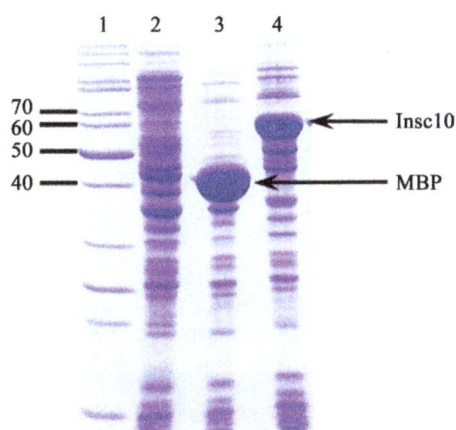

Figure 8. 12% SDS-PAGE showing pilot expression of Insc10 in BL21(DE3). Lanes 1: protein marker, 2: lysate of non-induced cells, 3: lysate of IPTG-induced cells expressing just MBP, 4: lysate of IPTG-induced cells expressing the MBP/C-terminal chimera Insc10.

order to minimize the flexibility between the two domains of the final protein [12]. **Figure 8** shows the pilot expression of this construct in *E. coli* BL21(DE3) cells induced with IPTG.

Insc10 was purified by affinity chromatography with amylose resin. The pooled fractions containing Insc10 were dialyzed and loaded into a MonoQ anionic exchange column. After several column-volume washes, the protein was eluted with a NaCl gradient. The two purest fractions were pooled, the protein concentrated and then applied to Superose 12 column for size-exclusion chromatography. This was a final purification step and it also helped to determine the oligomeric state of Insc10. Based on the elution volume (11.8 mL), it appeared that Insc10 is a monomer; it is worth mentioning that MBP is a monomer and it has been observed that its presence does not affect the quaternary structure of the other protein in the chimera [12]. **Figure 9** shows the electrophoresis analysis of the peak fractions and indicated that Insc10 is a mixture of truncated protein (although it contains more than 90% of the full protein). In order to avoid this problem, Insc11 was designed with a C-terminal His-tag. **Figure 10** shows the expression and purification results for this construct.

Fractions 3 and 4 after size-exclusion chromatography of Insc10 and pure Insc11 were concentrated up to 15 mg/mL for protein crystallization trials, and showed good solubility at that concentration. Crystallization could be feasible due to the designed linker between MBP and the C-terminal of Inscuteable [10]. Finally, since constructs with N-terminal MBP showed good expression levels, it was decided to apply it for Staufen as well (Stau2). **Figure 11** shows the pilot expression analysis, which indicates a very obvious new protein band at the expected size. A similar purification protocol used for Insc10 could be applied. It is worth mentioning that the quaternary structure of proteins is not disrupted by the presence of MBP, so, it may be possible to study Inscuteable-Staufen interactions both as chimeras with MBP [12].

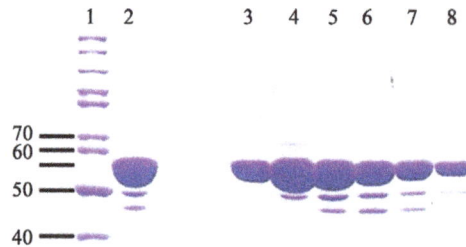

Figure 9. 12% SDS-PAGE electrophoresis analysis of Insc10 after Superose 12 size-exclusion chromatography. Lanes 1: protein marker, 2: Insc10 before chromatography, 3-8: main fractions from size exclusion chromatography.

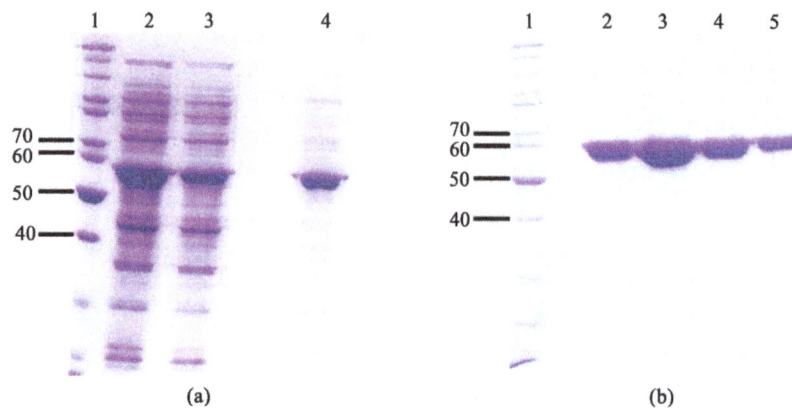

(a)   (b)

Figure 10. 12% SDS-PAGE electrophoresis analysis of Insc11 purification using metal-affinity and anion exchange chromatography.A Affinity chromatography, Lanes 1: protein marker, 2: Ins11 lysate, 3: lysate after incubation with the metal-affinity resin, 4: elution from the metal-affinity chromatography. B MonoQ anion exchange chromatography, Lanes 1: protein marker, 2-5: main fractions from elution.

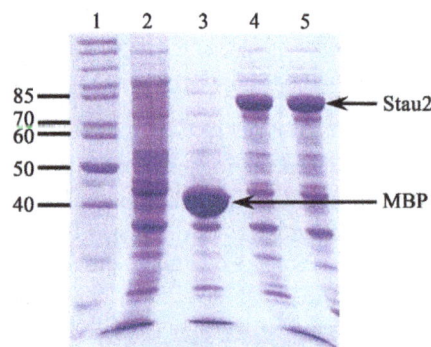

Figure 11. 12% SDS-PAGE electrophoresis analysis of the pilot expression of Stau2. Lanes 1: protein marker, 2: lysate of non-induced cells, 3: lysate of IPTG-induced cells expressing just MBP, 4-5: lysate of IPTG-induced cells expressing the MBP/Staufen chimera Stau2.

4. Conclusion

The generation of cellular diversity in developing organisms requires the synchronization of several molecular mechanisms; in *Drosophila* central nervous system, Inscuteable and Staufen play an important role for the asymmetric localization of cell fate determinants. This work provides a simple process for the production of large quantities of these proteins in *E. coli* for future structure-function studies. Gene synthesis was demonstrated to be a powerful tool for the expression of the *Drosophila* proteins Inscuteable and Staufen. The chimera design with maltose binding protein helped considerably to express high amounts of the C-terminal of both proteins, while the fusion protein glutathione-S-transferase allowed the soluble production of the main functional domain of Inscuteable.

References

[1]   Kucinich, R.E. (2005) Generating Neuronal Diversity in the *Drosophila* Central Nervous System: A View from the Ganglion Mother Cells. *Developmental Dynamics*, **232**, 609-616. http://dx.doi.org/10.1002/dvdy.20273

[2]   Chia, W. and Yang, X. (2002) Asymmetric Division of *Drosophila* Neural Progenitors. *Current Opinion in Genetics and Development*, **12**, 459-464. http://dx.doi.org/10.1016/S0959-437X(02)00326-X

[3]   Doe, C.Q., Chu-LaGraff, Q., Wright, D.M. and Scott, M.P. (1991) The Prospero Gene Specifies Cell Fate in the *Drosophila* Central Nervous System. *Cell*, **65**, 451-464. http://dx.doi.org/10.1016/0092-8674(91)90463-9

[4]   Yousef, M. and Matthews, B.W. (2005) Structural Basis of Prospero-DNA Interaction: Implications for Transcription Regulation in Developing Cells. *Structure*, **13**, 301-307. http://dx.doi.org/10.1016/j.str.2005.01.023

[5]   Shen, C.P., Knoblich, J.A., Chan, Y.M., Jiang, M.M., Jan, J.Y. and Jan, Y.N. (1998) Miranda as a Multidomain Adapter Linking Apically Localized Inscuteable and Basally Localized Staufen and Prospero during Asymmetric Cell Division in *Drosophila*. *Genes and Development*, **12**, 1837-1846. http://dx.doi.org/10.1101/gad.12.12.1837

[6]   Li, P., Yang, X., Wasser, M., Cai, Y. and Chia, W. (1997) Inscuteable and Staufen Mediate Asymmetric Localization and Segregation of *Prospero*RNA during *Drosophila* Neuroblast Cell Divisions. *Cell*, **90**, 437-447. http://dx.doi.org/10.1016/S0092-8674(00)80504-8

[7]   Kraut, R. and Campos-Ortega, J.A. (1996) Inscuteable, a Neural Precursor Gene of *Drosophila*, Encodes a Candidate for a Cytoskeleton Adaptor Protein *Developmental Biology*, **174**, 65-81. http://dx.doi.org/10.1006/dbio.1996.0052

[8]   Tio, M., Zavortink, M., Yang, X. and Chia, W. (1999) A Functional Analysis of Inscuteable and Its Role during *Drosophila* Asymmetric Cell Divisions. *Journal of Cell Science*, **112**, 1541-1551.

[9]   Withers-Martinez, C., Carpenter, E.P., Hackett, F., Ely, B., Sakid, M., Grainger, M. and Blackman, M.J. (1999) PCR-Based Gene Synthesis as an Efficient Approach for Expression of the A + T-Rich Malaria Genome *Protein Engineering*, **12**, 1113-1120. http://dx.doi.org/10.1093/protein/12.12.1113

[10] Smyth, D.R., Mrozkiewscz, M.K., McGrath, W.J.; Listwan, P. and Kobe, B. (2003) Crystal Structures of Fusion Proteins with Large Affinity Tags. *Protein Science*, **12**, 1313-1322. http://dx.doi.org/10.1110/ps.0243403

[11] Center, R.J., Kobe, B., Wilson, K.A., The, T., Howlett, G.J., Kemp, B.E. and Poumbourious, P. (1998) Crystallization of Trimeric Human T Cell Leukemia Virus Type 1 gp21 Octodomain Fragment as a Chimera with Maltose-Binding Protein. *Protein Science*, **7**, 1612-1619.

[12] Liu, Y., Manna, A., Li, R., Martin, W.E., Murphy, R.C., Cheung, A.L. and Zhang, G. (2001) Crystal Structure of the SarR Protein from *Staphylococcus aureus*. *Proceeding of the National Academy of Sciences United States of America*, **98**, 6877-6882. http://dx.doi.org/10.1073/pnas.121013398

21

Behavior and Viability of Blueberry Seeds through Germination and Tetrazolium Test

Ana Paula de Azevedo Pasqualini[1], Jessé Neves dos Santos[2], Ricardo Antonio Ayub[3*]

[1]Paraná Federal University, Curitiba, Brazil
[2]Ponta Grossa State University, Ponta Grossa, Brazil
[3]Phytotchny and Phytosanitary Department, Ponta Grossa State University, Ponta Grossa, Brazil
Email: anapauladeazevedo@gmail.com, jessens2@hotmail.com, *rayub@uepg.br

Abstract

Knowing the physiology of seeds and the elements that influence their germination is fundamental aspects in seminiferous propagation; important techniques are used to obtain genetic variability and development of new cultivars of blueberry. The aim of this study is to evaluate the germination behavior, as well as viability levels, through germination tests and tetrazolium, of *Vaccinium ashei* Reade seed cultivars Briteblue and Climax. Seeds treated or not with 5 M potassium hydroxide (KOH) were submitted to the germination test, on substrates, filter paper (SP) or solid culture medium with half of the salt concentration (MS/2), at temperatures of 10°C ± 2°C or 25°C ± 2°C. The maximum germination percentage of blueberry seeds was 40%. Both temperatures and substrates caused seed germination in the tested cultivars, and pretreatment with 5 M KOH for 5 minutes inhibited germination. Yet, the tetrazolium test, based on coloration of tissue, allowed the establishment of different levels of viability.

Keywords

Vaccinium ashei Reade, Germination Behavior, Dormancy

1. Introduction

Brazil has produced 59 t blueberry in 2012, in an area of 270 planted hectares in South and Southeast regions [1], and the most promising specie for regions of cold weather is the *Vaccinium ashei*; the species' cultivars that are better adapted to the Brazilian weather conditions are: Aliceblue, Bluebelle, Bluegem, Briteblue, Climax, Delite,

---
*Corresponding author.

Powderblue, Woodhard [2]. These cultivars are selected from other countries with different edaphoclimatic conditions displaying limitations for its cultivation, which drives the need for superior plants that are able to adapt to Brazil's particularities [3]. When searching for genetic variability and the development of new cultivars, the propagation of blueberry seeds becomes important [2].

In order to achieve satisfactory results in seminiferous propagation, it is imperative to know the seed's germinative behavior and physiologic quality, which can be evaluated through germination and force test, such as the tetrazolium test [4].

Seed's germination constitutes of a sequence of physical, biochemical and physiologic events, influenced by a variety of factors, which can act isolated or combined [5]. Those factors can be extrinsic, such as: light, temperature and humidity; and intrinsic, like: morphology, viability and dormancy [6].

As far as extrinsic factors are concerned, the *Vaccinium* seeds have its germination affected by temperature [7] and present positive and orthodox photoblastic behavior [8], meaning it germinates in the presence of light, and is able to maintain the viability even with low humidity content [9].

In a germination test, the substrate is another extern factor that influences seeds germination, due to its structure, aeration, water retention capacity and level of pathogens infestation [10]. The paper filter is the most utilized support in seeds germination, due to its capacity to simplify the test, increase the speed, and reduce costs [11]. However, the usage of MS medium culture [12] was cited as a substrate for seeds germination of Vaccinium meridionale Swartz since reduction of salts concentration in the culture medium allowed greater percentage of germination [13].

At the end of a germination test, the presence of non-germinated seeds may occur, because those died, were hard or in dormancy [14]. The seeds dormancy can be classified in: physiological, morphological, physical and combined (physical and physiological) dormancy [15]. The impermeability of the seed coat is a physical dormancy, which prevents the entry of water [16] and can be caused by the presence of substances such as suberin, lignin, cutin, tannins, pectins, as well as derivates of quinine [17]. Chemical studies indicate that the cuticle and pigment chains, present in the seed coat, are composed by insoluble polymeric materials, which can be depolymerized by alkaline hydrolysis [18] with KOH or NaOH [19].

Viable seeds that are considered in dormancy, present a temporary blockage to complete the germination [20]. The tetrazolium test is a fast method to determinate the viability of the seeds, based on the activity of dehydrogenase enzymes, present in living tissues [21].

The aim of the present work was to determinate the germinative behavior and levels of the viability of the seeds of *Vaccinium ashei* Reade cultivars Briteblue and Climax, by germination and tetrazolium tests.

2. Material and Methods

Site and biologic material

Blueberry seedlings from cvs. Climax and Briteblue experimentally planted in spacing of 4.0 m × 1.5 m in 2010, in Ponta Grossa—PR (25°05'35"S e 50°03'50"W e 950 m de altitude). The weather is classified as cfb [22], presenting well-defined dry seasons, frequent frosts during winter and soil Haplic cambisol dystrophic of clay texture, produced the fruits used in the experiments.

Experiment 1—Physiological quality and germinative behavior of *Vaccinium ashei* Reade seeds, cultivars Briteblue and Climax, treated or not with 5 M of potassium hydroxide (KOH), and submitted to the germination test in different substrates and temperatures

Ripe fruits, collected in February of 2012, were taken to the laboratory, selected, macerated and depulped in order to obtain the seeds. The seeds were washed with running water, spread on paper towels and shade dried at room temperature for 24 hours. Two experiments were performed, one using cv. Briteblue seeds, placed in the refrigerator for two months, and the other with fresh cv. Climax seeds. Furthermore, both cultivars' seeds were soaked in H_2O (control) or potassium hydroxide solution 5 M (KOH) for 5 minutes, and then submitted to the germination test, installed in Petri dishes, containing the autoclaved substrates: filter paper (SP) periodically moistened with sterile distilled water (2.5 times the weight of the substrate in water) or solid culture medium MS (6 g·L^{-1} of agar) with half of salt concentrations (MS/2). Then, the dishes were vetoed with plastic wrap and kept at constant temperatures of 10°C ± 2°C (BOD like cabinet) or 25°C ± 2°C (climatized room with air conditioning), with photoperiod of 16 hours of light, making the following treatments: T1) Control + SP + 10°C, T2) Control + SP + 25°C, T3) Control + MS/2 + 10°C, T4) Control + MS/2 + 25°C, T5) KOH + SP + 10°C, T6)

KOH + SP + 25°C, T7) KOH + MS/2 + 10°C, T8) KOH + MS/2 + 25°C. The experimental design was completely randomized with eight treatments, four replicates, and the experimental unit consisted of a Petri dish, containing 10 seeds. The effect of the use of KOH, substrates and temperatures on the germination performance of the seeds was evaluated by the percentage of germination and the first count of normal seedlings (presented all of the essential structures developed). After obtaining the first normal seedling, successive counts were performed every 7 days. The test finished when there was absence of germination after 30 days of test. The percentage data of germination were transformed to arcsen $\sqrt{P\%/100}$, submitted to the Barlett test, followed by analysis of variance, and when significant were compared by Duncan test ($p \leq 0.05$), using the SAS statistical package.

Experiment 2—Levels of seeds viability of *Vaccinium ashei* Reade cultivars Briteblue and Climax through the tetrazolium test

Fruits from cultivars Briteblue and Climax were collected in January of 2013, and were taken to the laboratory and placed under refrigaration. To separate the pulp from the seeds, fruits were placed together with water in a mixer. Next, the seeds were washed with running water, spread on paper towels and shade dried at room temperature for 3 days. To perform the tetrazolium test, three repetitions of 50 seeds were used; 150 seeds in total were pre-wetted by immersion in water for 24 hours, and the exposure of the fabrics to staining were made by a needle in the opposite side of the location of the embryo. Then, the seeds were immersed in a colorless solution of 2,3,4-triphenyl tetrazolium bromide 1% for 3 hours at 30°C and then kept 21 hours at room temperature in complete darkness condition [8]. After being immersed, the seeds were washed with running water and transversely cut with razor to visualize the embryo [14] in optic microscope (50× increase). The viability was scored by percentage of embryo color (0, 25, 50, 75 and 100%), and stained seeds with embryo above 50% were considered viable.

3. Results and Discussion

Experiment 1—Physiological quality and germinative behavior of *Vaccinium ashei* Reade seeds, cultivars Briteblue and Climax, treated or not with 5 M of potassium hydroxide (KOH), and submitted to the germination test in different substrates and temperatures

The disposal of the treatments T3 and T4 (Briteblue cultivar), and T2, T5 and T6 (Climax cultivar) was due to a contamination of the substrate by fungus. This interfered de evaluation of the factors: exposition or not to potassium hydroxide (KOH), substrates (SP and MS/2) and temperatures (10°C and 25°C) in factorial arrangement $2 \times 2 \times 2$.

Table 1 shows the results of analysis of variance of germination percentage transformed to arcsin $\sqrt{P\%/100}$, and the value of chi-square (χ^2) for the Barlett test, which showed homogeneity of variances of treatments.

For the germination test of blueberry seeds of Briteblue cultivar, in the treatments T1 and T2, which contained seeds immersed in water for 5 minutes and germinated on paper, had germination percentages of 30% and 40% respectively, yet they did not differ significantly as a function of temperature (10°C and 25°C). To the T6 and T8

Table 1. Test results of variance of germination percentage of data of the seeds of *Vaccinium ashei* Reade, Briteblue and Climax cultivars.

Vaccinium ashei Reade Briteblue Cultivar		
Variation Sources	G.L.	Mean Square
Treatments	5	0.00131635[*]
Error	18	0.00037102
Chi-Square (χ^2)		8.42603[*]
Vaccinium ashei Reade Climax Cultivar		
Variation Sources	G.L.	Mean Square
Treatments	4	0.00164518[*]
Error	15	0.00015064
Chi-Square (χ^2)		7.08139[*]

[*]significant at 5% of probability.

treatments, which contained seeds, treated with 5M of KOH and were germinated at a temperature of 25°C, had germination percentages of 7% and 10% respectively; these did not differ significantly in relation with the substrate (SP and MS/2).

The T2 treatment revealed to be statistically superior to the T6 treatment, showing that the exposure of the seeds to 5m of KOH inhibited the germination in relation with the seeds that were only immersed in water and germinated on paper at the same temperature (**Figure 1**).

For the Climax cultivar, the T1 and T4 treatments, presented the best statistical performance and germination percentages of 40% and 30% respectively, surpassing the other treatments. The T7 treatment, which had seeds treated with 5M of KOH and was germinated on the substrate MS/2 at 25°C, presented inferior statistical results than the other treatments, and germination percentage of 5% (**Figure 2**).

The first detached normal seedling was obtained after 46 days to Climax cv. treatment T8 (KOH + MS/2 + 25°C) and after 52 days to Briteblue cv. on treatment T5 (KOH + SP + 10°C). Seeds of *Vaccinium meridionale* Swartz, stored for a week in ambient conditions at a temperature of 18°C ± 2°C, using as substrate the culture medium MS with 1/3, 1/8 and 1/16 of the salt concentration, initiated the germination 42 days post the test installation [13].

The germination test of the seeds of *Vaccinium ashei* Reade Briteblue and Climax cultivars was terminated after 6 months of its installation, when the absence of germination was superior to 30 days. Evidence of dor-

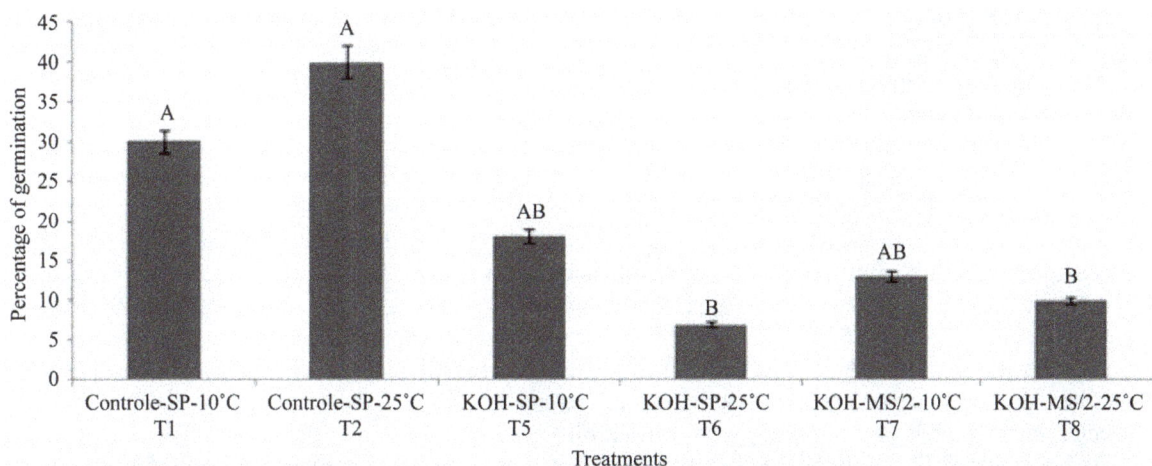

Figure 1. Germination of the blueberry seeds Briteblue cv. in different germinative treatments. Ponta Grossa, PR, 2012. *By the Duncan test p < 0.05, means followed by the same letter do not differ among themselves.

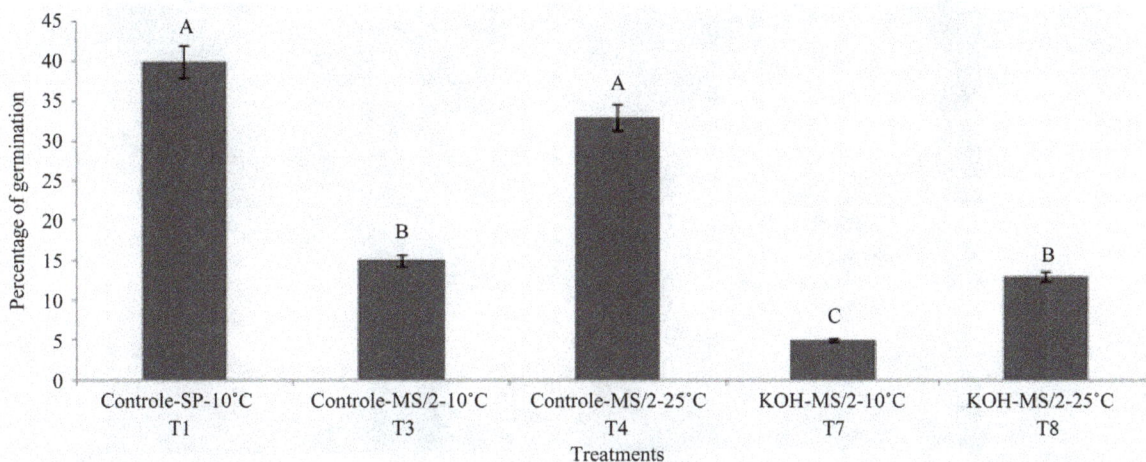

Figure 2. Germination of the blueberry seeds Climax cv. in different germinative treatments. Ponta Grossa, PR, 2012. *By the Duncan test P < 0.05, means followed by the same letter do not differ among themselves.

mancy in seeds of *Vaccinium spp.* are often manifested by the low and irregular germination, as observed in *V. angustifolium, V. ashei, V. canadense, V. corymbosum, V. macrocarpon* and *V. oxycoccus* [23]. In wild oat dormant seeds the use of KOH promoted significantly improvement on the germination [19]. Similarly seeds of *Vaccinium angustifolium* Ait. had maximum germination (approximately 80%) when treated with 5.3 M of KOH for 5 minutes [24]; results that were not seen in the tested cultivars of this study, showing that within the same genus, species have different behavior regarding KOH usage.

The usage of paper filter as a substrate has been observed in the seeds germination of *Vaccinium membranaceum* [25], *Vaccinium arctostaphylos* L. [26] and *Vaccinium parvifolium* Smith [7]. On the other hand, the culture medium MS with 1/3, 1/8 and 1/16 of the original salt concentration has been tested as a substrate [13] in the seeds germination of *Vaccinium meridionale*. Promoting germination percentages of 48, 63, 63.5% respectively. In the present experiment both substrates, SP and MS/2, promoted seeds germination of the *Vaccinium ashei* Reade for both tested cultivars.

Many temperature schemes can affect the germination of *Vaccinium sp* [7], fact that can be observed over the different temperatures used for germination of seeds of this genus. In order to simulate Canadian typical spring conditions, seeds of *Vaccinium angustifolium* Ait. were germinated at 10°C [24]. On the other hand, when working with *Vaccinium myrtillus* L. and *Vaccinium vitis-idaea*, obtained germination of 62% - 100% in presence of light and alternate temperatures of 20:10°C [27]. In the present study both constant temperatures of 10°C ± 2°C (BOD) and 25°C ± 2°C (SC) enabled the germination of *Vaccinium ashei* Reade seeds, although Climax cultivar, when variables were maintained constant, using or not KOH and substrate, at 25°C, showed germination percentage statistically superior than the one at 10°C.

Experiment 2—Levels of seeds viability of *Vaccinium ashei* Reade cultivars Briteblue and Climax through thetetrazolium test

In order to obtain satisfactory results in the tetrazolium test, is required that the 2,3,5, triphenil tetrazolium bromide solution is absorbed by the seeds. Thus, some species need preparation steps, such as puncture, cut and/or removal of the seed coat [21]. The puncture of the seed coat with a needle was essential so that the tetrazolium solution was able to act inside the seed, demonstrating impermeability by the tested cultivars. The class levels established in the tetrazolium test for the Briteblue and Climax cvs. seeds are represented in percentage staining red-orange of the embryo (**Figure 3** and **Figure 4**). Embryos *Vaccinium meridionale* Swartz, treated with tetrazolium solution at 1%, showed staining from light pink to dark pink, with shades of Orange [8]. Embryos with less than 50% of staining red-orange were considered not viable. The number of classes depends on the seed staining, morphological characteristics of the specie, and on the applied treatments, and for different species, distinct levels of classes can be proposed [28].

Accordingly with the levels of proposed classes, to Briteblue cv., were considered viable the seeds that presented 50%, 75% and 100% of staining red-orange of the embryo (**Figures 3(d)-(f)**), which represent respectively

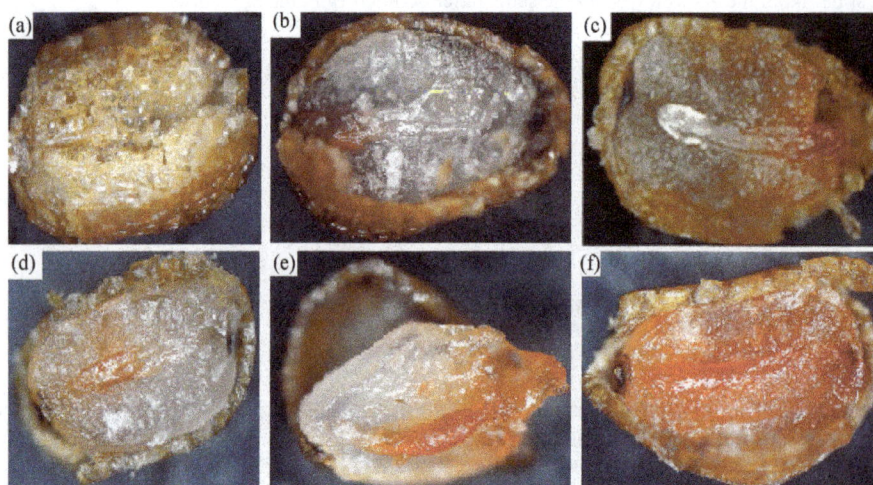

Figure 3. Embryo staining (a) dead (b) 0% (c) 25% (d) 50% (e) 75% and (f) 100% of blueberry seeds of Briteblue cv. treated with 2,3,5 tripenyl tetrazolium bromide solution. Ponta Grossa, PR, 2013.

13%, 7% and 9% of the evaluated seeds (**Figure 5(a)**). Seeds that showed red-orange embryo staining of approximately 25% (**Figure 3(c)**) corresponded to 20% of the sample and were not considered viable, as well as those that were not stained (dead or 0% of red-orange staining—(**Figure 3(a)** and **Figure 3(b)**), which represented 51% of the sample (**Figure 3(a)**), and from this percentage dead seeds correspond to 46%.

The same viability standards used to analyze the Briteblue cv. results, which were also used for Climax cv., were considered viable for the seeds that presented 50%, 75% and 100% of embryo staining red-orange (**Figures 4(d)-(f)**) and that represented respectively 8%, 13.5% and 28.5% of the sample (**Figure 4(b)**). Seeds that showed red-orange embryo staining of approximately 25% (**Figure 4(c)**) corresponded to 19% of the sample and were not considered viable. Those that were not stained (0% of red-orange staining or dead) represented 31% of the sample (**Figure 5(b)**).

Adding the percentage of the seeds that had the embryo stained red-orange over 50% in the tetrazolium test, 29% of the Briteblue cv. seeds and 50% of the Climax cv. seeds were considered viable. *Vaccinium meridionale* Swartz submitted to the tetrazolium test showed 84.2% of viability, and from those, 63% germinated and 21.2% did not germinate (remained dormant) [8]. The low percentage of germination of seeds can be related to elevated number of unviable seeds in the tetrazolium test, or viables that remained in latent state.

4. Conclusions

Briteblue and Climax cvs. of blueberry seeds, which were pre-treated with 5 M of KOH for 5 minutes and germi-

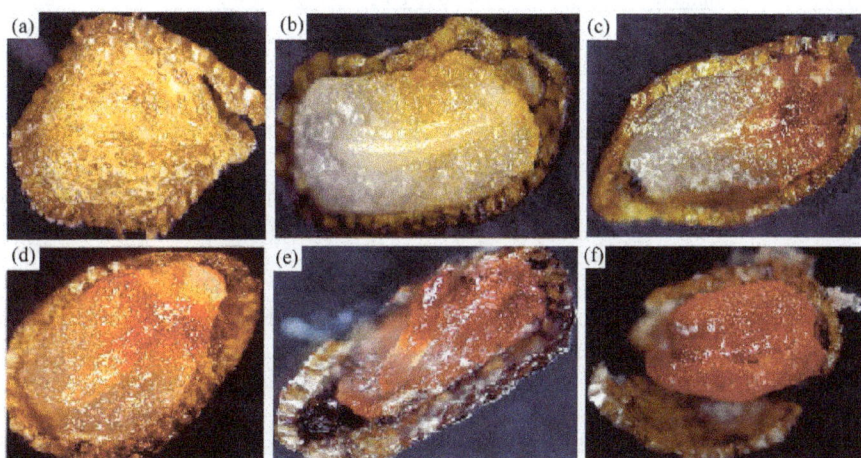

Figure 4. Embryo staining (a) dead (b) 0% (c) 25% (d) 50% (e) 75% and (f) 100% of blueberry seeds of Climax cv. treated with 2,3,5 tripenyl tetrazolium bromide solution. Ponta Grossa, PR, 2013.

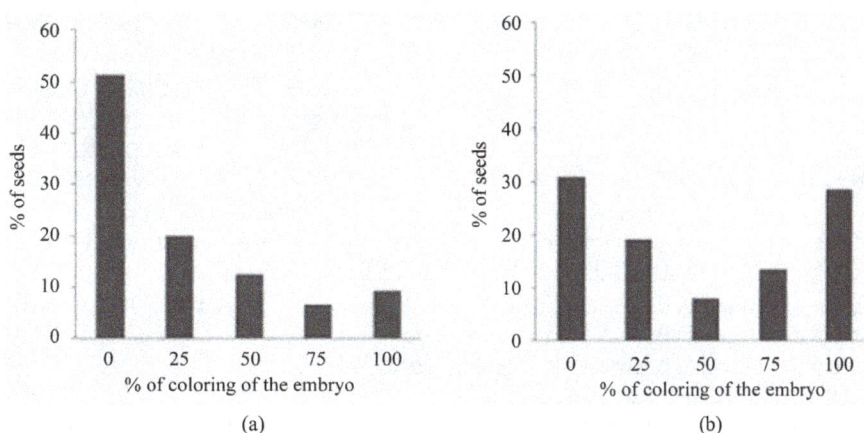

Figure 5. Percentage of staining of the blueberry seeds embryo (a) Briteblue cv. and (b) Climax cv., treated with 2,3,5 triphenyl tetrazolium bromide.

nated on the same substrate at the same temperature, had low germination percentage in relation with the non-treated seeds.

Blueberry seeds of both tested cultivars had their germinative behavior altered when the variables temperature (10°C and 25°C) and substrate (SP and MS/2) were combined.

Vaccinium ashei Reade blueberry seeds treated with combination of the variables: exposure or not to KOH, substrates (SP and MS/2) and temperature (10°C and 25°C) required up to 46 days to emit the first normal seedling and showed slow germination, and after 6 months of beginning the germination test, germination percentage dose did not exceed 40%.

The tetrazolium test, based on the staining of the fabrics, allows the establishment of different levels of viability, which is valuable for blueberry seeds.

Acknowledgements

Research support by CAPES, ARAUCÁRIA FOUNDATION, and UEPG.

References

[1]   Poll, H., Kist, B.B., Santos, C.E., Reetz, E.R., Carvalho, C. and Silveira, D.N. (2013) De bom tamanho: segmento das pequenas frutas constitui boa alternativa de renda para os produtores brasileiros e já cria oportunidades valiosas para o país. In: Gazeta Santa Cruz, Ed., *Anuário Brasileiro da Fruticultura*, Santa Cruz do Sul, 136 p. http://www.grupogaz.com.br/tratadas/eo_edicao/4/2013/04/20130401_e36fb3c90/pdf/3853_fruticultura_2013.pdf

[2]   Fachinello, J.C. (2008) Mirtilo. *Revista Brasileira de Fruticultura*, **30**, 285-576. http://dx.doi.org/10.1590/S0100-29452008000200001

[3]   de Oliveira Fischer, D.L., Fachinello, J.C., de Brum Piana, C.F., Bianchi, V.J. and Machado, N.P. (2014) Seleção de genótipos de mirtileiro obtidos a partir de polinização aberta. *Revista Brasileira de Fruticultura*, **36**, 221-231. http://dx.doi.org/10.1590/0100-2945-299/13

[4]   Carvalho, J.A., Pinho, É.V.R.V., Oliveira, J.A., Guimarães, R.M. and Bonome, L.T. (2002) Testes rápidos para avaliação da qualidade fisiológica de sementes de *Citromelo swingle*. *Revista Brasileira de Sementes (Online)*, **24**, 263-270. http://www.scielo.br/pdf/rbs/v24n1/v24n1a37.pdf

[5]   Amaro, M.S., Medeiros Filho, S., Guimarães, R.M. and Teófilo, E.M. (2006) Influência da temperatura e regime de luz na germinação de sementes de janaguba (*Himatanthus drasticus* (Mart.) Plumel.). *Ciência e Agrotecnologia*, **30**, 450-457. http://dx.doi.org/10.1590/S1413-70542006000300010 http://www.scielo.br/pdf/cagro/v30n3/v30n3a10.pdf

[6]   Biondi, D. and Leal, L. (2008) Tratamentos pré-germinativos em sementes de *Mimosa strobiliflora* Burkart. *Scientia Agraria*, **9**, 245-248. http://dx.doi.org/10.5380/rsa.v9i2.11013 http://dialnet.unirioja.es/servlet/articulo?codigo=2902758

[7]   Lopez, O.A., Barney, D.L., Shafii, B. and Price, W.J. (2008) Modeling the Effects of Temperature and Gibberellic Acid Concentration on Red Huckleberry Seed Germination. *Hortscience*, **43**, 223-228. http://dialnet.unirioja.es/servlet/articulo?codigo=2902758

[8]   Hernández, M.I.P., Lobo, M.A., Medina, C.I.C., Cartagena, J.R.V. and Delgado, O.A.P. (2009) Comportamiento de la germinación y categorización de la latencia en semillas de mortiño (*Vaccinium meridionale* Swartz). *Agronomía Colombiana*, **27**, 15-23. http://www.scielo.org.co/pdf/agc/v27n1/v27n1a03.pdf

[9]   Walck, J.L., Hidayati, S.N., Dixon, K.W., Thompson, K. and Poschlod, P. (2011) Climate Change and Plant Regeneration from Seed. *Global Change Biology*, **17**, 2145-2161. http://dx.doi.org/10.1111/j.1365-2486.2010.02368.x http://www.readcube.com/articles/10.1111%2Fj.1365-2486.2010.02368.x?r3_referer=wol&tracking_action=preview_c lick&show_checkout=1&purchase_referrer=onlinelibrary.wiley.com&purchase_site_license=LICENSE_DENIED

[10] Guedes, R.S., Alves, E.U., Gonçalves, E.P., Braga Júnior, J.M., Viana, J.S. and Colares, P.N.Q. (2010) Substratos e temperaturas para testes de germinação e vigor de sementes de *Amburana cearensis* (Allemão) A.C. Smith. *Revista Árvore*, **34**, 57-64. http://dx.doi.org/10.1590/S0100-67622010000100007

[11] Di Salvatore, M., Carafa, A.M. and Carratù, G. (2008) Assessment of Heavy Metals Phytotoxicity Using Seed Germination and Root Elongation Tests: A Comparison of Two Growth Substrates. *Chemosphere*, **73**, 1461-1464. http://www.sciencedirect.com/science/article/pii/S0045653508009831 http://dx.doi.org/10.1016/j.chemosphere.2008.07.061

[12] Murashige, T. and Skoog, F. (1962) A Revised Medium for a Rapid Growth and Bioassays with Tobacco Tissues Cultures. *Physiologia Plantarum*, **15**, 473-479. http://onlinelibrary.wiley.com/doi/10.1111/j.1399-3054.1962.tb08052.x/abstract

[13] Castro, C., Olarte, Y., Rache, L. and Pacheco, J. (2012) Development of a Germination Protocol for Blueberry Seeds (*Vaccinium meridionale* Swartz). *Agronomía Colombiana*, **30**, 196-203.
http://www.redalyc.org/articulo.oa?id=180325300005

[14] Ministério da Agricultura, Pecuária e Abastecimento (2009) Regras para análise de sementes. Secretaria de Defesa Agropecuária, Mapa/ACS, Brasília, DF, 395 p.
http://www.agricultura.gov.br/arq_editor/file/2946_regras_analise__sementes.pdf

[15] Baskin, J.M. and Baskin, C.C. (2004) A Classification System for Seed Dormancy. *Seed Science Research*, **14**, 1-16.
http://dx.doi.org/10.1079/SSR2003150

[16] D'Hondt, B., Brys, R. and Hoffmann, M. (2010) The Incidence, Field Performance and Heritability of Non-Dormant Seeds in White Clover (*Trifolium repens* L.). *Seed Science Research*, **20**, 169-177.
http://dx.doi.org/10.1017/S0960258510000152

[17] Costa, T.G., Dias, A.H. de S., Elias, T. de. F., Breier, T.B. and Abreu, H. dos S. (2011) Lignina e a dormência em sementes de três espécies de leguminosas florestais da Mata Atlântica. *Floresta e Ambiente*, **18**, 204-209.
http://www.floram.org/files/v18n2/v18n2a11.pdf

[18] Kolattukudy, P.E. (1981) Structure, Biosynthesis, and Biodegradation of Cutin and Suberin. *Annual Review of Plant Physiology*, **32**, 539-567. http://dx.doi.org/10.1146/annurev.pp.32.060181.002543
http://www.annualreviews.org/doi/abs/10.1146/annurev.pp.32.060181.002543

[19] Hou, J.Q. and Simpson, G.M. (1994) Effects of Immersing Dry Seeds in Alkaline Solutions on Seed Dormancy and Water Uptake in Wild Oat (*Avena fatua*). *Canadian Journal of Plant Science*, **74**, 19-24.
http://dx.doi.org/10.4141/cjps94-005 http://pubs.aic.ca/doi/pdf/10.4141/cjps94-005

[20] Debska, K., Krasuska, U., Budnicka, K., Bogatek, R. and Gniazdowska, A. (2013) Dormancy Removal of Apple Seeds by Cold Stratification Is Associated with Fluctuation in H_2O_2, NO Production and Protein Carbonylation Level. *Journal of Plant Physiology*, **170**, 480-488. http://dx.doi.org/10.1016/j.jplph.2012.11.018
http://www.ncbi.nlm.nih.gov/pubmed/23347818

[21] Costa, C.J. and dos Santos, C.P. (2010) Teste de tetrazólio em sementes de leucena. *Revista Brasileira de Sementes*, **32**, 66-72. http://dx.doi.org/10.1590/S0101-31222010000200008

[22] Köppen, W. (1948) Climatología: Com un estudio de los climas de la tierra. Fondo de Cultura Econômica, México, 479 p.

[23] Ellis, R.H., Hong, T.D. and Roberts, E.H. (1985) Chapter 36: Ericaceae: Handbook of Seed Technology for Genebanks. No. 3. In: Ellis, R.H., Hong, T.D. and Roberts, E.H., Eds., *Compendium of Specific Germination Information and Test Recommendations*, Vol. 2, International Board for Plant Genetic Resources, Rome, 456 p.
http://www.bioversityinternational.org/fileadmin/user_upload/online_library/publications/pdfs/52.pdf

[24] Gao, Y.P., Zheng, G.H. and Lawrence, V.G. (1988) Potassium Hydroxide Improves Seed Germination and Emergence in Five Native Plant Species. *Hortscience*, **33**, 274-276. http://hortsci.ashspublications.org/content/33/2/274.full.pdf

[25] Barney, D.L., Shafii, B. and Price, W.J. (2001) Cold Stratification Delays Germination of Black Huckleberry Seeds. *Hortscience*, **36**, 813. http://hortsci.ashspublications.org/content/36/4/813.full.pdf

[26] Shahram, S. (2007) Seed Dormancy and Germination of *Vaccinium arctostaphylos* L. *International Journal of Botany*, **3**, 307-311. http://dx.doi.org/10.3923/ijb.2007.307.311 http://www.researchgate.net/publication/26609984

[27] Baskin, C.C., Milberg, P., Andersson, L. and Baskin, J.M. (2000) Germination Studies of Three Dwarf Shrubs (*Vaccinium*, Ericaceae) of Northern Hemisphere Coniferous Forests. *Canadian Journal of Botany*, **78**, 1552-1560.
http://dx.doi.org/10.1139/b00-129 http://www.nrcresearchpress.com/doi/abs/10.1139/b00-129#.VmnBSHarTIU

[28] Sarmento, M.B., da Silva, A.C.S., Villela, F.A., dos Santos, K.L. and de Mattos, L.C.P. (2013) Teste de tetrazólio para avaliação da qualidade fisiológica em sementes de goiabeira-serrana (*Acca sellowiana* O. Berg Burret). *Revista Brasileira de Fruticultura*, **35**, 270-276. http://dx.doi.org/10.1590/S0100-29452013000100031

22

Exam Stress Induces Hormonal Changes amongst Students of the *Al-Haweeja* Technical College

Zainab A. Hassan[1], Ayoub A. Bazzaz[2]*, Noorhan A. Chelebi[2]

[1]Department of Clinical Analysis, Technical Institute, Kerkuk, Iraq
[2]Department of Basic Sciences, Faculty of Dentistry, University of Kerkuk, Kerkuk, Iraq
Email: *ayoubbazzaz@yahoo.co.uk

Abstract

Sixty students from both genders aged 19 - 22 years old at College of Technology undertaking half-term exams of 2012-2013 are encountered in this study. Blood samples were collected twice, *i.e.* before the exam inside the halls and during the rest time, to compare levels of some hormones, e.g. cortisol from all students, testosterone in male only and both estrogen and progesterone in female students. The female group was further subdivided into two subgroups, 15 each *i.e.* at the first half of the menstrual cycle (follicular phase) and second group at the second half of the menstrual cycle (luteal phase). The levels of cortisol had significantly ($p \leq 0.05$) raised from 12.3 ± 3.6 to 32.3 ± 4.2 ng/mL and from 11.6 ± 1.8 to 31.6 ± 7.3 ng/mL in both male and female students, respectively during exams in comparison with rest times. However, the levels of testosterone had significantly dropped ($p \leq 0.05$) from 6.63 ± 1.8 to 2.1 ± 0.4 ng/mL during the test-time. In female students, the levels of both estrogens and progesterone had significantly ($p \leq 0.05$) increased, *i.e.* in follicular from 202 ± 38 to 365 ± 22 and from 64.6 ± 8.0 to 160 ± 37 ng/mL at luteal phases, respectively and from 0.74 ± 0.03 to 1.5 ± 0.04 in follicular and 14.4 ± 2.4 to 29 ± 4.2 ng/mL at luteal phase, respectively in progesterone during the exam in comparison with rest times. These results indicate that all students had sustained stress during the exam-time which might have disturbed the regulation of various hormones in both genders consequently leading to further health effects.

Keywords

Stress, Hormones, Examination, Cortisol, Estrogen, Progesterone

---

1. Introduction

Stress, whether chronic or casual generated from different sources, could be experienced in our daily lives and may sometimes be a major life threatening *i.e.* severe illness. It can take a significant toll off us, both physically and emotionally leading to psychological sickness [1] due to its effects on immune system in the body [2]. Stress affects the body in other ways e.g. obviously, un-noticeable or uneasily detectable until it becomes more severe [1]. Stress hormones such as cortisol and epinephrine are released by the endocrine system in situations interpreted as potentially dangerous.

The mechanism of stress onsets by generating an alert which once recognized by the body then the adrenal hormone begins secretion, to increase body reflection to any unexpected situation leading to further increase in two systems in the body, the sympathetic and parasympathetic systems. The sympathetic system starts secreting noradrenaline or norepenefrin (NE) almost over 10 folds than at the normal status, *i.e.* more epenefrine (E) more than norepenefrine (NE). The stress or fear leads alert hypothalamus to receive signals from the brain which in turn activates the hormones to secrete. These hormones include cortisol, an anti-inflammatory factor that contributes to distribution of energy sources in the body which could be measured as an indicator to stress [3] [4]. Adrenal secretions of steroids are the main sources of responses to the stress which lasts a few hours and sometimes days in the blood in contrast to NE and E [5].

Cortisol is believed to affect the metabolic system while epinephrine plays a role in Attention Deficit Hyperactivity Disorder (ADHD) as well as depression and hypertension. Stress hormones are acted by mobilizing energy from storage to muscles, increasing heart rate, blood pressure and breathing rate and affect metabolic processes such as digestion, reproduction, growth and immunity [6]. Nevertheless, a subtle stress may sometimes be helpful in the life as it would activate both sympathetic and non-sympathetic nervous systems and organs to re-adapt ourselves with the so-called continuous changing environment and society [7]. Physiologically, stress, is controlled by chemicals in the body e.g. sex hormones which function to transmit certain messages from the producing glands to various body tissue through the blood circuit [8] [9]. This would lead, in turn, to certain reflections and overall behaviors. Accordingly, blood analysis could well be the most convenient tool to assess the hormonal changes in the body during various circumstances [10].

Sex hormones such as testosterone, estrogen, progesterone, oxytocin, and vasopressin are involved in sex motivation activity in most mammalian species which control the ability to engage in sexual behaviors [11]. Testosterone, the male hormone associated with the development of male characteristics, is present in both, but, in much smaller amounts in women than in men. It has a wide range of essential functions for a woman's health *i.e.* maintaining libido and sexual function [12] [13], maintain the normal growth and renewal of muscles, bone and other tissues [14], protect breast health at a cellular level [15]. Estrogen, produced by ovaries, is the second female hormone, makes cells grow, develops the uterus, breasts, periods, pregnancy and the egg within the ovary; however, an excess of it becomes toxic to the body. Progesterone is the sister hormone to estrogen, does work in a tandem to regulate and protect the health of the reproductive system throughout women fertile years. It also governs the second half of the menstrual cycle (Luteal phase) and is essential to maintain a pregnancy to term [12] [16]. It also plays a part in the regulation of blood sugar levels, has inherent calming properties, and protects breast, brain and bone health. When it comes to breast cancer prevention, the most important role of progesterone in the body is to balance estrogen [16]. The disturbance of these two hormones causes stress which could lead to many unexpected health defects in the life of females.

Stress could decrease the sexual stamina and weaken the sexual performance. In women it could also disturb the menstrual cycle and may decrease the fertility [17]. Over-increase in cortisol, in women, could completely stop the menstrual cycle [5] and could 50% increase the incidents of miscarriage and underweight newborns [18].

Neurophysiological and clinical studies have provided convincing evidence that altered stress hormone regulation that was frequently observed in depression and anxiety is caused by elevated secretion of the hypothalamic neuropeptides corticotrophin releasing hormone (CRH) and vasopressin produced a number of anxiety- and depression-like symptoms, which resulted in extensive validation of CRH_1 receptors as potential drug target [19]. Expression of neuronal tryptophan hydroxylase 2 (TPH2) mRNA and protein in midbrain serotonergic neurons was elevated, as well as increased in brain serotonin turnover in depressed suicide patients. The mechanisms underlying these changes are uncertain, but genetic influences, adverse early life experiences, or acute stressful life events, all of which can alter serotonergic neurotransmission and have been implicated in determining vulnerability to major depression [20]. Emerging evidence suggests that there are several different stress-related subsets of serotonergic neurons, each with a unique role in the integrated stress response [20].

The final-term exam stress is chosen as exams represent 60% of the final marks which imply determination of students fate in hunting best job opportunities. The objective of this research was to use these sexual hormones to assess the stress amongst students involved in final term-examinations at *Haweeja* Technical College.

2. Materials and Methods

Sixty University healthy students aged 20 - 22 years from both genders of College of Technology at *Haweeja* County were involved in this study during the sub-final tests (2012/2013) to assess the effects of stress on hormonal levels. The students have emphasized that the exams are their only worries at that stage. Five mL of blood samples were collected from all students at 09:00am inside the invigilating halls prior the onset of tests. Same amount of bloods were also collected from them during the rest time for comparison purposes. Levels of Cortisol, testosterone, estrogen and progesterone in the serum were measured according to ELISA obtained from Medical Supplied Enterprise in Iraq [21]. Biostatistics analysis was performed using two ways ANOVA Student-T-test.

3. Results

Levels of cortisol hormone were significantly increased ($p \leq 0.05$) in male students by almost three folds from 12.3 ± 3.6 ng/mL at the rest time to 32.3 ± 4.2 ng/mLin males and from 11.6 ± 3.8 ng/mL to 31.6 ± 7.3 ng/mL in female studentsin comparison with the final test times (**Table 1**). All students showed signs of anxiety and worries during the test time *i.e.* panic attacks and phobia. Interestingly, in male student, however, the levels of testosterone had significantly ($p \leq 0.05$) decreased three folds from 6.6 ± 1.8 ng/mL to 2.1 ± 0.4 ng/mL during the examination time.

In female students, the levels of estrogen hormone levels had significantly ($p \leq 0.05$) increased from 202 ± 38 to 365 ± 22 ng/mL at follicular phase and from 64.6 ± 8.0 ng/mL to 160 ± 37 ng/mL at luteal phase ($p \leq 0.05$). The levels of progesterone hormone had significantly ($p \leq 0.05$) doubled from 0.74 ± 0.03 to 1.5 ± 0.04 ng/mL for follicular phase and significantly ($p \leq 0.04$) from 14.1 ± 2.4 ng/mL to 29.0 ± 4.2 ng/mL at luteal phase. Surprisingly, the changes of these two hormones represented almost 60% for both phases during test times in comparison with the resting time.

Table 1. The mean values and standard deviations (±SD) of sex hormone levels in both male and female students during exams and a resting time; (n = 30 in each batch); Biostatistics used was two-ways ANOVA; (F) Follicular phase and (L) Luteal phase.

Hormonal values (mean & ± sd)	Cortisol (ng/mL)	Testosterone (ng/mL)	Estrogen (ng/mL)	Progesterone (ng/mL)
Male (Rest time)	12.3 ± 3.6	6.6 ± 1.8	-	-
Male (Stress time)	32.3 ± 4.2	2.1 ± 0.4	-	-
(Student T-test)	$p \leq 0.05$	$p \leq 0.05$	-	-
Female (Rest time) F	11.6 ± 1.8	-	202 ± 38	0.74 ± 0.03
Female (Stress time) F	31.6 ± 7.3	-	365 ± 22	1.5 ± 0.04
(Student T-test)	$p \leq 0.05$	-	$p \leq 0.05$	$p \leq 0.05$
Female (Rest time) L	-	-	64.6 ± 8	14.4 ± 2.4
Female (Stress time) L	-	-	160 ± 37	29.1 ± 4.2
(Student T-test)	-	-	$p \leq 0.04$	$p \leq 0.05$

4. Discussion

Although, the causes of stress are particular to the individual, it is difficult to make generalized statements about them; however, there are certain triggers that might cause stress [22]. The main cause, in this study, has been exam's hassle which seems to be common between students in both genders. Exam stress is normal but can give rise to anxiety which can interfere with performance of individual. Most students will have endured the trauma of exams at some stage in their lives, usually at school, college or university. Stress can be inevitable, however, the looming prospect of having to answer questions on most hated subjects often means that even the most laid-back of personalities will have experienced feelings of anxiety, worry and exam stress. Accordingly, body hormones will inevitably be changed dependent on degree of body response to the exam event.

Hormonal balance is an essential bio-factor in lives of both genders which its disturbance could cause many unexpected health consequence *i.e.* anxiety, stress, breast cancer, menstrual cycle [23]-[25]. By ignoring any other unexpected factors, the drop of cortisol hormone in both genders reflects the level of stress and anxiety the student were in during the examinations. Both genders, were at the same ages had undergone similar decline in cortisol levels. This may indicate the irrelevant correlation of stress and consequent hormonal imbalance to the gender. Exam stress and anxiety are a common phenomenon amongst Iraqi students, on almost all levels of academic studies as most students do not revise enough, on daily basis during the terms unless a few days prior to the exams. Accordingly, it could lead to personal self-under-trust amongst themselves in facing the exams due to mal-preparation. Such a phenomenon remains a must to be changed amongst the Iraqi students for a better performance.

Normally, the sex hormone disorders do occur during either an overproduction or underproduction of the hormones responsible for sexual characteristics and development. They eventually lead to motivation of individuals and vice versa [12] [26] [27]. In males, it is not produced enough they may experience a decline in libido (sex drive), erectile dysfunction, loss of muscle and loss of body hair [23]. Thetestosterone hormone had significantly declined during the exam. Assumingly, no sexual motivation did exist during the exam time due to such stress. However, stress in females could have caused un-overseen consequent effects *i.e.* disturbance in menstrual cycle, likelihood to carry female babes and could further lead to breast cancer in women [17]. The consequence of hormonal disturbance in students therefore, alerts the risk of blood toxicity and further changes in body performance, psychology and other un-overseen outcome both on short and long run.

The estrogen and progesterone, the two prominent hormones in woman, closely interrelated in many ways [28] are responsible for a lot of changes within the woman's body [13] [29] [30]. They both work in tandem, providing the body with necessary functions, balancing each other perfectly [31]. Exam stress in the present study could therefore, have disturbed their balance and caused serious changes in the body via misbalancing these two hormones on both short and long term. Unfortunately, the blood glucose levels have not been measured in current research with the hormone levels due to lack of finance. The blood glucose level could have provided important information regarding the correlation with hormone levels, under exam stress conditions. This would however be our next goal to achieve in similar work currently undertaken on other student samples at Kerkuk University.

As a response to stress which may exert many profound effects on human biological systems, the level of various hormones does change. Reactions to stress are associated with enhanced secretion of a number of hormones including glucocorticoids, catecholamines, growth hormone and prolactin, the effect of which is to increase mobilization of energy sources and to adapt the individual to its new circumstance [32].

5. Conclusion

It is concluded that cortisol could increase while sex hormones may well drop down due to stress generated from unnecessary extra anxiety during examination times amongst Iraqi students which, in turn, may lead to further harm to the body in both genders. Techniques that may help reduce stress include self-enhanced trust, regular revision of curricula, scheduling time off, relaxing, enough sleep, etc.

References

[1]   Martin, P. (1999) The Healing Mind: The Vital Links between Brain and Behavior, Immunity and Disease. Thomas Dunne Books, Publisher St. Martin's Griffin. http://www.goodreads.com/book/show/802181.The_Healing_Mind

[2]   Peled, R., Carmil, D., Siboni-Samocha, O. and Shoham-Vardi, I. (2008) Breast Cancer, Psychological Distress and Life Events among Young Women. *BMC Cancer*, **8**, 1-6. http://dx.doi.org/10.1186/1471-2407-8-245

[3]   Hellhammer, D.H., Wust, S. and Kudielka, B.M. (2009) Salivary Cortisol as a Biomarker in Stress Research. *Psychoneuroendocrinology*, **34**, 163-171.

[4]   Martinez, P.M., Martinz, L.R. and Ramos, R. (2009) Cortisol and Glucose: Reliable Indicators of Fish Stress. *Pan-American Journal of Aquatic Sciences*, **4**, 158-178.

[5]   Al-Abdulla, S. (2012) Nervous System, Structure and Organization. In: Al-Abdulla, S., Ed., *Physiology*, Dar Almassira Press, Amman, 115-138.

[6]   Sandman, C.A., Davis, E.P., Buss, C. and Glynn, L.M. (2011) Prenatal Programming of Human Neurological Function, *International Journal of Peptides*, **2011**, Article ID: 837596. http://www.ncbi.nlm.nih.gov/pmc/articles/PMC3133795/

[7]   McEwen, B. and Sapolsky, R. (2006) Stress and Your Health, the Hormone Foundation. https://scholar.google.com/scholar?q=McEwenandpolskyStress

[8]   Guber, H.A. and Farag, A.F. (2011) Evaluation of Endocrine Function. In: McPherson, R.A., Pincus, M.R., Eds., *Henry's Clinical Diagnosis and Management by Laboratory Methods*, Chap. 24, 22nd Edition, Elsevier Saunders, Philadelphia, 365-401.

[9]   Stewart, P.M. and Krone, N.P. (2011) The Adrenal Cortex. In: Melmed, S., Polonsky, K.S., Larsen, P.R., Kronenberg, H.M., Eds., *Williams Textbook of Endocrinolog*, Chap. 15, 12th Edition, Elsevier Saunders, Philadelphia, 479-545.

[10] Bazzaz, A.A. and Almanea, N.N. (2012) Assessment of Acute Effects of Diesel; Exhausted Emitted (DEE) in Blood Parameters in Guinea Pigs *Caviaporcellus*. *Journal of Environmental Sciences and Engineering*, **5**, 629-636.

[11] Wallen, K. (2001) Sex and Context: Hormones and Primate Sexual Motivation. *Hormone Behavior*, **40**, 339-357. http://dx.doi.org/10.1006/hbeh.2001.1696

[12] Johnson, D.F. and Phoenix, C.H. (1976) Hormonal Control of Female Sexual Attractiveness, Proceptivity, and Receptivity in Rhesus Monkeys. *Journal of Comparative and Physiological Psychology*, **90**, 473-483. http://dx.doi.org/10.1037/h0077216

[13] Spiteri, T., Musatov, S., Ogawa, S., Ribeiro, A., Pfaff, D.W. and Agmo, A. (2010) Estrogen-Induced Sexual Incentive Motivation, Proceptivity and Receptivity Depend on a Functional Estrogen Receptor α in the Ventromedial Nucleus of the Hypothalamus but Not in the Amygdala. *Neuroendocrinology*, **91**, 142-154. http://dx.doi.org/10.1159/000255766

[14] Harper, C.V., Barratt, C.L. and Publicover, S.J. (2004) Stimulation of Human Spermatozoa with Progesterone Gradients to Simulate Approach to the Oocyte. Induction of $[Ca^{2+}]_i$ Oscillations and Cyclical Transitions in Flagellar Beating. *The Journal of Biological Chemistry*, **279**, 315-325. http://dx.doi.org/10.1074/jbc.M401194200

[15] King, T.L. and Brucker, M.C. (2010) Pharmacology for Women's Health. Jones & Bartlett Publishers, Sudbury, 372-373.

[16] Burch, C. (2014) Your Hormone Balance, Hormone and Birth Consultant. http://yourhormonebalance.com/hormone-balance-basics/know-your-hormones/

[17] Ziegler, T.E. (2007) Female Sexual Motivation during Non-Fertile Periods: A Primate Phenomenon. *Hormones and Behavior*, **51**, 1-2. http://dx.doi.org/10.1016/j.yhbeh.2006.09.002

[18] Zutshi, T. (2005) Hormones in Obstetrics and Gynaecology. Jaypee Brothers Publishers, New Delhi, 74. http://dx.doi.org/10.5005/jp/books/10357

[19] Holsboer, F. and Ising, M. (2008) Central CRH System in Depression and Anxiety—Evidence from Clinical Studies with CRH_1 Receptor Antagonists. *European Journal of Pharmacology*, **583**, 350-357. http://dx.doi.org/10.1016/j.ejphar.2007.12.032

[20] Lowry, C.A., Hale, M.W., Evans, A.K., Heerkens, J., Staub, D.R., Gasser, P.J. and Shekhar, A. (2008) Serotonergic Systems, Anxiety, and Affective Disorder. *Annals of the New York Academy of Sciences*, **1148**, 86-94. http://dx.doi.org/10.1196/annals.1410.004

[21] Tietz, N.W. (1995) Clinical Guide to Laboratory Tests (ELISA). 3rd Edition, W.B. Saunders, Co., Philadelphia, 22-23.

[22] Scotte, E. (2014) Common Symptoms of Too Much Stress. http://stress.about.com/od/understandingstress/a/stress_symptoms.htm

[23] Gangestad, S.W. and Thornhill, R. (1998) Menstrual Cycle Variation in Women's Preferences for the Scent of Symmetrical Men. *Proceedings of the Royal Society of London*, **266**, 1913-1917. http://dx.doi.org/10.1098/rspb.1998.0380 http://www.ncbi.nlm.nih.gov/pmc/articles/PMC1690211/

[24] Gangestad, S.W., Thornhill, R. and Garver-Apgar, C.E. (2005) Adaptation to Ovulation Implications for Sexual and Social Behaviour. *Current Directions in Psychological Science*, **14**, 312-316. http://dx.doi.org/10.1111/j.0963-7214.2005.00388.x

[25] Jones, A., Dong, H.D.J., Duke, C.B., He, Y., Siddam, A., Miller, D.D. and Dalton, J.T. (2010) Nonsteroidal Selective Androgen Receptor Modulators Enhance Female Sexual Motivation. *The Journal of Pharmacology and Experimental Therapeutics*, **334**, 439-448. http://dx.doi.org/10.1124/jpet.110.168880

[26] Giles, J. (2008) Sex Hormones and Sexual Desire. *Journal for the Theory of Social Behavior*, **38**, 45-66.

[27] Blog, M. (2014) What Causes Stress. http://www.avogel.co.uk/health/stress-anxiety-low-mood/stress/causes/

[28] Scholten, A., Cisco, R.M., Vriens, M.R., Cohen, J.K., Mitmaker, E.J., Liu, C., Tyrrell, J.B., Shen, W.T. and Duh, Q.-Y.

(2013) Pheochromocytoma Crisis Is Not a Surgical Emergency. *Journal of Clinical Endocrinology & Metabolism*, **98**, 581-591. http://dx.doi.org/10.1210/jc.2012-3020

[29] Diamond, L.M. and Wallen, K. (2011) Sexual Minority Women's Sexual Motivation around the Time of Ovulation. *Archives of Sexual Behaviour*, **40**, 237-246. http://dx.doi.org/10.1007/s10508-010-9631-2

[30] Cummings, J.A. and Becker, J.B. (2012) Quantitative Assessment of Female Sexual Motivation in the Rat: Hormonal Control of Motivation. *Journal of Neuroscience Methods*, **204**, 227-233. http://dx.doi.org/10.1016/j.jneumeth.2011.11.017

[31] Iyer, S. (2011) How to Balance Estrogen and Progesterone. http://www.buzzle.com/articles/how-to-balance-estrogen-and-progesterone.html

[32] Ranabir, S. and Reetu, K. (2011) Stress and Hormones. *Indian Journal of Endocrinology and Metabolism*, **15**, 18-22. http://dx.doi.org/10.4103/2230-8210.77573 http://www.ncbi.nlm.nih.gov/pmc/articles/PMC3079864/

23

Myristoylated Alanine-Rich C Kinase Substrate Accelerates TNF-α-Induced Apoptosis in SH-SY5Y Cells in a Caspases-6 and/or -7-Dependent Manner

Atsuhiro Tanabe[1,2]*, Mitsuya Shiraishi[3], Yasuharu Sasaki[2]

[1]Laboratory of Biochemistry, Department of Bioscience and Engineering, Shibaura Institute of Technology, Saitama, Japan
[2]Laboratory of Pharmacology, School of Pharmacy, Kitasato University, Tokyo, Japan
[3]Department of Veterinary Pharmacology, Faculty of Agriculture, Kagoshima University, Kagoshima, Japan
Email: *tanabea@shibaura-it.ac.jp

Abstract

Cell proliferation, differentiation, and the elimination of unnecessary cells by apoptosis occur in the development of the nervous system. It is reported that brain dysplasia appears as the results of myristoylated alanine-rich C kinase substrate (MARCKS) knockout or the mutant mouse. We therefore expect that MARCKS participates in the development of the nervous system. However, the mechanism underlying such participation has not been identified. In this study, we observed the effects of the overexpression of MARCKS or unphosphorylatable MARCKS on cell proliferation and TNF-α-induced apoptosis in neuroblastoma SH-SY5Y cells. Furthermore, we restrained MARCKS expression by the RNAi method. In the results, MARCKS-overexpressing cells and not unphosphorylatable MARCKS-overexpressing cells showed increased cell proliferation rates. On the other hand, the RNAi decreased the proliferation of MARCKS-knocked down SH-SY5Y cells. These results indicated that MARCKS-overexpressing cells were more sensitive to TNF-α than normal SH-SY5Y cells. Moreover, in MARCKS-overexpressing cells TNF-α-induced apoptosis was inhibited by caspase-6 and -7 inhibitors but not by caspase-3 inhibitor. These results suggested that MARCKS participated in TNF-α-induced apoptosis in a caspase-6 and/or -7-dependent manner.

Keywords

Apoptosis, MARCKS, Caspase, Phosphorylation

---

*Corresponding author.

1. Introduction

Myristoylated alanine-rich C kinase substrate (MARCKS), which has a phosphorylated site domain (PSD), is cloned as a protein kinase C (PKC) substrate. It is known that several kinases, including ROCK, PKA, and MAPK, phosphorylate MARCKS *in vitro* [1]. Furthermore, we showed that PKC phosphorylated MARCKS not only directly but also through the RhoA/ROCK pathway in SH-SY5Y neuroblastoma cells [2].

MARCKS seems to be implicated in cell motility [3], secretion [4], membrane trafficking [5], and mitogenesis [6] through the regulation of cytoskeletal structure. Unphosphorylated MARCKS binds to actin directly and crosslinks F-actin, whereas phosphorylated MARCKS loses actin binding and polymerization activities [7]. Moreover, it is reported that MARCKS releases phosphatidylinositol-4,5-bisphosphate (PIP_2) through PSD in a phosphorylation-dependent manner [8] by which actin dynamics is regulated. Nevertheless, the exact mechanisms underlying the physiological roles of MARCKS are still unclear.

Stumpo and his colleagues reported that homozygous deletions of the MARCKS gene in mice led to abnormal brain development and perinatal death [9]. That report indicated that MARCKS played a pivotal role in the normal development processes of neurulation, hemisphere fusion, forebrain commissure formation, and the formation of cortical and retinal laminations. During nervous system development, neuronal cells not only proliferate but undergo a process of apoptosis [10]. We predicted that MARCKS was associated with apoptosis in the developing brain.

A lot of physiological phenomena are associated with apoptosis (e.g., germ cell development, elimination of tumor cells and DNA-damaged cells, and blood cell exchange). In these physiological phenomena, several apoptotic cascades are partially known. Tumor necrosis factor (TNF)-α and Fas ligand are examples of apoptosis-inducing factors [11]. These factors bind to specific receptors (so-called death receptors) on the cell surface, by which signals are transmitted into the cells. Although different factors are activated depending on the cause of the apoptosis, it is well known that caspases are the principal proteins in apoptosis.

In this report, we show that MARCKS is associated with both basal apoptotic levels and TNF-α-induced apoptosis in neuroblastoma cells. Moreover, we report that MARKCS-related apoptosis is independent from caspase-3 but dependent on caspase-6 and/or -7.

2. Materials and Methods

2.1. Cell Culture and Proliferation Assays

SH-SY5Y human neuroblastoma cells (nor-cells), GFP-expressing SH-SY5Y cells (GFP-OE-cells), GFP-fused wild-type MARCKS-overexpressing SH-SY5Y cells (wtMAR-OE-cells), and GFP-fused unphosphorylatable MARCKS-overexpressing cells (m3MAR-OE-cells) [12] were maintained in Dulbecco's modified Eagle's medium supplemented with 10% fetal bovine serum, 100 units/ml penicillin, and 100 μg/ml streptomycin at 37°C in a humidified atmosphere with 5% CO_2. Proliferation assays were performed by comparing the cell numbers between 2 and 5 days after the cells were plated in complete medium. All cell lines were plated on 24-well plates at a density of 5×10^3/well. The cells in each well (from 3 wells per cell line) were counted.

2.2. RNA Interference

MARCKS-specific double-stranded RNA oligonucleotides, each consisting of a 25-nucleotide sense sequence and a 25-nucleotide antisense sequence, were purchased from Invitrogen. The sequences for MARCKS siRNA were as follows: sense, 5'-uucgcugcggucuuggagaacuggg-3'; antisense, 5'-cccaguucuccaagaccgcagcgaa-3'.

SH-SY5Y cells were plated on 24-well plates at a density of 5×10^3 cells/well. After 24 h, 6.0 pmol of the MARCKS-specific or negative control siRNA (Invitrogen) was transfected to the cells using Lipofectamine2000 (Invitrogen).

2.3. Western Blotting Analysis

Trichloroacetic acid precipitants were subjected to Western blotting analyses as previously described [2]. The phosphorylation of MARCKS was detected with pS159-Mar-Ab [1] (1:5000 with Can Get Signal solution 1 (Toyobo)). Signals were detected by Chemi Lumi One (Nacalai Tesque) and Light Capture (ATTO). Densitometric analyses were performed using CS Analyzer (ATTO) software.

2.4. Apoptosis Detection

The cells were seeded on cover glasses in 35 mm dishes at a density of 1.0×10^5 cells/dish. After 2 days, the medium was changed to FBS-free DMEM with 1 μM TNF-α and incubated for 24 h. The cells were washed with ice-cold PBS (−) and fixed with 4% paraformaldehyde/4 mM EGTA/4% sucrose for 1 h at room temperature. The apoptotic cells were stained by the Apop Tag Red In Situ Apoptosis Detection Kit (Chemicon International), and the cover glasses were mounted on slide glasses with DAPI-containing Vectashield (Vector Laboratories) and a Biozero8000 fluorescence microscope (Keyence). The apoptotic cells were counted and the results were expressed as means ± SEM of the mean. The statistical differences for the response against TNF-α treatment in each cell line were determined by the two-sided Student's t-test. The data were analyzed using the Tukey-Kramer method for all pairwise comparisons between the TNF-α treatment mean. A difference of $p < 0.05$ was considered significant.

3. Results

3.1. MRACKS Expression's Effects on the Cell Proliferation Rate

Because it has been reported that MARCKS is associated with the proliferation of several cell types [13]-[15], we compared the growth speed among normal SH-SY5Y cells (nor-cells), GFP-expressing cells (GFP-OE-cells), GFP-fused wild-type MARCKS-overexpressing cells (wtMAR-OE-cells), and GFP-fused unphosphorylatable MARCKS (m3MARCKS)-overexpressing cells (m3MAR-OE-cells) [12]. We counted the cells at 2 days and 5 days after seeding.

Figure 1(a) shows the cell numbers at 5 days after seeding relative to those at 2 days after seeding. According to these results, the cell doubling times were calculated (nor-cell, 1.45 ± 1.21; GFP-OE-cell, 1.50 ± 0.20; wtMAR-OE-cell, 1.09 ± 0.31; m3MAR-OE-cell, 1.47 ± 0.17 days). The results indicated that MARCKS over-expression and its phosphorylation accelerate the growth rate of SH-SY5Y cells.

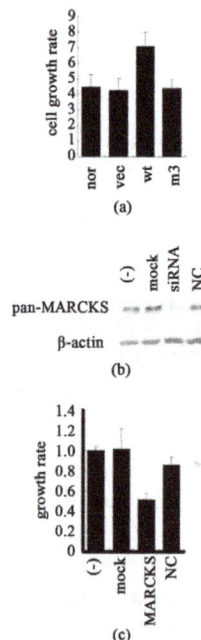

Figure 1. MARCKS involved in the growth of SH-SY5Y cells. (a) All cell lines (nor-, GFP-OE-, wtMAR-OE-, and m3MAR-OE-cells) were plated on 24-well plates at a density of 5×10^3/well. The cells per well at 2 and 5 days later were counted (n = 3). The cells at 5 days after were divided by those at 2 days after, and the growth rate was expressed as the mean ± SD. $^*p < 0.05$ compared with nor-cells; (b) SH-SY5Y cells were transfected with only transfection reagent (mock), MARCKS-specific siRNA (MARCKS), or negative control siRNA (NC). After 24 h, Western blotting assays were performed; (c) The MARCKS levels of (b) were quantified by densitometric analyses. The data for the nontreated cells were taken as 1; data represent mean ± SEM of four experiments. $^{**}p < 0.01$ compared with nontreated cells (nor: nor-cell; vec: GFP-OE-cell; wt: wtMAR-OE-cell; m3: m3MAR-OE-cells).

Next we observed the effects of MARCKS knockdown on cell growth (**Figure 1(b)**). At first we observed the effects of three MARCKS siRNA from Stealth RNAi (Invitrogen) and most effective sequence, that sequence is shown in Materials and Methods, is decided to use mainly hereafter (data not shown).

The growth of MARCKS-knocked down cells was slower than that of control cells (**Figure 1(c)**) and other siRNAs showed same but little slight effects. However, the cell-cycle phases did not differ among these four kinds of cells in flow cytometry (data not shown). These results showed that MARCKS is involved in the proliferation of SH-SY5Y cells.

3.2. MARCKS Is Involved in TNF-α-Induced Apoptosis of SH-SY5Y Cells

It is well recognized that some proteins are involved in both apoptosis and cell-cycle regulation [16]-[18]. Therefore, we observed the effects of MARCKS on apoptosis. In this report we used TNF-α, which is an important mediator of inflammation, apoptosis, and the development of secondary lymphoid structures, to induce apoptosis. The immunocytochemistry showed that, following 24 h TNF-α treatment, more than 60% of cells expanded and that the apoptotic cells were condensed (**Figure 2(a)** arrowheads). Next we investigated the difference in the apoptosis rate among these four cell types when we treated these cells with TNF-α. As we expected, TNF-α induced apoptosis in all of the kinds of cells we used. Especially in wtMAR-OE-cells, apoptosis was induced remarkably. On the other hand, the apoptosis rate of m3MAR-OE-cells was similar to those of nor- and GFP-OE-cells (**Figure 2(b)** and **Figure 2(c)**).

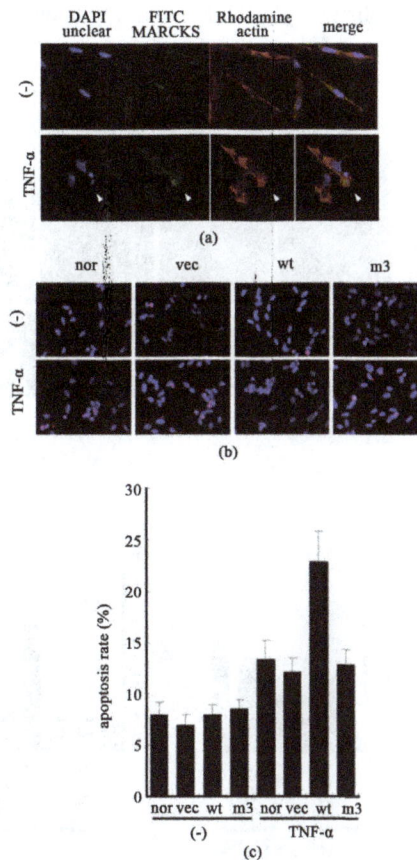

Figure 2. Phosphorylatable MARCKS accelerates TNF-α-induced apoptosis. (a) The nor-cells were exposed to TNF-α for 24 h. The nuclei, MARCKS, and actin were detected with DAPI, MARCKS-specific antibody and FITC-conjugated secondary antibody, and rhodamine-conjugated phalloidin, respectively. The arrow indicates apoptotic cell; (b) All cell lines (nor-, GFP-OE-, wtMAR-OE-, and m3MAR-OE-cells) were exposed to TNF-α for 24 h. The apoptotic cells were stained by the ApopTag Red In Situ Apoptosis Detection Kit (red) and the nuclei were stained with DAPI (blue); (c) The apoptotic cells in each cell line in B were counted, and the rates are shown. The data represent the mean ± SD of four experiments. $^{*}p < 0.05$ compared with TNF-α-treated nor-cells.

Next we knocked down MARCKS in SH-SY5Y cells with MARCKS-specific siRNA and treated the cells with TNF-α. The immunocytochemistry shows that MARCKS expression is repressed with MARCKS-specific siRNA but not with negative control siRNA (NC) (**Figure 3(a)**). The MARCKS knockdown increased the basal apoptosis rate while NC showed no effect. On the other hands, in the case of TNF-α-induced apoptosis MARCKS siRNA did not show significant differences with both mock and NC (**Figure 3(b)** left panel). As we thought in these sensitive conditions such adding damages to the cells by transfection regent and apoptosis-inducing regent even NC affects apoptosis rate, we calculated relative effects as fold (**Figure 3(b)** right panel). It shows that MARKCS siRNA almost completely inhibits TNF-α-induced apoptosis. These results strongly indicate that MARCKS is involved in apoptosis.

3.3. Phosphorylated MARCKS Is Involved in TNF-α-Induced Apoptosis

The outcome shown in **Figure 2** indicates that wild-type MARCKS accelerates TNF-α-induced apoptosis but m3MARCKS does not. Thus we thought that phosphorylated MARCKS is involved in the TNF-α-induced apoptosis cascade. To confirm our opinion, we first ascertained the pan-MARCKS expression levels in these cells. In nor- and GFP-OE-cells, the endogenous MARCKS (endo-MARCKS) amount increased slightly, although those of wtMAR-OE- and m3MAR-OE-cells did not change. On the other hand, the GFP-fused MARCKS (exo-MARCKS) amount in both wtMAR-OE- and m3MAR-OE-cells increased after TNF-α treatment (**Figure 4(a)** upper panel). Nonetheless, the amount of exo-MARCKS was smaller than that of endo-MARCKS. We also observed the phosphorylation level of MARCKS in TNF-α-treated cells. Without stimulation, exo-MARCKS in wtMAR-OE-cells was phosphorylated. In the early phase after TNF-α stimulation (0 - 1 h),

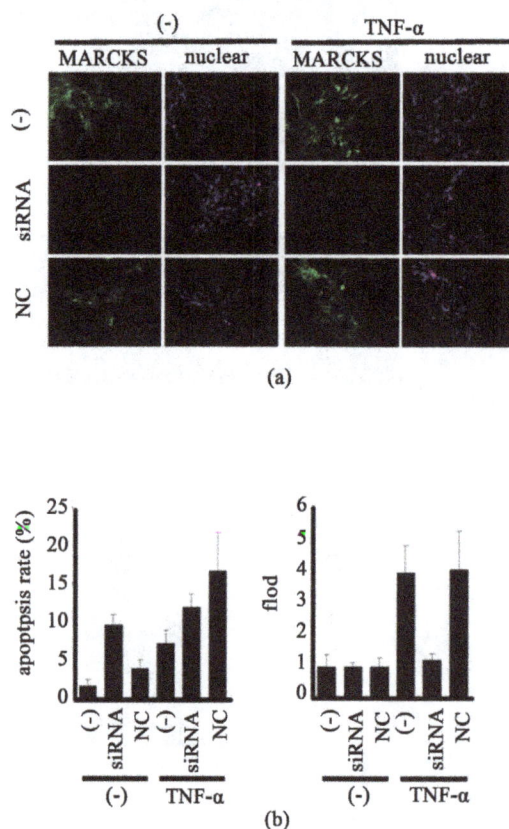

Figure 3. MARCKS knockdown reduces TNF-α-induced apoptosis. (a) MARCKS in the nor-cells was knocked down with MARCKS-specific siRNA. The cells were exposed to TNF-α for 24 h. The nuclei and MARCKS were detected with DAPI and with MARCKS-specific antibody and FITC-conjugated secondary antibody, respectively. The apoptotic cells were stained with the ApopTag Red In Situ Apoptosis Detection Kit; (b) left panel. The apoptotic cells in A were counted and the rates are shown. The data represent the mean ± SEM of three to six experiments. $^*p < 0.05$ compared with TNF-α-treated nor-cells (nor: nor-cell; vec: GFP-OE-cell; wt: wtMAR-OE-cell; m3: m3MAR-OE-cells).

(a)

(b)

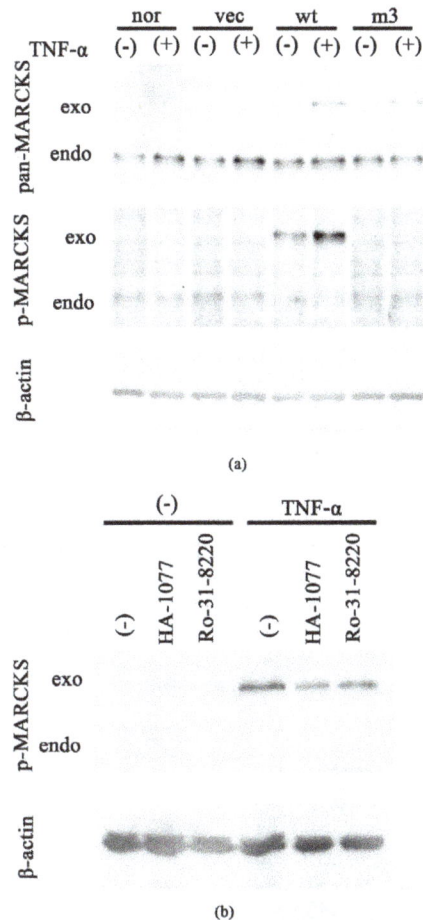

Figure 4. TNF-α induces MARCKS phosphorylation through ROCK and PKC. (a) All cell lines (nor-, GFP-OE-, wtMAR-OE-, and m3MAR-OE-cells) were exposed to TNF-α for 24 h. MARCKS, phosphorylated MARCKS, and β-actin were detected by Western blotting analyses using anti-MARCKS antibody, pS159-Mar-Ab, and anti-α-actin antibody, respectively; (b) wtMAR-OE-cells, pretreated with ROCK inhibitor HA-1077 or PKC inhibitor Ro-31-8220, were exposed to TNF-α for 24 h. Phosphorylated MARCKS and α-actin were detected by Western blot analyses using pS159-Mar-Ab and anti-α-actin antibody, respectively (nor: nor-cell; vec: GFP-OE-cell; wt: wtMAR-OE-cell; m3: m3MAR-OE-cells).

no increase in MARCKS phosphorylation was detected (data not shown). Nonetheless, 24 h after TNF-α treatment, only exo-MARCKS of wtMAR-OE-cells was phosphorylated (**Figure 4(a)**, middle panel). No change in the β-actin amount with TNF-α treatment was detected (**Figure 4(a)**, lower panel). The TNF-α-induced phosphorylation of exo-MARCKS in wtMAR-OE-cells was slightly reduced by pretreatment with either ROCK inhibitor HA-1077 or PKC inhibitor Ro-31-8220 (**Figure 4(b)**).

3.4. In wtMAR-OE-Cells, Caspase-6 and/or -7, but Not Caspase-3, Is Involved in TNF-α-Induced Apoptosis

Although the apoptosis cascade is not perfectly understood, caspase-3 is considered one of the most important proteins in apoptosis. Therefore, at first we used caspase-3 inhibitor to observe the participation of caspase-3 in the apoptosis of TNF-α-treated cells. In nor-, GFP-OE-, and m3MAR-OE-cells, TNF-α-induced apoptosis was decreased with a caspase-3 inhibitor z-DQMD-fmk. In wtMAR-OE-cells, on the other hand, TNF-α-induced apoptosis was not affected by z-DQMD-fmk (**Figure 5**).

Because caspases-6 and -7 are classified as effector caspases, as is caspase-3, we investigated the possibility that caspase-6 or -7 is involved in TNF-α-induced apoptosis in wtMAR-OE-cells. Before TNF-α treatment, we treated the cells with z-VEID-fmk or z-VEVD-fmk, which are caspase-6 and caspase-3/7 inhibitors, respectively. Both z-VEID-fmk and z-VEVD-fmk reduced the apoptosis rate in wtMAR-OE-cells (**Figure 6**). Another

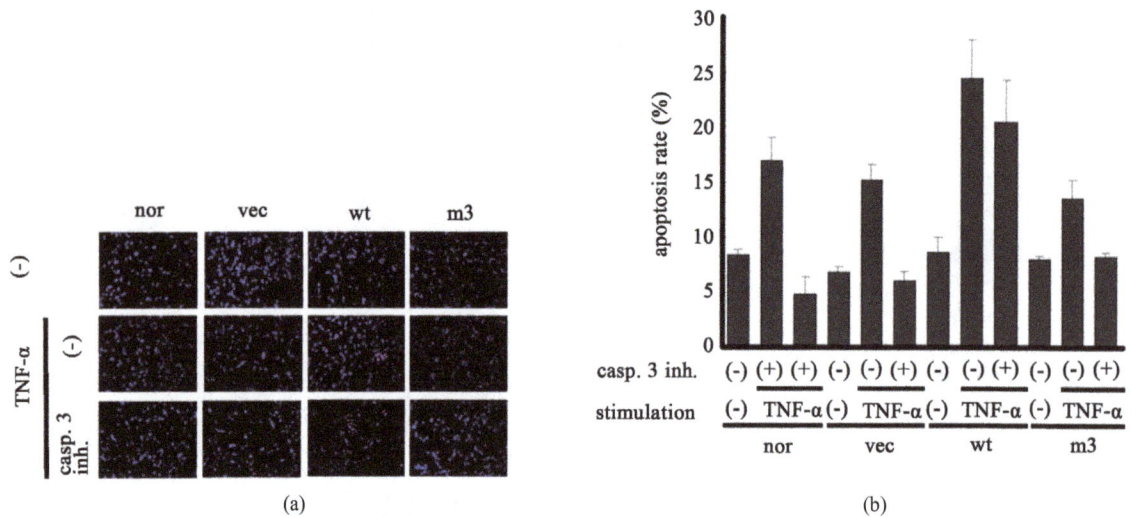

(a)   (b)

Figure 5. MARCKS upregulates the TNF-α induced apoptosis of caspase-3 independently. nor-cells pretreated with or without caspase-3 inhibitor z-DQMD-fmk were exposed to TNF-α for 24 h. The apoptotic cells were stained by the ApopTag Red In Situ Apoptosis Detection Kit (red), and the nuclei were stained with DAPI (blue); (b) The apoptotic cells in each cell line in A were counted, and the rate is shown. The data represent the mean ± SEM of three experiments. $^{**}p < 0.01$ and $^{*}p < 0.05$ compared with each TNF-α-treated cell line without caspase-3 inhibitor treatment (nor: nor-cell; vec: GFP-OE-cell; wt: wtMAR-OE-cell; m3: m3MAR-OE-cells).

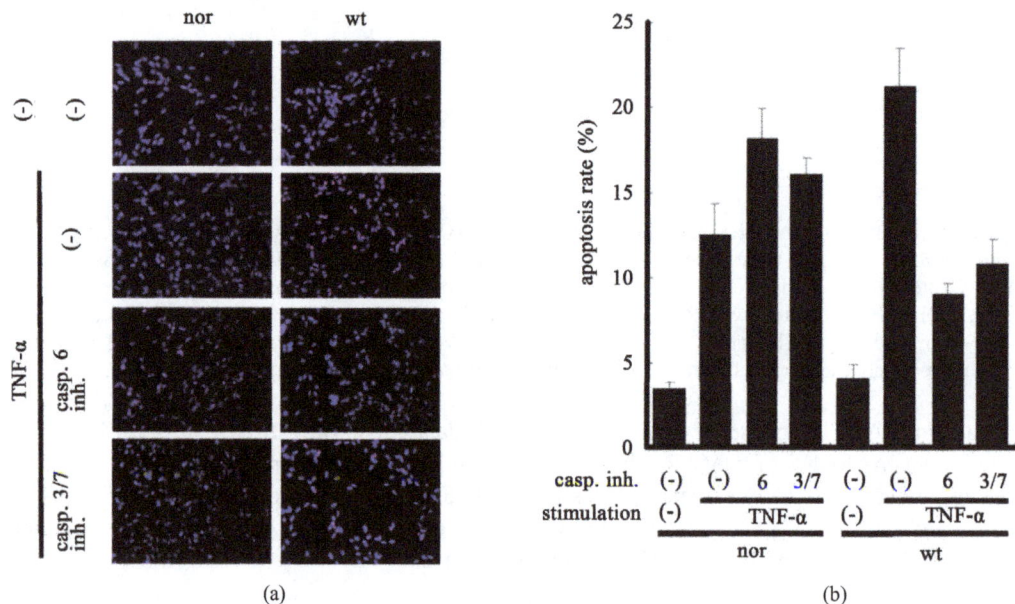

(a)   (b)

Figure 6. MARCKS upregulates TNF-α-induced apoptosis of caspase-6 and/or caspase-7 dependently. The nor-cells pretreated with or without caspase-3 inhibitor z-VEID-fmk or caspase-3/7 inhibitor z-VEVD-fmk were exposed to TNF-α for 24 h. The apoptotic cells were stained by the ApopTag Red In Situ Apoptosis Detection Kit (red), and the nuclei were stained with DAPI (blue); (b) The apoptotic cells in each cell line in A were counted, and the rate is shown. The data represent the mean ± SEM of three experiments. $^{**}p < 0.01$ compared with TNF-α-treated wtMAR-OE-cell line without caspase-3 inhibitor treatment (nor: nor-cell; wt: wtMAR-OE-cell).

caspase-3/7 inhibitor 5-[(S)-(-)-2-(Methoxymethyl)pyrrolidino] sulfonylisatin also inhibited the induction of apoptosis by TNF-α in wtMAR-OE-cells (data not shown) as well as z-VEVD-fmk. These results suggest that caspases-6 and/or 7 are involved in the apoptosis cascade in TNF-α-treated wtMAR-OE-cell.

Consequently, we consider that phosphorylated MARCKS is involved in TNF-α-induced apoptosis through caspases-6 and/or -7.

4. Discussion

The importance of MARCKS in the developing brain has been reported, and recently it was revealed that MARCKS modulates radial progenitor placement and proliferation in the developing brain [14]. It is well known that MARCKS is associated with proliferation in several types of cells besides glial cells [19] [20]. However, the relationship between MARCKS and neuronal cell proliferation has not been clarified. In this study, in SH-SY5Y cells, overexpression of MARCKS increased the proliferation rate, while MARCKS knockdown reduced it. Moreover, unphosphorylatable MARCKS overexpressed cells did not change the growth rate. Our findings suggest that MARCKS is involved in the growth of not only glial cells but also neuronal cells.

In the developing brain apoptosis, as well as cell proliferation, is an unavoidable cell response. In this study, we showed that wild-type MARCKS-overexpressing cells (wtMAR-OE-cells) upregulate TNF-α-induced apoptosis. It is well known that MARCKS is phosphorylated by PKC. Actually, TNF-α-induced MARCKS phosphorylation was partially inhibited by PKC inhibitor Ro-31-8220. Therefore, we thought that PKC plays an important role in phosphorylated MARCKS-involved apoptosis. Several PKCs are involved in TNF-α-induced apoptosis in several cell types [21]-[23]. In the future, an issue to examine is which PKC isoform is involved in TNF-α-treated SH-SY5Y and how MARCKS is associated with it.

In this study we showed that ROCK inhibitor HA-1077 also inhibited TNF-α-induced MARCKS phosphorylation at least partially. Besides PKC, ROCK is a candidate for kinase mediating MARCKS phosphorylation in the TNF-α-induced apoptosis. Because Mong and his colleagues reported that TNF-α activates JNK through the Rho/ROCK pathway [24], it is reasonable to infer that MARCKS is phosphorylated downstream from TNF-α. TNF-α also activates MAPK in several signal pathways [25]-[27] including the apoptotic cascade [28] [29]. It is reported that MAPK phosphorylates MARCKS *in vitro* [1] [30]. Moreover, Nwariaku and his colleagues indicated that ROCK inhibition attenuated TNF-α-induced MAPK activation [31]. Thus, ROCK is closely related to TNF-α. Previously we showed that PKC phosphorylates MARCKS through ROCK at least partially [2]. In the case of TNF-α-induced apoptosis, PKC may activate ROCK and would consequently induce the phosphorylation of MARCKS.

MARCKS knockdown also increased apoptosis without TNF-α treatment. In the case of muscle cell spreading, which is mediated by integrin, MARCKS was localized soon after cell adhesion to the focal adhesion site and was phosphorylated by PKCs, leading to translocation from the membrane to cytosol [32]. Cell adhesion involving integrin was associated closely with TNF-α-induced apoptosis [33]. It is possible that MARCKS regulates the basal apoptosis level through integrin-associated cell adhesion.

Talen *et al.* have reported that TNF-α induces the synthesis, myristoylation, and phosphorylation of MARCKS in neutrophils [34]. In our study, however, a slight increase in endo-MARCKS was recognized in nor- and GFP-OE-cells. On the other hand, although the exo-MARCKS amount was elevated by TNF-α, endo-MARCKS did not increase in wtMAR-OE- or m3MAR-OE-cells. We inserted MARCKS cDNA upstream from GFP, which is regulated under a cytomegalovirus (CMV) promoter in pEGFP-N1 to construct the GFP-MARCKS expression plasmid, and the CMV promoter is activated by TNF-α [35]. Therefore, TNF-α stimulation upregulates exo-MARCKS expression. Taken together, the results indicate that TNF-α raised the total MARCKS amount slightly in all types of cells. Although TNF-α elevated the exo-MARCKS amount, the increase was very small compared with the endo-MARCKS amount. However, the phosphorylated exo-MARCKS amount was several times larger than the phosphorylated endo-MARCKS amount. We cannot explain why exo-MARCKS is phosphorylated more easily than endo-MARCKS. However, from a different viewpoint we might be able to create a highly phosphorylated MARCKS expression system.

Caspase activation is a critical event in the onset of apoptosis [36]. Fourteen caspases have been identified to date [37]. According to the roles in apoptosis, caspases are divided roughly into two groups: initiator caspases and effector caspases. Caspases-2, -8, -9, -10, and -12 are initiator caspases. The apoptotic signals through the initiator caspases converge on effector caspases (caspases-3, 6, and 7). Although the roles of these caspases are clearly different, their activation mechanisms are remarkably similar [38]. All caspases recognize specific four-residue sequences and cleave peptide bonds located strictly after an Asp group. Inactive procaspases are digested by initiator caspases and activated. These active effector caspases digest their own substrates, so-called death substrates. In this study, we showed that inhibition of caspase-6 and -7, but not -3, reduced the TNF-α-induced apoptosis of wtMAR-OE-cells. However, the reduction of MARCKS amount which means MARCKS digestion was not detected. It remains unclear how MARCKS affects apoptosis.

5. Conclusion

Taken together, the present results show that MARCKS phosphorylation is involved in TNF-α-induced neuronal cell apoptosis. To our knowledge, this is the first report to show the association of MARCKS with apoptosis. This novel apoptosis pathway will lead to the elucidation of apoptosis mechanisms in neurons and in brain development.

Acknowledgements

We thanks Dr. Naoaki Saito for wtMARCKS and m3MARCKS express plasmid.

References

[1]   Nagumo, H., Ikenoya, M., Sakurada, K., Furuya, K., Ikuhara, T., Hiraoka, H. and Sasaki, Y. (2001) Rho-Associated Kinase Phosphorylates MARCKS in Human Neuronal Cells. *Biochemical and Biophysical Research Communications*, **280**, 605-609. http://dx.doi.org/10.1006/bbrc.2000.4179

[2]   Tanabe, A., Kamisuki, Y., Hidaka, H., Suzuki, M., Negishi, M. and Takuwa, Y. (2006) PKC Phosphorylates MARCKS Ser159 Not Only Directly but Also through RhoA/ROCK. *Biochemical and Biophysical Research Communications*, **345**, 156-161. http://dx.doi.org/10.1016/j.bbrc.2006.04.082

[3]   Zhao, Y., Neltner, B.S. and Davis, H.W. (2000) Role of MARCKS in Regulating Endothelial Cell Proliferation. *American Journal of Physiology—Cell Physiology*, **279**, C1611-C1620.

[4]   Blackshear, P.J. (1993) The MARCKS Family of Cellular Protein Kinase C Substrates. *Journal of Biological Chemistry*, **268**, 1501-1504.

[5]   McLaughlin, S. and Aderem, A. (1995) The Myristoyl-Electrostatic Switch: A Modulator of Reversible Protein-Membrane Interactions. *Trends in Biochemical Sciences*, **20**, 272-276. http://dx.doi.org/10.1016/S0968-0004(00)89042-8

[6]   Elzagallaai, A., Rose, S.D. and Trifaro, J.M. (2000) Platelet Secretion Induced by Phorbol Esters Stimulation Is Mediated through Phosphorylation of MARCKS: A MARCKS-Derived Peptide Blocks MARCKS Phosphorylation and Serotonin Release without Affecting Pleckstrin Phosphorylation. *Blood*, **95**, 894-902.

[7]   Hartwig, J.H., Thelen, M., Rosen, A., Janmey, P.A., Nairn, A.C. and Aderem, A. (1992) MARCKS Is an Actin Filament Crosslinking Protein Regulated by Protein Kinase C and Calcium-Calmodulin. *Nature*, **356**, 618-622. http://dx.doi.org/10.1038/356618a0

[8]   Wang, J., Arbuzova, A., Hangyás-Mihályné, G. and McLaughlin, S. (2001) The Effector Domain of Myristoylated Alanine-Rich C Kinase Substrate Binds Strongly to Phosphatidylinositol 4,5-Bisphosphate. *Journal of Biological Chemistry*, **276**, 5012-5019. http://dx.doi.org/10.1074/jbc.M008355200

[9]   Stumpo, D.J., Bock, C.B., Uttle, J.S. and Blackshear, P.J. (1995) MARCKS Deficiency in Mice Leads to Abnormal Brain Development and Perinatal Death. *Proceedings of the National Academy of Sciences of the United States of America*, **92**, 944-948. http://dx.doi.org/10.1073/pnas.92.4.944

[10] Cowan, V.M., Fawcett, J.W., O'Leary, D.D.M. and Stanfield, B.B. (1984) Regressive Events in Neurogenesis. *Science*, **225**, 1258-1265. http://dx.doi.org/10.1126/science.6474175

[11] Gruss, H.J. and Dower, S.K. (1995) Tumor Necrosis Factor Ligand Superfamily: Involvement in the Pathology of Malignant Lymphomas. *Blood*, **85**, 3378-3404.

[12] Tanabe, A., Shiraishi, M., Negishi, M., Saito, N., Tanabe, M. and Sasaki, Y. (2012) MARCKS Dephosphorylation Is Involved in Bradykinin-Induced Neurite Outgrowth in Neuroblastoma SH-SY5Y Cells. *Journal of Cellular Physiology*, **227**, 618-629. http://dx.doi.org/10.1002/jcp.22763

[13] McGill, C.J. and Brooks, G. (1997) Expression and Regulation of 80K/MARCKS, a Major Substrate of Protein Kinase C, in the Developing Rat Heart. *Cardiovascular Research*, **34**, 368-376. http://dx.doi.org/10.1016/S0008-6363(97)00041-2

[14] Weimer, J.M., Yokota, Y., Stanco, A., Stumpo, D.J., Blackshear, P.J. and Anton, E.S. (2009) MARCKS Modulates Radial Progenitor Placement, Proliferation and Organization in the Developing Cerebral Cortex. *Development*, **136**, 2965-2975. http://dx.doi.org/10.1242/dev.036616

[15] Manenti, S., Malecaze, F., Chap, H. and Darbon, J.M. (1998) Overexpression of the Myristoylated Alanine-Rich C Kinase Substrate in Human Choroidal Melanoma Cells Affects Cell Proliferation. *Cancer Research*, **58**, 1429-1434.

[16] Yang, M., Wu, S., Su, X. and May, W.S. (2006) JAZ Mediates G_1 Cell-Cycle Arrest and Apoptosis by Positively Regulating p53 Transcriptional Activity. *Blood*, **108**, 4136-4145. http://dx.doi.org/10.1182/blood-2006-06-029645

[17] Conzen, S.D., Gottlob, K., Kandel, E.S., Khanduri, P., Wagner, A.J., O'Leary, M. and Hay, N. (2000) Induction of Cell

Cycle Progression and Acceleration of Apoptosis Are Two Separable Functions of c-Myc: Transrepression Correlates with Acceleration of Apoptosis. *Molecular and Cellular Biology*, **20**, 6008-6018.
http://dx.doi.org/10.1128/MCB.20.16.6008-6018.2000

[18] Padmanabhan, J., Park, D.S., Greene, L.A. and Shelanski, M.L. (1999) Role of Cell Cycle Regulatory Proteins in Cerebellar Granule Neuron Apoptosis. *The Journal of Neuroscience*, **19**, 8747-8756.

[19] Monahan, T.S., Andersen, N.D., Martin, M.C., Malek, J.Y., Shrikhande, G.V., Pradhan, L., Ferran, C. and LoGerfo, F.W. (2009) MARCKS Silencing Differentially Affects Human Vascular Smooth Muscle and Endothelial Cell Phenotypes to Inhibit Neointimal Hyperplasia in Saphenous Vein. *The FASEB Journal*, **23**, 557-564.
http://dx.doi.org/10.1096/fj.08-114173

[20] McGill, C.J. and Brooks, G. (1997) Expression and Regulation of 80K/MARCKS, a Major Substrate of Protein Kinase C, in the Developing Rat Heart. *Cardiovascular Research*, **34**, 368-376.
http://dx.doi.org/10.1016/S0008-6363(97)00041-2

[21] Laouar, A., Glesne, D. and Huberman, E. (1999) Involvement of Protein Kinase C-β and Ceramide in Tumor Necrosis Factor-α-Induced but Not Fas-Induced Apoptosis of Human Myeloid Leukemia Cells. *The Journal of Biological Chemistry*, **274**, 23526-23534. http://dx.doi.org/10.1074/jbc.274.33.23526

[22] Chang, Q. and Tepperman, B.L. (2001) The Role of Protein Kinase C Isozymes in TNF-Alpha-Induced Cytotoxicity to a Rat Intestinal Epithelial Cell Line. *American Journal of Physiology—Gastrointestinal and Liver Physiology*, **280**, G572-G583.

[23] Comalada, M., Xaus, J., Valledor, A.F., López-López, C., Pennington, D.J. and Celada, A. (2003) PKC Epsilon Is Involved in JNK Activation That Mediates LPS-Induced TNF-Alpha, Which Induces Apoptosis in Macrophages. *American Journal of Physiology—Cell Physiology*, **285**, C1235-C1245.

[24] Mong, P.Y., Petrulio, C., Kaufman, H.L. and Wang, Q. (2008) Activation of Rho Kinase by TNF-Alpha Is Required for JNK Activation in Human Pulmonary Microvascular Endothelial Cells. *The Journal of Immunology*, **180**, 550-558.
http://dx.doi.org/10.4049/jimmunol.180.1.550

[25] Ho, A.W., Wong, C.K. and Lam, C.W. (2008) Tumor Necrosis Factor-Alpha Up-Regulates the Expression of CCL2 and Adhesion Molecules of Human Proximal Tubular Epithelial Cells through MAPK Signaling Pathways. *Immunobiology*, **213**, 533-544. http://dx.doi.org/10.1016/j.imbio.2008.01.003

[26] Okuma-Yoshioka, C., Seto, H., Kadono, Y., Hikita, A., Oshima, Y., Kurosawa, H., Nakamura, K. and Tanaka, S. (2008) Tumor Necrosis Factor-Alpha Inhibits Chondrogenic Differentiation of Synovial Fibroblasts through p38 Mitogen Activating Protein Kinase Pathways. *Modern Rheumatology*, **18**, 366-378.
http://dx.doi.org/10.3109/s10165-008-0069-5

[27] Lin, C.C., Tseng, H.W., Hsieh, H.L., Lee, C.W., Wu, C.Y., Cheng, C.Y. and Yang, C.M. (2008) Tumor Necrosis Factor-Alpha Induces MMP-9 Expression via p42/p44 MAPK, JNK, and Nuclear Factor-kappaB in A549 Cells. *Toxicology and Applied Pharmacology*, **15**, 386-398. http://dx.doi.org/10.1016/j.taap.2008.01.032

[28] O'Sullivan, A.W., Wang, J.H. and Redmond, H.P. (2009) The Role of P38 MAPK and PKC in BLP Induced TNF-Alpha Release, Apoptosis, and NFkappaB Activation in THP-1 Monocyte Cells. *Journal of Surgical Research*, **151**, 138-144. http://dx.doi.org/10.1016/j.jss.2008.02.031

[29] Iyer, C., Kosters, A., Sethi, G., Kunnumakkara, A.B., Aggarwal, B.B. and Versalovic, J (2008) Probiotic Lactobacillus Reuteri Promotes TNF-Induced Apoptosis in Human Myeloid Leukemia-Derived Cells by Modulation of NF-kappaB and MAPK Signalling. *Cellular Microbiology*, **10**, 1442-1452. http://dx.doi.org/10.1111/j.1462-5822.2008.01137.x

[30] Ohmitsu, M., Fukunaga, K., Yamamoto, H. and Miyamoto, E. (1999) Phosphorylation of Myristoylated Alanine-Rich Protein Kinase C Substrate by Mitogen-Activated Protein Kinase in Cultured Rat Hippocampal Neurons Following Stimulation of Glutamate Receptors. *The Journal of Biological Chemistry*, **274**, 408-417.
http://dx.doi.org/10.1074/jbc.274.1.408

[31] Nwariaku, F.E., Rothenbach, P., Liu, Z., Zhu, X., Turnage, R.H. and Terada, L.S. (2003) Rho Inhibition Decreases TNF-Induced Endothelial MAPK Activation and Monolayer Permeability. *Journal of Applied Physiology*, **95**, 1889-1895. http://dx.doi.org/10.1152/japplphysiol.00225.2003

[32] Disatnik, M.H., Boutet, S.C., Lee, C.H., Mochly-Rosen, D. and Rando, T.A. (2002) Sequential Activation of Individual PKC Isozymes in Integrin-Mediated Muscle Cell Spreading: A Role for MARCKS in an Integrin Signaling Pathway. *Journal of Cell Science*, **115**, 2151-2163.

[33] Fukushima, K., Miyamoto, S., Komatsu, H., Tsukimori, K., Kobayashi, H., Seki, H., Takeda, S. and Nakano, H. (2003) TNFα Induced Apoptosis and Integrin Switching in Human Extravillous Trophoblast Cell Line. *Biology of Reproduction*, **68**, 1771-1778. http://dx.doi.org/10.1095/biolreprod.102.010314

[34] Thelen, M., Rosen, A., Nairn, A.C. and Aderem, A. (1990) Tumor Necrosis Factor Alpha Modifies Agonist-Dependent Responses in Human Neutrophils by Inducing the Synthesis and Myristoylation of a Specific Protein Kinase C Sub-

strate. *Proceedings of the National Academy of Sciences of the United States of America*, **87**, 5603-5607. http://dx.doi.org/10.1073/pnas.87.15.5603

[35] Stein, J., Volk, H.D., Liebenthal, C.L., Krüger, D.H. and Prösch, S. (1993) Tumour Necrosis Factor α Stimulates the Activity of the Human Cytomegalovirus Major Immediate Early Enhancer/Promoter in Immature Monocytic Cells. *Journal of General Virology*, **74**, 2333-2338. http://dx.doi.org/10.1099/0022-1317-74-11-2333

[36] Mandal, D., Baudin-Creuza, V., Bhattacharyya, A., Pathak, S., Delaunay, J., Kundu, M. and Basu, J. (2003) Caspase 3-Mediated Proteolysis of the N-Terminal Cytoplasmic Domain of the Human Erythroid Anion Exchanger 1 (Band 3). *The Journal of Biological Chemistry*, **278**, 52551-52558. http://dx.doi.org/10.1074/jbc.M306914200

[37] Talanian, R.V., Dang, L.C., Ferenz, C.R., Hackett, M.C., Mankovich, J.A., Welch, J.P., Wong, W.W. and Brady, K.D. (1996) Stability and Oligomeric Equilibria of Refolded Interleukin-1β Converting Enzyme. *The Journal of Biological Chemistry*, **271**, 21853-21858. http://dx.doi.org/10.1074/jbc.271.36.21853

[38] Wilson, K.P., Black, J.A., Thomson, J.A., Kim, E.E., Griffith, J.P., Navia, M.A., Murcko, M.A., Chambers, S.P., Aldape, R.A. and Raybuck, S.A. (1994) Structure and Mechanism of Interleukin-1 Beta Converting Enzyme. *Nature*, **370**, 270-275. http://dx.doi.org/10.1038/370270a0

24

Obesity Effect on the Spine

Samir Zahaf[1], Bensmaine Mansouri[1], Abderrahmane Belarbi[1], Zitouni Azari[2]

[1]Department of Mechanical Engineering, University of Sciences and Technology, Oran, Algeria
[2]Laboratory of Biomechanics, Polymers and Structures, Ecole Nationale d'Ingénieurs de Metz, Metz, France
Email: zahafsamir1983@gmail.com, Smail_Mansouri@yahoo.fr, Belarbi_abd@yahoo.fr, azari@univ-metz.fr

Abstract

The objective of this work is to study the effect of obesity on the intervertebral discs and provide system analysis of the spine between multiple configurations of people, and know the risks due to this eccentric load (Big Belly). The results show that in all three previous loadings (obese people), the distributions of stress and strain are high in both D1 and D10 intervertebral discs. This shows that the distance between the point of load application and the spine axis has an important role in solicitation increasing and therefore its deformation: as main conclusion is that the concentration of the mass of stomach fat is a risk factor that leads to pain problems, deformation and herniated disc.

Keywords

Obes, Lumbar-Thoracic, Intervertebral Discs, Finite Element, Biomechanics, Von Mises Stress-Strain, Disc Degeneration

1. Introduction

The prediction of the mechanical behavior of the spine system is one of the major problems of biomechanics [1]. A better understanding of moving mechanisms of spine under different loads and stress distribution in this system is of fundamental importance in the advancement of technologies in the areas of spinal restorations inter vertebral prostheses, and osteopathic medicine bone [2].

The spine or rachis consists of a movable column of 24 free vertebrae and a fixed column formed of fused vertebrae: the sacrum and coccyx (**Figure 1**); it is the fixing strut of many essential muscles in the posture and locomotion and protects the spinal cord located in the vertebral canal; it supports the head and transmits the weight of the body to the hip joints; with a length of about 70 cm in men (60 cm in women), its reduction may reach 2 cm when standing [3].

The degeneration of the lumbar discs is definitely related to BMI; overweight and obesity are associated with a higher risk of problems of the spine. The MRI study [4] published in the January 30 edition of Arthritis and

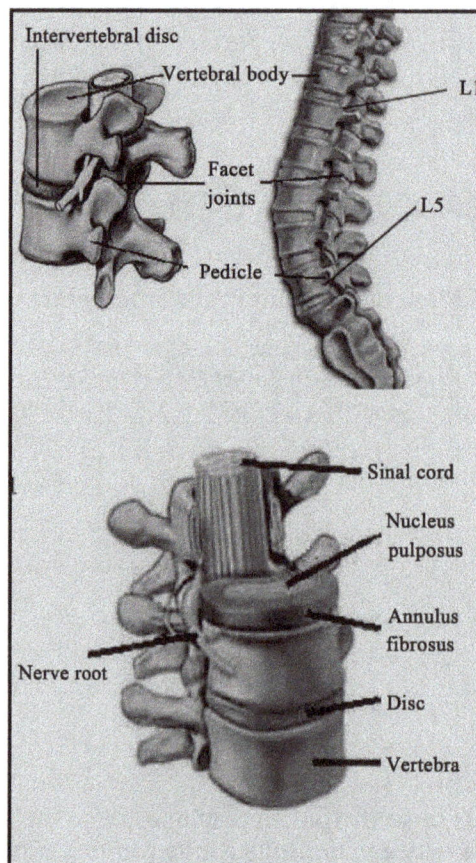

Figure 1. Anatomy of the lumbar spine [5].

Rheumatism, the Journal of the American College of Rheumatology, alerts on this effect of overweight in the development of degenerative disc disease and back pain and then herniated disc (**Figure 3**).

The term obesity is defined as an eccentric load (**Figure 2**); the load is represented by the mass of the belly (P1). In this work, the simulations of the disc degeneration, based on a finite element model of the spine depending on the mechanical properties were established; to define the boundary condition restriction on movements of translation and rotation of the spine has been applied in the frontal plane. We propose in this work to draw up a comprehensive study of stresses and deformations in the spinal discs distributions based on supported loads. The results showed that the level of degeneration increased in all inters vertebral discs but concentrated in the two disks D1 and D10.

Figure 3 shows two vertebrae of the spinal column with an intervertebral disc under the effect of a compound loading (compression P + bending moment P1). The compressive load P creates an internal pressure in the nucleus. This pressure will there after generate the disc degeneration or degenerative disc disease (**Figure 4**). As regards the forward flexion P1, if the load of the stomach increases, automatically distance between the point of load application and the axis of the spinal column increases, we see that the posterior portion of the annulus fibrosis is tensioned and the other front portion is compressed, that is to say the nucleus pulposus burst back (posterior compression), and this compression produced by disc protrusion comes into contact with a nerve root called herniated disc.

2. Material and Methods

The objective of this work is to provide an analysis between a geometric configuration of the spine system, to find the effect of eccentric loads on the latter and mainly on the inter vertebral discs by analyzing the stress distribution in this system using a 3D numerical simulation, based on the principles of the finite element method. The analysis of biomechanical problems includes several steps.

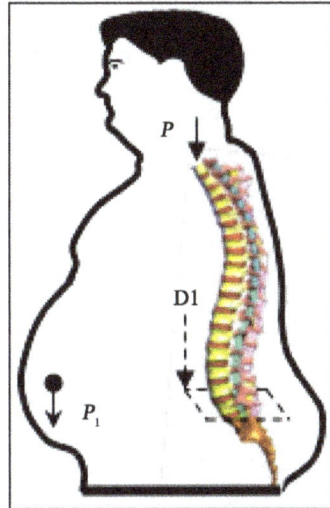

Figure 2. Constitution spine (obese person) at the disc depending on its condition.

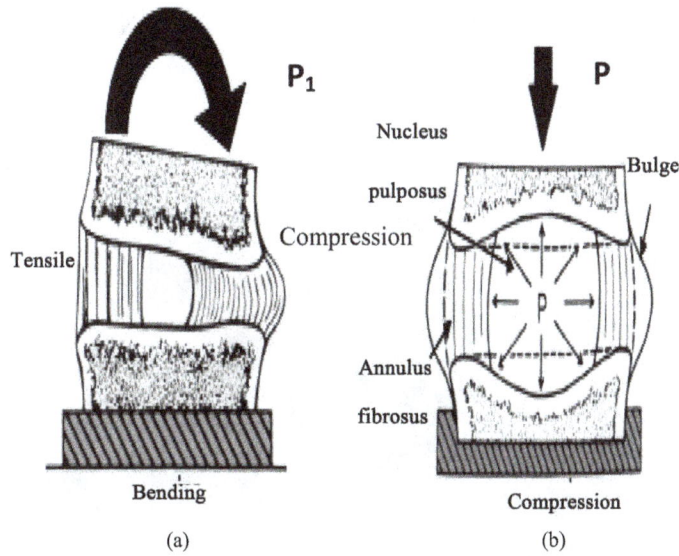

Figure 3. The intervertebral disc with (a) compression (b) bending [6].

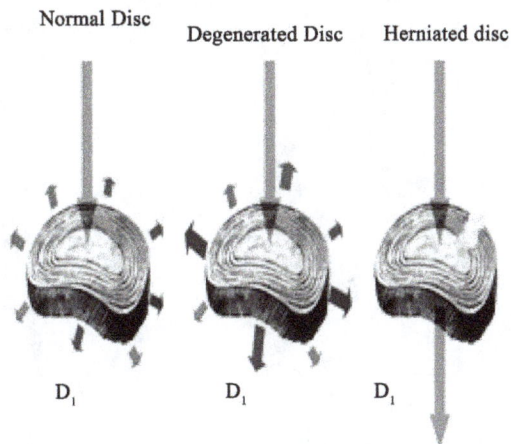

Figure 4. Load distribution at the disc D1 according to his [7] state.

The first is to study the form to define the geometrical configuration of the object, which allows the reconstitution of the vertebra, the ligament and bone using CAD programs.

The result is a 3D geometric model including these three components will then be prepared for use in finite element analyzes for the study of stresses and deformations distribution in the system.

The Steps to the execution of the 3D vertebra model (**Figure 5**).

a) Draw cortical bone that is the upper hinge and the lower hinge, and make the smoothing; this gives a solid body called the vertebral body.

b) Secondly, draw the posterior arch (blade + the pedicle) with the spinousprocess.

c) Finally we draw the transverse process.

The simulation of the disc degeneration is based on a finite element model of the healthy spine. **Figure 6** shows a spine model, this consists of five lumbar vertebrae (L1, L2, L3, L4 and L5) plus the sacrum, twelve thoracic vertebrae (TH1, TH2, TH3, TH4, TH5, TH6, TH7, TH8, TH9, TH10, TH11, TH12) and 17 inter vertebral discs between (S1-L5, L5-L4, L4-L3, L3-L2, L2-L1, L1-TH12 TH12-TH11, TH11, TH10, TH10-TH9, TH9-TH8, TH8-TH7, TH7-TH6, TH6-TH5, TH5-TH4, TH3-TH4, TH3-TH2 TH2-TH1) and various ligaments thoracic lumbar spine (anterior longitudinal ligament, posterior longitudinal ligament, ligament interspinous ligament supraspinatus, yellow ligament and capsular ligament).

In static loading conditions, the model of the reconstructed spine is used in an analysis for studying the role of the inter vertebral discs and the stress distribution in these disks as well as its supporting structures. The spine is reconstructed in 3D to study the system dimensions (IVD-ligament-bone) (**Figure 7**).

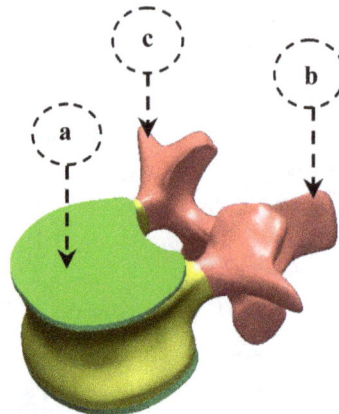

Figure 5. Lumbar vertebras.

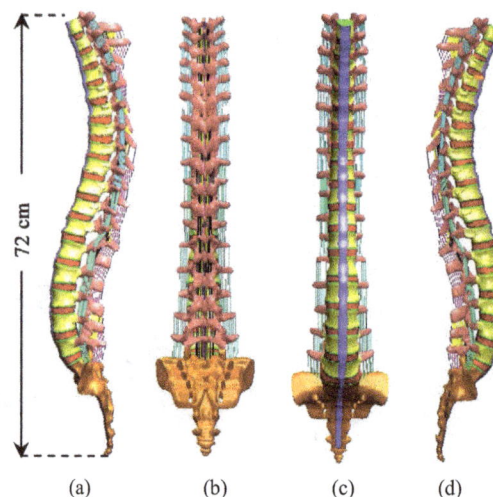

Figure 6. Spine studied. (a) Lateral (left) view; (b) Dorsal view; (c) Front view; (d) Lateral (right) view.

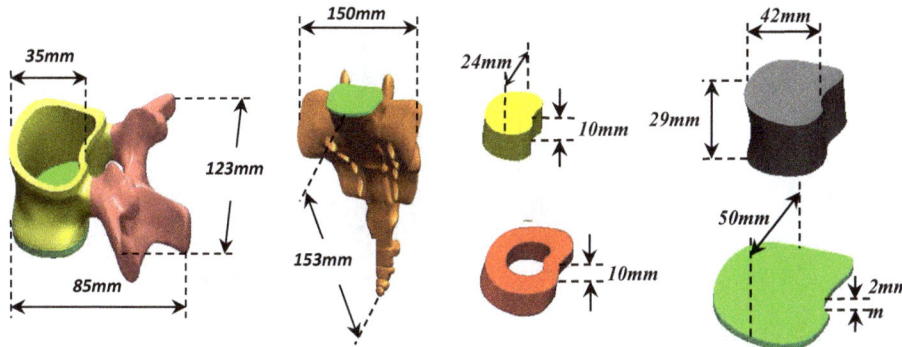

Figure 7. Vertebra and sacrum dimensions.

In order to define the boundary conditions, restriction on movements of translation and rotation of the spine has been applied in the lower plane, and defined as having zero displacements. Several charges in the anterior direction were applied as follows:

- The application of the load on the upper side of the thoracic vertebra TH1.
- The fixed part applied to the body of the sacrum.
- The interfaces between the different components of the system of the spine, the cortical bone, the inter vertebral disk and ligament are treated as perfectly bonded interfaces (**Figure 8**).

Figure 9 shows an isometric view of an exploded assembly of the spine and each component of the spine system is denoted by letters.

The selection of constitutive equations of the vertebral bone is defined as the part of the bone which carries the inter vertebral disc, composed of cortical bone, cancellous bone, the posterior arch, with a Young's modulus of about 12,000 MPa. It is recognized that cortical bone: has: better load capacity than the cancellous bone: cortical bone is considered as an isotropic material, and homogeneous linear elastic.

Table 1 shows the tensile strength of the structure annulus fibrosis different authors. These materials are in herently anisotropic and non-linear elastic.

The behavior of intertransverse ligament and inter-spinous ligament is nonlinear viscoelastic as in previous studies [8]; a linear elastic model is chosen to represent this behavior.

ANSYS WORKBENCH software was used for analyzing this geometry and generate the most suitable mesh. For the studied behavior, we used tetrahedral elements, type Solid187 conforming to defined parametric surfaces (**Figure 10**). It is necessary to mesh the components of the spine with small and confused elements to ensure optimum accuracy of the results of stresses and strains in the inter vertebral discs.

The material properties of the spine components were selected after a careful review of the published literature (**Table 2**) it was considered appropriate to define the cortical and cancellous bone as homogeneous and isotropic. The magnitudes of 12,000 MPa and 100 MPa (cortical and cancellous, respectively) were observed in all studies by various researchers. Since physiologically the nucleus is fluid filled, the elements were assigned low stiffness (1 MPa) values and near incompressibility properties (Poisson's ratio of 0.499). Biologically, the annulus fibrosus is comprised of layers of collagen fibers, which attributes to its non-homogenous characteristics. However, due to limitations in modeling abilities, the annulus was defined as a homogenous structure with a magnitude of 4.2 MPa. This was based on the modulus of the ground substance (4.2 MPa) and the collagen fibers reported in the literature, taking into account the volume fraction of each component.

The complete model of the spine (**Figure 10**) was realized by the *SOLIDWORKS SOFTWARE VERSION* 2014 and was then transferred to the software Calculates each element ends *ANSYS 14.5 WORKBENCHE* generated the default mesh then generated linear global custom mesh tetrahedra 10 nodes conform to surface.

The result ends element models of the spine contain a condensed mesh (**Figure 10**).

Figure 10 shows a complete model that consists of 11,762,783 elements and 17,150,901 nodes. L'os cortical contains (3,585,646 element and 5,132,199 nodes), cancellous bone in the cortical bone contains (2,496,448 element 3,471,929 nodes).

The posterior arch was modeled with tetrahedral elements to 10 nodes contains (2,377,091 element, 3,440,842 nodes), the nucleus pulposus in the annulus fibrosus were modeled with tetrahedral type elements 10 nodes (504,657 element 717,205 nodes), the annulus fibrosus were modeled with elements of type tetrahedral to 10

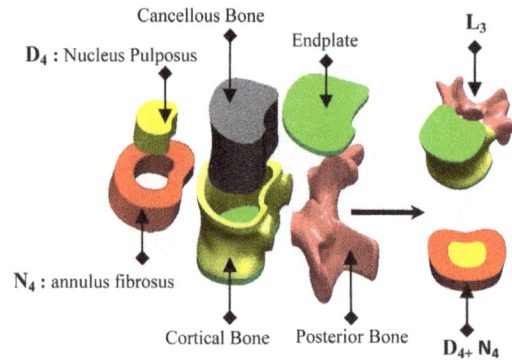

Figure 8. 3D modeling thoracic vertebra L3, D4 disc of the lumbar spine (SOLIDWORKS 2014 software).

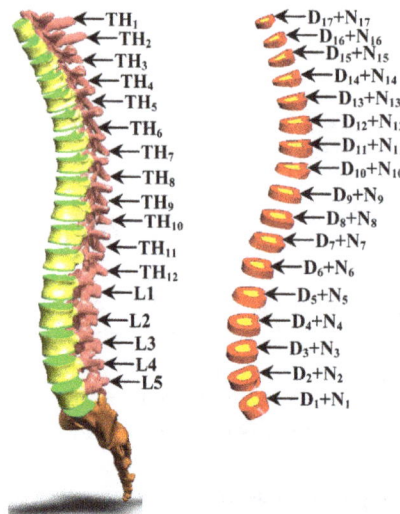

Figure 9. Assemblies in isometric perspective. Abbreviations: D_1: intervertebral disk upstairs one; N_1: nucleus in the intervertebral disc upstairs one; D_{10}: intervertebral disk upstairs ten; N_{10}: nucleus in the intervertebral disc upstairs ten; L_4: lumbar vertebra is on level four; D_4: intervertebral disk upstairs four; N_4: nucleus in the intervertebral disc upstairs four; MRI: Magnetic resonance imaging; FEM: Finite Element Method; IVD: inter vertebral disc.

Figure 10. Spine 3D finite element modeling (Ansys14.5 software).

Table 1. Mechanical characteristics of disc tissue [9].

Authors	σr (N/mm^2)
BROWN (axiale)	1.4
GALANTE (horizontale)	3.5 ± 0.3
GALANTE (sens fibre)	10.7 ± 0.9
WU	3.7

Table 2. Material properties specified in the model.

Material	Young modulus (MPa)	Poisson coefficient	References
Cortical bone	12,000	0.3	[10]-[21] [26] [35] [37]-[40] [42] [43]
Cancellous bone	100	0.2	[10] [13] [14] [16]-[18] [20] [21] [23]-[26] [35] [37]-[40] [42] [43]
Posterior bone	3500	0.25	[12]-[14] [17] [18] [20] [21] [23] [26] [27] [36]-[39]
Cartilage endplates	12,000	0.3	[20] [22] [24] [28]
Annulus ground substance	4.2	0.45	[10] [13] [16] [18]-[22] [24] [26] [30] [31] [34]-[39] [42] [43]
Nucleus pulposus	1	0.499	[11] [13]-[15] [17] [19] [20] [26] [29] [32]-[38] [42] [43]
Anterior longitudinal ligament	20	0.3	[13] [14] [16]-[18] [36] [37] [40] [41]
Posterior longitudinal ligament	20	0.3	[13] [14] [16] [17] [36] [37] [40] [41]
Ligamentum flavum	19.50	0.3	[13] [14] [16] [17] [36] [37] [40] [41]
Intertransverse ligament	58.7	0.3	[13] [14] [16] [17] [36] [37] [40] [41]
Inter-spinous ligament	11.6	0.3	[13] [14] [16] [17] [36] [37] [40] [41]
Supra-spinous ligament	15	0.3	[13] [14] [16] [17] [36] [37] [40] [41]
Capsular ligament	32.9	0.3	[13] [14] [16] [17] [36] [37] [40] [41]

nodes (1,434,546 element , 2,059,247 nodes). The gelatinous cartilage modeled with a tetrahedral element to 10 nodes (912,759 elements, 1,431,242 nodes). Finally the different types of ligaments generated by a tetrahedral mesh to 10 nodes (**Table 3**).

The diagram in (**Figure 11**) shows a normal person standing with a specific weight of 80 kg, the total mass (belly) is 13.25 kg representing the weight P1; the pressure load P is the mass of the top portion of the person's body (hands, forearms, arms, head) divided by the area of the thoracic vertebra TH1.

The length of the spine (thoracic + lumbar) is 72 cm and the distance between the specific weights of the belly which is the point of application of the load and the axis of the vertebral column (20 cm). For boundary conditions, the sacrum is fixed. (Embedding the sacrum (**Figure 12**).

We propose in this section to draw up a detailed study of Von Mises distributions constraints and deformations in the intervertebral discs as a function of supported loads.

Figure 12 shows three people in standing position, with respective specific bellies weight P1 (16.56 Kg, 19.88 kg, 24.85 kg).

The distance between the center of gravity of the belly and the axis of the vertebral column is between (30 cm/50 cm), the pressure load P is applied on the thoracic vertebra TH1.

To define the boundary conditions, restriction on movements of translation and rotation of the spine has been applied including frontal plane and defined as having zero displacements on the sacrum see **Figure 12**.

Consider the example of an anterior load (obese person), the colors represent the Von Mises stresses experienced by 17 intervertebral discs, blue represents the minimum stress and the red represents the maximum stress.

3. Results

Figure 13 shows clearly that the anterior load affects the disk D1 and D10 is the two most sought discs compared to other drives in the thoracic lumbar spine.

A load applied to the upper surface of the TH1 thoracic vertebra of the spinal column causes a high concentration of maximum Von Mises stresses in the anterior portion of the two discs D1 and D10 (red section) this is mentioned in **Figure 14**.

Moreover, the Von Mises stresses are minimal at the posterior part of the two intervertebral discs D1, D10 (blue contour) see **Figure 14**.

Figure 15 shows the Von Mises strain histogram in the inter vertebral discs for a normal person load 13.25 kg. We note that the Von Mises deformation values are highest in the two discs D1, D10 (2784, 2377) outline in red; this is mentioned in **Figure 16**.

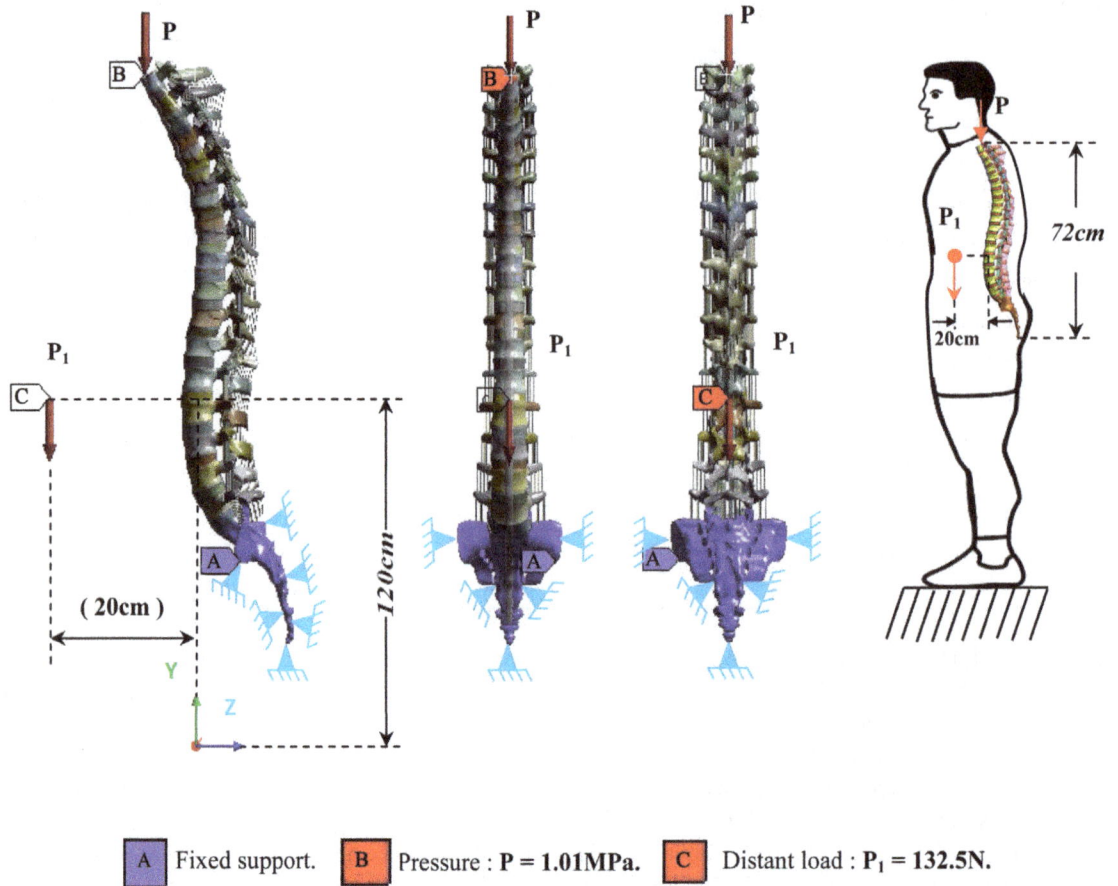

| A | Fixed support. | B | Pressure : **P = 1.01MPa.** | C | Distant load : **P₁ = 132.5N.** |

Figure 11. Mechanical model of the spine anterior load (normal person).

Table 3. Element and node numbers in the column vertebral system components.

Component	Nodes	Elements	Thickness
Cortical bone	5,132,199	3,585,646	1 mm
Cancellous bone	3,471,929	2,496,448	1 mm
Posterior bone	3,440,842	2,377,091	1 mm
Cartilage endplates	1,431,242	912,759	1 mm
Annulus ground substance	2,059,247	1,434,546	1 mm
Nucleus pulposus	717,205	504,657	1 mm
Anterior longitudinal ligament	227,078	128,365	1 mm
Posterior longitudinal ligament	158,748	92,426	1 mm
Ligamentum flavum	30,226	13,447	1 mm
Transverse ligament	285,328	131,648	1 mm
Inter-spinous ligament	28,968	13,158	1 mm
Supra-spinous ligament	17,833	8279	1 mm
Capsular ligament	51,816	24,072	1 mm
Total	**17,150,901**	**11,762,783**	**1 mm**

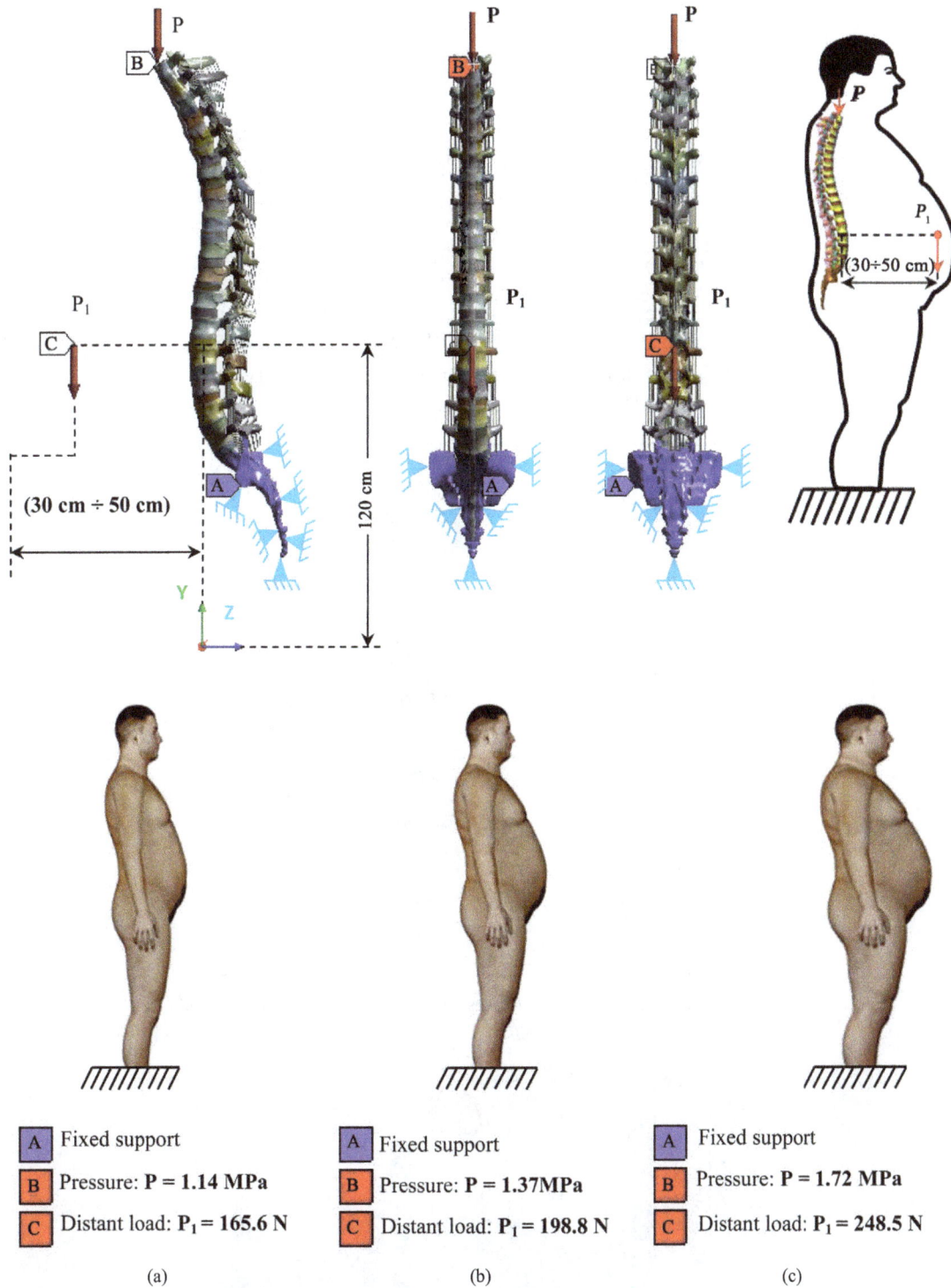

A	Fixed support	A	Fixed support	A	Fixed support
B	Pressure: P = 1.14 MPa	B	Pressure: P = 1.37MPa	B	Pressure: P = 1.72 MPa
C	Distant load: P_1 = 165.6 N	C	Distant load: P_1 = 198.8 N	C	Distant load: P_1 = 248.5 N
	(a)		(b)		(c)

Figure 12. Mechanical model of the spine [anterior load (fat person)].

Consider the example of an anterior load (obese person), the colors represent the Von Mises stresses experienced by 17 intervertebral discs, blue represents the minimum stress and the red represents the maximum stress. **Figure 17** shows the role of the intervertebral discs in the stress absorbing. Their distribution in the spine tends to be concentrated in the disc D1 on both sides, for the front side, a tracted portion to a maximum value of 15.216 MPa and the other has a minimum value of 0.035 MPa (**Figure 18**).

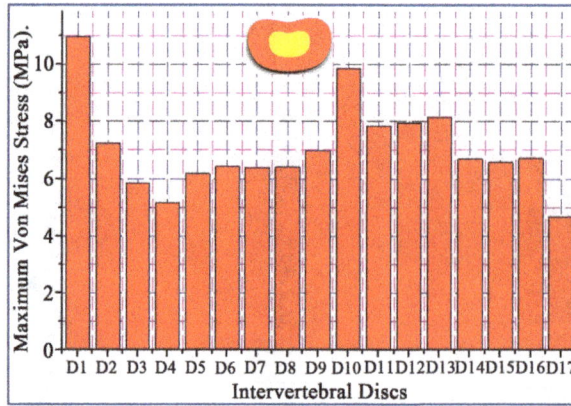

Figure 13. Histogram of the Von Mises stress in the IVD for a load of 13.25 kg.

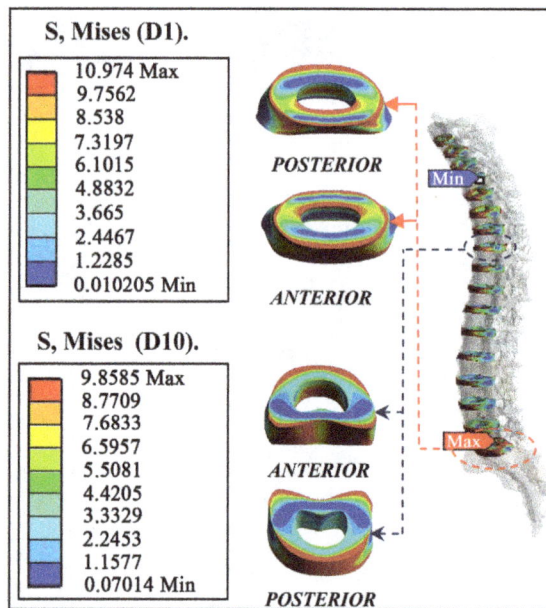

Figure 14. Von Mises stresses Distribution in the discs D1 and D10 for a load of 13.25 kg.

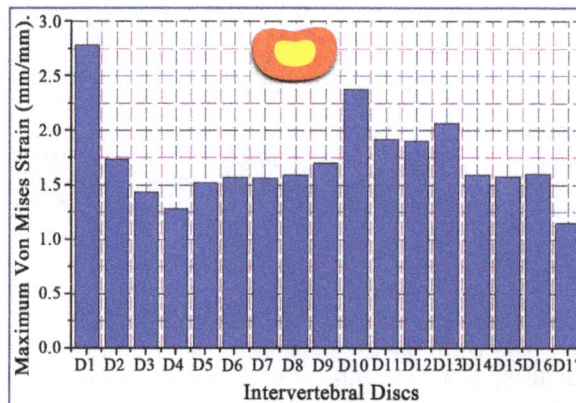

Figure 15. Histogram of the Von Mises strains in the IVD for a load of 13.25 kg.

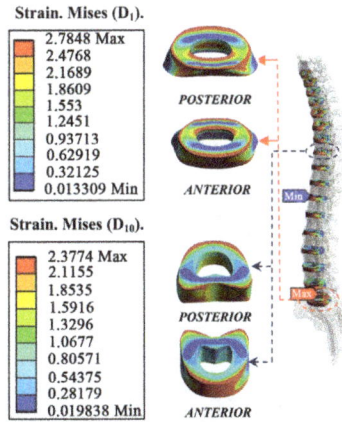

Figure 16. Von Mises strains Distribution in the discs D1 and D10 for a load of 13.25 kg.

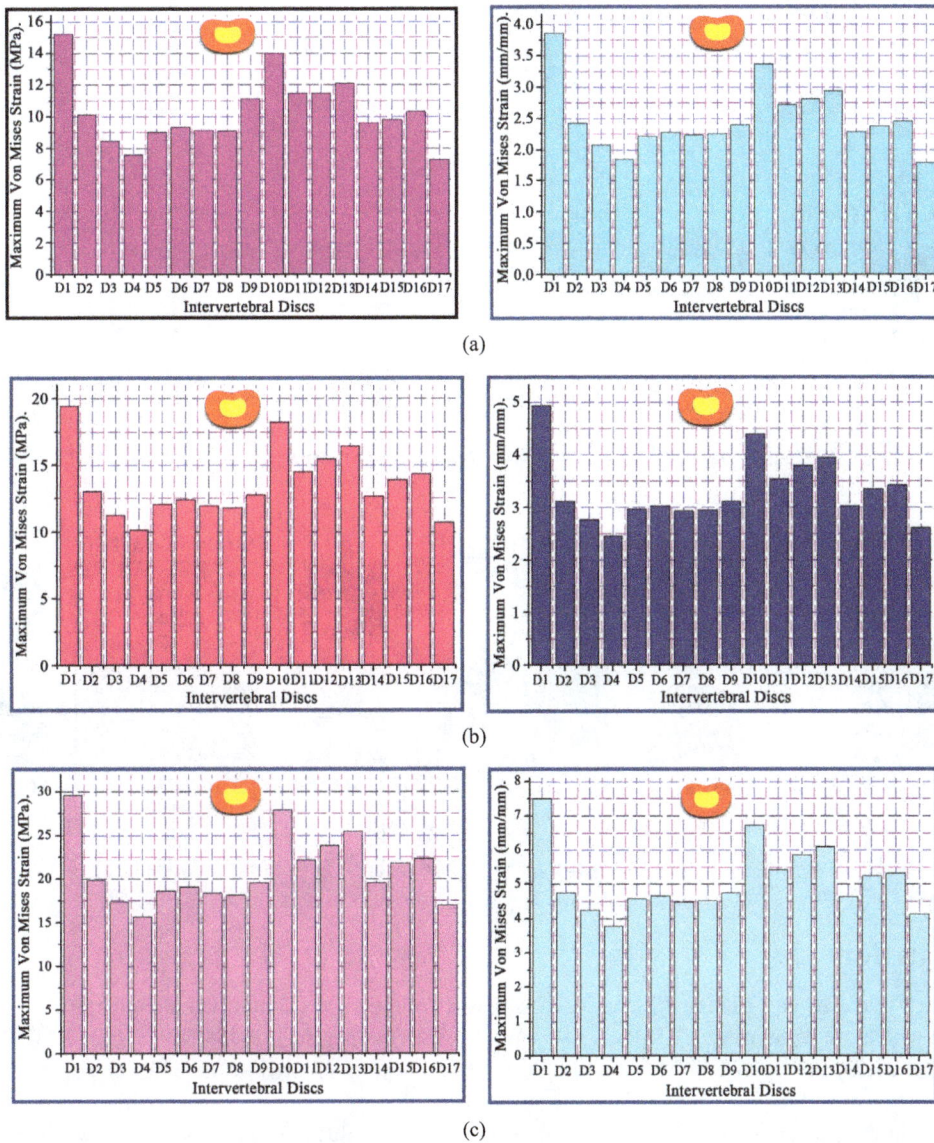

(a)

(b)

(c)

Figure 17. Histogram of the Von Mises stress and strains in the IVD for a Diffrent load. (a) 16.56 kg; (b) 19.88 kg; (c) 24.85 kg.

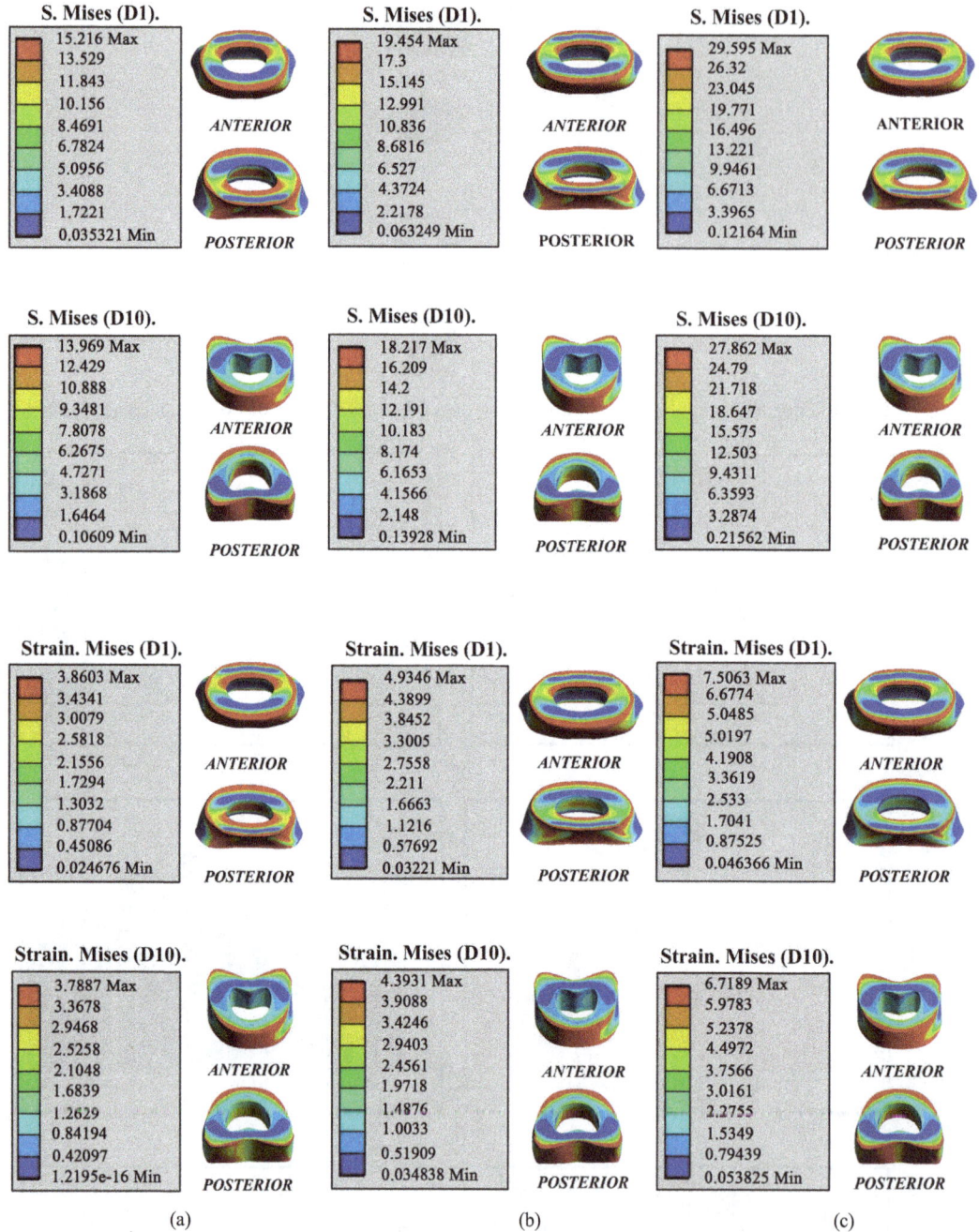

Figure 18. Histogram of the Von Mises stress and strains in the IVD for a Diffrent load. (a) 16.56 kg; (b) 19.88 kg; (c) 24.85 kg.

4. Discussion

We see in **Figure 18** and **Figure 19** Von Mises stresses reached a maximum value concentrated in the two discs D1, D10 which are equal to 15.21 MPa, 13.96 MPa for the first person, 19.454 MPa, 18.217 MPa for the second person and 29.595 MPa, 27.862 MPa for the third person who is located in the anterior portion of the disc (red part).

Regarding the deformations Von Mises, we notice that the values are greatest in the two intervertebral discs D1, D10 of 3.86, 3.788 for a load of 16.55 kg obesity and 4.934, 4.393 for an obesity load 19.88 kg and 7.506, 6.718 for an obesity load 29.85 kg compared to other discs of the vertebral column (**Figure 18**). Finally a normal person (without obesity), Von Mises stresses are concentrated in both discs D1 and D10 with values of 10.974 MPa, 9.858 MPa see **Figure 14**. In the case where the obesity of load augment that is to say the distance between the load application point and the axis of the spine augment (the most risky cases). These loads will cause according to the maximum distance of major constraints that will generated a phenomenon called disc degeneration (herniated disc) posterior side of the disc from its side will overwrite the spinal nerve (spinal cord) see **Figure 20(a)** and **Figure 20(b)**.

5. Conclusions

In sum, we concluded for the three cases of anterior load (obese persons) (**Figure 19**), that the distributions of

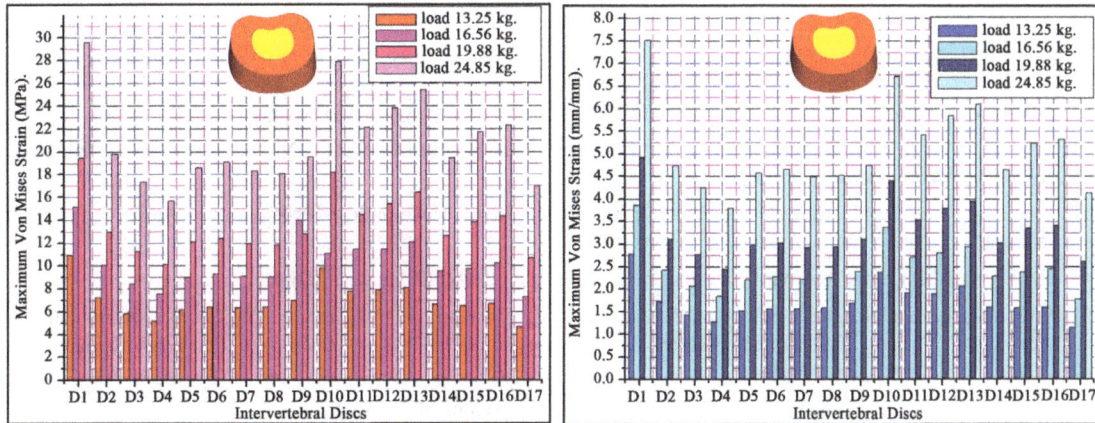

Figure 19. Histogram of the Von Mises stress and strains in the IVD for a Different loads.

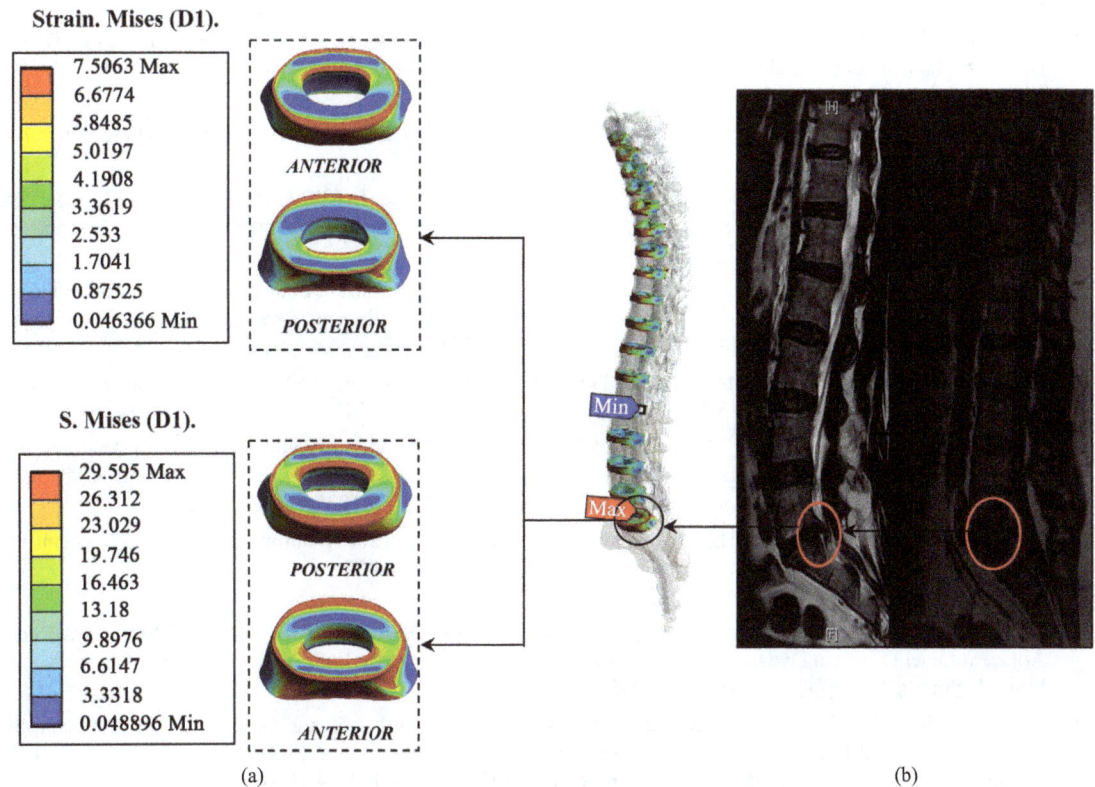

(a)   (b)

Figure 20. (a) Distribution of the Von Mises stress and strains in the Discs D1 for load 24.85 kg; (b) MRI showing a herniated disc.

Von Mises stresses are the highest in intervertebral discs and are concentrated in the disk D1, in contact with the L5 vertebra and the sacrum; **Figure 20(a)** clearly shows that the level of deformation of Von Mises is maximal in the disc D1, valued at 7.5063 mm/mm. This shows that the distance between the load application point and the axis of the spine has an important role in the increase of solicitation of this one and therefore of its deformation (**Figure 20(b)**).

Acknowledgements

The authors extend their appreciation to the Director of Scientific Research at LaBPS for funding the work through the Biomechanics Research Group.

References

[1]   Grillo, G., Koca, O., Eskitascioglu, G. and Usumez, A. (2005) Three Dimensional Finite Element Analysis of Functional Stresses in Different Bone Locations Produced by Implants Placed in the Maxillary Posterior Region of the Sinus Floor. *Journal of Prosthetic Dentistry*, **93**, 38-44. http://dx.doi.org/10.1016/j.prosdent.2004.10.001

[2]   Brunski, J.B. (1997) Biomechanics of Dental Implants. In: Block, M., Kent, J.N. and Guerra, L.R., Eds., *Implants in Dentistry*, W.B. Saunders, Philadelphia, 63-71.

[3]   Thomas, M. (2008) Contribution a l'analyse biomecanique et a l'evaluation des implants rachidiens. L'école nationale superieure d'arts et metiers specialite "biomecanique", Paris.

[4]   Dr Kassab, M. Centre Avicenne Médical, 2 Av Tahar Sfar, 2092, El Manar 2, Tunis.

[5]   Dietrich, M., Kedzior, K., Borkowski, P., Krzesinski, G., Skalski, K. and Zagrajek, T. (2005) A Nonlinear Analysis of the Human Vertebral Column and Medical Recommendations That Follow. *Bulletin of the Polish Academy of Sciences, Technical Sciences*, **53**, 179-194.

[6]   White Iii, A.A. and Panjabi, M.M. (1990) Clinical Biomechanics of the Spine.

[7]   Marcovschi Champain, S. (2008) Corrélations Entre Les Paramètres Biomécaniques Du Rachis Et Les Indices Cliniques Pour L'analyse Quantitative Des Pathologies Du Rachis Lombaire Et De Leur Traitement Chirurgical, Enam, Paris.

[8]   Starmans, F.J., Steen, W.H. and Bosman, F. (1993) A Three-Dimensional, Finite-Element Analysis of Bone Around Dental Implants in an Edentulous Human Mandible. *Archives of Oral Biology*, **38**, 491-496. http://dx.doi.org/10.1016/0003-9969(93)90185-O

[9]   Pr. Francois, L. (1997) Biomécanique Et Ostéosynthèse Du Rachis Ensm-Lbm Conférences D'enseignement De La Sofcot.

[10]   Ibarz, E., Más, Y., Mateo, J., Lobo-Escolar, A., Herrera, A. and Gracia, L. (2013) Instability of the Lumbar Spine Due to Disc Degeneration. A Finite Element Simulation. *Advances in Bioscience and Biotechnology*, **4**, 548-556. http://dx.doi.org/10.4236/abb.2013.44072

[11]   Song, M., Zhang, Z., Lu, M., Zong, J., Dong, C., Ma, K. and Wang, S. (2014) Four Lateral Mass Screw Fixation Techniques in Lower Cervical Spine Following Laminectomy: A Finite Element Analysis Study of Stress Distribution. *Biomedical Engineering Online*, **13**, 115. http://dx.doi.org/10.1186/1475-925X-13-115

[12]   Rundell, S.A., Isaza, J.E. and Kurtz, S.M. (2011) Biomechanical Evaluation of a Spherical Lumbar Interbody Device at Varying Levels of Subsidence. *SAS Journal*, **5**, 16-25.

[13]   Goel, V.K., Mehta, A., Jangra, J., Faizan, A., Kiapour, A., Hoy, R.W. and Fauth, A.R. (2007) Anatomic Facet Replacement System (AFRS) Restoration of Lumbar Segment Mechanics to Intact: A Finite Element Study and *in Vitro* Cadaver Investigation. *SAS Journal*, **1**, 46-54. http://dx.doi.org/10.1016/S1935-9810(07)70046-4

[14]   Holekamp, S., Goel, V., Kuroki, H., Huntzinger, J. and Ebraheim, N. (2007) Optimal Intervertebral Sealant Properties for the Lumbar Spinal Disc: A Finite-Element Study. *SAS Journal*, **1**, 68-73. http://dx.doi.org/10.1016/s1935-9810(07)70049-x

[15]   Castellvi, A.E., Huang, H., Vestgaarden, T., Saigal, S., Clabeaux, D.H. and Pienkowski, D. (2007) Stress Reduction in Adjacent Level Discs via Dynamic Instrumentation: A Finite Element Analysis. *SAS Journal*, **1**, 74-81. http://dx.doi.org/10.1016/S1935-9810(07)70050-6

[16]   López, E., Ibarz, E., Herrera, A., Mateo, J., Lobo-Escolar, A., Puértolas, S. and Gracia, L. (2014) Probability of Osteoporotic Vertebral Fractures Assessment Based on DXA Measurements and Finite Element Simulation. *Advances in Bioscience and Biotechnology*, **5**, 527-545. http://dx.doi.org/10.4236/abb.2014.56063

[17]   Kiapour, A., Kiapour, A.M., Kodigudla, M., Hill, G.M., Mishra, S. and Goel, V.K. (2012) A Biomechanical Finite Element Study of Subsidence and Migration Tendencies in Stand-Alone Fusion Procedures—Comparison of an *in Situ*

Expandable Device with a Rigid Device. *Journal of Spine*, **1**, 4.

[18] Zheng, S.-N., Yao, Q.-Q., Wang, L.-M., Hu, W.-H., Wei, B., Xu, Y. and Zhang, D.-G. (2013) Biomechanical Effects of Semi-Constrained Integrated Artificial Discs on Zygapophysial Joints of Implanted Lumbar Segments. *Experimental and Therapeutic Medicine*, **6**, 1423-1430.

[19] Byun, D.-H., Ah Shin, D., Kim, J.-M., Kim, S.-H. and Kim, H.-I. (2012) Finite Element Analysis of the Biomechanical Effect of Coflex™ on the Lumbar Spine, Laboratory Investigation. *Korean Journal of Spine*, **9**, 131-136. http://dx.doi.org/10.14245/kjs.2012.9.3.131

[20] Lan, C.-C., Kuo, C.-S., Chen, C.-H. and Hu, H.-T. (2013) Finite Element Analysis of Biomechanical Behavior of Whole Thoraco-Lumbar Spine with Ligamentous Effect. *The Changhua Journal of Medicine*, **11**, 26-41.

[21] Natarajan, R.N. and Andersson, G.B.J. (1997) Modeling the Annular Incision in a Herniated Lumbar Intervertebral Disc to Study Its Effect on Disc Stability. *Computers & Structures*, **64**, 1291-1297. http://dx.doi.org/10.1016/S0045-7949(97)00023-0

[22] Pitzen, T., Geisler, F.H., Matthis, D., Storz, H.M., Pedersen, K. and Steudel, W.I. (2001) The Influence of Cancellous Bone Density on Load Sharing in Human Lumbar Spine: A Comparison between an Intact and a Surgically Altered Motion Segment. *European Spine Journal*, **10**, 23-29. http://dx.doi.org/10.1007/s005860000223

[23] Polikeit, A. (2002) Finite Element Analysis of the Lumbar Spine: Clinical Application. Inaugural Dissertation, University of Bern, Bern.

[24] Denoziere, G. (2004) Numerical Modeling of a Ligamentous Lumber Motion Segment. Master's Thesis, Department of Mechanical Engineering, Georgia Institute of Technology, Georgia.

[25] Gwanseob, S. (2005) Viscoelastic Responses of the Lumbar Spine during Prolonged Stooping. PhD Dissertation, North Carolina State University, Raleigh.

[26] Sairyo, K., Goel, V.K., Masuda, A., Vishnubhotla, S., Faizan, A., Biyani, A., Ebraheim, N., Yonekura, D., Murakami, R.I. and Terai, T. (2006) Three-Dimensional Finite Element Analysis of the Pediatric Lumbar Spine. *European Spine Journal*, **15**, 923-929. http://dx.doi.org/10.1007/s00586-005-1026-z

[27] Rohlmann, A., Burra, N.K., Zander, T. and Bergmann, G. (2007) Comparison of the Effects of Bilateral Posterior Dynamic and Rigid Fixation Devices on the Loads in the Lumbar Spine: A Finite Element Analysis. *European Spine Journal*, **16**, 1223-1231. http://dx.doi.org/10.1007/s00586-006-0292-8

[28] Wilke, H.J., Neef, P., Caimi, M., Hoogland, T. and Claes, L.E. (1999) New *in Vivo* Measurements of Pressures in the Intervertebral Disc in Daily Life. *Spine*, **24**, 755-762. http://dx.doi.org/10.1097/00007632-199904150-00005

[29] Smit, T., Odgaard, A. and Schneider, E. (1997) Structure and Function of Vertebral Trabecular Bone. *Spine*, **22**, 2823-2833. http://dx.doi.org/10.1097/00007632-199712150-00005

[30] Sharma, M., Langrana, N.A. and Rodriguez, J. (1995) Role of Ligaments and Facets in Lumbar Spinal Stability. *Spine*, **20**, 887-900. http://dx.doi.org/10.1097/00007632-199504150-00003

[31] Lee, K. and Teo, E. (2004) Effects of Laminectomy and Facetectomy on the Stability of the Lumbar Motion Segment. *Medical Engineering & Physics*, **26**, 183-192. http://dx.doi.org/10.1016/j.medengphy.2003.11.006

[32] Rohlmann, A., Zander, T., Schmidt, H., Wilke, H.J. and Bergmann, G. (2006) Analysis of the Influence of Disc Degeneration on the Mechanical Behaviour of a Lumbar Motion Segment Using the Finite Element Method. *Journal of Biomechanics*, **39**, 2484-2490. http://dx.doi.org/10.1016/j.jbiomech.2005.07.026

[33] Ng, H.W. and Teo, E.C. (2001) Nonlinear Finite-Element Analysis of the Lower Cervical Spine (C4-C6) under Axial Loading. *Journal of Spinal Disorders*, **14**, 201-210. http://dx.doi.org/10.1097/00002517-200106000-00003

[34] Ng, H.-W., Teo, E.-C. and Zhang, Q.-H. (2004) Influence of Laminotomies and Laminectomies on Cervical Spine Biomechanics under Combined Flexion-Extension. *Journal of Applied Biomechanics*, **20**, 243-259.

[35] Gong, Z.Q., Chen, Z.X., Feng, Z.Z., Cao, Y.W., Jiang, C. and Jiang X.X. (2014) Finite Element Analysis of 3 Posterior Fixation Techniques in the Lumbar Spine. *Orthopedics*, **37**, E441-E448. http://dx.doi.org/10.3928/01477447-20140430-54

[36] Kim, H.-J., Kang, K.T., Chang, B.-S., Lee, C.-K., Kim, J.-W. and Yeom, J.-S. (2014) Biomechanical Analysis of Fusion Segment Rigidity upon Stress at Both the Fusion and Adjacent Segments: A Comparison between Unilateral and Bilateral Pedicle Screw Fixation. *Yonsei Medical Journal*, **55**, 1386-1394. http://dx.doi.org/10.3349/ymj.2014.55.5.1386

[37] Goel, V.-K., Kiapour, A., Faizan, A., Krishna, M. and Friesem, T. (2006) Finite Element Study of Matched Paired Posterior Disc Implant and Dynamic Stabilizer (360° Motion Preservation System). *SAS Journal*, **1**, 55-62. http://dx.doi.org/10.1016/S1935-9810(07)70047-6

[38] Tang, S. and Meng, X.Y. (2010) Does Disc Space Height of Fused Segment Affect Adjacent Degeneration in ALIF? A Finite Element Study. *Journal of Turkish Neurosurgery*, **3**, 296-303.

[39] Kim, K.-T., Lee, S.-H., Suk, K.-S., Lee, J.-H. and Jeong, B.-O. (2010) Biomechanical Changes of the Lumbar Segment after Total Disc Replacement: Charite®, Prodisc® and Maverick® Using Finite Element Model Study. *Journal of Korean Neurosurgical Society*, **47**, 446-453. http://dx.doi.org/10.3340/jkns.2010.47.6.446

[40] Agarwal, A., Agarwal, A.-K. and Goel, V.-K. (2013) The Endplate Morphology Changes with Change in Biomechanical Environment Following Discectomy. *International Journal of Clinical Medicine*, **4**, 8-17. http://dx.doi.org/10.4236/ijcm.2013.47A1002

[41] Zhong, Z.-C., Wei, S.-H., Wang, J.-P., Feng, C.-K., Chen, C.-S. and Yu, C.-H. (2006) Finite Element Analysis of the Lumbar Spine with a New Cage Using a Topology Optimization Method. *Medical Engineering & Physics*, **28**, 90-98. http://dx.doi.org/10.1016/j.medengphy.2005.03.007

[42] Sairyo, K., Goel, V.-K., Masuda, A., Vishnubhotla, S., Faizan, A., Biyani, A., *et al.* (2006) Three Dimensional Finite Element Analysis of the Lumbar Spine. *European Spine Journal*, **15**, 923-929.

[43] Goto, K., Tajima, N., Chosa, E., Totoribe, K., Kuroki, H., Arizumi, Y. and Arai, T. (2002) Mechanical Analysis of the Lumbar Vertebrae in a Three-Dimensional Finite Element Method Model in Which Intradiscal Pressure in the Nucleus Pulposus Was Used to Establish the Model. *Journal of Orthopaedic Science*, **7**, 243-246. http://dx.doi.org/10.1007/s007760200040

25

Changes in Human Hair Induced by UV- and Gamma Irradiation

Ervin Palma[1], David Gomez[1], Eugene Galicia[2], Viktor Stolc[3], Yuri Griko[3*]

[1]Evergreen Valley College, San Jose, CA, USA
[2]Carnegie Melon University, Moffett Field, CA, USA
[3]Division of Space Biosciences, NASA Ames Research Center, Moffett Field, CA, USA
Email: *Yuri.V.Griko@nasa.gov

Abstract

The effect of UV- and ^{137}Cs gamma radiation on the structural and chemical integrity of human hair was studied to determine the feasibility of using human hair as a non-invasive biomarker of radiation exposure to ionized gamma- and non-ionized UV-radiation. Steady state tryptophan (Trp) fluorescence and chemical analytical methods were used to evaluate the molecular integrity of Trp fluorophores and SH-groups in hair proteins and to assess the radiation induced damage quantitatively. It was found that human hair fibers were progressively damaged by exposure to both UV- and ionized gamma radiation. Damage to the hair was evidenced by a decrease in the fluorescence intensity as a result of observed depletion of the amino acid tryptophan as well as significant reduction in a number of free SH-groups in hair proteins. Hair damage was dose-dependent for exposures between 0 and 10.0 Gy and 0 - 20 J/cm^2 of UV-radiation. Additional results demonstrate that hair-fibers exposed to gamma rays, with much higher quantum energy than UV, undergo a smaller extent of changes in Trp fluorescence than when exposed to lower or equal energy of UV-irradiation. The stable Trp fluorophore appears to be extremely sensitive to UV-radiation in contrast to the ionized gamma radiation whose damage is originated from the reaction of free radicals and direct deposition of energy. We conclude that fluorescence spectroscopy represents a useful tool in the quantitative evaluation of the radiation exposure and could also be used for the rapid and non-invasive assessment of radiation dose *i.e.* biodosimeter. The approach is simple, non-invasive and appears to have considerable potential that enables quantitative evaluation of radiation dose exposure in a single hair fiber.

Keywords

Hair, Fluorescence, Radiation, Chemical Integrity

---

*Corresponding author.

1. Introduction

Chemical and physical properties of human hair are the subject of a remarkably wide range of scientific investigations due to their importance to the biomedical, cosmetics industry and forensic sciences. The principle protein component of hair is the cysteine rich keratin, which is composed of 18 amino acids and assembled into heavily melanized fibers that form up to 95% of hair fiber volume [1]. These protein components and structural organization of keratin contribute to most of the characteristic properties of hair.

Of the amino acids in keratin, cystine may account for as much as 24 percent. The numerous disulfide bonds formed by cystine are responsible for the great stability of keratin. On the other hand, keratin is very reactive, as cystine can easily be reduced, oxidized, and hydrolyzed [2].

Hair is very susceptible to chemical changes that occur with exposure to radiation [3]. Various abnormalities in the hair and hair follicles caused by radiation have been reported to be associated with structural re-arrangement and chemical modification in hair keratin [4]. UV-B radiation cleaves the disulfide bonds and decomposes tryptophan in hair [5]. We hypothesize that such changes might also occur upon exposure to ionizing gamma radiation and could be sensitive assays for radiation effects, and this approach could provide a basis for a non-invasive biological dosimeter [6]. While it is widely accepted that molecules of the tryptophan in the hair naturally fluoresce when illuminated, and also decompose when exposed to the ultraviolet light, their response to the ionized gamma radiation is undescribed. In contrast to skin or other cells in the body, hair fibre does not possess its own biological protective and repairing mechanisms against the impact of environmental effects. The absence of fibre regeneration makes it a potentially sensitive radiation dosimeter.

In this study, we attempt to evaluate sensitivity of Trp fluorophor to ionizing gamma radiation as well as identify key technical parameters and characteristics of fluorescent techniques that can quantify radiation-induced damage of hair. The fluorescence of this fluorophore was found to change in a predictable manner producing strong characteristic fluorescence when excited with ultraviolet light [7] [8]. This fact generates great interest for utilizing fluorescence spectroscopy to quantify damage from ionizing radiation. Such a non-invasive method that could provide a rapid and efficient assessment over time of the dose of radiation exposure would be of great value for the astronauts. Fluorescence spectroscopy is one such non-invasive technique for assessment of radiation dose utilizing hair as an accidental dosimeter following external irradiation.

2. Materials and Methods

2.1. Hair

Studies were carried out with scalp Caucasian dark brown hair collected from healthy volunteer. Single hair fiber (or bundle of two hairs) or 2 - 5 mg of hair was sufficient to make accurate quantitative analyses.

2.2. Fluorescence Spectroscopy

Fluorescence spectra in non-irradiated and in irradiated hairs were obtained with a Jobin Yvon FluoroMax-2 fluorescence spectrophotometer (Edison, NJ), equipped with a 150 W Xenon lamp, double monochromators on the excitation and emission, a photomultiplier detector (H5783P-04, Hamamatsu, Hamamatsu City, Japan). The individual hair fibers or a bundle of 2 hairs were positioned diagonally in a 1.0 × 1.0-cm polystyrene fluorometric cuvette as shown in **Figure 1** and were placed in the sample compartment of a FluoroMax-2 spectrofluorometer.

Excitation-emission spectra are constructed by measuring fluorescence emission at various excitation wavelengths, from 270 nm to 450 nm in increments of 5 nm. Emission spectra were collected starting 15 nm higher than the excitation wavelength to generate a total emission scan of 200 nm. Each fluorescence measurement consisted of a set of two serial emission spectra collected by the same positioning the hair holder. Care was taken to fix position of the hair bundle within the holder between measurements after irradiation to different dose. All fluorescence spectra were corrected for instrument response.

2.3. Gamma Irradiation

A dose response curve for low LET γ-radiation has been developed from in vitro irradiation of human hairs using [137]Cs Shepherd and Associates Mark I model 30 Irradiator available at NASA Ames Research Center. A

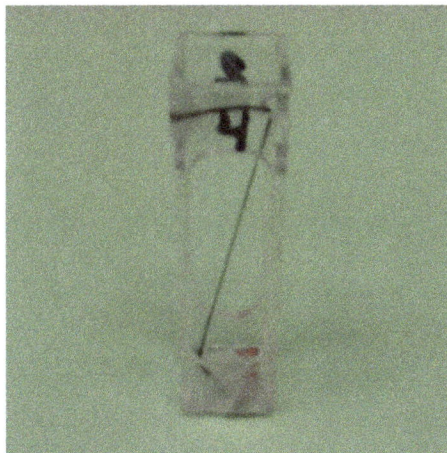

Figure 1. Positioning of hair bundle in fluorescence cuvette used in experiments. Bundle of 2 human hairs was placed in a fixed position for radiation treatment and fluorescence experiments.

dose rate was determined and exposure time was calculated from each experiment for the corresponding dose rate at one of three turntable positions with exposure rates of 0.411 Gy/min, 0.176 Gy/min, and 0.116 Gy/min respectively. The samples for the dose points were prepared in a total of 3 experiments, each with a control.

A bundle of two hairs were arranged into a disposable cuvette, which was customized for fast and reliable positioning of hair. Fluorescence characteristics of fixed hairs were first measured as non-irradiated control, and then the sample was exposed gamma radiation and changes in fluorescence spectra of irradiated sample were recorded again without changing in position of hair in cuvette. Data points for hair samples exposed to fractionated dose of radiation up to 10 Gy at 2 Gy increments were plotted to generate dose response curve.

2.4. UV-Light Irradiation

The Ultra-Violet Products Ltd., transilluminator consisting of eleven eight watt metal halide gas plasma lamps with reflecting elements has been used for UV irradiation of hair. The spectral frequencies of the light source were 279 and 295 nm. Hairs randomly positioned on the Petri dish were exposed to light from the halogen lamp at room temperature with fluencies ranging from 0 to 20 J/cm^2 at the irradiance of 7000 mkW/cm^2. The measured temperature rise in the medium was less than 3°C during exposure to an irradiation fluency of 20 J/cm^2. UVB monitor (UVB-500C, National Biological Corp.) was used to measure the UV dose rate. A Scientech 362 power energy meter (Scientech Co., Boulder, Colorado) was utilized to measure the power of the UV-source.

2.5. Determination of Sulfhydryl Groups

Previously described sensitive assay utilizing fluorescein mercuric acetate (FMA) for quantitative assessment of SH-groups in proteins was used to measure changes in content of sulfhydryl groups of hair keratin after exposure to UV- and gamma-radiation [9]. FMA was purified from the mono mercuric derivatives on G50 Sephadex (1.5 × 32 cm) column in solution of 10 mM [10]. Stock solution of FMA with concentration 5 × 10^{-4} M was stored in 10 mM NaOH at room temperature and can be used in experiments without additional purification over two weeks. Concentration of FMA was measured spectrophotometrically at 499 nm using molar extinction coefficient $E = 7.8 \times 10^{-4}$ M^{-1}·cm^{-1} [9].

3. Results and Discussion

3.1. Effect of Gamma Radiation on Fluorescence Signal of Hair

Figure 2 shows fluorescence spectra of human hair unexposed to gamma irradiation (a) along with fluorescence

(a)

(b)

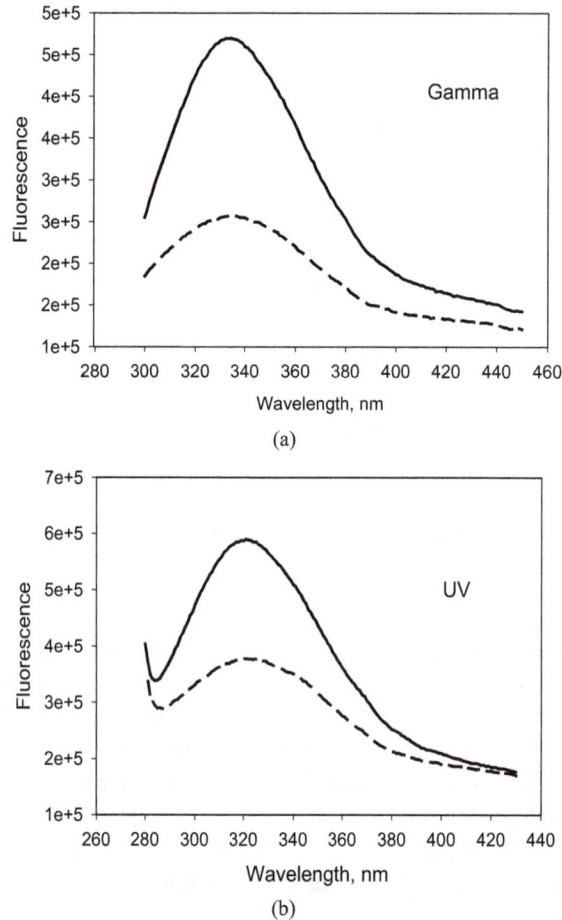

Figure 2. Changes in fluorescence spectra of hair after exposure to 10 Gy gamma and 10 J/cm² of UV-radiation. Solid line: non-irradiated; Dashed line: 10 J/cm² and 10 Gy. Resulting mean fluorescence spectra from different individuals (scalp area) before and after the irradiation.

from hair exposed to 10 Gy radiation (b). The damage to the exposed hair is demonstrated by a significant decrease in fluorescence intensity.

These changes in fluorescence are associated with tryptophan that absorbs UV light at 290 nm and emits fluorescence at 340 nm, indicating a degradation of this amino acid. For both irradiated and non-irradiated hairs, the signal changed during spectra recording was noticed and the exposure of the samples to light was therefore minimized.

These results demonstrate that hair may serve as a robust indicator of radiation exposure and impact of the radiation on hair can be quantified with fluorescence spectroscopy. A calibration curve can be established to determine the degree of damage from various dose of radiation.

Figure 3 shows data points for hair samples exposed to fractionated dose of radiation up to 10 Gy at 2 Gy increments plotted to generate dose response curve.

The average changes in hairs fluorescence induced by gamma irradiation, obtained by pooling the fluorescence data of 3 hair samples, are presented as a function of radiation dose and the error bars represent standard deviations within the studied population.

To evaluate the equation of dose-response curves, the number of hair samples was examined at different doses, and dose-response curve of the induced MN was obtained by fitting the linear-quadratic model $y = a + bD + cD^2$, where y is the changes in fluorescence of hair, a is the spontaneous yield, b is the coefficient of the one-track component, c is the coefficient of the two-track component, and D is the dose in Gy. When plotting on a linear

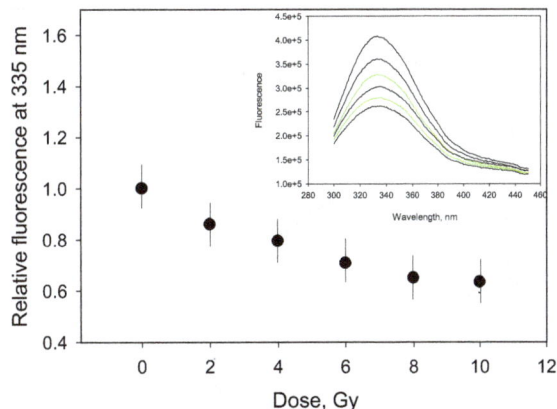

Figure 3. Dose dependence of hair fluorescence in response to gamma irradiation as measured at 335 nm. Insert show changes in fluorescence spectra of human hair exposed to different doses of gamma irradiation (0; 2; 4; 6; 8; 10 Gy).

scale against radiation dose, the line of the best fit was $Y = 0.004 + (1.882 \times 10^{-2} \pm 9.701 \times 10^{-5})$ D $+ (1.43 \times 10^{-3} \pm 1.571 \times 10^{5})$ D^2, (R^2 = 0.9996) after the gamma exposure. There was a significant relationship between the change in tryptophan fluorescence of hair and dose. The dose-response curves were linear-quadratic. These data show trends towards decreasing of hair's fluorescence with increasing dose.

3.2. Effect of UV-Radiation on Fluorescence Signal of Hair

Figure 4 shows changes in the fluorescence intensity of human hair after exposure to different doses of UV-radiation. Damage to hair fibers from various exposure time of UV radiation presented as dose response curve is indicated by decrease in fluorescence intensity with increasing radiation dose. After a non-linear, preliminary stage, hair shows a linear dose-response curve to UV-radiation in a range from about 10 J/cm^2 to 25 J/cm^2, although the signal is already detectable below 1 J/cm^2. The best fit of the entire dose response curve shows that it exhibits linear-quadratic behavior.

Repetitive recording of the fluorescence spectra of the hair progressively resulted in decrease spectra intensity providing evidence on accumulative degradation of Trp after each spectra recording.

In order to confirm differences between impact of UV- and gamma irradiation on structural and chemical properties of hair as obtained from fluorescence measurements, quantitative analysis of keratin's SH-groups was performed. **Figure 5** shows changes in the content of sulfohydryl group in keratin exposed to different dose of UV- and gamma radiation.

Exposure to radiation resulted in a progressive increase in the number of free SH-groups on UV irradiated hair in amount proportional to the decrease in fluorescence intensity. In contrast to UV-irradiation, the ionized radiation has little effect on the number of free SH-groups in hair. This is in agreement with the tryptophan fluorescence measurements, which reveals that hair responds differently to UV- and gamma radiation in terms of chemical modification of tryptophan. The UV-sensitivity of sulfur groups and tryptophan of hair keratin is higher than that for the gamma radiated hairs. Although our results are consistent with the general understanding of the effect of the gamma ray and UV-light exposure on hair, we find that damaging impact of radiation do not correlated with considering amount of absorbed radiation energy. A gamma ray emitted by an atom of Cesium 137 has an energy of 0.662 million electron-volts or 10.592×10^{-14} J. (1 MeV is 1×10^{6} eV and 1 eV = 1.6022×10^{-19} J). The energy of light with a wavelength of 280 nm is 7.099×10^{-19} J as calculated using Equation E$_\lambda$ = hc. This is almost 5 orders of magnitude less than energy of gamma radiation. In the case of UV radiation of hair we have shown that changes in tryptophan fluorescence and number of SH-group are disproportionally larger in a case of UV radiation than an equal energy of gamma radiation. It might be caused by the presence of the sensitive targeting hot spots, which are less sensitive to the energy of radiation but more sensitive to the density of the radiation energy as it might be in the case of UV- and gamma-radiation. The effectiveness of the produced changes depends upon the structure of the polymer and the experimental conditions of irradiation such as energy and fluence.

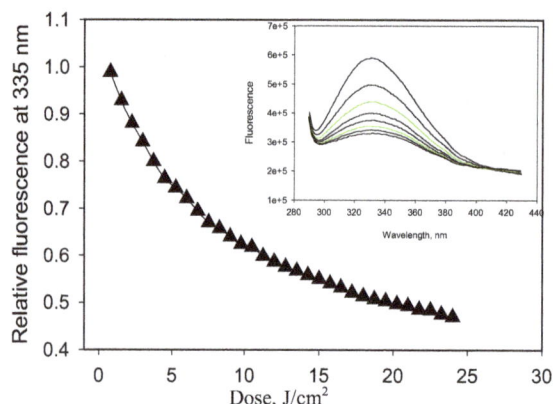

Figure 4. Dose dependence of hair fluorescence in response to UV-irradiation as measured at 335 nm. Insert show changes in fluorescence spectra of human hair exposed to different doses of UV- irradiation (0.225; 4.5; 6.75; 9.0; 11.28; 13.5 J/cm^2).

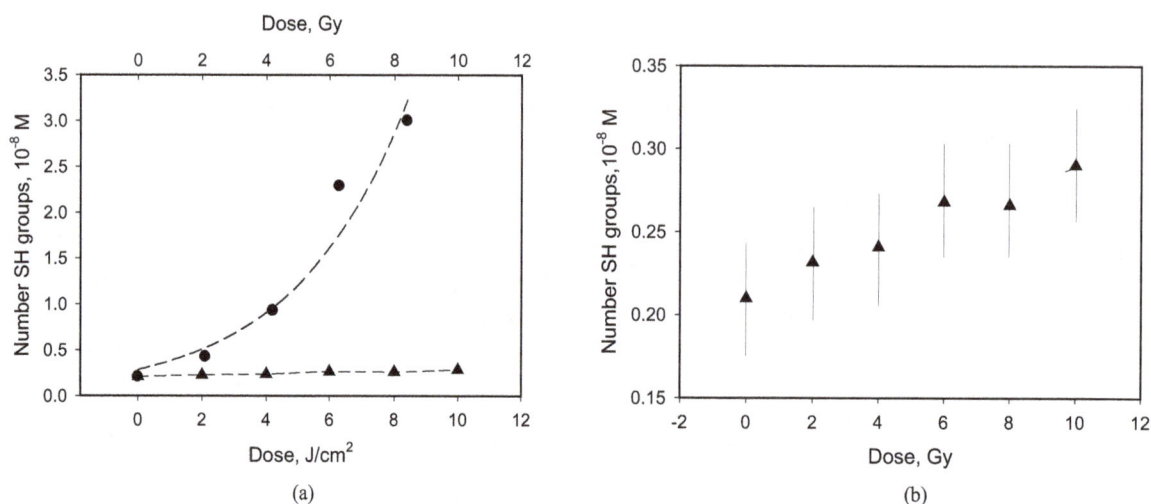

(a)

(b)

Figure 5. Changes in content of SH- groups of human hair exposed to different doses of gamma-irradiation (a). Comparison of changes in content of SH- groups induced by UV- and gamma irradiation (b).

Another difference between gamma radiation and UV is that UV radiation has predominant damaging effect on the wavelength at which material best absorbs while gamma radiation is non-specific. In the case of UV, the damage is done by photons absorbed directly by tryptophan. Photons absorbed by other components of the target do not contribute to the critical damage. In contrast, the damage by gamma for such specific target as tryptophan is less effective per amount of energy deposited but it can affect other targets in hair's keratin. In addition, up to 70% of damaging effect of ionizing radiation arrives from the secondary effect of the radiation, therefore, damage to hair's tryptophan is not mostly due to direct deposition of gamma energy, but to free radicals generated through ionization of water or other components of the target. Therefore the damaging effect of ionizing radiation is significantly reduced in dehydrated biological materials such as hair that has only between 10% - 30% of water compared to 70% of water in biological tissues [11].

The difference in character of the dose response curves for UV- and gamma radiation can be explained by not only the specific energy absorption by tryptophan but also by limited number of available targets for non-penetrated UV radiation (tryptophan) in hair keratin. Previous studies of the fluorescence characteristics of the hair suggested that the fluorescence from the tryptophane contributed significantly to the 340 nm band as a correlation was observed between the fluorescence intensity of that band and the number of tryptophanes in UV-damaged hair [12]. What makes the dose response curve steeper at the low UV-radiation doses, is the sufficient number of intact tryptophans per the targeting surface of hair. This number is progressively decreases with time (or dose) of radiation exposure and the dose response curve becomes less steeper at the high UV doses when no more intact tryptophans are available. In contrast to non-penetrated UV-radiation, penetrated gamma radiation targets not only tryptophans located on the surface of hair but also the tryptophans hidden inside, so "absorption per unit mass" is appropriate. When significantly higher number of tryptophans is damaged, this leads to depletion of the tryptophane fluorescence signal at higher doses of gamma radiation exposure. It can be predicted that at the sufficiently high doses of UV- and gamma radiation when all tryptophans will be damaged, dose response will become flat, and not responsive to detect further radiation-induced damage.

While the depletion of tryptophane fluorescence in response to ionized radiation closely resembles that of the UV irradiated samples, UV- and gamma radiation may damage hair through different mechanisms. It is reasonable to expect that changes in tryptophan fluorescence associated with UV radiation is due to its specific absorption in the UV light, while the radiation induced damage to the tryptophane residues in human hair due to the gamma irradiation is less specific.

In addition to the depletion of the amino acid tryptophan, we have consistently detected a progressive increase in the number of free SH-groups on UV irradiated hair in amounts proportional to the decrease in fluorescence intensity. The increasing amount of free sulfhydryl groups (-SH-) in keratin can originate from disruption of the disulfide cysteine bonds (-SS-) induced by radiation. The breaking of all of these disulfide bonds in cysteine, may results in the amount of cysteine equals 720 µmol/g of feathers [13]. Data presented in **Figure 5** show that after exposure to 10Gy gamma radiation a sulfohydril group content in hair keratin change insignificantly from 0.21 µmol/g in non-radiated hair to 0.29 µmol/g in irradiated hair, which indicates that only less than 10% of the disulfide bonds were broken. Therefore, according to these criteria the native keratin was not significantly modified. Exposure to the UV-radiation change sulfhydryl group content from 0.21 µmol/g in non-radiated hair to 3.0 µmol/g in hair exposed to 10 J/cm^2, which is substantially higher than in a case of gamma irradiation. Therefore, in contrast to UV irradiation, the ionized radiation has little or no effect on the number of free SH groups in hair. The disruption of the disulfide bridge nearby the tryptophan residue by UV light excitation of this aromatic residue has been reported in previous studies [14]. Tryptophan excitation energy disrupts a neighboring disulfide bridge, which in turn leads to altered structural integrity and stability. While the Trp residues are certain "hot spot" for the UV radiation, facilitating disruption of S-S bridges, this might be not the case for the ionized gamma radiation whose damage originates from the reaction of free radicals and direct deposition of energy. In fact, it has been demonstrated that disruption of S-S crosslinks in hair induced by radiation may occur not with SH groups as an end product, but rather through the oxidation leading to the formation of cysteic acid, $CySO_3H$. This may explain the lack of reduced SH-groups in the gamma-irradiated hair in contrast to the UV-irradiation.

The discrepancy between delivered dose and damaging effects of ionizing and non-ionizing radiations on hair may also be derived from the difficulties involved in relating UV doses to gamma ray. At the same time, there are several publications demonstrating that biological effect of UV-radiation in dose range 10 - 20 J/m^2 can be significantly more damaging than of 7 or 10 Gy gamma irradiation [15] [16].

3.3. Potential Application

Since we have observed a direct correlation between the total delivered dose of ionized radiation and specific features of the fluorescence spectra it would be desirable to identify features leading to application of this property for the potential design of a sensitive biological dosimeter. The ability of a human hair to predict radiation dose throughout a window of time represents an important advance in the development of non-invasive dosimetry. In a large-scale radiologic emergency, estimates of exposure doses and radiation injury would be required for individuals without physical dosimeters. Current methods are inadequate for the task, so we are developing simple fluorescence-based approach for radiation dosimetry. This approach could provide both an estimate of physical radiation dose and an indication of the extent of individual injury or future risk.

We demonstrated that a portable fluorometer based on Ocean Optics Inc. software and fiberoptics is capable of quantitatively evaluating the dose of radiation exposure in one hair fiber, and could be used for the rapid and non-invasive assessment of radiation dose, *i.e.* bio-dosimeter. We anticipate doing further investigation to determine if tryptophan can also serve as a sensitive marker for the ionizing radiation to measure the total dose of radiation exposure using fluorescence spectroscopy.

We have repeated some of the florescence measurements on human hair utilizing both bench-top FluoroMax-2 fluorescence spectrophotometer and miniaturized optical unit produced by Ocean Optics (**Figure 6**).

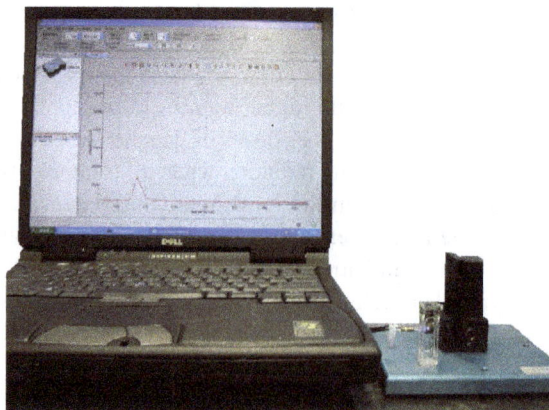

Figure 6. Ocean Optics Inc. Fiberoptics Prototype for portable device estimating dose exposure of hair using fluorescence marker. The picture represents an actual miniaturized fluorometer attached to a Dell laptop. In order to take a radiation exposure dose reading, a human hair could be fixed in to a cuvette after which the cuvette would be inserted in to the fluorometer where it would be read in live time.

Results on estimation of absorbed dose were consistent with those obtained using bench-top fluorometer demonstrating that portable miniaturized version of fluorometer can be used for assessment of radiation dose received by astronauts during space mission without using a sophisticated radiation dosimeter.

Our use of human hair for evaluating the biological effects of radiation could be helpful in predicting the extent of DNA damages in different scenarios, namely after an occupational, environmental, accidental or medical exposure. However, it may depend on gender, age and life style. Our particular interest is the follow-up study of astronauts during long-term spaceflight mission on International Space Station (ISS). With this approach it would be possible to easily detect biological effects of doses in case of unexpected solar and galactic radiation events. Such capability could be crucial in making life saving medical decisions for exposed astronauts and for determining other health consequences.

Future efforts in the area of applied hair-based bio-dosimetry will be on developing customization to market available miniaturized fluorometers and selecting the proper instrument for hair measurement such as solid state fluorescence that may fill in the gaps of the current assay for spaceflight applications. Two particular parameters are important in selecting the proper instrument for this hair evaluation: a) data should be collected quickly, to be able to process a large number of samples; b) Instrument should be compact, fully automated, and simple for users to operate. It would be advantageous if the instrument can be equipped with a fiber-optic accessory, so human hair can be positioned and fixed for examination.

Apart from steady state fluorescence spectrometry used in this study other hair analyzing fluorescence techniques such as solid-state fluorescence should be evaluated and compared for consistency of signal in hair analyses.

Further studies are needed to better control the uncertainties involved. The fluorescence signal dependence with hair qualities such as color, thickness, growth rate and treatments should, for example, be addressed. Variation based on preparation of the hair prior to analysis, *i.e.* the positioning procedure, also needs to be investigated further. To our knowledge, there have been no previous studies on the fluorescence measurements of hair following gamma radiation exposure. Hence, the dynamics behind dose of exposure and subsequent fluorescence signal are, to a large extent, unknown.

Acknowledgements

The research reported here was funded by the NASA Human Research Program (HRP), San Jose State University Bridges Program, and the NASA contract to the Carnegie Mellon University Silicon Valley # NNX08- AB13A. We would like to thank Dr. Chad Paavola for using Fluromax-2 machine. Especially indebted to Rebecca Ruf, thank you for the support and collaboration.

References

[1] Wolfram, L.J. and Lindemann, M.K.O. (1971) Some Observations on the Hair Cuticle. *Journal of the Society of Cosmetic Chemistry*, **22**, 839-850.

[2] Dowling, L.M., Crewther, W.G. and Parry, D.A.D. (1986) Secondary Structure of Component 8c-1 of α-Keratin. *Biochemical Journal,* **236**, 705-712. http://dx.doi.org/10.1042/bj2360705

[3] Signori, V. (2004) Review of the Current Understanding of the Effect of Ultraviolet and Visible Radiation on Hair Structure and Options for Photoprotection. *International Journal of Cosmetic Science*, **26**, 217-219. http://dx.doi.org/10.1111/j.0142-5463.2004.00223_7.x

[4] Nogueira, A.C.S., Dicelio, L.E. and Joekes, I. (2006) About Photo-Damage of Human Hair. *Photochemical & Photobiological Sciences*, **5**, 165-169. http://dx.doi.org/10.1039/B504574F

[5] Robbins, C.R. (2012) Chemical and Physical Behavior of Human Hair. 5th Edition, Springer, Berlin, 74. http://www.beck-shop.de/fachbuch/leseprobe/9783642256103_Excerpt_001.pdf http://dx.doi.org/10.1007/978-3-642-25611-0

[6] Potten, C.S., Burt, P.A., Roberts, S.A., Deshpande, N.A., Williams, P.C. and Ramsden, J. (1996) Changes in the Cellularity of the Cortex of Human Hairs as an Indicator of Radiation Exposure. *Radiation and Environmental Biophysics*, **35**, 121-125. http://dx.doi.org/10.1007/BF02434035

[7] Chandrashekara, M.N. and Ranganathaiah, C. (2010) Chemical and Photochemical Degradation of Human Hair: A Free-Volume Microprobe Study. *Journal of Photochemistry and Photobiology B: Biology*, **101**, 286-294. http://dx.doi.org/10.1016/j.jphotobiol.2010.07.014

[8] Kollias, N., Gillies, R., Moran, M., Kochevar, I.E. and Anderson, R.R. (1998) Endogenous Skin Fluorescence Includes Bands That May Serve as Quantitative Markers of Aging and Photoaging. *The Journal of Investigative Dermatology*, **111**, 776-780. http://dx.doi.org/10.1046/j.1523-1747.1998.00377.x

[9] Morenkova, S.A. and Nagler, L.G. (2005) Fluorometric Method for Determination of Keratin SH-Groups in Human Epidermis. *Biomedical Chemistry (Russian)*, **51**, 220-223.

[10] Karush, F., Klinman, N.R. and Marks, R. (1964) An Assay Method for Disulfide Groups by Fluorescence Quenching. *Analytical Biochemistry*, **9**, 100-114. http://dx.doi.org/10.1016/0003-2697(64)90088-0

[11] Elsner, P., Berardesca, E. and Maibach, H.I. (1994) Bioengineering of the Skin: Water and the Stratum Corneum. CRC Press, Boca Raton

[12] Longo, V.M., Pinheiro, A.C., Sambrano, J.R., Angell, J.A.M., Longo, E. and Varela, J.A. (2013) Towards an Insight on Photodamage in Hair Fibre by UV-Light: An Experimental and Theoretical Study. *International Journal of Cosmetic Science*, **35**, 539-545. http://dx.doi.org/10.1111/ics.12054

[13] Schrooyen, P.M., Dijkstra, P.J., Oberthür, R.C., Bantjes, A. and Feijen, J. (2000) Partially Carboxymethylated Feather Keratins. 1. Properties in Aqueous Systems. *Journal of Agricultural and Food Chemistry*, **48**, 4326-4334. http://dx.doi.org/10.1021/jf9913155

[14] Neves-Petersen, M.T., Gryczynski, Z., Lakowicz, J., Fojan, P., Pedersen, S., Petersen, E. and Petersen, S.B. (2002) High Probability of Disrupting a Disulphide Bridge Mediated by an Endogenous Excited Tryptophan Residue. *Protein Science*, **11**, 588-600. http://dx.doi.org/10.1110/ps.06002

[15] Zeng, X., Keller, D., Wu, L. and Lu, H. (2000) UV but Not Gamma Irradiation Accelerates p53-Induced Apoptosis of Teratocarcinoma Cells by Repressing MDM2 Transcription. *Cancer Research*, **60**, 6184-6188.

[16] Deacon, D.H., Hogan, K.T., Swanson, E.M., Chianese-Bullock, K.A., Denlinger, C.E, Czarkowski, A.R., Schrecengost, R.C., Patterson, J.W., Teague, M.W. and Slingluff, C.L. (2008) The Use of Gamma-Irradiation and Ultraviolet-Irradiation in the Preparation of Human Melanoma Cells for Use in Autologous Whole-Cell Vaccines. *BMC Cancer*, **8**, 360. http://dx.doi.org/10.1186/1471-2407-8-360

26

Characterization of Islets from Chronic Calcific Pancreatitis Patients of Tropical Region with Distinct Phenotype

P. Pavan Kumar[1], M. Sasikala[1*], K. Mamatha[1], G. V. Rao[2], R. Pradeep[2], R. Talukdar[1], D. Nageshwar Reddy[1,2]

[1]Asian Healthcare Foundation, Hyderabad, India
[2]Asian Institute of Gastroenterology, Hyderabad, India
Email: *aigres.mit@gmail.com

Abstract

Background and Objective: Islet autotransplantation is performed to preserve endocrine function in patients undergoing pancreatic resections for painful chronic pancreatitis. We characterized islets isolated from chronic pancreatitis patients (CP) of tropical region. Patients and Methods: Pancreatic tissues were obtained from CP patients with and without diabetes undergoing pancreatic resections (n = 35) and brain-dead multi organ donors (n = 6; considered as controls). Islets isolated were assessed for yield, purity, viability and in vitro islet function (Glucose stimulated insulin release, GSIR) as per standard protocols. Results: Islets from CP patients without diabetes were similar to controls in yield (control 4120 - 6100 IE/g, CP 3550 - 5660 IE/g), purity (control 78% ± 12%, CP 70% ± 8.2%) and viability (control 85% ± 8%, CP 81% ± 10%) and islets from CP patients with diabetes showed decreases in yield (3002 - 2300 IE/g), purity (61% ± 16%) and viability (62% ± 21%). Islets measuring 50 - 200 μ were similar in abundance in controls (94.74% ± 3.2%) and CP patients with and without diabetes, 86.31% ± 4.9%, 91.03% ± 3.8%. GSIR of islets from CP patients and controls were similar at 5.5 mM glucose (2.8 - 3.1 μU/ml). However, GSIR at 16.5 mM glucose was decreased in CP patients (control 18.5 ± 0.6, CP without diabetes 11.8 ± 0.3, CP with diabetes 4.3 ± 0.3 μU/ml). Conclusion: Our results demonstrate suitability of islets isolated from CP patients of tropical region for autotransplantation.

Keywords

Islets, Tropical Chronic Pancreatitis, Islet Functions, Transplantation

---

*Corresponding author.

1. Introduction

Islet autotransplantation (IAT) is an accepted treatment option for chronic pancreatitis (CP) patients undergoing pancreatic resections [1]-[12]. Total pancreatectomy followed by islet autotransplantation (TPIAT) is performed in many western countries to relieve intractable pain, preserve endocrine function and improve quality of life in patients with chronic pancreatitis [13], with autografts being more durable than allografts [14]. About one third of the adult patients with CP receiving IAT are insulin independent, with another one third being partially independent and requiring minimal dose of insulin. CP patients who did not achieve insulin independence after IAT had partial graft function and did not experience morbidity due to brittle diabetes [1]. A recent meta analysis of TPIAT by Wu *et al.* has indicated that islet autotransplantation is safe for CP patients requiring pancreatic resections for pain relief [15]. However, several issues including timing of TPIAT, identification of patients who may benefit from TPIAT, predictable islet isolation outcomes, etc., need to be addressed to improve long term functions of transplanted islets [1] [15]. Despite such limitations, improvements in isolation techniques to increase islet yield for better clinical outcomes have been reported [16].

Prevalence of chronic pancreatitis in South East Asian countries, more so in South India, is comparatively higher (114 - 200/100,000 populations in southern India) than in western countries (10 - 15/100,000) [17]. Presence of large dense pancreatic calculi, ductal dilatation and early onset endocrine dysfunction in nonalcoholic younger generation renders CP a distinct phenotype in India [18]. Etiology of CP in western countries is mainly due to alcohol consumption, while ≈ 60% - 70% of CP in India and China are reported to be idiopathic in nature [17]. Though the etiopathogenesis of CP is still not clear, genetic, nutritional and inflammatory factors have been implicated in the disease that leads to abdominal pain, maldigestion and diabetes. Importantly, diabetes in these CP patients is ketosis-resistant, often brittle and difficult to attain normoglycemia with conventional treatment requiring multiple doses of insulin [19].

In pursuit of initiating islet autotransplantation to benefit CP patients from tropical region affected with severe form of the disease at young age, we isolated islets from resected pancreatic tissues obtained from these patients as well as from pancreata obtained from brain dead multi organ donors (MODs; considered as controls) and assessed the suitability for autotransplantation. Our results demonstrate presence of viable islet mass secreting insulin in response to glucose challenge in CP patients with and without diabetes.

2. Materials and Methods

2.1. Pancreas Procurement and Preservation

The protocols were approved by the institutional review board and all the patients had given informed consent. Patients with documented CP, as evidenced by pancreatic calcifications on CT scan, ductal or parenchymal abnormalities on secretin stimulated magnetic resonance cholangiopancreatography (MRCP) and histopathological confirmation, and undergoing either lateral pancreaticojejunostomy or distal pancreatectomy or Whipples procedure as a surgical treatment modality were included (n = 310). Severity of CP was evaluated on parenchymal assessment by endoscopic ultrasonography (EUS) and ductal morphology by endoscopic retrograde cholangiopancreatography (ERCP). Based on Cambridge classification [20] and histopathological grading of fibrosis [21] [22] patients were categorized either as patients with mild or with severe CP. Based on fasting blood glucose levels CP patients were further categorized into those with and without diabetes. Tissue resected from tail region (10 - 20 g) of pancreata (n = 35) upon distal pancreatectomy was used for islet isolation and characterization. CP patients below 18 years of age and those with acute exacerbation of chronic pancreatitis, pancreatic cancer and inability to give informed consent were excluded.

Pancreata (n = 6) were obtained from brain dead MODs with the consent of the relatives and permission from Jeevan Daan committee located in Hyderabad. Pancreata were retrieved as per standard protocol by the organ retrieval committee while retrieving other organs. Immediately after pancreatic resection the specimen was immersed in cold University of Wisconsin solution and sent to islet isolation facility. Pancreata from brain dead donors were used for standardizing islet isolation for clinical transplantation and comparing islet functions with that obtained from CP pancreata before initiating islet transplantation in CP patients.

2.2. Islet Isolation

Islets were isolated from the pancreatic tissue (10 g from tail region) of brain dead MODs as well as from pancreatic specimens resected (10 g, distal pancreatectomy specimens) from CP patients. All the procedures were conducted within the islet isolation facility, established as per cGMP requirements. The pancreatic specimens were cleaned of surrounding fat, muscle, connective tissues and washed thrice with RPMI 1640 nutrient medium.

Islets were isolated from MOD pancreas as per the semi-automated Ricordi isolation method [23]. The main pancreatic duct was cannulated with a 24 gauge catheter and infused with collagenase V solution (2.5 mg/gm of pancreas, Sigma/Roche, USA) using a peristaltic pump under recommended conditions of pressure (60 - 180 mmHg), flow rate (>30 ml/min), temperature (6°C - 16°C) and time (~15 minutes). For isolating islets from pancreatic tissue obtained from CP patients, visible ducts in the resected tissue were located and infused with collagenase V solution (2 - 3 mg/gm of resected pancreas) manually using a needle and 20 cc syringe over a period of 15 minutes. For samples with extensive parenchymal fibrosis and calcifications, interstitial perfusion was performed manually by injecting the enzyme solution in to pancreatic tissue. The tissue was then chopped into small pieces and digested with collagenase V (20 - 25 mg collagenase/10gms pancreas) at 37°C for 10 - 30 minutes in a Ricordi chamber; longer periods of incubation were resorted for digestion of severely fibrosed tissue. Progression of digestion was continuously monitored by microscopic examination of aliquots for release of islets by staining with dithizone. When islets were free from acinar tissue, digestion was arrested by dilution and cooling of the enzyme solution with large volumes (4 - 5 L) of Hank's solution at 4°C. The resultant, dispersed pancreatic islets were centrifuged using opti prep based density gradient centrifugation for purification [24]. Isolated islets were assessed for viability and functional efficacy as described below apart from evaluating the sample for its sterility.

2.3. Histological Examination of Isolated Islets

Morphological and histological characters of islets within pancreatic tissue from tail region were assessed (before subjecting the tissue for islet isolation) by staining with hematoxylin-eosin, Masson's trichrome and immnuostaining employing standard protocols with minor modifications [25] [26]. In addition, morphological integrity and endocrine cell composition of isolated islets was determined after culturing islets on a cover slip and immnuostaining with guinea pig anti- insulin, mouse anti-glucagon, rabbit anti-somatostatin and rabbit anti-pancreatic polypeptide antibodies and fluorescent tagged secondary antibodies (goat anti guinea pig alexa 488 for insulin, anti-mousealexa 546 for glucagon, anti-rabbitalexa 546 for somatostatin and pancreatic polypeptide; Santacruz Biotechnologies, USA). All fluorescent images were captured using a CARV II bioimager (BD Bio Sciences, USA).

Islet yield was estimated after isolation by counting dithizone (DTZ)-stained islets in 10 μl of homogenous suspension using Olympus microscope (CX41, Tokyo, Japan). Briefly, islets were stained with 10 μl 0.1% DTZ solution made in Krebs Ringer bicarbonate buffer (pH 7.4) containing 10mM HEPES and incubated at 37°C for 10 to 15 min. The stained islets were counted under an inverted microscope (Olympus CX 41; Olympus Corporation, Tokyo, Japan) immediately after isolation and expressed as islet equivalents (IEQ/g). Islet purity was obtained upon counting the relative number of DTZ-stained cells among all the pancreatic cells and expressed as percentage [24].

Islet viability was determined in all the samples using calcein AM, propidium iodide double fluorescence membrane integrity assay to assess the amount of live (green colored) versus dead (red colored) cells [27]. Islets were initially incubated with 0.5 mM calcein-AM at 37°C for 45 min following by incubation with 10 μl PI (Propidium Iodide) for 15 min at room temperature in the dark. The stained cells were then washed with PBS and visualized under fluorescent microscope (Olympus CX 41; Olympus Corporation, Tokyo, Japan) attached with digital ProgResC5 cool camera and Capture Pro software (version 2.5).

In **Figure 3** the islet size was measured on pancreatic tissue histology. Five blocks per pancreas were prepared from the tail region of each pancreatic specimen. 20 serial sections of 5 μ thickness were sliced from each block. Sections 50 μ apart were used for counting islets. Islet size was measured by capturing islet images in five regions in each section using 4x objective to count islet numbers. One hundred islets were marked for measuring the islet size using 20x objective. Islet diameter in μm was measured by drawing a line across the islet as mentioned earlier Balamurugan *et al.*, using Olympus microscope employing ProgResC5 cool camera and CapturePro software (version 2.5). Care was taken to avoid measuring same islet. Islets of different sizes (50 - 150 μm, 150 - 200 μm and >200 μm) were counted separately and represented in percentages (mean ± SD).

2.4. Functional Assessment of Isolated Islets; Glucose Stimulated Insulin Secretion

Islet equivalents measuring approximately 150 μm in diameter (n = 50) from controls and CP patients were subjected to static insulin secretory functions on exposure to basal and high glucose concentrations as described earlier [28] following the method by Bottino *et al.* [29]. Insulin secreted into the medium by isolated islets in response to basal (5.5 mmol/L) and high (16.5 mmol/L) glucose stimulation, was measured in triplicates by enzyme-linked immunosorbent assay (Mercodia, Uppsala, Sweden) and reported in terms of micro units of insulin released per minute/islet. Insulin content and islet DNA content were measured once after isolation as recommended [29].

3. Statistical Analysis

The data are expressed as mean ± SD and were analyzed by Analysis of variance (ANOVA). Pair wise comparison was made by Scheffe's test to compare between the groups using SPSS, 20th version. $P < 0.05$ was considered significant.

4. Results

4.1. Patient Characteristics

A total of 41 subjects (35 CP patients and 6 MODs) were included in the study to characterize islet functions. All the CP patients (22 males, age 34.9 ± 13) showed intra ductal calcifications, dilated ducts and pancreatic atrophy on radiological examination and Inter lobular acinar atrophy, fibro collagenous tissue with intact islets on histopathological examination. Pancreatic specimens obtained from MODs (5 males, age 25 ± 23) during 2012-2013 were considered as controls in this study. Brain death in the MODs was mainly due to road accidents except for one donor with cardio respiratory failure (female, 30 years). Patients with chronic pancreatitis undergoing pancreatic resections during 2009-2013 (n = 310) were considered for this study and 248 of them were diagnosed to have CP with involvement of the main pancreatic duct. Pancreatic tissue from 192 patients undergoing pancreaticojejunostomy and 21 patients undergoing other surgeries were not used for islet isolation. Pancreatic tissue (10 - 20 g, from tail region of the pancreas) was obtained from CP patients undergoing distal pancreatectomy (n = 35, **Figure 1**). Among these, patients with focal lesions on CT scan, parenchymal heterogeneity observed on EUS and intralocular fibrosis were considered to be patients with mild CP (n = 12) and all of these patients were non diabetic. Patients with increased ductal wall echoes on CT scan, dilatation, calcifications and calculi in the main pancreatic duct as well as inter-acinar fibrosis were considered to be having severe pancreatitis (n = 23). Out of the 35 CP patients included for characterization of islets, 12 patients who had fasting blood glucose 114 ± 16 mg/dl, decreased C- peptide 0.91 ± 0.52 ng/ml and HbA1C 8.63% ± 1.87% were diagnosed to have severe CP with diabetes (type3C); the remaining 23 patients with fasting blood glucose 81 ± 17 mg/dl, C-peptide 1.6 ± 0.95 ng/ml and HbA1C 6.11% ± 0.95% levels were considered to be CP patients without

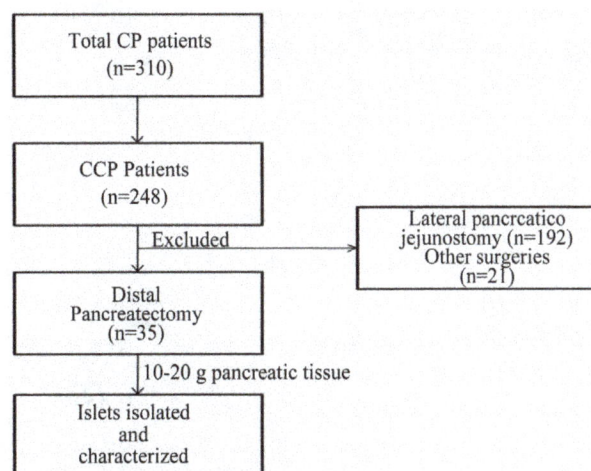

Figure 1. Patient disposition.

diabetes (mild CP without diabetes 12, severe CP without diabetes 11). Seven out of 12 CP patients with diabetes were treated with insulin and five were treated with oral hypoglycemics.

4.2. Islet Yield

Islet yield in CP patients without diabetes was not significantly different (Islets/g 4655 ± 550; IEQ/g 5786 ± 98) in comparison to controls (Islets/g 5186 ± 885; IEQ/g 5653 ± 211). However, the yield was significantly lower (Islets/g 2348 ± 385; IEQ/g 2464 ± 52) in CP patients with diabetes (P < 0.01). Purification by density gradient centrifugation resulted inislets with 78% ± 12% purity in controls, 70% ± 8.2% in CP patients without diabetes and 61% ± 16.5% in CP patients with diabetes (**Table 1**).

4.3. Islet Size

Measurement of islet size from controls as well as CP patients revealed small (50 - 200 μm) and large (>200 μm) categories of islets. Islets measuring below 50 μm were not considered. Islets measuring 50-200 μm were similar in numbers and represented 94.74% ± 3.2% in controls, 86.31% ± 4.9% in CP patients with diabetes and 91.03% ± 3.8% in CP patients without diabetes. However, islets measuring >200 μm were present in higher numbers in CP patients with diabetes (7.0% ± 2.4%) and in CP patients without diabetes (9.08% ± 3.1%) in comparison to controls (5.16% ± 1.40%) as shown in **Table 1**.

4.4. Histological Examination and Islet Viability

Histological examination and immunostaining of isolated islets showed the presence of α (glucagon+), β (insulin+), δ (somatostatin+) and pancreatic polypeptide (PP) secreting cells in the islets isolated from control as well as CP patients with and without diabetes. (**Figure 2** panel B). A significant decrease in viability of islets was observed in CP patients with diabetes (62% ± 21%; P > 0.05) as compared to those without diabetes (81% ± 10%) and controls (85.5% ± 8.9%) as shown in **Figure 3**.

4.5. Islet Functions

Similar to our earlier report on GSIR of islets in CP, Glucose stimulated insulin secretion by the islets from controls, CP patients with and without diabetes was similar at basal glucose concentration (2.8 - 3.1 μU/min/50islet). In comparison to controls (18.5 ± 0.6 μU/min/50islet), insulin secretion in response to high glucose in culture

Table 1. Characteristics of islets isolated from control and CP patients.

Parameters	Controls (n = 6)	CP Patients (n = 35)	
		Without Diabetes (n = 23)	With Diabetes (n = 12)
Islet yield (IEQ/g)	5653 ± 211	5786 ± 98	*2464 ± 52
Islet size (% of islets in each category)			
50 - 100 μ	21.30 ± 0.70	16.68 ± 1.45	16.54 ± 1.10
100 - 150 μ	36.24 ± 1.32	32.63 ± 3.32	47.36 ± 1.56
150 - 200 μ	37.20 ± 2.83	40.00 ± 2.65	27.13 ± 2.30
>200 μ	5.16 ± 1.40	9.08 ± 3.10	7.00 ± 2.40
Purity %	78 ± 12	70 ± 8.2	61 ± 16.5
Viability%	85.5% ± 8.9	81 ± 10	*62 ± 21
GSIR μU/ml	18.5 ± 0.6	11.8 ± 0.3	*4.3 ± 0.3
SI	6.1 ± 1.4	5.6 ± 1.0	*1.8 ± 1.1

*P < 0.05.

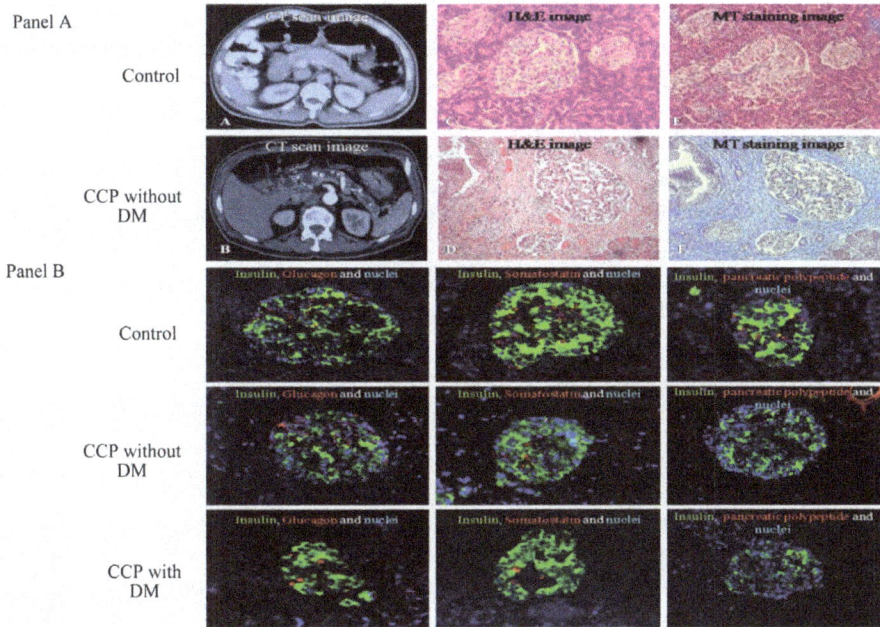

Figure 2. CT scan images and Histological features of islets in control and CP patients. Panel A: Images of MOD (control) with brain injury and calcific chronic pancreatitis patients. (A) CT scan image of control (B) CT scan image of CP patient with calcifications in the main pancreatic duct. Hematoxylin and Eosin (H&E) staining of tissue sections showing islets (C) acinar tissue and islets in controls (D) islets and peri islet fibrosis in patients with CP. Masson's trichrome (MT) staining in control and CP patient, amidst profound fibrosis (E) & (F). Panel B: Immunofluorescent staining of islets from control (upper lane) and CP patients without diabetes (middle lane) and CP patient with diabetes (basaler lane) depicting insulin (green) and glucagon (red), insulin (green) and somatostatin (red) and insulin (green) pancreatic polypeptide (red) secreting cells.

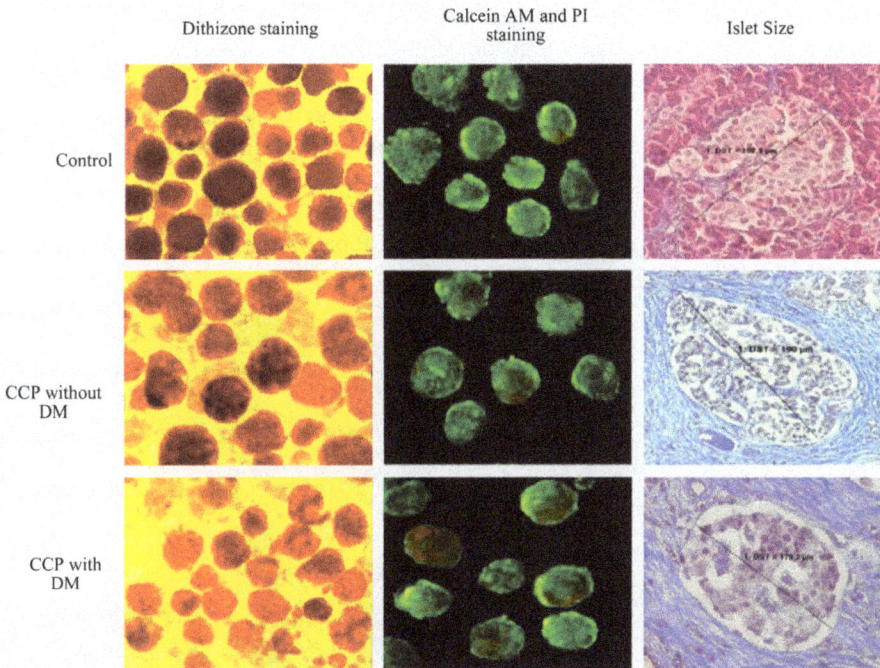

Figure 3. Depicts islets isolated from controls and CP patients with and without diabetes. Islets were stained with dithizone for purity (left column), Calcein AM and propidium iodide for viability (middle column); Green indicates viable mass and red indicates nonviable mass. Viable islets are seen in controls and CP patients although viability is decreased in CP patients with diabetes. Islet size in the three groups is depicted in the last column.

medium was decreased by islets from CP patients without diabetes (11.8 ± 0.3 in mild and 8.4 ± 0.5 μU/min/50 islet in severe CP). Islets from CP patients with diabetes showed very minimal response to high glucose challenge (4.3 ± 0.3 μU/min/50islet) as shown in **Figure 4**.

GSIR of islets isolated from control and CP patients with DM and CP patients without DM shows a progressive decrease in GSIR of islets from control to mild, mild to severe and control to CP with diabetes and CP without diabetes to CP with diabetes. $P < 0.001$

Insulin content for smaller sized islets (50 - 150 μm) was 420 ± 130 μU/islet in CP and 460 ± 158 μU/islet in controls with the DNA contents being 19 ± 8 ng/islet in controls and 17 ± 9 ng/islet in CP. Stimulation index (SI), as expressed by dividing basal insulin over high glucose insulin release was 6.1 ± 2.3 in controls, 5.3 ± 1.9 in CP without diabetes and 1.8 ± 1.1 in CP with diabetes.

5. Discussion

Earlier studies have shown the beneficial effects of islet autotransplantation in chronic pancreatitis patients who undergo total/near total/partial pancreatectomy for pain relief. Such attempts are not made in India, wherein the phenotype of CP is distinct and patients present with large ductal stones, severe fibrosis and early onset of diabetes. This study was conducted primarily to characterize the islets isolated from CP patients with and without diabetes. We assessed islet viability, evaluated function of islets, compared with islets isolated from MOD sand demonstrate suitability of islets isolated from CP patients for autotransplantation.

It is well established that glycemic control in islet graft recipients depends upon the islet mass and quality of islets infused. Islet yield obtained in CP patients without diabetes was not significantly different from that of control (**Table 1**), while the yield was significantly decreased in CP patients with severe pancreatitis. Though the number of islets isolated in CP patients with diabetes was significantly decreased, islets could be isolated even in CP patients with diabetes, which is akin to the observations of a recently published data [30]. In addition, Takita *et al*. have demonstrated that glycemic control was excellent in both mild and advanced chronic pancreatitis patients who received total pancreatectomy with islet autotransplantation even when the yield was decreased [31]. Researchers at University of Minnesota demonstrated that of the CP patients receiving 2500 - 5000 IEQ/kg, nearly one fifth were insulin independent and one eighth of patients among those receiving <2500 IEQ/kg achieved insulin independence at the end of 3 years [1]. The majority of CP patients receiving IAT after pancreatectomy, were shown to have islet graft function (C-peptide positive) and hemoglobin A1C levels in the range of <7% [1]. Therefore, islet yield obtained from CP patients with and without diabetes seems to be adequate for use in IAT. However, due caution is to be exercised during islet isolation to clear the calcifications before enzyme infusion into the duct, since the calcifications might cause mechanical injury and loss of function of islets.

The majority of islets in CP patients were between 50 - 200 μm in size; 94.74% ± 3.2% in controls, 86.31% ± 4.9% in CP patients with diabetes and 91.03% ± 3.8% in CP patients without diabetes, although the number of

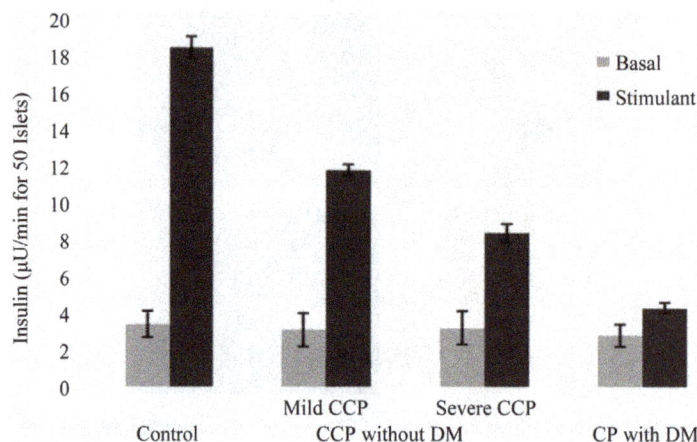

Figure 4. Glucose stimulated insulin secretory function of islets in control and CP patients.

islets measuring >200 μ were increased. A study by Suszynski *et al.* involving 58 CP patients established that those who receive marginal dose of islets were likely to achieve insulin independence if the graft transplanted had greater number of smaller islets [32]. Since smaller islets contribute 80% - 90% of the isolate in CP patients, function of the graft could likely be efficient.

Presence of viable islet mass in CP patients with and without diabetes (62% ± 21% and 81% ± 10% respectively) indicated that the islets if transplanted could help stabilize glycemic control. These findings are similar to a proof of concept study which demonstrated that significant islet mass could be isolated from CP patients with c peptide positive diabetes. Even if glycemic control cannot be completely achieved, these studies have demonstrated that C peptide secreting viable islet mass could at least avoid microvascular complications of diabetes [30]. Islet cluster integrity as evidenced by the presence of α (glucagon+), β (insulin+), δ (somatostatin+) and polypeptide+ cells indicate that the tight coordination of insulin gene expression can be maintained in CP patients.

Glucose stimulated insulin release function of islets from control and CP patients at basal glucose concentration indicate presence of viable mass secreting insulin at basal level. However, decreased response by islets from CP patients without diabetes and minimal response by islets from CP patients with diabetes at high glucose concentration indicate β cell dysfunction. This result is in accordance with our earlier study and could be due to long standing exposure to the inflammatory microenvironment that is prevalent in chronic pancreatitis. HbA1C levels of 6.11% ± 0.95% in CP patients without diabetes indicate existence of glucose intolerance. Decreased *in vitro* GSIR (~50% - 60% of control; **Figure 4**) by islets from these patients could result in glucose intolerance. This result corroborates with our earlier finding demonstrating β cell dysfunction even in non-diabetic CP patients [33]. Gradual decrease in GSIR by islets from controls to CP without diabetes (mild, severe) and to CP with diabetes clearly demonstrates a progressive dysfunction/damage of the islets as the disease progresses. Presence of viable islet mass with a progressive decrease in response to high glucose demonstrates the effect of inflammation on the function of islets. Therefore moving the viable islets from chronic inflammatory milieu to a non inflamed environment before the islets are damaged might improve the high glucose response of the islets from CP patients with and without diabetes. Takita *et al.* and others have demonstrated that CP patients with preexisting diabetes receiving IAT also maintained good glycemic control [31]. This is probably due to shifting the viable islets from inflamed pancreas to non inflamed environment in the liver. To date there are no data demonstrating permanent damage to the islets in CP. Many investigators also demonstrated that C peptide positive viable β cell mass maintain functions of transplanted islets to the extent that microvascular complications caused by diabetes can be avoided [34]. This study describes the characteristic of islets isolated from CP patients and further studies are focused on assessing the glucose homeostasis in CCP patients after autologous islet transplantation.

6. Conclusion

In conclusion, our results demonstrate presence of significant viable and functional islet mass in CP patients with and without diabetes and their suitability for autotransplantation.

Acknowledgements

The authors are thankful to all the multi organ donor relatives and Jeevan Dhaan committee for consenting to donate pancreas.

References

[1]   Sutherland, D.E., Radosevich, D.M., Bellin, M.D., Hering, B.J., Beilman, G.J., Dunn, T.B., Chinnakotla, S., Vickers, S.M., Bland, B., Balamurugan, A.N., Freeman, M.L. and Pruett. T.L. (2012) Total Pancreatectomy and Islet Autotransplantation for Chronic Pancreatitis. *Journal of American College of Surgeons*, **214**, 409-426. http://dx.doi.org/10.1016/j.jamcollsurg.2011.12.040

[2]   Blondet, J.J., Carlson, A.M., Kobayashi, T., Jie, T., Bellin, M., Hering, B.J., Freeman, M.L., Beilman, G.J. and Sutherland, D.E. (2007) The Role of Total Pancreatectomy and Islet Autotransplantation for Chronic Pancreatitis. *Surgical Clinics of North America*, **87**, 1477-1501. http://dx.doi.org/10.1016/j.suc.2007.08.014

[3]   Bellin, M.D. and Sutherland. D.E. (2010) Pediatric Islet Autotransplantation: Indication, Technique, and Outcome. *Current Diabetes Reports*, **10**, 326-331. http://dx.doi.org/10.1007/s11892-010-0140-4

[4]   Clayton, H.A., Davies, J.E., Pollard, C.A., White, S.A., Musto, P.P. and Dennison, A.R. (2003) Pancreatectomy with Islet Autotransplantation for the Treatment of Severe Chronic Pancreatitis: The First 40 Patients at the Leicester General Hospital. *Transplantation*, **76**, 92-98. http://dx.doi.org/10.1097/01.TP.0000054618.03927.70

[5]   Garcea, G., Weaver, J., Phillips, J., Pollard, C.A., Ilouz, S.C., Webb, M.A., Berry, D.P. and Dennison, A.R. (2009) Total Pancreatectomy with and without Islet Cell Transplantation for Chronic Pancreatitis: A Series of 85 Consecutive Patients. *Pancreas*, **38**, 1-7. http://dx.doi.org/10.1097/MPA.0b013e3181825c00

[6]   Morgan, K., Owczarski, S.M., Borckardt, J., Madan, A., Nishimura, M. and Adams, D.B. (2012) Pain Control and Quality of Life after Pancreatectomy with Islet Autotransplantation for Chronic Pancreatitis. *Journal of Gastrointestinal Surgery*, **16**, 129-134. http://dx.doi.org/10.1007/s11605-011-1744-y

[7]   Cameron, J.L., Mehigan, D.G., Broe, P.J. and Zuidema, G.D. (1981) Distal Pancreatectomy and Islet Autotransplantation for Chronic Pancreatitis. *Annals of Surgery*, **193**, 312-317. http://dx.doi.org/10.1097/00000658-198103000-00010

[8]   Takita, M., Naziruddin, B., Matsumoto, S., Noguchi, H., Shimoda, M., Chujo, D., Itoh, T., Sugimoto, K., Onaca, N., Lamont, J.P., Lara, L.F. and Levy, M.F. (2010) Variables Associated with Islet Yield in Autologous Islet Cell Transplantation for Chronic Pancreatitis. *Proceedings (Baylor University. Medical Center)*, **23**, 115-120.

[9]   Argo, J.L., Contreras, J.L., Wesley, M.M. and Christein, J.D. (2008) Pancreatic Resection with Islet Cell Autotransplant for the Treatment of Severe Chronic Pancreatitis. *American Journal of Surgery*, **74**, 530-536.

[10] Berney, T., Rüdisühli, T., Oberholzer, J., Caulfield, A. and Morel, P. (2000) Long-Term Metabolic Results after Pancreatic Resection for Severe Chronic Pancreatitis. *American Board of Surgery*, **135**, 1106-1111.

[11] Dixon, J., DeLegge, M., Morgan, K.A. and Adams, D.B. (2008) Impact of Total Pancreatectomy with Islet Cell Transplant on Chronic Pancreatitis Management at a Disease-Based Center. *American Journal of Surgery*, **74**, 735-738.

[12] Ahmad, S.A., Lowy, A.M., Wray, C.J., D'Alessio, D., Choe, K.A., James, L.E., Gelrud, A., Matthews, J.B. and Rilo, H.L. (2005) Factors Associated with Insulin and Narcotic Independence after Islet Autotransplantation in Patients with Severe Chronic Pancreatitis. *Journal of the American College of Surgeons*, **201**, 680-607.

[13] Drewes, A.M. (2013) Understanding and Treatment of Chronic Pancreatitis. *World Journal of Gastroenterology*, **14**, 7219-7221. http://dx.doi.org/10.3748/wjg.v19.i42.7219

[14] Sutherland, D.E., Gruessner, A.C., Carlson, A.M., Blondet, J.J., Balamurugan, A.N., Reigstad, K.F., Beilman, G.J., Bellin, M.D. and Hering, B.J. (2008) Islet Autotransplant Outcomes after Total Pancreatectomy: A Contrast to Islet Allograft Outcomes. *Transplantation*, **27**, 1799-1802. http://dx.doi.org/10.1097/TP.0b013e31819143ec

[15] Wu, Q., Zhang, M., Qin, Y., Jiang, R., Chen, H., Xu, X., Yang, T., Jiang, K. and Miao, Y. (2015) Systematic Review and Meta-Analysis of Islet Autotransplantation after Total Pancreatectomy in Chronic Pancreatitis Patients. *Endocrine Journal*, **62**, 227-234. http://dx.doi.org/10.1507/endocrj.EJ14-0510

[16] Balamurugan, A.N., Bellin, M.D. and Papas, K. (2011) Maximizing Islet Yield from Pancreata with Chronic Pancreatitis for Use in Islet Auto-Transplantation Requires a Modified Strategy from Islet Allograft Preparations. *Pancreas*, **40**, 1312.

[17] Braganza, J.M., Lee, S.H., McCloy, R.F. and McMahon, M.J. (2011) Chronic Pancreatitis. *The Lancet*, **377**, 1184-1197. http://dx.doi.org/10.1016/S0140-6736(10)61852-1

[18] Balakrishnan, V. and Rajesh, G. (2012) Chronic Pancreatitis: A South India Perspective. *Medicine Update*, **22**, 445-449.

[19] Unnikrishnan, R. and Mohan, V. (2015) Fibrocalculous Pancreatic Diabetes (FCPD). *Acta Diabetologica*, **52**, 1-9. http://dx.doi.org/10.1007/s00592-014-0685-9

[20] Ahmed, S.A., Wray, C., Rilo, H.L., Choe, K.A., Gelrud, A., Howington, J.A., Lowy, A.M. and Matthews, J.B. (2006) Chronic Pancreatitis: Recent Advances and Ongoing Challenges. *Current Problems in Surgery*, **43**, 127-238. http://dx.doi.org/10.1067/j.cpsurg.2005.12.005

[21] Ammann, R.W., Heitz, P.U. and Klöppel, G. (1996) Course of Alcoholic Chronic Pancreatitis: A Prospective Clinicomorphological Long-Term Study. *Gastroenterology*, **111**, 224-231. http://dx.doi.org/10.1053/gast.1996.v111.pm8698203

[22] Chong, A.K., Hawes, R.H., Hoffman, B.J., Adams, D.B., Lewin, D.N. and Romagnuolo, J. (2007) Diagnostic Performance of EUS for Chronic Pancreatitis: A Comparison with Histopathology. *Gastrointestinal Endoscopy*, **65**, 808-814. http://dx.doi.org/10.1016/j.gie.2006.09.026

[23] Ricordi, C., Lacy, P.E. and Scharp, D.W. (1989) Automated Islet Isolation from Human Pancreas. *Diabetes*, **38**, 140-142. http://dx.doi.org/10.2337/diab.38.1.S140

[24] Latif, Z.A., Noel, J. and Alejandro, R. (1988) A Simple Method of Staining Fresh and Cultured Islets. *Transplantation*, **45**, 827-830. http://dx.doi.org/10.1097/00007890-198804000-00038

[25] Bancroft, J. and Stevens, A. (1982) Theory and Practice of Histological Techniques. 2nd Edition, Churchill Living-

stone, New York, 131-135.

[26] Balamurugan, A.N., Chang, Y., Fung, J.J., Trucco, M. and Bottino, R. (2003) Flexible Management of Enzymatic Digestion Improves Human Islet Isolation Outcome from Sub-Optimal Donor Pancreata. *American Journal of Transplantation*, **3**, 1135-1142. http://dx.doi.org/10.1046/j.1600-6143.2003.00184.x

[27] Lorenzo, A., Razzaboni, B., Weir, G.C. and Yankner, B.A. (1994) Pancreatic Islet Cell Toxicity of Amylin Associated with Type-2 Diabetes Mellitus. *Nature*, **21**, 756-760. http://dx.doi.org/10.1038/368756a0

[28] Mitnala, S., Pondugala, P.K., Guduru, V.R., Rabella, P., Thiyyari, J., Chivukula, S., Boddualli, S., Hardikar, A.A. and Reddy, D.N. (2010) Reduced Expression of PDX-1 Is Associated with Decreased Beta Cell Function in Chronic Pancreatitis. *Pancreas*, **39**, 856-862. http://dx.doi.org/10.1097/MPA.0b013e3181d6bc69

[29] Bottino, R., Balamurugan, A.N., Bertera, S., Pietropaolo, M., Trucco, M. and Piganelli, J.D. (2002) Preservation of Human Islet Cell Functional Mass by Anti-Oxidative Action of a Novel SOD Mimic Compound. *Diabetes*, **51**, 2561-2567. http://dx.doi.org/10.2337/diabetes.51.8.2561

[30] Bellin, M.D., Beilman, G.J., Dunn, T.B., Pruett, T.L., Chinnakotla, S., Wilhelm, J.J., Ngo, A., Radosevich, D.M., Freeman, M.L., Schwarzenberg, S.J., Balamurugan, A.N., Hering, B.J. and Sutherland, D.E. (2013) Islet Autotransplantation to Preserve Beta Cell Mass in Selected Patients with Chronic Pancreatitis and Diabetes Mellitus Undergoing Total Pancreatectomy. *Pancreas*, **42**, 317-321. http://dx.doi.org/10.1097/MPA.0b013e3182681182

[31] Takita, M., Naziruddin, B., Matsumoto, S., Noguchi, H., Shimoda, M., Chujo, D., Itoh, T., Sugimoto, K., Onaca, N., Lamont, J., Lara, L.F. and Levy, M.F. (2011) Implication of Pancreatic Image Findings in Total Pancreatectomy with Islet Autotransplantation for Chronic Pancreatitis. *Pancreas*, **40**, 103-108. http://dx.doi.org/10.1097/MPA.0b013e3181f749bc

[32] Suszynski, T.M., Wilhelm, J.J., Radosevich, D.M., Balamurugan, A.N., Sutherland, D.E., Beilman, G.J., Dunn, T.B., Chinnakotla, S., Pruett, T.L., Vickers, S.M., Hering, B.J., Papas, K.K. and Bellin, M.D. (2014) Islet Size Index as a Predictor of Outcomes in Clinical Islet Autotransplantation. *Transplantation*, **27**, 1286-1291. http://dx.doi.org/10.1097/01.TP.0000441873.35383.1e

[33] Sasikala, M., Talukdar, R., Pavan Kumar, P., Radhika, G., Rao, G.V., Pradeep, R., Subramanyam, C. and Nageshwar Reddy, D. (2012) *β*-Cell Dysfunction in Chronic Pancreatitis. *Digestive Diseases and Science*, **57**, 1764-1772. http://dx.doi.org/10.1007/s10620-012-2086-7

[34] Ali, M.A. and Dayan, C.M. (2009) The Importance of Residual Endogenous Beta Cell Preservation in Type 1 Diabetes. *British Journal of Diabetes and Vascular Diseases*, **9**, 248-253.

27

Hydrogen Sulfide in Proliferating and Differentiated Cells in Primary Cultures of Juvenile Brain of Masu Salmon *Oncorhynchus masou*

Evgeniya V. Pushchina[1], Sachin Shukla[2], Anatoly A. Varaksin[1]

[1]Zhirmunsky Institute of Marine Biology, Far East Branch, Russian Academy of Sciences, Vladivostok, Russia
[2]Prof. Brien Holden Eye Research Centre, L.V. Prasad Eye Institute, Hyderabad, India
Email: puschina@mail.ru

Abstract

Analysis of proliferative activity and the ability to neuron differentiation was performed in cultured cells of the brain and spinal cord of juvenile masu salmon *Oncorhynchus masou*. Proliferating cell nuclear antigen (PCNA) was used as a proliferative marker, while the markers of neuronal differentiation—a neuron protein HuCD, and a neuron-specific transcriptional factor with two DNA-binding sites Pax6—detected neurons. The results showed that cell proliferation occurred mainly in the suspension cell fraction. In monolayer, a few cells were only found to express PCNA. The results of morphological and immunohistochemical analysis allow us to conclude that proliferative activity in primary cultures from the *O. masou* brain is mainly connected with the suspension fraction of small cells. In contrast, a positive correlation between the cells expressing cystathionine β-synthase (CBS), a marker of H_2S synthesis, and the cells expressing PCNA in the monolayer, indicates the participation of H_2S in proliferative activity of neurons in primary cultures. The data obtained suggest that the hydrogen sulphide is also involved in the process of differentiation.

Keywords

Cell Culture, Fish, PCNA, HuCD, Hydrogen Sulfide, Pax6, Regeneration, Adult Neurogenesis

1. Introduction

After injury of the central nervous system, in particular of the spinal cord, the locomotor activity of the fish can effectively recover [1]. It concerns the ability of the central projection neurons to regenerate damaged axons, the emergence of new cells in the zone of injury, and the occurrence of high proliferative activity in nearby neurogenic niches and proliferative areas of the brain [2]. Both of these processes, axon regeneration and neurogenesis, contribute to anatomical and functional recovery in the injured CNS of adult fish. Its intrinsic growth and repair capacity, combined with its experimental amenability, present the salmon fishes as a good model for investigations of CNS restoration [3]. However, mechanisms such as high reparative activity in the nervous tissues of fish, including both anatomical and functional regeneration remain poorly explored. Some *in vitro* model systems have demonstrated their potential for studies on CNS injury and repair [4]. But, so far a model system of adult salmon primary brain neurons has not been described. Here, we report on an *in vitro* model system of neural cells from masu salmon *O. masou*.

We present morphological data and some cellular characterization supporting the use of this novel *in vitro* tool in investigations of neurochemical properties, axonal growth and neurogenesis in CNS. For a more detailed study of the properties of cells expressing CBS, their relationship with the cells with different neurochemical specialization in the central nervous system of fish, their characteristics of the processes of proliferation and differentiation, and the features of participation of hydrogen sulfide (H_2S) in reparative neurogenesis, the primary culture of brain and spinal cord from the *O. masou* was set up and properties of proliferation and differentiation were analyzed with the help of specific markers (HuCD, Pax6, CBS, and PCNA).

2. Materials and Methods

2.1. Animals

The fishes were kept in the standard environment in the aquarium (salinity: 5‰, temperature: 15°C - 17°C). Five specimens each of the masu salmon, *Oncorhynchus masou* (in age of one year and four months with the average body length 20 - 22 cm and average body weight 25 - 30 g) were taken from Rasan fish hatchery (Primorisky region, Russia).

The conditions of the animals were monitored by the aquarium equipment room: the room temperature in the aquarium was maintained by air conditioners. Salinity in water aquariums was created by adding fresh water to sea water with salinity equal to the formation of 5‰. The level of salinity of sea water was monitored using a refractometer (Refractometer "AtagoATS-S/Mill-E", Japan).

2.2. Primary Culture

The fishes were sacrificed by decapitation; brain and spinal cord were dissected out aseptically and washed in sterile PBS. The minced tissues were transferred to a sterile 15 ml tube and washed thrice with PBS. In each wash, the pieces were allowed to settle down for some time and the supernatant was discarded. The tissues were then treated with trypsin (0.25% and 0.025%) and collagenase (28 U and 56 U), incubated in water bath at 28°C for 15 minutes, transferred to a 50 ml sterile tube and suspended in a complete growth medium (five times to the volume of the trypsin used): Leibovitz's L-15 medium containing 10% fetal bovine serum and 0.4% (v/v) penicillin/streptomycin antibiotic cocktail (Gibco, Invitrogen, USA). A single cell suspension was prepared by disaggregation. The resulting suspension was allowed to stay in the centrifuge tube for 5 minutes, followed by careful aspiration of the floating cell clumps with the aid of pipette. The suspension was centrifuged at 200 × g for 5 minutes, the supernatant was discarded and the pellet was resuspended in the complete L-15 medium. The resulting cell suspensions from brain and spinal cord were seeded in the small (35 mm) specially coated duplex dishes and maintained in the incubator at 28°C for 3 - 4 days for further proliferation and differentiation. Cells were regularly monitored under the microscope (Axiovert Apotom 200 M, Carl Zeiss, Germany).

2.3. Immunocytochemistry

After 4 days in culture, the cells from suspension were centrifuged at 250 × g for 5 min and fixed with 4% paraformaldehyde for 30 min at room temperature, followed by a PBS wash. To quench the endogenous peroxidase activity, the cells were incubated with 3% hydrogen peroxide for 5 min and washed with PBS. To reduce the background staining, cells were incubated with 10% normal serum in PBS and then with primary antibodies for 1 hour at 37°C against PCNA (Santa Cruz Biotechnology; USA; 1:200); Pax6 (Chemicon; USA; 1:300); HuCD (Invitrogen; USA; 1:200); CBS (Abcam; ab54883, UK; 1:200). Cells were gently washed with PBS and incubated for 10 min with diluted biotinylated secondary antibody, followed by a PBS wash. Further staining was carried out as described in Vectastain Elite Kit (Vector Laboratories, Burlingame, USA). The microscopic analysis was carried out in duplex dishes under inverted microscope AxiovertApotom 200 M, (Carl Zeiss, Germany).

The cells from suspension population were treated by similar procedure on the glass slidescoated with poly-L-lysine. After above mentioned procedure of immunocytochemical labeling, cells from suspension fraction were incubated with peroxidase substrate solution during 10 min. To identify the reaction products, the slices were incubated in a substrate for detection of peroxidase (VIP Substrate Kit; Vector Laboratories, Burlingame, USA); the process of staining was controlled under a microscope. Then, the glass slides were washed out in three changes of phosphate buffer, dried at room temperature, dehydrated using a standard technique, and embedded in medium BioOptica (Italy). To estimate the specificity of the immunocytochemical reaction, we used negative control. The cells were incubated in a medium containing 1% nonimmune horse serum (instead of primary antibodies) for 2 h, and then all procedures were performed as was described above. In all control experiments, the immunopositivity in the studied cells was absent.

2.4. Statistical and Morphometric Analysis

For statistical and morphometric analysis we used software of microscope research class Axiovert 200 M with module Apotome. To do this, were obtained micrograph cell monolayer (for each version of the marking were removed and analyzed 10 non-overlapping fields of view with zoom lens X20). Densitometric investigation of the optical density of immunolabeled cells was performed using software Axiovision in microscope Axiovert. The optical density of immunoprecipitate marked cells were studied on the samples from 50 - 100 cells. Analysis of morphometric parameters of cell culture is given in **Table**. Data are expressed as the mean ± S.E.M. and were analyzed with an ANOVA followed by post *hoc* Tukey's tests unless otherwise stated. P-values < 0.05 were considered to be statistically significant.

On the basis of morphometric analysis was allocated 5 morphological types of cells in accordance with generally accepted neurohistological classification [5]. The first class was attributed to cells the diameter of which was more than 40 μm. In second class were cells with the diameter from 20 to 40 μm, in third class cells with diameter of 15 to 20 μm, fourth from 10 to 15 μm, and the fifth class with diameter of less than 10 μm. Morphometric parameters and correlation study between some of the parameters were analyzed by Microsoft Excel 2010.

3. Results and Discussion

Morphometric analysis of dissociated salmon brain and spinal cord cells, grown on laminin coated duplex dishes, revealed 5 morphological types of cells in accordance with classification [5]. In the resulting cultures after four days, most cells were round with a small one, or two processes (**Figure 1(A)**). We observed that our salmon primary cell culture was heterotypic (consisting of five main types of cells), representing the cellular composition of the adult salmon's brain and spinal cord in the monolayer (**Figure 1(A)**) and suspension (**Figure 1(B)**) fractions. Morphometric characteristics of cells are described in **Table**. Most of the cells in monolayer were round or oval in shape, while some cells were at the initial stage of formation of outgrowths.

In the present study, cells were isolated from the juvenile fishes brain, which, under proper culture conditions, formed neurospheres (**Figure 1(C)**), contained PCNA-immunolabeled cells (**Figure 1(D)**) and were self-renewing. We hypothesized that, these cells can be considered as intrinsic stem cells [6]. The intrinsic stem cells are likely to have originated from the proliferation zones in the brain regions contained proliferative zones [7]. *In vivo*, cells within these proliferation zones in *O. masou* persist during adulthood to undergo mitotic divisions and to produce new neurons [8].

Existing reports about the presence of endogenous sulphides in the brain and about the role of H_2S in facilitating the modulation of neuronal activity [9] lead us to analyze the CBS activity in masu salmon's brain.

CBS activity was identified both in cells of suspension fraction and in monolayer (**Figure 1(E)**). Morphometric parameters of CBS-positive and -negative cells from brain of masu salmon are presented in **Table**. The vast majority of CBS-positive cells present small cells of the 5th (84.1%) and 4th (14.9%) types. Thus, the primary culture of the brain and spinal cord of masu salmon contained a heterogeneous population of cells expressing CBS. Cells of the 4th and 5th types in suspension fraction of brain and spinal cord were PCNA-positive.

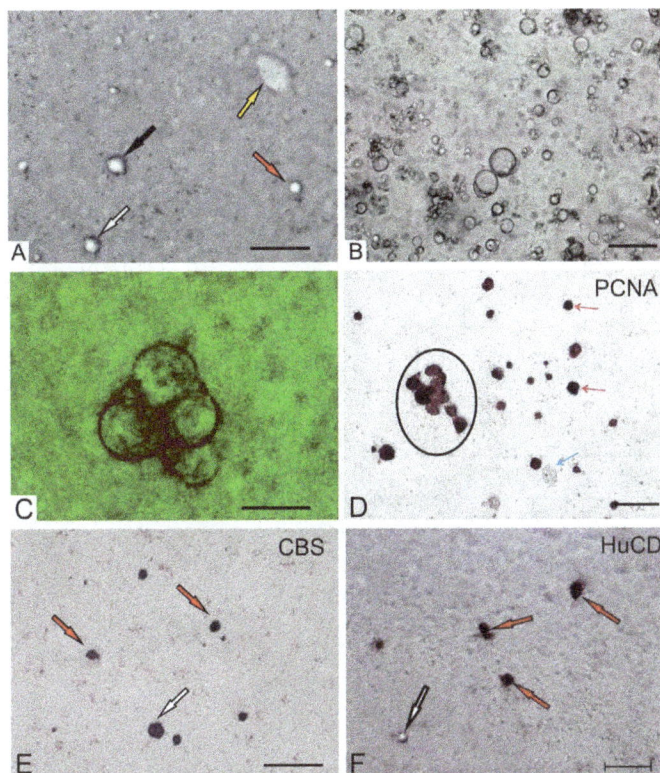

Figure 1. Primary culture of brain cells of masu salmon *Oncorhynchus masou*. (A) Cells in a monolayer (colored arrows indicate different types of cells); (B) Cells in suspension; (C) Neurosphere; (D) PCNA-immunopositive cells (red arrows) and negative (blue arrow) in suspension, oval contoured conglomerate of PCNA-immunopositive cells (neurosphere); (E) CBS-immunopositive cells of 4th (white arrow) and 5th (red arrows) types; (F) HuCD-immunopositive (red arrows) and negative (white arrow) cells in monolayer. The scale bar: (A), (B), (D)-(F) 50 μm; (C) 10 μm.

The data also testify the participation of the transcription factor Pax6 in facilitating proliferation of the cells of the salmon's brain, and suggest H$_2$S-dependent mechanisms for such participation (**Figure 2**). The level of CBS activity in these cells is also high. Thus, the predominant cell type characterized by the highest level of expression of CBS, which were large and medium cells of 1 - 3 types deprived of outgrowths and tended to be grouped into clusters.

Figure 2 shows the results of a quantitative analysis of cells of 4th and 5th types in monolayer immunolabeled with CBS, Pax6 and PCNA that indicate high quantitative correspondence to these types of cells, which indirectly suggest that the small cells, which expressed CBS and Pax6 may be proliferating, and/or out from mitosis cells belonging to the same type. Based on these data, we concluded the existence of a positive correlation between small cells expressing CBS, Pax6 and PCNA in *O. masou* brain cells.

We then explored the identity of the remaining 39% non-neuronal (HuCD negative) cells in salmon brain and spinal cord cultures. While neuronal differentiation requires exit from the cell cycle, non-neuronal brain cells, including glia and stem/progenitor cells, are potentially mitotically active. Proliferating cell nuclear antigen (PCNA) is expressed in mitotic cells throughout the cell cycle with prominent nuclear expression [10]. Furthermore, some examples of cytoplasmic immunoreactivity have been reported [11]. Immunocytochemical analysis with anti-PCNA revealed subpopulations of PCNA positive, as well as PCNA negative cells (**Table**; **Figure 1(D)**). Thus, the results of morphological and immunohistochemical analysis allow to conclude that: 1) Proliferative activity in primary culture cells in the brain of juvenile masu salmon is mainly present in suspension cell population of 4th and 5th type; 2) There is a positive correlation between cells expressing CBS, Pax6 and PCNA, which indicates the participation of hydrogen sulfide in proliferative activity of cells in postembryonic neurogenesis of masu salmon.

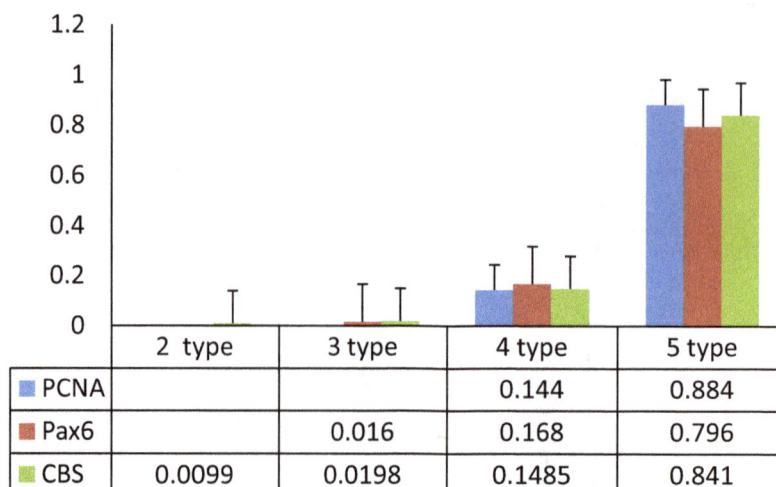

Figure 2. Correlation between expression of HuCD, Pax6 and CBS in cells of monolayer in *Oncorhynchus masou* brain. Data are shown as M ± m. x—types of cells, y—number of immunopositive cells (%).

Table. Morphometric characteristic, number and optical density of immunoreactivity of PCNA-, CBS- and HuCD-positive and negative cells (M ± m) in monolayer of primary culture of masu salmon *Oncorhynchus masou*.

	CBS-positive and negative cells						
	CBS-negative cells			**CBS-positive cells**			
Types of cells	Large diameter µm	Small diameter µm	Percentage %	Large diameter µm	Small diameter µm	Percentage %	Optical density (E ± e)
1	-	-	-	45.45 ± 2.68	38.38 ± 2.89	15.9	142.76 ± 14.74
2	24.28 ± 2.02	22.11 ± 2.89	13.30	31.30 ± 4.93	26.40 ± 5.40	66.0	160.06 ± 12.48
3	17.27 ± 1.22	14.46 ± 2.32	18.30	18.18 ± 1.78	14.98 ± 2.60	11.36	144.25 ± 14.21
4	12.57 ± 1.20	10.91 ± 1.78	48. 3	12.8	9.85	2.27	120.9 ± 10.2
5	8.75 ± 0.84	7.69 ± 0.82	18.30	5.9 ± 1.27	5.5 ± 0.70	4.50	136.2 ± 11.03
	PCNA-positive and negative cells						
	PCNA-negative cells			**PCNA-positive cells**			
1	-	-	-	-	-	-	-
2	24.66 ± 1.88	21.14 ± 2.11	6.7	-	-	-	-
3	17.02 ± 1.52	15.09 ± 2.03	15.20	16.35 ± 0.86	14.61 ± 1.52	-	118.3 ± 44.74
4	12.11± 1.36	10.71 ± 1.71	47.40	12.55	9.73	-	89.95 ± 5.72
5	8.80 ± 0.90	7.93 ± 0.95	30.50	7.55 ± 1.44	5.98 ± 0.25	-	82.6 ± 7.91
	HuCD-positive and negative cells						
	HuCD-negative cells			**HuCD-positive cells**			
1	-	-	-	-	-	-	-
2	22.42 ± 1.78	19.09 ± 3.05	5.90	21.92 ± 1.07	16.86 ± 1.99	12.50	121.69 ± 22.2
3	17.40 ± 1.57	12.91 ± 2.72	11.90	17.07 ± 1.88	13.56 ± 2.69	31.20	120.38 ± 24.3
4	11.75 ± 1.16	9.79 ± 1.44	44.70	12.54 ± 1.09	9.69 ± 1.56	56.20	102.22 ± 13.7
5	8.17 ± 1.40	6.85 ± 1.25	37.30	-	-	-	-

Expression of HuCD, a marker of neuro differentiation, in primary culture has shown that this marker is expressed by cells of 2 - 4 types in the masu salmon's brain (**Figure 1(F)**). Our results indicate that cells exhibit various degrees of maturity in culture. Data suggest that the majority of HuCD positive cells are neuronally committed stem cells or differentiated neurons. In the suspension fraction, this marker was found among small cells that may indicate early neuro differentiation of cells released from the proliferative cycle. Investigation of the correlation relationships between HuCD, Pax6 and CBS-producing cells in the monolayer of masu salmon's brain, showed inverse correlation between the expressions of these markers (**Figure 3**). Thus, among differentiated populations of large cells of the 2nd type revealed the largest number of CBS- and Pax6-producing cells and the least amount of HuCD-immunopositive cells. Conversely, in the monolayer among small cells of the 4th type, the largest percentage of HuCD-immunopositive and the lowest percentage Pax6 and CBS-producing cells were observed.

Thus, the process of cell differentiation in obtaining primary culture is more characteristic for monolayer. The data obtained suggest that the hydrogen sulphide (H_2S) is also involved in the process of neuronal differentiation; however, the nature of its participation appears to be nonlinear and inversely proportional in terms of mathematical correlations.

An important advantage of our heterogeneous culture system is that it simulates the diverse cellular composition of the juvenile masu salmon brain. Therefore, it may allow cell-cell interactions to occur that play important roles in physiological events such as development or restoration *in vivo*. However, to answer certain research questions, it may be necessary to isolate specific cellular populations from dissected brain utilizing purification techniques such as fluorescence-activated cell sorting (FACS) or immunopanning. It will be interesting to explore the mechanism behind the participation of H_2S in neuronal proliferation and differentiation in correlation with the characteristic markers like PCNA, Pax6 and HuCD. Our further efforts will be directed to understand the underlying mechanism. The potential of our culture system is diverse as it could be used to investigate axonal growth and regeneration, neuronal/glia biology, or stem cell behavior. The neurons in our culture exhibit typical growth and morphological features. Our juvenile masu salmon brain cell culture provides a valuable tool to study numerous developmental and restorative events. Previously, mammalian neuronal cultures provided insight into axon growth mechanisms [12], improved our understanding of regeneration-associated signaling pathways. Due to the intrinsic regenerative capacity of masu salmon CNS neurons *in vivo*, our system provides opportunity to explore axonal growth and regeneration utilizing a powerful genetic system and the controlled conditions of the culture.

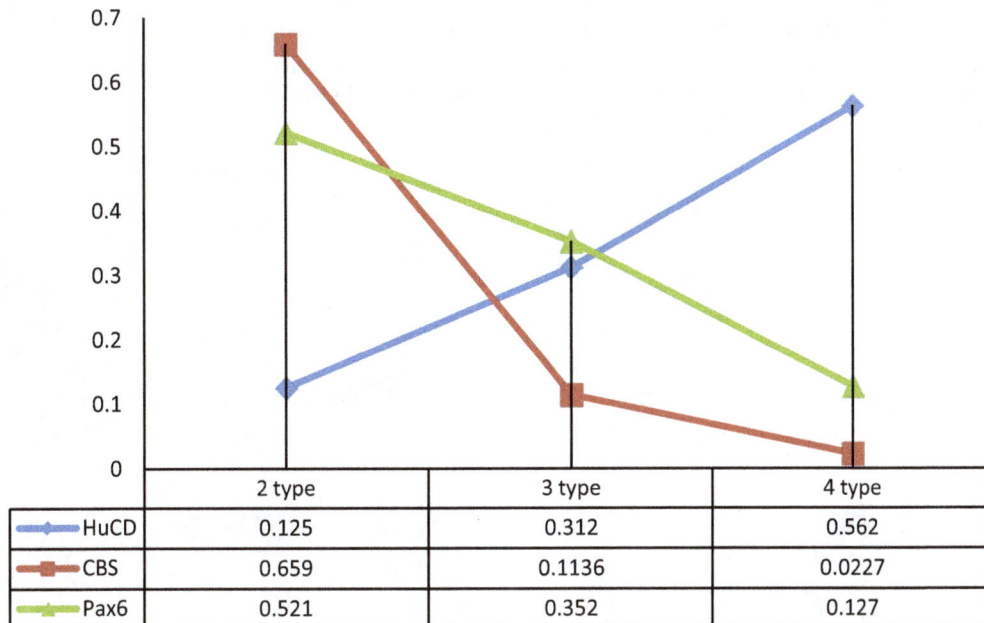

	2 type	3 type	4 type
HuCD	0.125	0.312	0.562
CBS	0.659	0.1136	0.0227
Pax6	0.521	0.352	0.127

Figure 3. Correlation between expression of PCNA, Pax6 and CBS in cells of monolayer in *Oncorhynchus masou* brain. x—types of cells, y—number of immunopositive cells (%).

Acknowledgements

This work was supported by the Grant of President of Russian Federation (MD-4318.2015.4) and DST-INSPIRE Faculty Grant of Government of India (IFA14-LSBM-104).

References

[1]   Kizil, C., Kaslin, J., Kroehne, V. and Brand, M. (2012) Adult Neurogenesis and Brain Regeneration in Zebrafish. *Developmental Neurobiology*, **72**, 429-461. http://dx.doi.org/10.1002/dneu.20918

[2]   Becker, T., Wullimann, M.F., Becker, C.G., *et al.* (1997) Axonal Regrowth after Spinal Cord Transection in Adult Zebrafish. *Journal of Comparative Neurology*, **377**, 577-595. http://dx.doi.org/10.1002/(SICI)1096-9861(19970127)377:4<577::AID-CNE8>3.0.CO;2-#

[3]   Candal, E., Anadon, R., DeGrip, W.J. and Rodriguez-Moldes, I. (2005) Patterns of Cell Proliferation and Cell Death in the Developing Retina and Optic Tectum of the Brown Trout. *Developmental Brain Research*, **154**, 101-119. http://dx.doi.org/10.1016/j.devbrainres.2004.10.008

[4]   Sakowski, S.A., Lunn, J.S., Busta, A.S., *et al.* (2012) A Novel Approach to Study Motor Neurons from Zebrafish Embryos and Larvae in Culture. *Journal of Neuroscience Methods*, **205**, 277-282. http://dx.doi.org/10.1016/j.jneumeth.2012.01.007

[5]   Arévalo, R.R., Alonso, J.R., Garcia-Ojeda, E., *et al.* (1995) NADPH-Diaphorase in the Central Nervous System of the Tench (*Tinca tinca* L., 1758). *Journal of Comparative Neurology*, **352**, 398-420. http://dx.doi.org/10.1002/cne.903520307

[6]   Doe, C.Q., Fuerstenberg, S. and Peng, C.-Y. (1998) Neural Stem Cells: From Fly to Vertebrates. *Journal of Neurobiology*, **36**, 111-127. http://dx.doi.org/10.1002/(SICI)1097-4695(199808)36:2<111::AID-NEU2>3.0.CO;2-4

[7]   Zupanc, G.K. and Sîrbulescu, R.F. (2013) Teleost Fish as a Model System to Study Successful Regeneration of the Central Nervous System. *Current Topics in Microbiology and Immunology*, **367**, 193-233. http://dx.doi.org/10.1007/82_2012_297

[8]   Pushchina, E.V., Obukhov, D.K. and Varaksin, A.A. (2013) Features of Adult Neurogenesis and Neurochemical Signaling in the Cherry Salmon *Oncorhynchus masou* Brain. *Neural Regeneration Research*, **8**, 13-23.

[9]   Wang, R. (2012) Physiological Implication of Hydrogen Sulfide: A Whiff Exploration That Blossomed. *Physiological Reviews*, **92**, 791-896. http://dx.doi.org/10.1152/physrev.00017.2011

[10] Waseem, N.H. and Lane, D.P. (1990) Monoclonal Antibody Analysis of the Proliferating Cell Nuclear Antigen (PCNA). Structural Conservation and the Detection of a Nucleolar Form. *Journal of Cell Science*, **96**, 121-129.

[11] Vriz, S., Lemaitre, J.M., Leibovici, M., *et al.* (1992) Comparative Analysis of the Intracellular Localization of c-Myc, c-Fos, and Replicative Proteins during Cell Cycle Progression. *Molecular and Cellular Biology*, **12**, 3548-3555.

[12] Blackmore, M.G., Moore, D.L., Smith, R.P., *et al.* (2010) High Content Screening of Cortical Neurons Identifies Novel Regulators of Axon Growth. *Molecular and Cellular Neuroscience*, **44**, 43-54. http://dx.doi.org/10.1016/j.mcn.2010.02.002

28

Changes in Salivary and Plasma Markers during and Following Short-Term Maximal Aerobic Exercise Assessed during Cognitive Assessment

Christine Lo Bue-Estes[1,2]*, Peter J. Horvath[2]

[1]Department of Sports Medicine, Mercyhurst University, University at Buffalo, Buffalo, USA
[2]Department of: Exercise and Nutrition, University at Buffalo, Buffalo, USA
Email: *lobue007@hotmail.com

Abstract

This study assessed multiple salivary and plasma markers before and after incremental short-term maximal aerobic exercise and in a non-exercising control in conjunction with cognitive testing. Subjects: Apparently healthy 18 - 30 years old low CVD risk females participated (n = 19). Methods: Subjects completed two conditions: 1) exercise: short maximal treadmill exercise and cognitive assessment pre- and post-exercise and, 2) non-exercise: with cognitive assessment timed to match testing in the exercising condition. Non-stimulated, timed salivary samples and venous blood were collected before and after exercise and after recovery. Results: Saliva: Over time α-amylase increased in both exercise and non-exercising conditions. Exercise had increases in α-amylase at time matched control points up to 36% greater than the non-exercising conditions. Following exercise and recovery from exercise α-amylase increased compared to baseline (ranging from 47% to 290%). Baseline cortisol was 33% higher than post-exercise and 59% higher than recovery irrespective of exercise. Plasma: NEFA was 50% higher at post-exercise and recovery compared to baseline without exercise and 36% higher at post-exercise and recovery compared to baseline with exercise. Glucose and lactate were, 18% and 50% higher respectively, after exercise compared to baseline and recovery with exercise. Post-exercise glycerol was 11% higher than recovery. Differences between Conditions: Post-exercise glucose and lactate were 20% and 40% higher respectively with exercise. Glycerol was 11% lower after exercise. Conclusions: We demonstrated that acute exercise coupled with cognitive task increased α-amylase levels, but not cortisol, potentially due to a differential stress response, but most likely due to the timing of sample collection.

---

*Corresponding author.

Keywords

α-Amylase, NEFA, Cortisol, Exercise, Salivary Biomarker

1. Introduction

Aerobic exercise affects every system in the human body. Blood flow, metabolism, and multiple chemical levels all change acutely as a result of aerobic exercise; some changes have fleeting effects, and others are long lasting. Chronic adaptations also occur to numerous systems as a result of repetitive aerobic training over time. Authors have demonstrated that gene expression in the brain is up-regulated and that performance on cognitive tasks is altered by chronic aerobic exercise [1]. The mechanisms responsible for these changes in the brain however have yet to be identified.

Changes occurring in the periphery impact the brain, sometimes directly, sometimes indirectly. Assessing changes in hormones, metabolites, and other chemicals in the body during varying exercising conditions may help us to identify potential chemical mechanisms involved in altering brain structure and function. Aerobic exercise has the potential to affect brain function in many ways. Over time aerobic training results in increased blood volume and the ability to increase cardiac output [2] [3]. Increased cardiac output allows for increased delivery of fuel and oxygen and removal of substrates in the brain. If chemical changes observed during and following acute short-term maximal aerobic exercise bouts alter how the brain works, via changes in fuel availability and delivery, these changes could be part of the mechanism(s) responsible for changes in cognitive function during and following exercise and as such need to be assessed.

Saliva is an easily collected, non-invasive, biological sample that is capable of providing valuable information on acute changes in hormones and metabolites, without some of the potentially negative consequences of other biological samples such as venous and arterial blood or tissue biopsies. Salivary Alpha-amylase [4]-[8] and cortisol [9]-[11] have both been assessed in numerous studies assessing the impact of both physical and psychological stressors on the levels of these hormones. Changes in glucose, lactate, glycerol, and NEFA levels during and following exercise are all potential indicators of altered fuel source availability [12]-[26]. Changes in fuels sources could play an important role in fatigue. The purpose of the current study was to assess changes in multiple salivary and plasma factors as assessed during cognitive testing before and after incremental short-term maximal aerobic exercise in young healthy women.

2. Methods

2.1. Subjects

Apparently healthy 18 - 30 years old females, currently using pharmaceutical birth control were recruited to participate in this experiment (no minors were allowed to participate). Subjects were deemed "apparently healthy" when it was determined that they were free of signs and symptoms of cardiovascular and pulmonary disease and met the criteria for the American College of Sports Medicine (ACSM) low risk stratification for coronary artery disease [27]. Additional exclusionary criteria included: diagnosed learning disability, a concussion in the preceding six months (since both learning disabilities and concussion could be a confounding factor when assessing cognitive function), and lastly, the use of medication that could influence cognitive performance, including painkillers and antidepressants. Nineteen subjects completed the study.

Only females actively taking birth control were included to avoid potentially confounding results based on phase of the menstrual cycle since hormone levels vary based on the phase of the menstrual cycle (even though phase of menstrual cycle has been shown to have insignificant impact on cognitive performance) [28]. Testing was done after the first 3 days in a pill or birth control patch cycle and before the last three days of the birth control cycle. Prior to any participation in this study, including screening, a copy of the informed consent approved by the University at Buffalo's Institutional Review Board, had to be signed. Subjects completed three laboratory visits; all visits were completed at the same time of day, in the morning to avoid changes based on circadian variations. The day before each lab visit subjects: 1) refrained from strenuous exercise, 2) did not drink alcoholic beverages for the entire day, 3) and did not eat or drink anything after midnight.

2.2. Screening Lab Visit

After subjects read and signed the approved consent form, a questionnaire documenting the subject's demoFigureics, medical history, and other pertinent information was completed. Fasting cholesterol (TC and HDL) and blood glucose were measured using a Cholestech LDX (Hayward, CA) via a finger stick blood sample. Resting heart rate, blood pressure, height and weight were measured. Body composition was assessed via 4-site skin caliper measurements, using Lange® skin calipers (Beta Technology Inc., Santa Cruz, CA), and then calculated using equations derived by Jackson and Pollock [29]. These results were used to determine whether a subject met the ACSM low risk stratification for coronary artery disease. Subjects also provided a urine sample, collected immediately prior to testing, that was used for pregnancy screening, hCG ACON laboratories Inc. (San Diego, CA) to assure that only non-pregnant women participated.

Following the successful completion of ACSM low CVD risk screening, each subject completed a discontinuous Modified Bruce Treadmill VO_2 max treadmill test on a Landice L7 treadmill using a Vacumed CPX Mini system (Ventura, California), which was used to perform breath by breath gas analysis. This served to determine each subject's level of aerobic fitness and obtain an objective measure of exercise to exhaustion for each subject. Subject's heart rate was continually monitored throughout testing with a Polar© heart rate monitor (Woodbury, NY). Rate of perceived exertion was measured at the end of each exercise interval, using the Borg Scale 6 - 20.

2.3. Second & Third Lab Visits

The second and third visits were assigned in counterbalanced order to prevent an effect of learning. The "**Exercising Visit**" involved computerized neuropsychological assessment before, and after short-term aerobic exercise to exhaustion. The "**Non-Exercising Visit**" involved computerized neuropsychological assessment over time without any exercise, where cognitive assessments occurred at time points that mirrored the cognitive testing in the exercise protocol.

Cognitive Testing: During each of these two visits subjects were instructed how to take Automated Neuropsychological Assessment Metrics (ANAM), the computerized software program used to test cognitive variables for this study. ANAM is a windows/PC based, mouse operated software program designed to assess various aspects of cognitive performance; it has strong correlations to traditional neuropsychological tests and was created by the US Department of Defense as a rapid, reliable, easily repeatable neuropsychological test [30] [31].

The full ANAM full battery takes approximately 12 to 15 minutes to complete and has been used to evaluate simple reaction time and both the speed and accuracy of other cognitive functions, including information processing, visual spatial memory, continual processing (attention), code substitution (short term memory), and working memory [30] [31]. We used the 2001 version of ANAM which consisted of seven modules: 1) simple reaction time, 2) code substitution, 3) procedural reaction time, 4) spatial processing, 5) visual spatial memory, 6) working memory, and, 7) code substitution delayed [30] [31]. For a more in depth review of these subtests please consult (Lo Bue-Estes et al., 2008). Each module is preceded by written on screen instructions that explain the specific sub-test, and all but the delayed memory test are followed by several practice problems.

2.4. Exercising Visit

Once the subject was instructed how to take ANAM, the subject took one practice test to familiarize herself with the test. The subject then had a five-minute rest between the **practice ANAM** and her **baseline ANAM**. Then each subject participated in a treadmill VO_2 max test that was customized for her based on her performance on the Modified Bruce test in the first lab visit. By customizing the workloads based on each person's fitness level the amount of time each subject spent on the treadmill was very consistent between subjects. **Non-Exercise Visit:** Once the subject was instructed how to take ANAM, the subject took one practice tests to become familiar with the test. The subjects had a five-minute rest between her practice and baseline ANAM test. Then there was a break that mirrored the time a subject would spend exercising during the "Exercising Visit" followed by another ANAM, a 30-minute break following the completion of that ANAM and then the last full ANAM.

2.5. Blood & Saliva Collection

At three of the data collection time points, **baseline**, **post maximal exercise**, and **recovery**, a non-stimulated, timed salivary sample and a 15 mL venous blood sample were taken. The salivary sample was collected into a 5 mL polypropylene cryovial (to avoid hormone binding) via passive drool collected through a short section of common drinking straw. The venous blood sample was collected using 21G 3/4 Vacutainer® brand Safety-Lok™ blood collection set into in a BD Vacutainer EDTA K_2 sterile tube and a BD Vacutainer Serum tube). (*Note: Midway through the study, after completion of the 1st seven subjects, BD the maker of our vacutainers switched from glass vials to plastic.*) Blood samples were spun for 30 minutes using a Sorvall RT6000B centrifuge, Wilmington, DE (at 3000 rev/min (approximately 6 1/3 on the speed dial)) at 4 degrees Celsius. Salivary samples were weighed and then vortexed using a Vortex-Genie prior to aliquotting, while still in the cryovial used for sample collection. Salivary samples were then aliquoted using glass Fisher Brand Pasteur into 2 mL polypropylene cryovials.

2.6. Serum Sample Processing

Serum samples for all times (1 = Baseline, 2 = Post, and 3 = Recovery) and conditions (1 = Non-exercise Visit, 2 = Exercise Visit) were processed using a COBAS FARA II (Basel, Switzerland). Chemicals from Wako Chemicals USA Inc. (Richmond, VA) were used in all serum and plasma testing. None of the samples had undergone any freeze-thaw cycles prior to this round of analysis. Samples were allowed to thaw for 60 minutes at room temperature in the morning, and then placed into a refrigerator to thaw the rest of the way overnight. The following morning 300 μL samples were pipetted using Gilson Pipetman pipettes into Fara Cups and loaded into numbered racks. Racks were placed in the refrigerator until processing began the following morning. Samples were refrigerated when not actively being assessed. Twenty-nine plate wells were used to run samples. Total Cholesterol, Glucose, Glycerol, and Non-esterified fatty acids (NEFA) were all measured in duplicate.

2.7. Saliva Sample Processing

Saliva samples for all times (1 = Baseline, 2 = Post and 3 = Recovery) and conditions (1 = Non-exercise, 2 = Exercise) were packaged and sent for analysis to: Salimetrics, LLC (State College, PA). Salivary estradiol was assessed in duplicate using a high-sensitivity enzyme immunoassay (Cat. No. 1-3702/1-3712, Salimetrics LLC, State College PA). The test used 100 ul of saliva per determination, has a lower limit of sensitivity of 1.0 pg/mL, standard curve range from 1.0 pg/mL to 32.0 pg/mL, an average intra-assay coefficient of variation of 7.1% and an average inter-assay coefficient of variation 7.5%. α-Amylase assay was completed using a commercially available kinetic reaction assay (Salimetrics LLC, State College PA) following Granger *et al.* [32]. Intra-assay variation (CV) computed for the mean of 30 replicate tests was less than 7.5%. Inter-assay variation computed for the mean of average duplicates for 16 separate runs was less than 6%. Salivary cortisol in assays was performed in duplicate using a highly sensitive enzyme immunoassay (Salimetrics, PA). The test uses 25 ul of saliva per determination, has a lower limit of sensitivity of 0.003 ug/dl, standard curve range from 0.012 to 3.0 ug/dl, and average intra- and inter-assay coefficients of variation 3.5% and 5.1% respectively. Method accuracy, determine by spike and recovery, and linearity, determined by serial dilution are 100.8% and 91.7%.

2.8. Statistics

All statistical calculations for salivary and plasma variables were assessed using Sigma Stat 3.5 (Jandel). Two Way Repeated Measure ANOVAs, were run to assess the effects of time, condition, and time by condition. All subject data are presented as mean ± SD, plasma and salivary samples are presented as mean ± SEM.

3. Results

Subjects were 21.8 ± 2.7 years old, 23.5 ± 4.9 percent body fat, 176.3 ± 8.9 cm tall, 61.6 ± 8.7 kg, with an average VO_2 max of 51.3 ± 6.8 ml/kg/min.

3.1. Salivary Results

Estradiol was assessed only at the baseline time point in both conditions to determine if there was a difference in hormone levels between the experimental conditions. No difference in salivary estradiol was present (Non-Exercise = 2.4 ± 0.4 pg/mL, Exercise = 2.5 ± 0.3 pg/mL), with both conditions having estradiol levels that most closely reflect the follicular phase. α-Amylase was higher post and recovery than baseline in both conditions (ranging from 47% to 290%) and was also 35.5% higher at the post-exercise assessment in the Exercise condition compared to the Non-Exercise condition (see **Figure 1**). Cortisol at baseline was 33% higher than post-exercise and 59% higher than recovery (see **Figure 2**) but was not different at any time point between conditions.

3.2. Plasma Results

Non-Exercising condition: Glucose (**Figure 3**) and Glycerol (**Figure 4**) levels were not different at any time point. Lactate was 61% (**Figure 5**) higher at post compared to baseline and recovery. NEFA (**Figure 6**) was 50% higher at post and recovery compared to baseline. *Exercising condition*: Glucose and lactate were, 18% and 50%

Alpha Amylase Before & After Exercise

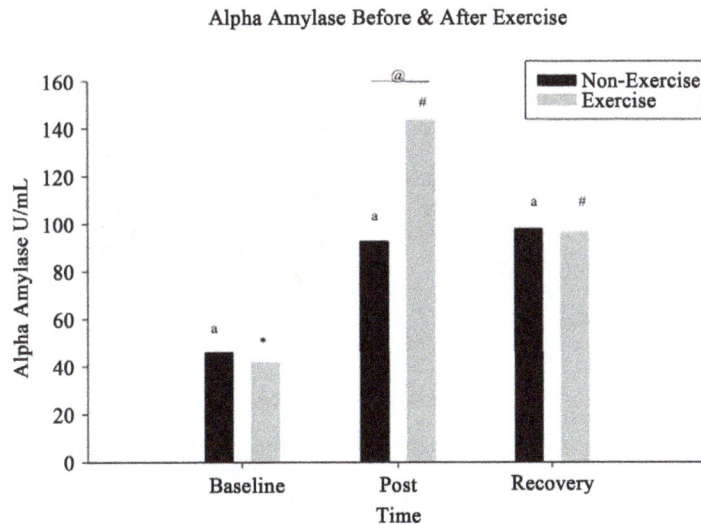

Figure 1. Values are Mean ± SEM, non-exercising time points not sharing a common letter different $P \leq 0.05$, exercise time points not sharing a common symbol different $P \leq 0.05$, time points with symbol over both non-exercise & exercise data different between conditions $P \leq 0.05$.

Cortisol before and after exercise

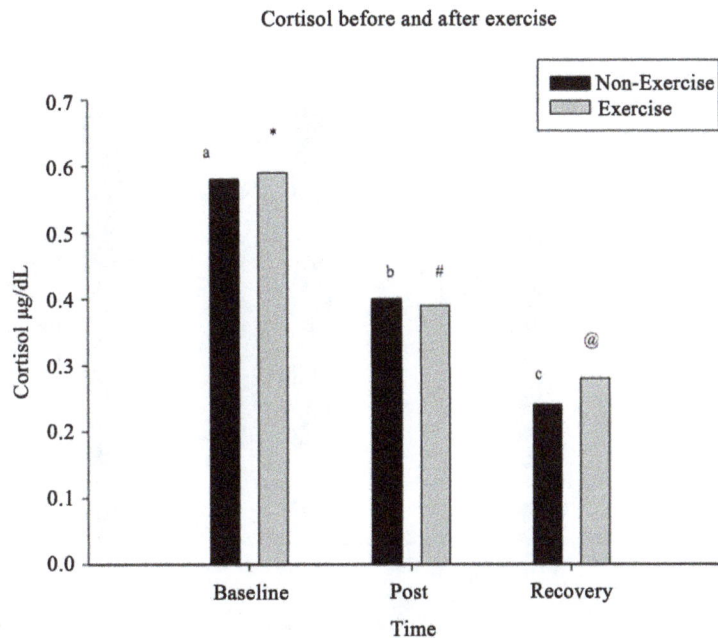

Figure 2. Values are Mean ± SEM, non-exercising time points not sharing a common letter different $P \leq 0.05$, exercise time points not sharing a common symbol different $P \leq 0.05$, time points with symbol over both non-exercise & exercise data different between conditions $P \leq 0.05$.

Glucose before and after exercise

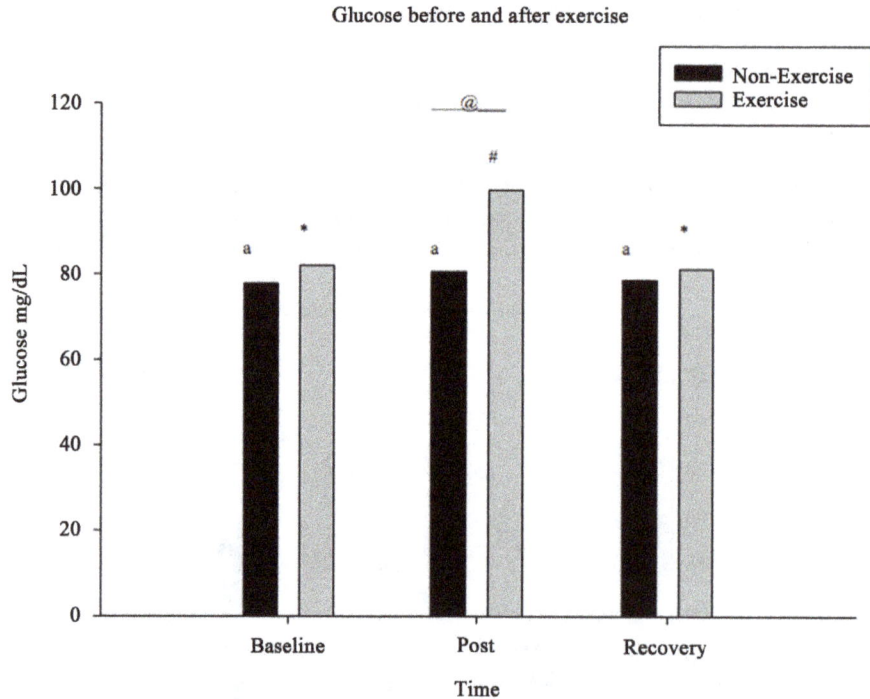

Figure 3. Values are Mean ± SEM, non-exercising time points not sharing a common letter different $P \leq 0.05$, exercise time points not sharing a common symbol different $P \leq 0.05$, time points with symbol over both non-exercise & exercise data different between conditions $P \leq 0.05$.

Glycerol before and after exercise

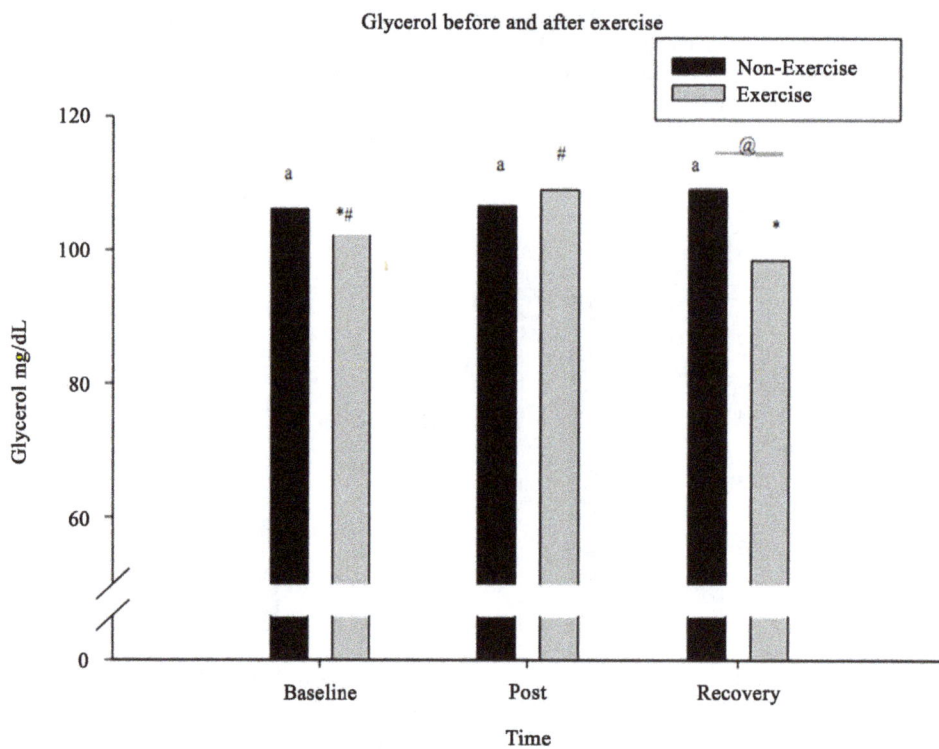

Figure 4. Values are Mean ± SEM, non-exercising time points not sharing a common letter different $P \leq 0.05$, exercise time points not sharing a common symbol different $P \leq 0.05$, Time points with symbol over both non-exercise & exercise data different between conditions $P \leq 0.05$.

Lactate before and after exercise

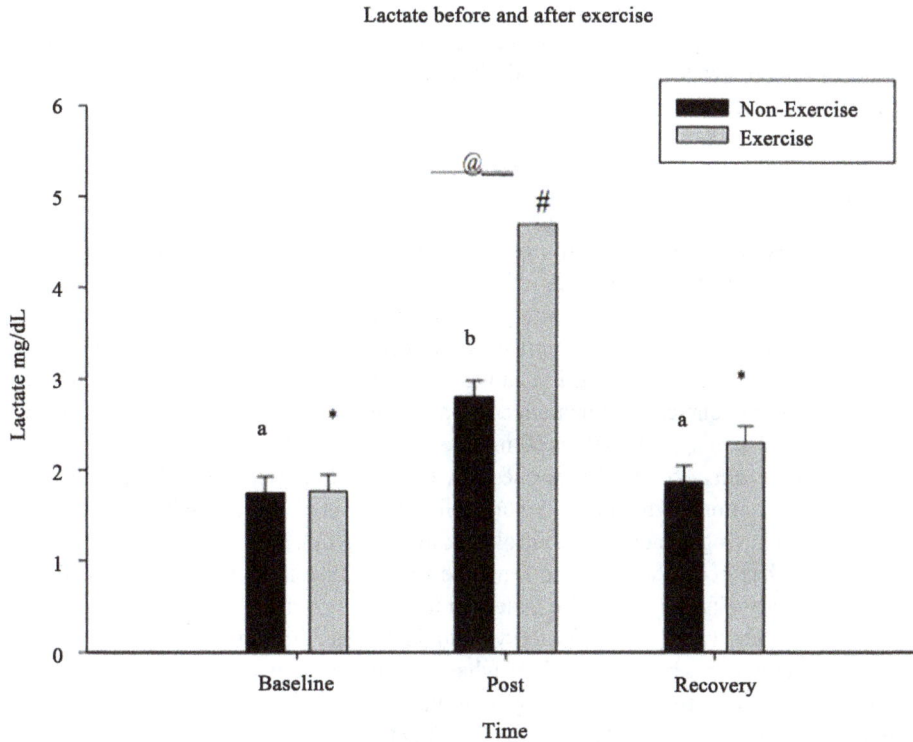

Figure 5. Values are Mean ± SEM, non-exercising time points not sharing a common letter different $P \leq 0.05$, exercise time points not sharing a common symbol different $P \leq 0.05$, Time points with symbol over both non-exercise & exercise data different between conditions $P \leq 0.05$.

NEFA before and after exercise

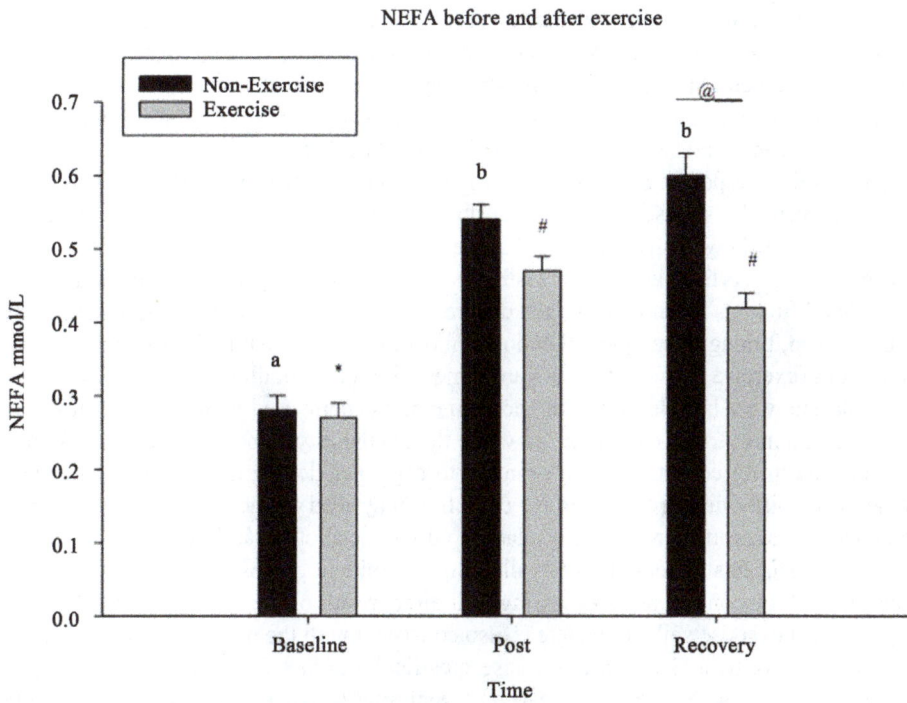

Figure 6. Values are Mean ± SEM, non-exercising time points not sharing a common letter different $P \leq 0.05$, exercise time points not sharing a common symbol different $P \leq 0.05$, time points with symbol over both non-exercise & exercise data different between conditions $P \leq 0.05$.

higher respectively, post exercise compared to baseline and recovery. Post exercise glycerol is 11% higher than recovery. NEFA was 36% higher at post and recovery compared to baseline. *Differences between conditions*: post exercise glucose and lactate were, 20% and 40% higher respectively in the Exercise condition. NEFA was 15% higher in the Non-Exercise condition at both post exercise and recovery. Glycerol was 11% higher in the Non-Exercise Condition.

4. Discussion

The methodological restriction of using only females actively taking birth control during certain phases of their medication cycle proved useful as the levels of estradiol did not vary between conditions, thus removing that hormone as a potential covariate. Salivary α-amylase is considered an indicator of stress induced changes in the sympathic adrenomedullary system [4]-[6] and salivary cortisol is an indicator of hypothalamus pituitary adrenocortical system [5]. In addition to being indicative of stresses within different systems these salivary markers also exhibit different time courses of appearance/disappearance. The time course of appearance and disappearance of salivary α-amylase is approximately 10 minutes, whereas cortisol's peak takes approximately 20 min as cortisol diffuses into the saliva from other tissues [4]. It is also important to understand that these markers are affected by circadian variations (which is why controlling for time of day and taking into account time of day in interpreting results and having an accurate control condition are key).

Salivary α-amylase has been shown to be sensitive to both physical and psychological stressors and is secreted directly by salivary glands [4] [5]. We observed no change over time in the levels of α-amylase during the non-exercise condition, showing that level of psychological stress did not change in our control (non-exercise) condition. This finding is different than much of what is currently in press [5] [7] however it may be due to the fact that much of the previous research used psychological tasks that involved a socially stressful component as well. Our findings that the exercise intervention significantly increased α-amylase levels compared to baseline levels and increased the α-amylase levels between conditions with the exercise condition being higher than the non-exercising condition are consistent with much of what is in press [33] [34].

The timing of our saliva collection occurred during the optimal window to detect changes in alpha amylase, but not in the optimal window to detect cortisol changes. Some studies show correlations between α-amylase and cortisol and others do not [35], this disparity may be 2 fold, partly because of the difference in the timing of appearance/disappearance for these markers and or because they are indicative of the stress responses in different systems. Some interventions may affect one stress system and not the other or affect the systems to varying degrees. A recent meta analysis by Dickerson *et al.* [11] reviewed 208 articles and found that while in general that cortisol increases in response to psychological stress, but that the type of stress matters and that not all stressors elicit a cortisol response. Stressor that involve uncontrollable aspects especially when coupled with social stresses elicit cortisol responses more so than those stressors that are not uncontrolled, or are not coupled with social stresses. We showed changes in cortisol levels consistent with circadian changes over the course of a morning, with subjects' cortisol levels decreasing as the morning progresses (prior to noon); but contrary to many other authors' findings we did not show a difference between our exercise and non-exercise conditions. As previously mentioned, timing of sample collection is most likely responsible for a lack of detected changes both between conditions (exercise vs. non-exercise) and times within the conditions.

Glucose and lactate were both higher at the post time points compared to all others, but significantly more so in the Exercise than non-exercise condition. So while the availability of both of these fuel sources changed over time in the non-exercising condition, they changed to a greater degree during the exercising condition. This change in fuel availability during the Exercise condition may lead to increased ability to deliver more fuel to metabolically active tissues. If more fuel is available to drive metabolism an individual may be able to "do more" than they would without this increased fuel availability. A spike in glucose following short-term maximal exercise is consistent with gluconeogenesis occurring as a direct result of the increased fuel demand occurring peripherally [3]. The post exercise spike in lactate is also consistent with the normal alterations in lactate following a maximal aerobic exercise bout [3]. Also in the case specifically of lactate, since there is an approximate time lag of about 30 minutes from when lactate appears in the plasma to when it crosses the blood brain barrier [36], changes in brain function/performance may experience a time delay as the lactate takes time to be delivered. Increased availability of lactate has also been shown to increase lactate transport across the blood brain barrier [36].

Increases in NEFA levels in the non-exercise condition are consistent with lypolysis and changes in LPL activity in healthy fasted individuals over time [22]. The increased amount of NEFA present at the recovery time point in the non-exercising compared to Exercising condition is most likely due to increased uptake of fatty acids post-exercise to replenish energy stores used during the exercise bout. Our NEFA findings are consistent with Marion-Latard et al. showing increased oxidation of NEFA during recovery from exercise [24]. Glycerol, which is a measure of triglyceride breakdown and mobilization from both adipose and muscle tissues, similarly to NEFA showed increased levels at the recovery time point in the non-exercise compared to the exercise conditions. It was also however increased during the exercising condition post exercise compared to recovery. This disparity between the non-exercising and exercising conditions again similarly to NEFA is most likely a result of increase peripheral uptake post-exercise to replenish energy stores. The increase in glycerol levels post exercise compared to recovery is also most likely a result of two phenomena, 1) increased fuel mobilization during exercise to fuel the increased demands of intense exercise and, 2) the previously mentioned peripheral increased fuel uptake post exercise to replenish energy stores used during exercise. Wee et al. [37] showed an increase in glycerol following a 20 minute bout of submaximal exercise (cycling at 70% of VO_2 max) followed by a sharp decline in the initial recovery phase, hours into recovery from exercise they observe additional increases in glycerol levels. The magnitude of increase that they observed is much greater than ours, but this is most likely due to the fact that our initial post exercise assessment occurs significantly later than theirs.

5. Conclusion

We demonstrated that acute exercise coupled with cognitive task in our protocol increases α-amylase levels, but not cortisol, potentially due to a differential stress response, but most likely due to the timing of sample collection. Additional research with sample collection occurring at time points that would maximize the ability to detect changes in both salivary α-amylase and cortisol is warranted to see if this combination of physical and psychological stresses taxes both of the representative stress systems equally or in a similar fashion.

References

[1]   Colcombe, S.J., et al. (2004) Cardiovascular Fitness, Cortical Plasticity, and Aging. Proceedings of the National Academy of Sciences of the United States of America, 101, 3316-3321. http://dx.doi.org/10.1073/pnas.0400266101

[2]   Foss, M.L. and Keteyian, S.J. (1998) Fox's Physiological Basis for Exercise and Sport. 6th Edition, McGraw-Hill, Boston.

[3]   Mougios, V. (2006) Exercise Biochemistry. Matz. K., Ed., Human Kinetics, Champaign, Illinois, 332.

[4]   Granger, D.A., et al. (2007) Salivary Alpha-Amylase in Biobehavioral Research: Recent Developments and Applications. Annals of the New York Academy of Sciences, 1098, 122-144. http://dx.doi.org/10.1196/annals.1384.008

[5]   Takai, N., et al. (2007) Gender-Specific Differences in Salivary Biomarker Responses to Acute Psychological Stress. Annals of the New York Academy of Sciences, 1098, 510-515. http://dx.doi.org/10.1196/annals.1384.014

[6]   Nater, U.M., et al. (2006) Stress-Induced Changes in Human Salivary Alpha-Amylase Activity—Associations with Adrenergic Activity. Psychoneuroendocrinology, 31, 49-58. http://dx.doi.org/10.1016/j.psyneuen.2005.05.010

[7]   Nater, U.M., et al. (2005) Human Salivary Alpha-Amylase Reactivity in a Psychosocial Stress Paradigm. International Journal of Psychophysiology, 55, 333-342. http://dx.doi.org/10.1016/j.ijpsycho.2004.09.009

[8]   Rohleder, N., et al. (2004) Psychosocial Stress-Induced Activation of Salivary Alpha-Amylase: An Indicator of Sympathetic Activity? Annals of the New York Academy of Sciences, 1032, 258-263. http://dx.doi.org/10.1196/annals.1314.033

[9]   Ben-Aryeh, H., et al. (1989) Effect of Exercise on Salivary Composition and Cortisol in Serum and Saliva in Man. Journal of Dental Research, 68, 1495-1497. http://dx.doi.org/10.1177/00220345890680110501

[10] Chicharro, J.L., et al. (1998) Saliva Composition and Exercise. Sports Medicine, 26, 17-27. http://dx.doi.org/10.2165/00007256-199826010-00002

[11] Dickerson, S.S. and Kemeny, M.E. (2004) Acute Stressors and Cortisol Responses: A Theoretical Integration and Synthesis of Laboratory Research. Psychological Bulletin, 130, 355-391. http://dx.doi.org/10.1037/0033-2909.130.3.355

[12] Dalsgaard, M.K., Ide, K., Cai, Y., et al. (2002) The Intent to Exercise Influences the Cerebral O_2/Carbohydrate Uptake Ratio in Humans. Journal of Physiology, 540, 681-689. http://dx.doi.org/10.1113/jphysiol.2001.013062

[13] Dalsgaard, M.K., Ide, K., Cai, Y., et al. (2003) Cerebral Metabolism Is Influenced by Muscle Ischaemia during Exer-

cise in Humans. *Experimental Physiology*, **88**, 297-302. http://dx.doi.org/10.1113/eph8802469

[14] Ide, K. and Secher, N.H. (2000) Cerebral Blood Flow and Metabolism during Exercise. *Progress in Neurobiology*, **61**, 397-414. http://dx.doi.org/10.1016/S0301-0082(99)00057-X

[15] Ogoh, S., Dalsgaard, M.K., Yoshiga, C.C., *et al.* (2005) Dynamic Cerebral Autoregulation during Exhaustive Exercise in Humans. *American Journal of Physiology—Heart & Circulatory Physiology*, **288**, H1461-H1467. http://dx.doi.org/10.1152/ajpheart.00948.2004

[16] Grego, F., Vallier, J.M., Collardeau, M., *et al.* (2004) Effects of Long Duration Exercise on Cognitive Function, Blood Glucose, and Counter Regulatory Hormones in Male Cyclists. *Neuroscience Letters*, **364**, 76-80. http://dx.doi.org/10.1016/j.neulet.2004.03.085

[17] Owen, G., Turley, H. and Casey, A. (2004) The Role of Blood Glucose Availability and Fatigue in the Development of Cognitive Impairment during Combat Training. *Aviation, Space, and Environmental Medicine*, **75**, 240-246.

[18] Peters, A., Schweiger, U., Pellerin, L., *et al.* (2004) The Selfish Brain: Competition for Energy Resources. *Neuroscience & Biobehavioral Reviews*, **28**, 143-180. http://dx.doi.org/10.1016/j.neubiorev.2004.03.002

[19] Baker, J.S., Morgan, R., Hullin, D., *et al.* (2006) Changes in Blood Markers of Serotoninergic Activity Following High Intensity Cycle Ergometer Exercise. *Research in Sports Medicine*, **14**, 191-203. http://dx.doi.org/10.1080/15438620600854744

[20] Hargreaves, M. (2004) Muscle Glycogen and Metabolic Regulation. *Proceedings of the Nutrition Society*, **63**, 217-220. http://dx.doi.org/10.1079/PNS2004344

[21] Layden, J.D., Malkova, D. and Nimmo, M.A. (2004) During Exercise in the Cold Increased Availability of Plasma Nonesterified Fatty Acids Does Not Affect the Pattern of Substrate Oxidation. *Metabolism: Clinical & Experimental*, **53**, 203-208. http://dx.doi.org/10.1016/j.metabol.2003.09.018

[22] Stich, V. and Berlan, M. (2004) Physiological Regulation of NEFA Availability: Lipolysis Pathway. *Proceedings of the Nutrition Society*, **63**, 369-374. http://dx.doi.org/10.1079/PNS2004350

[23] Bulow, J., Gjeraa, K., Enevoldsen, L.H., *et al.* (2006) Lipid Mobilization from Human Abdominal, Subcutaneous Adipose Tissue Is Independent of Sex during Steady-State Exercise. *Clinical Physiology & Functional Imaging*, **26**, 205-211. http://dx.doi.org/10.1111/j.1475-097X.2006.00664.x

[24] Marion-Latard, F., Crampes, F., Zakaroff-Girard, A., *et al.* (2003) Post-Exercise Increase of Lipid Oxidation after a Moderate Exercise Bout in Untrained Healthy Obese Men. *Hormone & Metabolic Research*, **35**, 97-103. http://dx.doi.org/10.1055/s-2003-39051

[25] Marion-Latard, F., De Glisezinski, I., Crampes, F., *et al.* (2001) A Single Bout of Exercise Induces Beta-Adrenergic Desensitization in Human Adipose Tissue. *American Journal of Physiology—Regulatory Integrative & Comparative Physiology*, **280**, R166-R173.

[26] Stich, V., de Glisezinski, I., Berlan, M., *et al.* (2000) Adipose Tissue Lipolysis Is Increased during a Repeated Bout of Aerobic Exercise. *Journal of Applied Physiology*, **88**, 1277-1283.

[27] American College of Sports Medicine (2000) ACSM's Guidelines for Exercise Testing and Prescription. 6th Edition, Lippincott Williams & Wilkins, Philadelphia, 368.

[28] Gordon, H.W. and Lee, P.A. (1993) No Difference in Cognitive Performance between Phases of the Menstrual Cycle. *Psychoneuroendocrinology*, **18**, 521-531. http://dx.doi.org/10.1016/0306-4530(93)90045-M

[29] Jackson, A.S., Pollock, M.L. and Ward, A. (1980) Generalized Equations for Predicting Body Density of Women. *Medicine & Science in Sports & Exercise*, **12**, 175-181. http://dx.doi.org/10.1249/00005768-198023000-00009

[30] Bleiberg, J., Garmoe, W.S., Halpern, E.L., *et al.* (1997) Consistency of Within-Day and Across-Day Performance after Mild Brain Injury. *Neuropsychiatry Neuropsychology and Behavioral Neurology*, **10**, 247-253.

[31] Bleiberg, J., Kane, R.L., Reeves, D.L., *et al.* (2000) Factor Analysis of Computerized and Traditional Tests Used in Mild Brain Injury Research. *Clinical Neuropsychologist*, **14**, 287-294. http://dx.doi.org/10.1076/1385-4046(200008)14:3;1-P;FT287

[32] Granger, D.A. and Kivlighan, K.T. (2003) Integrating Biological, Behavioral, and Social Levels of Analysis in Early Child Development: Progress, Problems, and Prospects. *Child Development*, **74**, 1058-1063. http://dx.doi.org/10.1111/1467-8624.00590

[33] Li, T.-L. and Gleeson, M. (2004) The Effect of Single and Repeated Bouts of Prolonged Cycling and Circadian Variation on Saliva Flow Rate, Immunoglobulin A and Alpha-Amylase Responses. *Journal of Sports Sciences*, **22**, 1015-1024. http://dx.doi.org/10.1080/02640410410001716733

[34] Walsh, N.P., Blannin, A.K., Clark, A.M., *et al.* (1999) The Effects of High-Intensity Intermittent Exercise on Saliva IgA, Total Protein and Alpha-Amylase. *Journal of Sports Sciences*, **17**, 129-134. http://dx.doi.org/10.1080/026404199366226

[35] Kivlighan, K.T. and Granger, D.A. (2006) Salivary Alpha-Amylase Response to Competition: Relation to Gender, Previous Experience, and Attitudes. *Psychoneuroendocrinology*, **31**, 703-714. http://dx.doi.org/10.1016/j.psyneuen.2006.01.007

[36] Aubert, A., Costalat, R., Magistretti, P.J. and Pellerin, L. (2005) Brain Lactate Kinetics: Modeling Evidence for Neuronal Lactate Uptake upon Activation. *Proceedings of the National Academy of Sciences of the United States of America*, **102**, 16448-16453. http://dx.doi.org/10.1073/pnas.0505427102

[37] Wee, J., Charlton, C., Simpson, H., *et al.* (2005) GH Secretion in Acute Exercise May Result in Post-Exercise Lipolysis. *Growth Hormone & Igf Research*, **15**, 397-404. http://dx.doi.org/10.1016/j.ghir.2005.08.003

29

Genotoxicity of Some Essential Oils Frequently Used in Aromatherapy

Nadya Mezzoug[1], Mohamed Idaomar[1], Dominique Baudoux[2], Pascal Debauche[2], Véronique Liemans[2], Abdesselam Zhiri[2*]

[1]Laboratory of Biology and Health, Faculté des Sciences, Université Abdelmalek Essaâdi, Tétouan, Morocco
[2]S.A. PRANAROM International, Ghislenghien, Belgium
Email: *azhiri@pranarom.com

Abstract

Genotoxic properties of the essential oils extracted from *Artemisia dracunculus* (tarragon), *Ocimum basilicum* (basil), *Cinnamomum loureirii* (cinnamon), *Laurus nobilis* (laurel), *Satureja montana* (savory) and *Rosmarinus officinallis* (rosemary) are studied by *Drosophila melanogaster* Somatic Mutation and Recombination Test (SMART). The high bioactivation crossed with a high cytochrome P450-dependent bioactivation capacity is used. This assay is principally based on the loss of heterozygosity of the suitable recessive markers' multiple wing hairs (*mwh*) and *flare*-3 (*flr³*) which can lead to the formation of mutant clones of larval cells, and which are then going to be expressed as spots on the wings of adult flies. Third-instar *larvae* are treated for 48 hr with different concentrations of the essential oils dissolved in Tween-80 at 0.2% or 2%. The wings of the emerging adults are analyzed for the occurrence of different types of mutant spots. No statistically significant differences in spot frequencies between negative controls and treated series are observed. These results suggest that the six essential oils at concentrations tested are not genotoxic towards somatic cells of *D. melanogaster*.

Keywords

Essential Oils, Genotoxicity, Somatic Mutation, Mitotic Recombination, *Drosophila melanogaster*

1. Introduction

Essential oils are natural volatile substances found in a variety of plants' particular fragrance. They are products of the secondary metabolism of plants, and generally are fragrant volatile materials consisting of a complex

---

*Corresponding author.

mixture of which the most common constituents are the monoterpene hydrocarbons, oxygenated monoterpenes, sesquiterpene hydrocarbons, oxygenated sesquiterpenes, benzene derivatives, and non-isoprenoid components including alcohols, aldehydes and ketones. The flavor-imparting essential oils' content of the spices, herbs and leafy vegetables is important and can represent more than 5% of their fresh mass [1]. The essential oils can be isolated from several parts of plant usually by using the distillation method.

Essential oils have been widely used in traditional medicine, as food additives or food preservatives as well as in aromatherapy and in the industries of perfumes and cosmetics. Over the last couple of years, there has been an increasing interest in the use of the essential oils as a large number of them have been investigated for their biological activities. Indeed, the antimicrobial properties of essential oils and their constituents against some bacteria and fungi are described in more than 500 reports [2]. Moreover, the antiphlogistic, cough-relieving and spasmolytic effects of essential oils have been observed in experimental animals [3]. Also the essential oils can play a significant role as antioxidant [4], insecticide [5] anti-inflammatory and antinociceptive [6]. Furthermore, the hepatoprotective [7] and anticarcinogenic activities of specific essential oils were revealed [8]. However, only few papers contain data on their toxicity, and less about their chronic toxicities such as teratogenesis, carcinogenesis and mutagenesis. Though, it must be noted that studies on genotoxicity of individual components of essential oils are much more abundant, about 30 constituents of essential oils, mainly monoterpenes and alkenylbenzenes, have been tested for their genotoxicity. About one-third of them have shown a genotoxic effect in one or several genotoxicity tests [9].

Based on this preliminary information, and led by our consistent interest to assess the genotoxic profile of our natural essences [10]-[12], the aim of the present study was to investigate the genotoxicity of six essential oils frequently used in aromatherapy by the Somatic Mutation and Recombination Test (SMART) in *Drosophila melanogaster*.

Genotoxicity tests were developed to detect genotoxic substances and to assess the genetic hazard of chemicals to humans. The tests in *Drosophila melanogaster* present undoubted advantages: they are an *in vivo* system that uses a eukaryotic organism with metabolic machinery similar to that found in mammalians cells [13] and which the genome sequencing has shown that half of the identified protein sequences are similar to the mammalian proteins [14]. On the other hand, those assays are also characterized by their rapidity as compared with that of prokaryotic or unicellular *in vitro* tests [15].

The SMART is a sensitive short-term assay for the detection of somatic mutation and mitotic recombination [13]. It has been widely used in both version based on eye or wing marker [16] [17]. In both cases loss of heterozygosity leads to uncovering and expression of the recessive marker gene(s) in the larval imaginal disc cells. The SMART assay has been extensively validated and a hundred of compounds and complex mixtures have been analysed since the assay was developed [18].

The SMART system using wing tissue is based on the markers *mwh* (multiple wing hairs) and *flr* (*flare*) and takes advantage of the possibility to expose and analyze a large number of cells: approximately 25,000 cells in one wing [19]. If a genetic alteration occurs in *larvae*, this alteration will form a clone of mutant cells that can be detected as a spot of mutant cells on the body surface of the adult flies [19]. Single spots, either *mwh* or *flr*, can be produced by somatic point mutations or deletions and may also result from aneuploidy or chromosome loss [13] [20]. Single *mwh* spots also arise from recombination between the *mwh* and *flr* loci. Twin spots, consisting of both *mwh* and *flr* subclones are originated exclusively from mitotic recombination [13] [20]. All of these mutational events detected by the wing SMART assay well permit to measure mutagenic damage that could be induced by the essential oils tested.

2. Material and Methods

2.1. Essential Oils and Plant Material

Oils of *Cinnamomum loureirii* (cinnamon), *Artemisia dracunculus* (tarragon), *Laurus nobilis* (laurel), *Ocimum basilicum* (basil), *Rosmarinus officinallis* (rosemary), and of *Satureja montana* (savory) are provided by PRANAROM International Company, Belgium. The plants were collected from different geographic areas in the world and are listed in **Table 1** by plant species, common names as well as part used.

2.2. Extraction and Identification of Essential Oils

The essential oils were extracted by steam distillation. This was performed at a low pressure without chemical

Table 1. Botanical names, location and part used of selected aromatic plant species.

Botanical names	Common names	Location	Part used
Cinnamomum loureirii	Cinnamon	Vietnam	Bark
Artemisia dracunculus	Tarragone	Iran	Flowering tops
Laurus nobilis	Laurel	Slovenia	Leaf
Ocimum basilicum	Basil	Madagascar	Flowering tops
Rosmarinus officinallis	Rosemary	France	Flowering tops
Satureja montana	Savory	France	Flowering tops

descalers. The essential oils analyses were carried out by GC/MS using a Hewlett-Packard GCD system. HP-INNOWAX capillary column (60 m × 0.25 mm, 0.5 µm film thickness) was used with helium as carrier gas with flow 22 - 25 psi. GC oven temperature was held at 50°C for 6 min, then programmed at 2°C/min to 250°C and then held at this temperature for 20 min. The injector and detector temperatures were 250°C and 280°C, respectively, injection in split mode, volume injected 1 µl of a solution 5/100 in hexane of the oil. Automatic calibration of the masses by autotuning was used in MS. Mass range was from *m/z* 30 to 350. Library search was carried out using the combination of NKS library with 75,000 spectra and a personnel aromatic library. **Table 2** lists the oils major components obtained by gas chromatography analysis.

2.3. Mutagenicity Assay

For mutagenicity testings, the essentials oils were dissolved in the Tween-80 at 0.2% or 2% depending on the solubility of the oils. The essential oils were administered to *Drosophila larvae* at different concentrations, ranging from 0.025% to 0.3%. Solutions were always freshly prepared immediately before use. The solvents are also used as negative controls. Urethane (CAS registry number: 51-79-6) was used as positive control.

2.3.1. *Drosophila* Stocks and Cross

The high bioactivation cross was used when *NORR/NORR; NORR/NORR; flr³/In* (3LR) *TM3, ri p^p sep l(3)89Aa bx^{34e} e Bd^s* (*flr³/TM3*) females are crossed with *NORR/NORR; NORR/NORR; mwh/mwh* males. *NORR* strains (New ORR) has chromosomes 1 and 2 from DDT-resistant which are responsible for a high constitutive level of cytochrome P450 [21]. The *mwh* strain is homozygote for the wing cell marker multiple wing hairs (*mwh*, 3 - 0.3). The *flr³/TM3* strain contains the wing cell marker allele *flare*3 (*flr³*, 3 - 38.8) and the balancer chromosomes TM. More detailed information on genetic markers and descriptions are given by Lindsley and Zimm [22].

2.3.2. Experimental Procedures

Eggs from the high bioactivation were collected for 8h in culture bottles containing the live fermenting yeast. After 72 ± 4 h, the *larvae* were collected off the food with a 20% NaCl solution. The *larvae* were transferred to individual vials containing 1.5 g of food prepared from *Drosophila* Instant Medium (Carolina Biological Supply) hydrated by 5 ml of the essential oils dissolved in Tween-80 at different concentrations. The *larvae* were fed on this medium for the rest of their development which corresponds to a chronic treatment of approximately 48h until pupation. Negative solvent controls were included in all treatments. All experiments were conducted at 25°C and 65% relative humidity.

The hatched flies were collected from the treatment vials and flies of the trans-heterozygous (*mwh flr + /mwh + flr*) genotype were stored in 70% ethanol. The wings of adult flies were mounted on slides and scored under 400× magnification for the presence of cell clones showing malformed wing hairs. Such spots appeared as single spots expressing either the multiple wing hair (*mwh*) or flare (*flr*) phenotype or as twin spots with adjacent *mwh* and *flr* areas.

2.4. Data Evaluation and Statistical Analysis

For the evaluation of the recorded genotoxic effects, the frequencies of spots per fly of a treated series were compared to its concurrent negative control. Statistical analyses have been conducted using a Chi-square test.

Table 2. Major chemical components of *Cinnamomum loureirii*, *Artemisia dracunculus*, *Laurus nobilis*, *Ocimum basilicum*, *Rosmarinus officinallis* and *Satureja montana* essential oils.

	C. loureirii	A. dracunculus	L. nobilis	O. basilicum	R. officinallis	S. montana
α-Pinene	—	1.19	4.13	0.24	26.48	1.10
Camphene	—	—	0.41	—	6.21	0.69
β-Pinene	—	—	3.48	0.41	2.66	0.42
P-Menthatriene	—	—	—	—	1.48	—
Sabinene	—	—	6.53	0.15	—	—
Benzene propanal	1.32	—	—	—	—	—
β-Myrcene	—	0.10	0.94	0.20	1.50	0.89
1,8-Cineole	—	—	38.73	4.02	10.71	1.32
Limonene	—	3.04	1.47	0.32	3.61	0.95
Cis-β-ocimene	—	8.49	—	0.13	0.16	0.35
Cinnamaldehyde	81.66	—	—	—	—	—
Trans-β-ocimene	—	10.05	—	1.47	—	0.10
γ-Terpinene	—	—	0.88	0.13	1.35	7.01
P-Cymene	—	—	0.53	0.09	1.98	9.85
Terpinolene	—	—	0.40	—	0.73	0.16
α-Copaene	1.55	—	—	—	—	—
β-Caryophyllene	0.66	—	0.50	0.56	0.32	1.20
Coumarine mw = 146	2.63	—	—	—	—	—
Cinnamyle acetate	3.66	—	—	—	—	—
Camphre	—	—	—	0.24	2.82	1.28
Linalol	—	—	7.77	1.89	2.84	2.42
Bornyle acetate	—	—	0.53	0.45	13.19	0.49
Carvacrol methyl ether	—	—	—	—	—	4.88
Terpinene-4-ol	—	—	3.49	0.60	1.18	0.91
Methyl chavicol	—	75.23	0.08	77.41	0.13	—
α-Terpineol	—	—	3.21	0.20	1.32	1.88
Terpenyle acetate	—	—	12.34	—	—	—
Borneol	—	—	0.12	0.10	4.45	3.41
Verbenone	—	—	—	—	7.27	—
Carvone	—	—	—	—	0.11	0.83
δ-Cadinene	0.89	—	0.13	—	—	0.50
Geraniol	—	—	—	—	1.33	—
Methyleugenol	—	0.71	6.79	3.75	0.11	0.54
Eugenol	—	0.31	1.97	0.14	—	0.19
T-Cadinol	—	—	—	1.66	—	—
Thymol	—	—	—	—	—	10.59
Carvacrol	—	—	—	—	—	36.67

Accordingly we distinguished small single spots (one or two cells affected), large single spots (more than two cells affected) and twin spots.

3. Results

Before starting the genotoxicity experiments, toxicity is evaluated in the first instance in order to determine the concentrations that are going to be tested in the wing spot test. The toxicity shown in **Table 3** was measured by determining the fraction of the *larvae* developing to adulthood in comparison with negative controls. All six essential oils tested present toxicity in 48h larval feeding experiments at different degrees. At a concentration of 0.3%, the essential oils of *Artemisia dracunculus*, *Ocimum basilicum* and *Cinnamomum loureirii* were shown to be very toxic for *larvae* and no adult flies survived. The essential oil from *Satureja montana* was less toxic at this concentration. For their part, *Laurus nobilis* and *Rosmarinus officinalis* presented weak toxicity at 0.3%.

The results of the genotoxicity study obtained with the six essential oils in the *Drosophila* wing spots test after chronic exposure are shown in **Table 4**. The spot data for small single spots, large single spots and twin spots

Table 3. Toxicity of the essentials oils tested expressed in % of lethality of the treated larvae.

Oil	Dose (%)	Toxicity (%)
Artemisia dracunculus		
	0.3	100
	0.2	90 - 100
	0.1	25 - 50
	0.05	5 - 10
Ocimum basilicum		
	0.3	100
	0.2	90 - 100
	0.1	50
	0.05	5 - 10
Cinnamomum loureirii		
	0.3	100
	0.2	80 - 100
	0.1	50 - 75
	0.05	25 - 50
Satureja montana		
	0.3	80 - 100
	0.2	50
	0.1	10 - 25
	0.05	5 - 10
Laurus nobilis		
	0.3	25
	0.2	10 - 20
	0.1	5 - 10
Rosmarinus officinallis		
	0.3	25
	0.2	10 - 20
	0.1	5 - 10

Table 4. Wing spots data obtained after chronic exposure with the essential oils.

Compounds concentration (%)	Number of wings	Spots per wing (number of spots)			
		Small single spots	Large single spots	Twin spots	Total spots
Cinnamomum loureirii					
Tween-80 0.2%	71	0.28 (20)	0.06 (04)	0.00 (00)	0.34 (24)
0.025	40	0.27[a] (11)	0.00[a] (00)	0.00[a] (00)	0.27[a] (11)
0.05	40	0.40[a] (16)	0.02[a] (01)	0.00[a] (00)	0.42[a] (17)
0.1	40	0.35[a] (14)	0.00[a] (00)	0.00[a] (00)	0.35[a] (14)
Artemisia dracunculus					
Tween-80 0.2%	71	0.28 (20)	0.06 (04)	0.00 (00)	0.34 (24)
0.025	40	0.35[a] (14)	0.00[a] (00)	0.00[a] (00)	0.35[a] (14)
0.05	42	0.45[a] (19)	0.10[a] (04)	0.00[a] (00)	0.55[a] (23)
0.1	25	0.32[a] (08)	0.20[a] (05)	0.00[a] (00)	0.52[a] (13)
Laurus nobilis					
Tween-80 2%	158	0.52 (82)	0.04 (06)	0.00 (00)	0.56 (88)
0.1	40	0.43[a] (17)	0.02[a] (01)	0.00[a] (00)	0.45[a] (18)
0.2	32	0.34[a] (11)	0.09[a] (03)	0.00[a] (00)	0.43[a] (14)
0.3	39	0.41[a] (16)	0.03[a] (01)	0.00[a] (00)	0.44[a] (17)
Ocimum basilicum					
Tween-80 2%	158	0.52 (82)	0.04 (06)	0.00 (00)	0.56 (88)
0.025	40	0.32[a] (13)	0.00[a] (00)	0.00[a] (00)	0.32[a] (13)
0.05	40	0.50[a] (20)	0.05[a] (02)	0.00[a] (00)	0.55[a] (22)
0.1	40	0.30[a] (12)	0.02[a] (01)	0.00[a] (00)	0.32[a] (13)
Rosmarinus officinallis					
Tween-80 2%	158	0.52 (82)	0.04 (06)	0.00 (00)	0.56 (88)
0.1	40	0.45[a] (18)	0.10[a] (04)	0.00[a] (00)	0.55[a] (22)
0.2	40	0.35[a] (14)	0.02[a] (01)	0.00[a] (00)	0.37[a] (15)
Satureja montana					
Tween-80 2%	158	0.52 (82)	0.04 (06)	0.00 (00)	0.56 (88)
0.05	37	0.68[a] (25)	0.08[a] (03)	0.00[a] (00)	0.76[a] (28)
0.1	33	0.45[a] (15)	0.06[a] (02)	0.00[a] (00)	0.51[a] (17)
Urethane					
Distilled water	40	0.30 (12)	0.03 (01)	0.00 (00)	0.33 (13)
5 mM	40	2.53[d] (101)	0.65[d] (26)	0.27[c] (11)	3.45[d] (138)

[a]$p > 0.05$; [c]$p < 0.01$; [d]$p < 0.001$.

together with the total number of spots are presented. Different concentrations of each compound were assessed using *larvae* of the high bioactivation cross. The compounds were dissolved in Tween-80 0.2% or Tween-80 2%. Urethane was dissolved in distilled water. For each essential oil, the treated series were compared with the negative control corresponding to the pooled results of the spontaneous mutations detected for the corresponding solvent used. The spontaneous frequency of total Spots obtained in Tween-80 2% (0.56) was higher than the value in the Tween-80 0.2% (0.34) and statistically different. Moreover, there was no significant difference between mutations detected with the Tween-80 0.2% and water (0.33). Urethane at 5 mM increased significantly ($p < 0.001$) the small single spots, large simple spots, and the total of spots. The induction of the small size clone was more important than that of the large clones. Also the frequency of twin spots was increased ($p < 0.01$) in the presence of this promutagen.

Around and/or low concentrations than LC50 were chosen for all the six essential oils to conduct the genotoxicity experiments. The number of spots as well as their type and size were recorded. Two basic types of spots, single and twin, could be observed in the SMART assay. In our treatments, small single spots predominated; large single spots were rare whereas twin spots were absent. Most *mwh* clones were small sizes; however, clones that are larger than 32 cells were absent. Among single spots in six different series, only few *flr* spots were observed; all other single spots showed the *mwh* phenotype partly because the majority of the clones detected are small sizes and a mutational event at the *flr+* does not express itself in clone smaller than a certain size [23]. Moreover, *flr³* probably arises from relatively rare events like point mutations at the locus, interstitial deletions and perhaps double crossing-over [24]. Twin spots are absent, partly because only rare mitotic recombination events which take place between the *flr* locus and the centromere produce this type of spot [13].

From the statistically treated data summarized on **Table 4** it can be pointed out that, the essential oils tested do not induce a significant increase in the frequency on any of the three categories of spots. Although an increase of the frequencies of mutations was observed with *Satureja montana* and *Artemisia dracunculus* essential oils in comparison with the negative solvent control; but this increase in mutant frequency was not considered biologically significant as there was no evidence of a dose-response effect. Also a weak increase was detected with the essential oil from *Cinnamomum loureirii* at 0.05% that remains not statistically or biologically significant. However, there was no increase of frequencies for other essential oils tested. In addition, a reduction in the rate of spontaneous mutations was observed for these oils, this effect being dose dependant for *Rosmarinus officinallis* contrarily to *Laurus nobilis* and *Ocimum basilicum* oils.

4. Discussion

After the exposition of the *larvae* of *Drosophila melanogaster* to the studied essential oils, a significant toxicity effect was observed. Many of plant secondary products are known for their high toxicity, and they are involved in plant defence mechanisms against herbivores as well as insects. This is the case of essential oils of which the insecticidal action was already demonstrated [5] [25]. In addition, the strains used in SMART assays are characterized by a high level of P450 [21], which is known for its role in the metabolism of several insecticides and plant toxins [26].

All the six essential oils tested are not genotoxic. These results are in agreement with the results previously shown by other authors using different assays, which demonstrated the absence of the genotoxicity of a great number of essential oils and stated out that only few essential oils are genotoxic [10] [12] [27]-[32]. But it must be noted that the positive results demonstrating the genotoxic effect of many essential oils were also found [9]. Positive and negative results about some of the oils tested in our study were reported; the genotoxic properties of *Artemisia dracunculus* and *Satureja montana* essential oils were studied with *Bacillus subtilis* Rec-assay and *Salmonella*/microsome reversion assay; only the oil of *Artemisia dracunculus* can be active in the Rec-assay but not in the *Salmonella* test [29]. With respect to our results, the Tarragon essential oil did not confirm its genotoxic potential in the *Drosophila* wing spot test even if the frequencies of spots detected in the treatments with this essential oil were weakly enhanced, but without statistical significance. Cinnamon bark oil was studied in the Ames *Salmonella* reversion assay, in the *Bacillus subtilis* DNA-repair test (Rec-assay) without S9 and in the *Escherichia coli* WP2 uvrA reversion test, and showed negative results in the 3 microbial test systems [27]. This negative effect of Cinnamon essential oil was confirmed by SMART assay in our study. However, a genotoxic potential of the Basil essential oil in rat hepatocytes *in vitro* and in rat liver in an *in vivo* test was demonstrated [33] but not with SMART assay according to our results.

On the other hand, the main components of essential oils have been tested for their mutagenicity by a range of genotoxicity tests, and produced contradictory results. The negative mutagenic effect, which can confirm our results, was found in the Ames *Salmonella* reversion assay and in the *Escherichia coli* WP2 uvrA reversion test with methyl chavicol, which is present in both *Artimisia dracunculus* essential oil at 75.23% and *Ocimum basilicum* oil at 77.41%, and with cinnamaldehyde, the main compound of *Cinnamomum loureirii* essential oil (81.66%) [27] [34]. Cinnamaldehyde did not cause any DNA damage in the SOS-chromotest [35]. However, this compound exhibited a weak mutagenic response in TA100 *Salmonella* strain with mouse liver S9 [36] and gave a positive response in the *E. coli* DNA repair test [35]. It has been reported that cinnamaldehyde was positive in a *Drosophila* sex-linked recessive lethal mutation test and in a chromosomal aberration test with Chinese hamster fibroblasts [37]. A positive effect was also detected in the *Bacillus subtilis* DNA-repair test (Rec-assay) with cinnamaldehyde and with methyl chavicol [27]. 1,8-Cineole, an important molecule found in Laurel oil and even more present in Rosemary essential oil, did not show any mutagenic effect by the *Salmonella* assay [32]. Thymol was screened for mutagenic activity using the same test; no effect was detected with this important savory oil compound [34]. Thymol was also tested *in vitro* on human pulp fibroblasts without any genotoxic effect [38]. However, weak significant genotoxic effect was observed in the DNA repair test with thymol and also with carvacrol which is the major constituent of the savory essential oil [35]. In the SOS-chromotest, none of the carvacrol and thymol was positive [35].

The genotoxicity tests were negative for d-limonene which is present in all essential oils at different proportions [39]. Also, no evidence of myrcene-induced clastogenicity was observed using the rat bone marrow cytogenetic *in vivo* assay [40] and no mutagenic effect in *Salmonella* typhimurium with borneol was detected [34]. However, the Terpineol present in many essential oils tested caused a slight but dose-related increase in the number of his + revertants with TA102 *Salmonella* tester strain both without and with addition of S9 mixture [32]. For their part, methyl eugenol and eugenol did not show mutagenicity in the Ames assay and in the *Escherichia coli* WP2 uvrA reversion test [27] [34]. In the human pulp fibroblasts *in vitro*, eugenol did not show any genotoxic effect [38]. But using the bone marrow micronucleus assay in mice, eugenol showed a significant induction of micronucleus in 400 and 600 mg/kg doses [41]. Mutagenic capacity of eugenol was also demonstrated by *in vivo* eukaryotic assays on mice [42].

Thus, the evaluation of the genotoxicity of the components of essential oils showed data that can vary according to the organism and the genotoxic assay used. Moreover, the negative results obtained in the present study suggest that the studied essential oils for the tested concentrations are not genotoxic in *D. melanogaster*. This is not always in accordance with the data of some of their constituents. This can be explained by the antagonistic phenomena when some antimutagenic compounds can be present in the oils tested and oppose the mutagenic effect of other components of the mixture of essential oil. Further experiments are suggested to evaluate possible antigenotoxic properties of these oils or some of their constituents.

Acknowledgements

This work was supported by PRANAROM International Company and the Ministry of Walloon Region, Belgium (Direction générale des technologies, de la recherche et de l'énergie, Promotion de l'innovation technique, Jambe-Namur).

References

[1]   Achinewhu, S.C., Ogbonna, C.C. and Hart, A.D. (1995) Chemical Composition of Indigenous Wild Herbs, Spices, Nuts and Leafy Vegetables Used as Food. *Plant Foods for Human Nutrition*, **48**, 341-348. http://dx.doi.org/10.1007/BF01088493

[2]   Kalemba, D. and Kunicka, A. (2003) Antibacterial and Antifungal Properties of Essential Oils. *Current Medicinal Chemistry*, **10**, 813-829. http://dx.doi.org/10.2174/0929867033457719

[3]   Inouye, S., Yamaguchi, H. and Takizawa, T. (2001) Screening of the Antibacterial Effects of a Variety of Essential Oils on Respiratory Tract Pathogens, Using a Modified Dilution Assay Method. *Journal of Infection and Chemotherapy*, **7**, 251-254. http://dx.doi.org/10.1007/s101560170022

[4]   Miguel, M.G. (2010) Antioxidant and Anti-Inflammatory Activities of Essential Oils: A Short Review. *Molecules*, **15**, 9252-9287. http://dx.doi.org/10.3390/molecules15129252

[5]   Liu, Z.L., Chu, S.S. and Liu, Q.R. (2010) Chemical Composition and Insecticidal Activity against *Sitophilus zeamais* of the Essential Oils of *Artemisia capillaris* and *Artemisia mongolica*. *Molecules*, **15**, 2600-2608. http://dx.doi.org/10.3390/molecules15042600

[6]   Leite, B.L., Bonfim, R.R., Antoniolli, A.R., Thomazzi, S.M., Araújo, A.A., Blank, A.F., Estevam, C.S., Cambui, E.V.,

Bonjardim, L.R., Albuquerque Júnior, R.L. and Quintans-Júnior, L.J. (2010) Assessment of Antinociceptive, Anti-Inflammatory and Antioxidant Properties of *Cymbopogon winterianus* Leaf Essential Oil. *Pharmaceutical Biology*, **10**, 1164-1169. http://dx.doi.org/10.3109/13880200903280000

[7]   Samojlik, I., Lakić, N., Mimica-Dukić, N., Daković-Svajcer, K. and Bozin, B. (2010) Antioxidant and Hepatoprotective Potential of Essential Oils of Coriander (*Coriandrum sativum* L.) and Caraway (*Carum carvi* L.) (*Apiaceae*). *Journal of Agricultural and Food Chemistry*, **58**, 8848-8853. http://dx.doi.org/10.1021/jf101645n

[8]   Edris, A.E. (2009) Anti-Cancer Properties of *Nigella* spp. Essential Oils and Their Major Constituents, Thymoquinone and Beta-Elemene. *Current Clinical Pharmacology*, **4**, 43-46. http://dx.doi.org/10.2174/157488409787236137

[9]   Lazutka, J.R., Mierauskiene, J., Slapsyte, G. and Dedonyte, V. (2001) Genotoxicity of Dill (*Anethum graveolens* L.), Peppermint (*Menthaxpiperita* L.) and Pine (*Pinus sylvestris* L.) Essential Oils in Human Lymphocytes and *Drosophila melanogaster*. *Food and Chemical Toxicology*, **39**, 485-492. http://dx.doi.org/10.1016/S0278-6915(00)00157-5

[10] Idaomar, M., El Hamss, R., Bakkali, F., Mezzoug, N., Zhiri, A., Baudoux, D., Munoz-Serrano, A., Liemans, V. and Alonso-Moraga, A. (2002) Genotoxicity and Antigenotoxicity of Some Essential Oils Evaluated by Wing Spot Test of *Drosophila melanogaster*. *Mutation Research*, **513**, 61-68. http://dx.doi.org/10.1016/S1383-5718(01)00287-X

[11] El Hamss, R., Idaomar, M., Alonso-Moraga, A. and Munoz Serrano, A. (2003) Antimutagenic Properties of Bell and Black Peppers. *Food and Chemical Toxicology*, **41**, 41-47. http://dx.doi.org/10.1016/S0278-6915(02)00216-8

[12] Mezzoug, N., Elhadri, A., Dallouh, A., Amkiss, S., Skali, N.S., Abrini, J., Zhiri, A., Baudoux, D., Diallo, B., El Jaziri, M. and Idaomar, M. (2007) Investigation of the Mutagenic and Antimutagenic Effects of *Origanum compactum* Essential Oil and Some of Its Constituents. *Mutation Research*, **629**, 100-110. http://dx.doi.org/10.1016/j.mrgentox.2007.01.011

[13] Graf, U., Würgler, F.E., Katz, A.J., Frei, H., Juon, H., Hall, C.B. and Kale, P.G. (1984) Somatic Mutation and Recombination Test in *Drosophila melanogaster*. *Environmental Mutagenesis*, **6**, 153-188. http://dx.doi.org/10.1002/em.2860060206

[14] Rubin, G.M., Yandell, M.D., Wortman, J.R., Gabor Miklos, G.L., Nelson, C.R., Hariharan, I.K., *et al.* (2000) Comparative Genomics of the Eukaryotes. *Science*, **287**, 2204-2215. http://dx.doi.org/10.1126/science.287.5461.2204

[15] Vogel, E.W. (1992) Tests for Recombinagens in Somatic Cells of *Drosophila*. *Mutation Research*, **284**, 159-175. http://dx.doi.org/10.1016/0027-5107(92)90030-6

[16] Vogel, E.W. and Zijlstra, J.A. (1987) Mechanistic and Methodological Aspects of Chemically-Induced Somatic Mutation and Recombination in *Drosophila melanogaster*. *Mutation Research*, **182**, 243-264. http://dx.doi.org/10.1016/0165-1161(87)90010-0

[17] Vogel, E.W. and Nivard, M.J. (1993) Performance of 181 Chemicals in a *Drosophila* Assay Predominantly Monitoring Interchromosomal Mitotic Recombination. *Mutagenesis*, **8**, 57-81. http://dx.doi.org/10.1093/mutage/8.1.57

[18] Delgado-Rodriguez, A., Ortiz-Marttelo, R., Villalobos-Pietrini, R., Gomez-Arroyo, S. and Graf, U. (1999) Genotoxicity of Organic Extracts of Airborne Particles in Somatic Cells of *Drosophila melanogaster*. *Chemosphere*, **39**, 33-43. http://dx.doi.org/10.1016/S0045-6535(98)00586-4

[19] Guzmán-Rincón, J. and Graf, U. (1995) *Drosophila melanogaster* Somatic Mutation and Recombination Test as a Biomonitor. In: Butterworth, F.M., Corkun, L.D. and Guzmán-Rincón, J., Eds., *Biomonitors and Biomarkers of Environment Change*, Plenum Press, New York, 169-181.

[20] Graf, U., Juon, H., Katz, A.J., Frei, H.J. and Wurgler, F.E. (1983) A Pilot Study on a New *Drosophila* Spot Test. *Mutation Research*, **120**, 233-239. http://dx.doi.org/10.1016/0165-7992(83)90095-7

[21] Frölich, A. and Würgler, F.E. (1989) New Tester Strains with Improved Bioactivation Capacity for the *Drosophila* Wing-Spot Test. *Mutation Research*, **216**, 179-187. http://dx.doi.org/10.1016/0165-1161(89)90003-4

[22] Lindsley, D.L. and Zimm, G.G. (1992) The Genome of *Drosophila melanogaster*. Academic Press, San Diego, 1133.

[23] Szabad, J., Soos, I., Polgar, G. and Hejja, G. (1983) Testing the Mutagenicity of Malondialdehyde and Formaldehyde by the *Drosophila* Mosaic and the Sex-Linked Recessive Lethal Tests. *Mutation Research*, **113**, 117-133. http://dx.doi.org/10.1016/0165-1161(83)90224-8

[24] Spano, M.A., Frei, H., Würgler, F.E. and Graf, U. (2001) Recombinagenic Activity of Four Compounds in the Standard and High Bioactivation Crosses of *Drosophila melanogaster* in the Wing Spot Test. *Mutagenesis*, **16**, 385-394. http://dx.doi.org/10.1093/mutage/16.5.385

[25] Isman, B.M. (2000) Plant Essential Oils for Pest and Disease Management. *Crop Protection*, **19**, 603-608. http://dx.doi.org/10.1016/S0261-2194(00)00079-X

[26] Scott, J.G., Liu, N. and Wen, Z. (1998) Insect Cytochromes P450: Diversity, Insecticide Resistance and Tolerance to Plant Toxins. *Comparative Biochemistry and Physiology, Part C, Pharmacology, Toxicology and Endocrinology*, **121**, 147-155. http://dx.doi.org/10.1016/S0742-8413(98)10035-X

[27] Sekizawa, J. and Shibamoto, T. (1982) Genotoxicity of Safrole-Related Chemicals in Microbial Test Systems. *Mutation Research*, **101**, 127-140. http://dx.doi.org/10.1016/0165-1218(82)90003-9

[28] Andersen, P.H. and Jensen, N.J. (1984) Mutagenic Investigation of Flavourings: Dimethyl Succinate, Ethyl Pyruvate and Aconitic Acid Are Negative in the *Salmonella*/Mammalian-Microsome Test. *Food Additives and Contaminants*, **1**, 283-288. http://dx.doi.org/10.1080/02652038409385855

[29] Zani, F., Massimo, G., Benvenuti, S., Bianchi, A., Albasini, A., Melegari, M., Vampa, G., Bellotti, A. and Mazza, P. (1991) Studies on the Genotoxic Properties of Essential Oils with *Bacillus subtilis* Recassay and *Salmonella*/Microsome Reversion Assay. *Planta Medica*, **57**, 237-241. http://dx.doi.org/10.1055/s-2006-960081

[30] Franzios, G., Mirotsou, M., Hatziapostolou, E., Kral, J., Scouras, Z.G. and Mavragrani-Tsipidou, P. (1997) Insecticidal and Genotoxic Activities of Mint Essential Oils. *Journal of Agricultural and Food Chemistry*, **45**, 2690-2694. http://dx.doi.org/10.1021/jf960685f

[31] Karpouhtsis, I., Pardali, E., Feggou, E., Kokkini, S., Scouras, Z.G. and Mavragani-Tsipidou, P. (1998) Insecticidal and Genotoxic Activities of Oregano Essential Oils. *Journal of Agricultural and Food Chemistry*, **46**, 1111-1115. http://dx.doi.org/10.1021/jf970822o

[32] Padilha de Paula, J., Gomes-Carneiro, M.R. and Paumgartten, F.J. (2003) Chemical Composition, Toxicity and Mosquito Repellency of *Ocimum selloi* Oil. *Journal of Agricultural and Food Chemistry*, **88**, 253-260. http://dx.doi.org/10.1016/S0378-8741(03)00233-2

[33] Muller, L., Kasper, P., Muller-Tegethoff, K. and Petr, T. (1994) The Genotoxic Potential *in Vitro* and *in Vivo* of the Allyl Benzene Etheric Oils Estragole, Basil Oil and Transanethole. *Mutation Research*, **325**, 129-136. http://dx.doi.org/10.1016/0165-7992(94)90075-2

[34] Azizan, A. and Blevins, R.D. (1995) Mutagenicity and Antimutagenicity Testing of Six Chemicals Associated with the Pungent Properties of Specific Spices as Revealed by the Ames Salmonella/Microsomal Assay. *Archives of Environmental Contamination and Toxicology*, **28**, 248-258. http://dx.doi.org/10.1007/BF00217624

[35] Stammati, A., Bonsi, P., Zucco, F., Moezelaar, R., Alakomi, H.L. and Von Wright, A. (1999) Toxicity of Selected Plant Volatiles in Microbial and Mammalian Short-Term Assays. *Food and Chemical Toxicology*, **37**, 813-823. http://dx.doi.org/10.1016/S0278-6915(99)00075-7

[36] Dillon, D., Combes, R. and Zeiger, E. (1998) The Effectiveness of *Salmonella* Strains TA100, TA102 and TA104 for Detecting Mutagenicity of Some Aldehydes and Peroxides. *Mutagenesis*, **13**, 19-26. http://dx.doi.org/10.1093/mutage/13.1.19

[37] Ishidate, M., Sofuni, T., Yoshikawa, K., Hayashi, M., Nohmi, T., Sawada, M. and Matsuoka, A. (1984) Primary Mutagenicity Screening of Food Additives Currently Used in Japan. *Food and Chemical Toxicology*, **22**, 623-636. http://dx.doi.org/10.1016/0278-6915(84)90271-0

[38] Chang, Y.C., Tai, K.W., Huang, F.M. and Huang, M.F. (2000) Cytotoxic and Non-Genotoxic Effects of Phenolic Compounds in Human Pulp Cell Cultures. *Journal of Endodontics*, **26**, 440-443. http://dx.doi.org/10.1097/00004770-200008000-00002

[39] Whysner, J. and Williams, G.M. (1996) D-Limonene Mechanistic Data and Risk Assessment: Absolute Species-Specific Cytotoxicity, Enhanced Cell Proliferation, and Tumor Promotion. *Pharmacology and Therapeutics*, **71**, 127-136. http://dx.doi.org/10.1016/0163-7258(96)00065-4

[40] Zamith, H.P., Vidal, M.N., Speit, G. and Paumgartten, F.J. (1993) Absence of Genotoxic Activity of Beta-Myrcene in the *in Vivo* Cytogenetic Bone Marrow Assay. *Brazilian Journal of Medical and Biological Research*, **26**, 93-98. http://www.ncbi.nlm.nih.gov/pubmed/8220273

[41] Ellahuene, M.F., Perez-Alzola, L.P., Orellana-Valdebenito, M., Munoz, C. and Lafuente-Indo, N. (1994) Genotoxic Evaluation of Eugenol Using the Bone Marrow Micronucleus Assay. *Mutation Research*, **320**, 175-180. http://dx.doi.org/10.1016/0165-1218(94)90044-2

[42] Woolverton, C.J., Fotos, P.G., Mokas, M.J. and Mermigas, M.E. (1986) Evaluation of Eugenol for Mutagenicity by the Mouse Micronucleus Test. *Journal of Oral Pathology*, **15**, 450-453. http://dx.doi.org/10.1111/j.1600-0714.1986.tb00656.x

Abbreviations

SMART: Somatic Mutation and Recombination Test
C. loureirii: *Cinnamomum loureirii*
A. dracunculus: *Artemisia dracunculus*
L. nobilis: *Laurus nobilis*
O. basilicum: *Ocimum basilicum*
R. officinallis: *Rosmarinus officinallis*
S. montan: *Satureja montana*
GC/MS: Gas Chromatography/Mass Spectrometry

Permissions

All chapters in this book were first published in ABB, by Scientific Research Publishing; hereby published with permission under the Creative Commons Attribution License or equivalent. Every chapter published in this book has been scrutinized by our experts. Their significance has been extensively debated. The topics covered herein carry significant findings which will fuel the growth of the discipline. They may even be implemented as practical applications or may be referred to as a beginning point for another development.

The contributors of this book come from diverse backgrounds, making this book a truly international effort. This book will bring forth new frontiers with its revolutionizing research information and detailed analysis of the nascent developments around the world.

We would like to thank all the contributing authors for lending their expertise to make the book truly unique. They have played a crucial role in the development of this book. Without their invaluable contributions this book wouldn't have been possible. They have made vital efforts to compile up to date information on the varied aspects of this subject to make this book a valuable addition to the collection of many professionals and students.

This book was conceptualized with the vision of imparting up-to-date information and advanced data in this field. To ensure the same, a matchless editorial board was set up. Every individual on the board went through rigorous rounds of assessment to prove their worth. After which they invested a large part of their time researching and compiling the most relevant data for our readers.

The editorial board has been involved in producing this book since its inception. They have spent rigorous hours researching and exploring the diverse topics which have resulted in the successful publishing of this book. They have passed on their knowledge of decades through this book. To expedite this challenging task, the publisher supported the team at every step. A small team of assistant editors was also appointed to further simplify the editing procedure and attain best results for the readers.

Apart from the editorial board, the designing team has also invested a significant amount of their time in understanding the subject and creating the most relevant covers. They scrutinized every image to scout for the most suitable representation of the subject and create an appropriate cover for the book.

The publishing team has been an ardent support to the editorial, designing and production team. Their endless efforts to recruit the best for this project, has resulted in the accomplishment of this book. They are a veteran in the field of academics and their pool of knowledge is as vast as their experience in printing. Their expertise and guidance has proved useful at every step. Their uncompromising quality standards have made this book an exceptional effort. Their encouragement from time to time has been an inspiration for everyone.

The publisher and the editorial board hope that this book will prove to be a valuable piece of knowledge for researchers, students, practitioners and scholars across the globe.

List of Contributors

Sunita Kumari Yadav, Vibhuti Sharma and Aparna Dixit
School of Biotechnology, Jawaharlal Nehru University, New Delhi, India

Carmelita N. Marbaniang
RNA Biology Group, Institute for Molecular Infection Biology, University of Würzburg, Würzburg, Germany

Rewaida Abdel-Gaber, Manal El Garhy and Kareem Morsy
Zoology Department, Faculty of Science, Cairo University, Cairo, Egypt

Jorge Parodi and Pamela Olivares
Laboratorio Fisiología de la Reproducción, Escuela de Medicina Veterinaria, Núcleo de Investigaciónen Producción Alimentaria, Facultad de Recursos Naturales, Universidad Católica de Temuco, Temuco, Chile

Viviana Chavez
Laboratorio de Investigación y Educación Tonalli Ltda, Temuco, Chile

Matías Peredo-Parada
Departamento de Ingenieríaen Obras Civiles, Universidad de Santiago de Chile, Santiago, Chile Plataforma de Investigaciónen Ecohidrología y Ecohidráulica, EcoHyd Ltda, Santiago, Chile

Shaiful Azuar Mohamad, Mat Rasol Awang, Rusli Ibrahim, Mohd Yusof Hamzah, Rosnani Abdul Rashid, Sobri Hussein and Khairuddin Abdul Rahim
Agrotechnology and Biosciences Division, Malaysian Nuclear Agency, Bangi, Malaysia

Choong Yew Keong
Herbal Medicine Research Centre, Institute For Medical Research, Kuala Lumpur, Malaysia

Fauzi Daud, Aidil Abdul Hamid and Wan Mohtar Wan Yusoff
Faculty of Science and Technology, Universiti Kebangsaan Malaysia, Bangi, Malaysia

Jennifer Anne Northmore, Marie Leung and Simon Dich Xung Chuong
Department of Biology, University of Waterloo, Waterloo, Canada

Khadidja Hassaballah
Department of Biology, Faculty of Exact and Applied Sciences, University of N'Djamena, N'Djamena, Chad

Vounparet Zeuh
Livestock Polytechnic Institute of Moussoro, Moussoro, Chad

Raman A. Lawal and Olivier Hanotte
School of Life Sciences, The University of Nottingham, University Park, Nottingham, UK

Mbacké Sembene
Department of Animal Biology, Faculty of Sciences and Techniques, Cheikh Anta Diop University of Dakar, Dakar, Senegal

Felicia N. Anike, Omoanghe S. Isikhuemhen and Dietrich Blum
Mushroom Biology and Fungal Biotechnology Laboratory, North Carolina A&T State University, Greensboro, USA

Hitoshi Neda
Forestry and Forest Products Research Institute (FFPRI), Tsukuba, Japan

Sairong Fan, Junjun Wang, Yingge Mao, Yuan Ji, Liqin Jin, Xiaoming Chen and Jianxin Lu
Key Laboratory of Laboratory Medicine, Ministry of Education of China, School of Laboratory Medicine & Life Science, Wenzhou Medical University, Wenzhou, China

Junjun Wang, Yingge Mao, Yuan Ji, Liqin Jin, Xiaoming Chen and Jianxin Lu
Institute of Glycobiological Engineering, School of Laboratory Medicine & Life Science, Wenzhou Medical University, Wenzhou, China

Maria Luiza P. de Oliveira and Ed Stover
USDA-ARS Subtropical Insects and Horticulture Research Unit, Fort Pierce, USA

Gloria Moore
Horticultural Science Department, Institute of Food and Agricultural Science, University of Florida, Gainesville, USA

James G. Thomson
USDA-ARS Crop Improvement and Utilization, Albany, USA

Md. Ismail and Hossain Uddin Shekhar
Department of Biochemistry and Molecular Biology, University of Dhaka, Dhaka, Bangladesh

Md. Faruk Hossain
Department of Biological Sciences, St John's University, New York, USA

Md. Fazlul Karim
Department of Biological Sciences, Eastern Illinois University, Charleston, USA

Mohsen Arooni-Hesari, Reza Talaei-Hassanloui and Qodrat Sabahi
Department of Plant Protection, College of Agriculture and Natural Resources, University of Tehran, Karaj, Iran

Rodrigo Sanchez, Pamela Olivares and Jorge Parodi
Laboratorio Fisiología de la Reproducción, Escuela de Medicina Veterinaria, Núcleo de Investigación en Producción Alimentaria, Facultad de Recursos Naturales, Universidad Católica de Temuco, Temuco, Chile

Hector Herrera and Rodrigo Sanchez
Empresa Vitapro Chile, Castro, Región de Los Lagos, Chile

Erico Carmona
Grupo de Genotoxicología, Escuela de Medicina Veterinaria, Núcleo de Estudios Ambientales, Facultad de Recursos Naturales, Universidad Católica de Temuco, Temuco, Chile

Allisson Astuya
Laboratory of Cell Culture and Marine Genomics, Marine Biotechnology Unit, Faculty of Natural and Oceanographic Sciences, University of Concepcion and Sur-Austral COPAS Program, University of Concepcion, Concepción, Chile

Gaku Matsunaga, Syuuji Karasuda, Ryo Nishino, Hideto Fukushima and Masahiro Matsumiya
Department of Marine Science and Resources, College of Bioresource Sciences, Nihon University, Kanagawa, Japan

Ebtesam M. Abd-El-Basset
Department of Anatomy, Faculty of Medicine, Kuwait University, Kuwait city, Kuwait

Steingrimur Stefansson, Stephen P. Bruttig and David H. Ho
HeMemics Biotechnologies Inc., Rockville, MD, USA

David S. Chung
Division of Biology & Medicine, Brown University, Providence, RI, USA

Jamie Yoon
Earl Warren College, UC San Diego, CA, USA

Won Seok Yoo
Department of Chemistry, Michigan University, Ann Arbor, MI, USA

Young Wook Park
Seoul International School, Seongnam, South Korea

Sung-Jae Chung, George Kim and David Hahn
Fuzbien Technology Institute, Rockville, MD, USA

Huyen Le
Nauah Solutions, Mclean, VA, USA

Sung-Jae Chung
Marymount University School of Arts & Science, Arlington, VA, USA

Yue Zhao
Department of Radiation Oncology, University of Texas Southwestern Medical Center, Dallas, USA

Ruby A. Ynalvez and Carmen G. Cruz
Department of Biology and Chemistry, Texas A & M International University, Laredo, Texas, USA

Marcus A. Ynalvez
Department of Social Sciences, Texas A & M International University, Laredo, Texas, USA

Mohamed A. M. Kadry and Mohamed A. S. Marie
Zoology Department, Faculty of Science, Cairo University, Cairo, Egypt

Eman M. E. Mohallal
Ecology Unit of Desert Animals, Desert Research Center, Cairo, Egypt

Doha M. M. Sleem
Zoology Department, Faculty of Science, Ain Shams University, Cairo, Egypt

Olena M. Nedukha
Department of Cell Biology and Anatomy, Institute of Botany, National Academy of Sciences of Ukraine, Kiev, Ukraine

Xristo Zárate, Teresa Vargas-Cortez, Jéssica J. Gómez-Lugo, Claudia J. Barahona, Elena Cantú-Cárdenas and Alberto Gómez-Treviño
Universidad Autónoma de Nuevo León, Facultad de Ciencias Químicas, Cd. Universitaria, San Nicolás de los Garza, México

Megan M. McEvoy
Department of Chemistry and Biochemistry, University of Arizona, Tucson, USA

Ana Paula de Azevedo Pasqualini
Paraná Federal University, Curitiba, Brazil

Jessé Neves dos Santos
Ponta Grossa State University, Ponta Grossa, Brazil

Ricardo Antonio Ayub
Phytotchny and Phytosanitary Department, Ponta Grossa State University, Ponta Grossa, Brazil

Zainab A. Hassan
Department of Clinical Analysis, Technical Institute, Kerkuk, Iraq

Ayoub A. Bazzaz and Noorhan A. Chelebi
Department of Basic Sciences, Faculty of Dentistry, University of Kerkuk, Kerkuk, Iraq

Atsuhiro Tanabe
Laboratory of Biochemistry, Department of Bioscience and Engineering, Shibaura Institute of Technology, Saitama, Japan

Yasuharu Sasaki and Atsuhiro Tanabe
Laboratory of Pharmacology, School of Pharmacy, Kitasato University, Tokyo, Japan

Mitsuya Shiraishi
Department of Veterinary Pharmacology, Faculty of Agriculture, Kagoshima University, Kagoshima, Japan

Samir Zahaf, Bensmaine Mansouri and Abderrahmane Belarbi
Department of Mechanical Engineering, University of Sciences and Technology, Oran, Algeria

Zitouni Azari
Laboratory of Biomechanics, Polymers and Structures, Ecole Nationale d'Ingénieurs de Metz, Metz, France

Ervin Palma and David Gomez
Evergreen Valley College, San Jose, CA, USA

Eugene Galicia
Carnegie Melon University, Moffett Field, CA, USA

Viktor Stolc and Yuri Griko
Division of Space Biosciences, NASA Ames Research Center, Moffett Field, CA, USA

P. Pavan Kumar, M. Sasikala, K. Mamatha, R. Talukdar and D. Nageshwar Reddy
Asian Healthcare Foundation, Hyderabad, India

G. V. Rao, R. Pradeep and D. Nageshwar Reddy
Asian Institute of Gastroenterology, Hyderabad, India

Evgeniya V. Pushchina and Anatoly A. Varaksin
Zhirmunsky Institute of Marine Biology, Far East Branch, Russian Academy of Sciences, Vladivostok, Russia

Sachin Shukla
Prof. Brien Holden Eye Research Centre, L.V. Prasad Eye Institute, Hyderabad, India

Christine Lo Bue-Estes
Department of Sports Medicine, Mercyhurst University, University at Buffalo, Buffalo, USA

Christine Lo Bue-Estes and Peter J. Horvath
Department of: Exercise and Nutrition, University at Buffalo, Buffalo, USA

Nadya Mezzoug and Mohamed Idaomar
Laboratory of Biology and Health, Faculté des Sciences, Université Abdelmalek Essaâdi, Tétouan, Morocco

Dominique Baudoux, Pascal Debauche, Véronique Liemans and Abdesselam Zhiri
S.A. PRANAROM International, Ghislenghien, Belgium